Regional Handbooks of Economic Development

The China Handbook
The India Handbook
The Japan Handbook
The CIS Handbook
The Central and Eastern Europe Handbook
The Southeast Asia Handbook

Forthcoming

The South America Handbook

The Southeast Asia Handbook

Edited by
Patrick Heenan and Monique Lamontagne

Advisers

Greg Bankoff
University of Auckland

Michael Haas
California State University at Fullerton

FITZROY DEARBORN PUBLISHERS

LONDON • CHICAGO

Copyright 2001 by
FITZROY DEARBORN PUBLISHERS

All rights reserved including the right of reproduction in whole or in part
in any form. For information write to:

FITZROY DEARBORN PUBLISHERS
919 North Michigan Avenue
Chicago, Illinois 60611
USA
or
310 Regent Street
London W1B 3AX
UK

British Library Cataloguing in Publication Data and
Library of Congress Cataloging in Publication Data are available.

ISBN 1-884964-97-4

First published in the USA and UK 2001
Typeset by Florence Production Ltd, Stoodleigh, Devon
Printed in Great Britain by The Bath Press

Contents

Editors' Note		vii
Map: National Frontiers and Capital Cities in Southeast Asia		x

History and Context

One	Entrepot of Trade: Southeast Asia until the 1870s *Stephen Dobbs*	3
Two	Imperial Frameworks: Southeast Asia from 1870 to 1957 *Adrian Vickers*	12
Three	World Economies: Southeast Asia since the 1950s *John Minns*	24

Nation States

Four	Thailand *Michael Kelly Connors*	41
Five	Malaysia *Fiona Yap*	56
Six	Singapore *Michael Haas*	67
Seven	Brunei *Geoffrey C. Gunn*	78
Eight	The Philippines *Raul Pertierra*	87
Nine	Indonesia *William Case*	97
Ten	East Timor *Damien Kingsbury*	110
Eleven	Vietnam *Athar Hussain*	121
Twelve	Cambodia *Sue Downie*	132

Thirteen	Laos *Gerald W. Fry and Manynooch Nitnoi Faming*	145
Fourteen	Burma *Clark D. Neher*	157

Facing the New Century

Fifteen	Nation-building, Ethnicity, and Politics *David Martin Jones and Kirsten E. Schulze*	167
Sixteen	Environment, Resources, and Hazards *Greg Bankoff*	179
Seventeen	Agriculture *John Minns*	193
Eighteen	Financial Services and the Asian Crisis *Yoichiro Sato*	207

Relations with the Wider World

Nineteen	Relations with Northeast Asia *Stephen Hoadley*	219
Twenty	Worlds in Collision: Southeast Asia and the West *Mark Beeson*	231

Appendices

Chronology	243
Glossary	285
Personalities	287
Political and Economic Institutions	301
Ethnic Groups	306
Bibliography	317
Index	323

Editors' Note

Over the decades since the end of World War II, Southeast Asia has undergone perhaps more far-reaching economic, political and social changes than any other region. In 1945, every country in Southeast Asia, apart from Thailand, was still formally part of one or another western empire. Officials and soldiers from the United Kingdom, France, the Netherlands, Portugal, and the United States were returning to take the reins from the defeated Japanese, and only the Philippines had been given any guarantee that there would be a transition to independence. All the economies of the region, apart from the port city of Singapore, were overwhelmingly agricultural, and their societies were dominated by traditional rulers, whose collaboration with the colonial powers had long been crucial to securing stability.

Within 20 years, however, all except the two smallest countries, East Timor and Brunei, had become independent sovereign states. In the cases of Vietnam, Cambodia, and Laos – and, to lesser degrees, Burma, Malaysia, and the Philippines too – independence came at the cost of deepening divisions, partly but not wholly along Cold War lines, and long-lasting struggles for unification. All the countries of the region had also begun to feel the effects of nascent industrialization, and of opening up to foreign trade and investment. They did so at widely varying speeds, and with very different impacts on their domestic politics and international relations, whether with the former colonial powers or with the new superpowers that were becoming increasingly engaged in the region.

Twenty years later still, in the mid-1980s, Vietnam, Laos, Cambodia, and Burma remained isolated from the world economy, controlled by Stalinist governments, and crushed by the burden of decades of internal warfare. Meanwhile, the rest of the region was about to reach the peak of economic growth and social upheaval that became widely known, with the World Bank's blessing, as the "East Asian miracle." The economic achievements of the 1980s and 1990s made it seem possible to many observers that these countries, all tellingly presented as if they fitted a single pattern, could maintain prosperity and growth without political reform, or that growth would somehow lead to reform in any case. Those few, both within the region and beyond it, who predicted further upheavals to come tended to be ignored. So, too, did those who tried to draw the world's attention to the violent oppression going on in Burma and East Timor, or to the varying degrees of repression in every other country in the region.

Since 1997, the financial crisis that has swept through the region, and beyond it, has exposed the fragility of the economic achievements of recent decades, as well as the defects of rule by authoritarian regimes. It has also exposed, once again, the lasting differences among the nation-states of Southeast Asia. Indonesia, which alone accounts for around 40% of the region's population, is beginning to recover from the crisis over the "loss" of East Timor, which has become formally independent, though under UN auspices, and still bears the scars of more than two decades of brutal repression. It is not at all clear, however, that Indonesia can remain united in the face of interreligious and interethnic conflicts, tensions between the center and the provinces, and the continuing overall strains of rapid modernization. Malaysia, the Philippines, and Thailand are also confronted with the challenge of finding unity and stability in an era of very rapid change and widespread dissatisfaction. The unique city state of Singapore and the microstate of Brunei are very different from each other, and from their much larger neighbors, but both face similar hard choices. Vietnam, Cambodia, and Laos, now also members of ASEAN, are beginning to catch up with the other members, but not without transitional difficulties and debates over whether and how to retain at least some elements of the social and political structures created in past decades. Finally, Burma remains deeply divided, under the contested control of an extremely harsh military regime, and it is clearly not easy to maintain an international consensus on how best to deal with it. The region as a whole still faces major difficulties in reviving economic growth, securing social stability, furthering intraregional cooperation, and reforming political systems that can no longer be justified, if they ever could have been, by the existence of domestic emergencies or international confrontation.

Given this background of underlying complexity and contemporary change, our goal here has been, not to say all that could be said about the region, but to discuss a defined range of facts and issues, as concisely, accessibly, and evenhandedly as possible. We hope that the book will have captured some of the sheer diversity of the region, and of each of the countries within it, and also indicated its importance for the rest of the world. Some of the recently fashionable predictions about the coming of the "Asian Century" – or, even more vaguely, the "Asia-Pacific Century" – may well prove to have been overstated, but there can be little doubt that Asia in general, and Southeast Asia in particular, will have an increasing influence on the global economy, and on non-Asian cultures, over the course of the 21st century.

We would like to thank all those who have helped in the making of this book. Our contributors have been patient and cooperative throughout; their contributions include not just the chapters in their names, but large parts of the appendices. Our academic advisers did useful work, recommending contributors and helping to shape the structure of the book, and we are also grateful to Professor Colin Mackerras of Griffith University, Professor Hal Hill of the Australian National University, Richard Powell and his colleagues, Dr Graham Field, and many others who have taken an interest in the project. Finally, Roda Morrison has once again supplied valuable editorial support at every stage.

Linguistic Conventions

We have generally opted to use the country and place names that are still the most familiar and most widely accepted in English: for example, "Burma" rather than "Myanmar," "Celebes" rather than "Sulawesi." Alternative place names are listed in Appendix 2; the official names of the countries of the region are given in Appendix 4.

The naming of the western portion of the island of Papua New Guinea has presented special problems. Known as West (or Western) New Guinea under Dutch colonial rule, and renamed Irian Barat and then Irian Jaya under Indonesian control, it has since become known in Indonesia simply as Papua. However, we have elected to call it West Papua, chiefly in order to distinguish it from the independent state of Papua New Guinea, which occupies the eastern portion of the island. As for its capital city, formerly known as Hollandia, Kota Baru, and Sukarnopura, its current name, Jayapura, is used throughout. It is not yet clear which names may be officially adopted – let alone which names may pass into general use outside the region – if the territory becomes an independent state in its turn.

Currency Symbols

AU$	Australian dollar	B	baht (Thailand)
M$	ringgit (Malaysia)	B$	Bruneian dollar
NK	new kip (Laos)	BK	kyat (Burma)

P	Philippine peso	CR	new riel (Cambodia)
Rp	Indonesian rupiah	D	dong (Vietnam)
S$	Singapore dollar	US$	US dollar (interim official currency, East Timor)

Abbreviations

The following abbreviations are used in more than one chapter of this book, and/or in the appendices:

ASEAN	Association of Southeast Asian Nations
EU	European Union
GDP / GNP	Gross Domestic Product / Gross National Product
IMF	International Monetary Fund
UK	United Kingdom (when used as an adjective)
UN	United Nations
US	United States (when used as an adjective)

In other cases, the meaning of each abbreviation or adjective is stated at its first appearance within each chapter.

National Frontiers and Capital Cities in Southeast Asia

History and Context

HISTORY AND CONTEXT

Chapter One

Entrepot of Trade: Southeast Asia until the 1870s

Stephen Dobbs

Southeast Asia has long had a reputation for being a region rich in natural resources, and a dynamic zone of commerce and trade. From ancient times, speculation abounded outside the region over the plenitude and wealth of this part of the world. Cloves, nutmeg, aromatic woods, camphor, and resins from the region were available in the markets of the Roman empire and Han China more than 2,000 years ago (Reid 1993 p. 1). In ports and trading centers from the Atlantic to the China Sea, ancient mariners discussed, speculated and fantasized about what they called "The Land of Gold" (Shaffer p. 2). The rare assortment of condiments, woods, lacquers, and other products collected in Southeast Asia, and exchanged along ancient trade routes, had no match elsewhere in the world, and encouraged the belief that this was an exotic domain of mystery and incredible wealth (see Savage). The Alexandrian astronomer and geographer Ptolemy, in his *Geographike Hyphegesis* (Guide to Geography), pointed to Southeast Asia as the location of the fabled peninsula of the "Golden Khersonese" (Savage pp. 28–33). According to the Romans, the area was a mythical place inhabited by "wild men" and was rich in gold, animals, and other resources considered valuable across the ancient world. Although the early chronicles contain exaggerated, even fanciful, accounts, their references to the area's wealth and abundant natural resources were a direct consequence of its dynamic and active role in world trade networks from ancient times.

The Geographical Setting

A combination of geographical and historical factors has given Southeast Asia an "international orientation" (Kathirithamby-Wells p. 1). Readily accessible by sea and strategically located to command the principal maritime routes between China, India, the Middle East, and Europe, it played the role of warehouse to the world for centuries. Not surprisingly for a region that encompasses such a vast amount of ocean, its own maritime tradition developed early. Southeast Asia had skilled sailors who had already developed techniques for transoceanic sailing, centuries before Christopher Columbus or Vasco da Gama embarked on their epic voyages at the end of the 15th century. They discovered early the push-pull effect of the monsoon seasons and used this knowledge to become nomads of the Southern Ocean (Hall p. 186). During Central Asia's cold winter months, the northeast monsoon drove their craft outward through the archipelago to explore, trade, and populate new areas. Return journeys were timed to coincide with the southwest monsoon, when the heat of Central Asia's summer months sucked in the cool air of the Southern Ocean like a magnet pulling toward continental Asia. Without compasses or charts, intrepid sailors from Southeast Asia traversed thousands of kilometers of open water, and could find their way around the archipelago and beyond with almost pinpoint accuracy. They navigated using the winds and by observing the stars, ocean currents and swells, and cloud formations, as well as the behavior of marine and bird life (Hall p. 185). They developed rigging and techniques such as the "balance-lug sail," which allows vessels to sail into the wind at an angle or to tack a zigzag path directly into the wind. Such techniques went on to be used universally by sailing craft (Shaffer p. 13).

"Malay" sailors were known in China as early as 300 BCE, and there is evidence that they were trading and settling along the east coast of Africa by the first century CE (Hall p. 186). They had established permanent communities on the Malagasy coast even earlier, perhaps before Roman times (Shaffer p. 15). In the years before there was an overland Silk Road, "Malay" traders were shipping cinnamon from South China to the east coast of Africa, from where it was traded Northward through Ethiopia and on into Europe (Shaffer p. 16). In fact, everywhere they ventured the sailors of Southeast Asia carried the produce of one region to the next. It is widely recognized that bananas, coconuts, cocoyam (taro), and a variety of other plants were transported from Southeast Asia to Africa by these early sojourners (Shaffer p. 15). The broader implications of the region's pioneering efforts in long-distance trade were global. From at least the third century until the 19th century, Southeast Asia was at the center of world trade (Kathirithamby-Wells p. 1). The region's importance to international exchange was such that from around 1000 CE until the industrial era, the traffic in spices out of the archipelago determined the amount and extent of all international commerce (Hall p. 183).

Trading Centers: Funan, Srivijaya, and Majapahit

Trade, and the connections that it brought with international commercial routes, played an important role in the development of Southeast Asia's first states by creating wealth and introducing new artistic, religious and political ideas. Throughout both maritime and continental Southeast Asia, "international emporiums" sprung up to service, in the first instance, the trade between western Asia and China, and, later, the trade between Europe and China (Kathirithamby-Wells p. 1). These emporiums linked the region's own dynamic commerce to world trade routes. While the significance of individual ports and states rose and declined depending on a range of domestic and commercial factors, the region's importance in global trade grew steadily from the first centuries of the common era. Demand for Chinese silk in the markets of India, the Middle East, and the Mediterranean world of the Roman empire first brought foreign merchants and trade into the region en route to China. The presence of traders from India, China, the Middle East, and, later, Europe provided opportunities to introduce a variety of regional produce into the trade between China and the lands to the West. Scented gharuwood and sandalwood became valued items used in the making of incense. A variety of resins from Southeast Asia's vast rain forests were also successfully introduced into this global trade. Camphor from Barus, a port on the northwest coast of Sumatra, became highly prized for use in medicines, varnish, and incense. The fine spices, cloves, nutmeg and mace, of the Moluccas proved to be the most significant of the local commodities introduced into international trade.

By the first century of the common era, Indian traders were passing through Southeast Asia as they sought the source of silk in China. With the Roman empire at its zenith in the West, and Han China in the East, a new and truly international trade route developed, incorporating significant maritime stretches. Goods from China made their way westward through mainland Southeast Asia, India, and the Middle East, and from there into the Mediterranean, and vice versa (Whitmore p. 141). The use of this route avoided the necessity of sailing through the archipelago of Southeast Asia by transporting goods across the narrow Isthmus of Kra (around 56 kilometers wide), at the northern tip of what is now Malaysia. While this meant more handling of the goods, with unloading and then reloading on the other side of the land bridge, sailing around the Malay Peninsula would have added around 2,500 kilometers to the journey. Then there were the dangerous currents and shoals at its southern tip to contend with, not to mention the pirates believed to frequent these waters (Shaffer, p. 20). The main items of trade along this route were luxury goods such as Middle Eastern aromatics (frankincense and myrrh), drugs, and the Chinese silk for which there was a considerable demand in Roman markets (Hall p. 192).

Southeast Asia provided one of the principal points of exchange along this route and the trade nurtured the first recorded polity and mainland entrepot to emerge in the region. Situated on the eastern shore of the Gulf of Thailand, Funan, as it was called by the Chinese, was an important stopover on the route to China and a major market for Asian goods, particularly silk (Hall, p. 192). Its main port, located at Oc-eo on the western edge of the Mekong Delta, in what is now Vietnam, had to be reached via a network of canals that linked the Gulf of Thailand with this vast waterway (Shaffer, p. 22). It provided traders from China, India, and the Middle East with a convenient and sheltered weighpoint where they could rest and goods could be exchanged or supplies taken on for the rest of the journey. Initially, it was mainly Indian merchants making their way to China who frequented Oc-eo, but over time it also attracted merchants from the Middle East, and even Greece, seeking to purchase Chinese silk. As mariners were dependent on the monsoon winds, their stay in Funan could stretch out to as much as five months as they waited for the right season and winds so they could continue on their journey. The rich fertile lands of the Mekong Delta provided sufficient food and surplus for the development of Funan as a polity, and to support the seasonal traders and sailors on their way to or from China (Hall p. 193).

Funan's success attracted trade from other ports in Southeast Asia and provided an outlet for a range of local goods. Produce from the region was brought in to supply the inhabitants of Funan as well as to be exchanged for foreign trade goods. Among the first local products to find their way onto this international trade route at Funan were a variety of resins (pine resins, benjamin gum) and aromatic woods (gharuwood, sandalwood) from as far afield as Sumatra and Timor. These goods were, in part at least, introduced as substitutes for frankincense and myrrh, which were much in demand in China but had to be transported from East Africa and Arabia (Shaffer p. 23). Other local products, such as camphor, tortoiseshell, and the feathers of exotic birds also found a place in the growing list of commodities available on this route. Moluccan spices (cloves, nutmeg, and mace) also began to be traded at Funan. These "fine spices" were valued for their flavors and their supposed medicinal qualities (Hall pp. 195–96).

By the sixth century CE, Funan's role in the commerce between the West and China was being eclipsed as more and more trade bypassed it. From around the fifth century, the international trade route had begun to pass through the Malacca Straits, no longer using the narrow neck of the Malay Peninsula as a short cut between the Bay of Bengal and the Gulf of Thailand. The main commercial interest of merchants in the sixth century was still in the exchange of Middle Eastern aromatics and drugs for Chinese silk (Whitmore p. 142), but the infusion of Sumatran resins and other products from the archipelago into this long-distance trade was reason enough to form closer links with the islands. The new sea passage offered a range of benefits. Sumatran ports functioned as collection centers for local goods as well as products from the eastern archipelago. In addition, they also provided vessels, crews, and port facilities for those involved in trading operations (Whitmore p. 142). The flow of trade through the Malacca and Sunda Straits ushered in a new era of commerce centered in the archipelago world. In later centuries, mainland polities such as Angkor (Cambodia), Pagan (Burma), and Ayutthaya (Thailand) flourished and participated in the burgeoning intraregional and international trade of Southeast Asia, although they did not develop international emporiums or centers of the type that Funan had been. Nevertheless, these early states fused increasingly large areas of hinterland and coast together in what Kenneth R. Hall has called "hierarchical political and economic networks" (Hall, p. 270).

By the seventh century, Srivijaya, located on the Musi River, on the northeastern coast of

Sumatra, where Palembang stands today, had become the principal port and polity on the new maritime route through Southeast Asia. Having subdued rival ports along the coast and through the Malacca Straits, Srivijaya became the main link between the wealth of the archipelago and international trade until the 11th century. Its hegemony extended from the upper Malay Peninsula to Java and southwestern Borneo. A well-equipped harbor provided excellent facilities for refitting or repairing ships and taking on fresh supplies. Additionally, there was ample opportunity to trade, especially in the valued resins, camphor, and spices of the archipelago, and several thousand itinerant peddlers and traders from many lands thronged its markets (Cady p. 69).

Srivijaya's preeminent position as a center of commerce lasted for around 600 years. By the 11th century, changes in the international and regional situation resulted in the trade route moving further to the Southeast. To begin with, East Java was becoming increasingly active in developing its own trade and tapping into the wealth of the Java Sea region (Whitmore p. 144). Indian and Chinese traders who had frequented the Sumatran ports now wanted to establish more direct contact with the sources of their supplies. They became increasingly active on regional sea routes, sailing further into the archipelago. For Srivijaya there was now competition from the developing polities and ports in East Java, around the delta of the Brantas River. Between 990 and 1007, the competition to attract international trade and secure food resources from Central Java led to open hostilities and periodic war between the two competing polities (Hall p. 210). Srivijaya's demise as a regional maritime power and center for international trade was completed following a devastating attack by the Indian kingdom of Chola in 1025 (Shaffer p. 74).

A series of kingdoms located in East Java dominated the regional and international trade routes in the archipelago between 927 and 1528. The largest and most enduring of East Java's polities was Majapahit (1293–1528). With its royal city located around 80 kilometers upriver from Surabaya, on the Brantas River, Majapahit became the major political and economic power in the region. Its founder, having foiled attempts by the Mongols to subjugate the kingdom in 1292–93, went on to extend its authority throughout the archipelago. At its peak, Majapahit counted as part of its realm the entire island world that makes up contemporary Indonesia, and controlled the all-important Malacca Straits and the spice islands (Hall p. 215). After the defeat of the Mongols, trade relations were reestablished with China, while the growing demand for spices in western Europe meant more trade with the West than ever before. The demand for spices in Europe had burgeoned in the wake of an agricultural and commercial revolution. Europeans had taken to eating more meat and the demand for spices to flavor salted and preserved meat and vegetables brought about a rapid growth in the trade (Hall p. 217). Majapahit was well-situated to benefit from the additional commerce.

Majapahit's rulers worked to maximize the benefits of the spice trade to their realm and they had several advantages over earlier emporiums such as Srivijaya. Majapahit, like earlier polities in East Java, had direct trade links with the spice islands, instead of having to rely on intermediaries. Additionally, the kings of East Java controlled vast rice supplies and this placed them in an ideal situation to control the trade of the archipelago. They were able to stockpile large amounts of this grain, which could be traded throughout the region for forest produce and, of course, spices too. So close were the ties that Majapahit's rulers formed with the spice merchants that they often appear as little more than agents of trade for the rulers (Hall p. 218). The kings were paid a fee for the use of the kingdom's ports, they had control over the flow of bullion and other luxury goods that passed through the realm, and they received a share of the money made from international and regional commerce (Hall p. 218).

However, the central role of Majapahit and East Java in regional and international trade was in decline by the 15th century. The century had barely begun when Majapahit experienced a civil war (1405–06) that threatened its hold over the surrounding lands. By this time, the spice trade that had fueled the region's growth had become so large that it was all but impossible for the weakened Javanese rulers to control. The Sumatran pepper ports had become so important to Ming China that it sent out a military force to secure the Malacca Straits by clearing out a den of Chinese pirates operating from Palembang. During the same expedition, an alliance was formed with the port of Malacca (or Melaka), which had been founded in 1402. China's protective umbrella ensured Malacca's security and growth for decades (Hall p. 227). It became the center of the spice trade and even trade in Javanese goods as it formed links with that island's trading communities. By the end of the 15th century, Malacca was the preeminent port in the islands. Across and beyond the region controlled from Majapahit and East Java, coastal communities found a new independence in the course of the international trade boom, and were able deprive the realm of much of its income from trade. Ultimately, this independence led to their adoption of Islam, a further indication of how important new overseas links were. By the end of the 14th century, Islam had spread throughout the island world of Southeast Asia. It provided the coastal communities and ports with a new sense of autonomy that allowed them to question the legitimacy of the Majapahit rulers – an activity that culminated in 1528 with the core of the realm falling to a coalition of Moslem-Javanese coastal communities (Hall p. 229).

The Cultural Impact of Trade

External trade brought material goods and wealth for those who could control it. However, at least equally important were the new religious and political ideas that were adopted and integrated with local beliefs and traditions. Before the coming of Islam, the most notable ideas brought in from outside the region were the cosmologies of India and the Middle East. Hinduism and Buddhism played key roles in both the mainland and the maritime courts of Southeast Asia from around the second century CE. Over centuries of contact, and in an uneven fashion, Indian religious and cultural practices were introduced to the region in a process that has been described (at risk of oversimplification) as "Indianization." Significantly, however, there was no mass migration of Indians into Southeast Asia: the bearers of these new modes of thought were the traders and priest-scholars who followed early trade routes into the region (Osborne p. 24). Local chiefs or rulers adopted Indian political and religious ideas to enhance, embellish and legitimize their authority (Cady p. 43), for they provided structured and well-developed doctrines for sustaining religious and political power.

While Hindu and Buddhist influences were a feature of the early states in Southeast Asia, it was Islam that ultimately dominated the archipelago. Like Hinduism and Buddhism, Islam was introduced to the region via the maritime trade routes that linked India, Southeast Asia, and China. It is widely held that Sufis – mystics who arrived with traders from India – brought Islam to Southeast Asia (Shaffer p. 101). By the end of the 13th century, Samudra-Pasai, on the north coast of Sumatra, was the seat of the first Muslim polity in the region. After the rulers of Malacca were converted to Islam, in the 15th century, the religion quickly spread throughout the archipelago. Its rise was due, in part at least, to Malacca's commercial domination of the region's maritime trade from around the middle of the 15th century (De Casparis and Mabbett p. 330). In addition, successive rulers actively promoted, persuaded and compelled neighboring states to adopt the new faith. By embracing Islam, the rulers of Malacca added to their prestige and power, linking themselves to a resurgent Islam that had seen the Ottoman Turks take Constantinople (Istanbul) in 1453 and transform it into a center of Islamic culture and learning. Middle Eastern accoutrements were adopted, such as the title "Sultan," in order to emphasize these rulers' position and power over and above the princes of nearby and possibly competing polities (Andaya p. 517).

There had been two notable exceptions to the process of "Indianization": Vietnam, and the islands of the Philippines (Osborne, p. 24). Geographically, the Philippine islands were distant from the early trade routes linking India and China. As a result, these islands were not exposed to the religious and political influences observed elsewhere in Southeast Asia. It was Christianity, the last world religion to reach Southeast Asia, that did most to shape the ontological world of the Philippines, beginning with the arrival there of Ferdinand Magellan, in 1521. Vietnam's development was also distinctive, mainly because for more than 1,000 years, between the second century BCE and the tenth century CE, it was under Chinese rule. Its years as a dominion of China coincided with the period in which Indian influences were strongest throughout the rest of Southeast Asia. In particular, Mahayana Buddhism entered Vietnam via China – in contrast to the Hinayana or Theravada form of Buddhism that spread through its neighbors – and Vietnam's rulers long looked to Chinese models for workable systems of government. By the 12th century, the Vietnamese had their own emperors, ruling what was called, significantly, a "southern" (*Nam*) realm (Taylor p. 147).

While the peoples of Southeast Asia adopted and used religious and political ideas from other cultures as the basis for building more powerful administrative structures, they were not without their own cultures. Rather than simply replacing these older traditions, they took whatever was useful to them and adapted it to local conditions, while discarding those elements that would not fit within their established cultural matrices. For example, while Hinduism had a considerable influence on the region's first polities, the caste system of India was never adopted anywhere in the region: caste terminology was used to describe the courts of Southeast Asia's rulers but was never given a broader application. Similarly, while Sanskrit was taken up as the language of government and religion, it did not go on to be used more widely, and ultimately it was discarded, although scripts based on Indian models continued to be used to render indigenous languages. In a number of important areas, such as agriculture and metallurgy, there was certainly nothing that outsiders could offer to supplement Southeast Asia's own achievements. The peoples of the region developed their own techniques for wet rice cultivation, independent of any imported technology or ideas (Osborne pp. 24–25). Indeed, their success with rice production was crucial in the development of large state systems with dense populations, and also in supporting trading activity. The refining of metals was also well-advanced in Southeast Asia before trade brought in outsiders: the region began producing bronze and copper from as early as 2000 BCE (Shaffer, p. 11). The numerous elements borrowed from the cultures of the region's nearest neighbors never subsumed the essential indigenous characteristics that distinguished the region (Reid 1988 p. 6), and whatever influence India, China, and even Islam had largely followed in the wake of trade links.

It is clear that the peoples of Southeast Asia themselves played an active role in developing outward-looking and trade-oriented networks of ports and polities. These did not arise simply as a result of outside factors and demands, but rather as an expression of early statehood (Kathirithamby-Wells p. 1) in increasingly sophisticated and complex societies that recognized the value of trade exchanges. Within the region itself, commercial activity played an important role by connecting distant populations to each other socially, culturally and economically. Until the "trade revolution" of the 17th century, maritime links within Southeast Asia had more influence on the way people lived than any outside forces, and they held the region together as a socioeconomic unit (Reid 1988 p. 6). Arguably, then, it was the peoples of the region who instigated international demand for Southeast Asia's produce in the first place, by distributing it to foreign shores. They took advantage of new commercial opportunities, as they arose, to further extend state power and wealth.

Continuity and Change from around 1600 to 1870

Until 1600, cloves were found only on a handful of volcanic islands along the western coast of Halmahera – Ternate, Tidore, Motir, Makian, and Batjan – while the nutmeg tree, also the source of mace, was restricted to fewer than ten islands in the Banda Sea (Hall p. 209). Locating the spice islands and controlling the trade in these valuable condiments was the driving force behind western Europe's advance into Southeast Asia from the 16th century. The Portuguese conquest of Malacca in 1511 was the first attempt by a western European power to control the lucrative spice trade. However, the efforts of these outsiders to dominate the spice trade and extend their influence in Southeast Asia had only limited success for the next three centuries. Any control they had was largely confined to coastal areas and commercial ventures, and, for the most part, the interlopers operated within established economic and political frameworks. It was not until the 19th century that they emerged as a force capable of reshaping Southeast Asia's political, social and economic fabric.

Nevertheless, the arrival of outsiders from western Europe began a process by which Southeast Asia was opened to greater direct trade with the outside world. In a clear break with the past, the Portuguese went directly to the source of the spices, bypassing intermediary suppliers and ports. The immediate effect was that the Javanese control of the spice trade that had been built up over more than 1,000 years was broken (Whitmore p. 147). However, the Portuguese were not able to monopolize the trade. Rather, the opening up of the archipelago resulted in a variety of groups, Asian and European, competing to exploit the region's wealth. Commercial activity continued to be largely in Asian hands. Aceh became the center of the Sumatra pepper and gold trade, while Johore handled the trade of the eastern archipelago. On the mainland, Pegu (Burma) and Ayutthaya (Thailand) thrived because of the new foreign contacts and the opening of the maritime realm (Whitmore p. 147).

When the Dutch entered Southeast Asia, in the 17th century, they too were forced to operate within this competitive environment and had only limited success in controlling the region's trade. Operating from their base at Batavia (now Jakarta), which they founded in 1619, they attempted to restrict access to the Molucca islands and dominate the Java Sea and the archipelago. They also became actively engaged with the continental cities and ports of Burma, Ayutthaya, Laos, and Vietnam. Trading on the mainland, they found themselves in direct competition with the Japanese for a short period in the 1630s, before the Japanese withdrew into semi-isolation, and then later they had to compete with the Chinese (Whitmore p. 148). The Chinese had managed to establish themselves in almost all Southeast Asia's key trading centers before the Europeans arrived (Whitmore p. 148), and they became important agents in the European centers of trade such as Malacca, Batavia, Manila, and Penang. Thus, while Europeans played a role in opening up the trade of Southeast Asia, to a large extent they were simply another element within an already complex trade network. Indigenous polities and ports were actively engaged with the Europeans, and continued to maintain a high level of autonomy, particularly in the political and cultural spheres, until the 19th century.

International commerce in the 19th century ushered in an era of unprecedented economic and social change for Southeast Asia. The region's longstanding connections with international trade had played a central part in shaping its political, economic and social development from at least the beginning of the common era. However, the western European powers were now on the verge of radically transforming the economic relationship that had prevailed for centuries. The old mercantile order, involving the collection and transportation of goods produced by others for sale at a profit on the international market, was about to give way to new economic and social imperatives. The development of western capitalism and, in particular, the industrial revolution in western Europe provided the impetus for this change over the course of the 19th century. The industrial revolution armed western Europe with a range of new powers that gave it the capability to undertake direct and forceful intervention in the political, economic and social structures of Southeast Asia. Advances in communications, weaponry, and transportation, among other areas, gave it an edge over the indigenous polities and the other Asian states operating in the region (Elson p. 141). China and India were increasingly subjected to European power, and effectively eliminated from any balancing role in the region.

Rivalries among the powers of western Europe resulted in a process of territorial annexation in Southeast Asia from around

1850. In this environment, trading posts gave way to colonies, which could supply raw materials to feed the industries of western Europe. They were also functioned as markets for the processed goods that the factories of the industrializing world produced in ever greater abundance (Osborne p. 77). Clear lines of demarcation between areas of influence were quickly drawn, in a process that ultimately created the modern states of Southeast Asia. By the mid-19th century, Britain had established itself as the predominant power in the Malacca Straits and on the Malay Peninsula, most notably with its acquisition of the island of Singapore in 1819. It also annexed Lower Burma, in 1852, and forced the Burmese kingdom to open its doors to trade. The Dutch, meanwhile, were extending their hold over the archipelago, proclaiming it to be the Dutch East Indies, and in the Philippines the Spanish continued to exert and extend their authority. The French also entered the race for colonies in Southeast Asia, beginning with acquisitions around the Mekong Delta between 1859 and 1862 – which became Cochin China (southern Vietnam) – and later extending their control to include Cambodia, Laos, and the rest of Vietnam (the provinces of Annam and Tonkin). Siam (Thailand) was the only state in Southeast Asia that avoided succumbing to the imperial ambitions of western Europeans, although in 1855 King Mongkut signed the Bowring Treaty, which granted huge concessions to the colonial powers, including the acceptance of extraterritoriality (meaning that citizens of European powers were beyond his state's jurisdiction) and the abolition of state trade monopolies. Aware of Britain's military actions against China (the Opium War of 1839–42) and Burma (1852), Mongkut was under no illusions as to Siam's position.

Under European direction and control, the region became a major supplier of rice, rubber, tin, and other commodities for the world market. In the process Southeast Asia's physical, political, economic and social landscape was transformed. Jungle gave way to plantation agriculture and mining, while the influx of tens of thousands of Chinese and Indian laborers to work in these colonial enterprises forever changed the region's social and cultural composition. By 1870, the European powers were poised to complete their transformation of Southeast Asia.

Further Reading

Andaya, Barbara Watson, and Yoneo Ishii, "Religious Developments in Southeast Asia, c. 1500–1800," in Nicholas Tarling (editor), *The Cambridge History of Southeast Asia*, Cambridge, New York, and Singapore: Cambridge University Press, 1992

This chapter of an authoritative work provides an excellent discussion of indigenous beliefs in those parts of Southeast Asia that were not greatly influenced by the world religions, but it also examines the spread of Islam and Christianity in Southeast Asia.

Cady, John F., *Southeast Asia: Its Historical Development*, New York: McGraw-Hill, 1964

This general history of Southeast Asia provides an extended discussion of the early influences that India and China exerted on the region.

De Casparis, J.G., and I.W. Mabbett, "Religion and Popular Beliefs of Southeast Asia c.1500," in Tarling, as cited under Andaya and Ishii

This informative study of early religious beliefs in Southeast Asia is primarily concerned with the pre-Islamic period.

Elson, Robert, "International Commerce, the State and Society: Economic and Social Change," in Tarling, as cited under Andaya and Ishii

Elson provides a detailed survey of the economic and social changes brought about in Southeast Asia from the 19th century onward, as European powers extended their influence in the region. It focuses primarily on the late 19th century.

Hall, Kenneth R., "Economic History of Early Southeast Asia," in Tarling, as cited under Andaya and Ishii

In this excellent examination of economic development in Southeast Asia up to the 16th century, Hall highlights the region's role in international trade from the beginning of the common era.

Kathirithamby-Wells, J., "Introduction: An Overview," in Kathirithamby-Wells, J., and John Villiers (editors), *The Southeast Asian Port and Polity: Rise and Demise*, Singapore: Singapore University Press, 1990

This is the first in a collection of essays detailing the evolution of the port-polity in Southeast Asia, highlighting the close links between the growth of trade and statehood.

Osborne, Milton, *Southeast Asia: An Introductory History*, second edition, Sydney, London, and New York: Allen and Unwin, 1983; seventh edition, Chiang Mai: Silkworm, 1997

A general study of Southeast Asia that examines the region's traditional past and the impact of colonialism

Reid, Anthony, *Southeast Asia in the Age of Commerce 1450–1680*, Volume I, *The Lands Below the Winds*, New Haven, CT, and London: Yale University Press, 1988

In both volumes of this book, Reid examines in detail the development of Southeast Asia from the 15th century to the 17th century. Volume I explores the cultural, physical and material world of Southeast Asia during this period, with a focus on the role that the region's peoples played in their own development.

Reid, Anthony, *Southeast Asia in the Age of Commerce 1450–1680*, Volume II, *Expansion and Crisis*, New Haven, CT, and London: Yale University Press, 1993

In this volume Reid details the impact of international trade, in particular the spice trade, and examines changes in state structure, urbanization, and religion brought about by the global trade boom of the period covered.

Savage, Victor, *Western Impressions of Nature and Landscape in Southeast Asia*, Singapore: Singapore University Press, 1984

A detailed study of European attitudes and ideas about Southeast Asia from ancient times through to colonial times

Shaffer, Lynda, *Maritime Southeast Asia to 1500*, Armonk, NY: M.E. Sharpe, 1996

This study traces the major changes that took place in the maritime realm of Southeast Asia up to the end of the 16th century.

Taylor, Keith W., "The Early Kingdoms," in Tarling, as cited under Andaya and Ishii

Taylor traces the political development of Southeast Asia's earliest states, giving detailed coverage of continental as well as maritime states up to the time of Majapahit.

Whitmore, J. K., "The Opening of Southeast Asia: Trading Patterns through the Centuries," in Karl L. Hutterer (editor), *Economic Exchange and Social Interaction in Southeast Asia: Perspectives from Prehistory, History, and Ethnography*, Ann Arbor, MI: University of Michigan Center for South and Southeast Asian Studies, 1977

This study provides an outline of Southeast Asia's engagement with global trade from early in the common era until the 19th century.

Dr Stephen Dobbs is Tutor in Asian Studies at Murdoch University, Perth, Western Australia.

HISTORY AND CONTEXT

Chapter Two

Imperial Frameworks: Southeast Asia from 1870 to 1957

Adrian Vickers

What did imperialism do in Southeast Asia? Its evils were long taken for granted, made into political rhetoric, and used by nationalist movements in their struggles against European and US domination. This rhetoric continued during the Cold War when "neocolonialism" was challenged in forums such as the Nonaligned Movement (established in 1961 but prefigured in the Asia-Africa Conference held in Bandung, Indonesia, in April 1955). The 1980s and 1990s saw a revival of the debates about imperialism through the arrival of post-colonial theory and its obverse: nostalgia for the "good old days" of life in the colonies. Anti-imperialist language continues in the 21st century in evocations of "Asian values" against "the West." Yet gauging the effects of imperialism on the everyday lives of people in Southeast Asia requires an analysis that goes beyond the assumptions underlying this rhetoric.

To begin with, identifying imperialism with a specific historical period is often difficult, despite such assertions as that of President Sukarno of Indonesia, that Dutch imperialism lasted for 350 years in his country. The majority of writers on the subject identify the period 1870–1942 as the period of "real" or "high" (western) imperialism. In the case of British Malaya, this era did not end until the granting of independence in 1957. The period was inaugurated by the final bursts of nationalist unification in Europe. It was characterized by a combination of bureaucracy-building, aggressive military policies, promotion of liberal trade regimes, and the development of a bourgeois culture characterized by notions of respectability and race. It was abruptly terminated in Southeast Asia by the creation of the Japanese "Greater East Asia Co-Prosperity Sphere," which demonstrated that imperialism was not new, not exclusively western, and might best be handled by those who actually had emperors. In strict terms, imperialism's "empire building" is not identical to "colonialism," or the processes of setting up dependent colonies through the presence of settlers from the home country. However, for the period 1870–1957 the two terms describe different aspects of a single phenomenon.

Further, not all of Southeast Asia experienced imperialism as incorporation into western empires. Thailand (known as Siam until 1938, and again from 1946 to 1949) remained independent, Brunei was only nominally under British "protection," and there were lots of loose ends in the tidying up of the map of Southeast Asia undertaken at the end of the 19th century. The period up to 1870 had seen a process of uneven colonization in the region, based around ports and the support of trade bases rather than the acquisition of territory for territory's sake. The Portuguese had been in the region longer than any other Europeans, but had the least interest in doing anything about their tiny bases, the most notable of which was the eastern part of the island of Timor. The Spanish did not occupy the Philippines for land, but simply as a door to China, as it was for the United States when it took over the archipelago at the end of the 19th Century. The Dutch had been in Java and other parts of "their" East Indies since the early

16th century, but they did not take over large parts of that archipelago – for example, Aceh, parts of Sumatra, Borneo, Celebes, South Bali or West Papua – until after 1900. The British acquisitions on the Malay Peninsula added up to a loose collection of "Federated" and "Unfederated" Malay sultanates and separate states taken over between the 1780s and the end of the 19th century. The French colonies that constituted Indochina were a collection of protectorates and territorial acquisitions built up from the 1860s to the 1890s. They were to have difficulty forming into three cohesive nations in the period immediately after independence, especially given the barriers that France and then the United States put up against that independence.

Given this diversity of experience, it seems difficult to say that there was a single, common experience of imperialism in Southeast Asia, or that imperialism and colonialism had similar effects across the region. Most commentators see imperialism as inaugurating forms of modernity, and giving those forms of modernity a distinctively western pattern. This pattern is usually seen as adding up to capitalism and western forms of governmentality, reinforced by cultures of dominance that have been carried over into the postcolonial period. The three elements were not always even in their effects or in their distribution across the region, but still the forms of colonialism and capitalist relations, particularly in the late 19th and early 20th centuries, had much in common. This commonality derives from the shared models and ways of action adopted by western governments and companies. In particular, the transnational forms of capitalism, from small family-run plantations to large multinational enterprises, produced common ways of working across the region.

Capitalism in Southeast Asia

The numerous writers on imperialism mostly agree that there is a close relationship between it and capitalism, but beyond that there is very little consensus about how the relationship worked. The idea of self-sustaining closed societies violently penetrated by capitalism, thanks to imperialism, makes for a convenient idealized image, but this is not a picture applicable in Southeast Asia. The scholarship of Asian trade has demonstrated the ubiquity of forms of money and large-scale buying and selling throughout the region for nearly 2,000 years. Those considered remote or "primitive" were involved in exchanges of exotic goods. Papuans were linked in trading chains to the Dutch and Portuguese in eastern Indonesia. Marriages among the Alorese in what is now eastern Indonesia involved elephant tusks and Indian textiles. Dayaks in the forests of Borneo made use of large Chinese pots that were known as "Martavans" because they had been carried by foreign traders along routes through the port of Martaban in Burma. By the latter part of the 19th century, Malay states were already engaged in capitalist forms of production before they were taken over by the British (see Burns), while Siam/Thailand's transformation into a modern economy took place without direct western control (see Hong).

J.A. Hobson and V.I. Lenin produced perhaps the two most influential versions of the argument that European imperialism was produced or necessitated by the need of capital to expand: the colonies provided resources, markets, and sites for the investment of surplus capital. The many reassessments of their arguments have pointed both to the lack of direct relationship between economic imperatives and the territories acquired, and to the lack of profitability of many colonies. One particularly telling argument is that private capital did not consistently make substantial investments into their own country's colonies, especially in Asia (see Fieldhouse pp. 386–92, Haviden and Meredith pp. 308–09, and Porter pp. 41–42). Derek Fieldhouse has put it more simply by dismissing the myth that "modern empires... were great machines deliberately constructed by Europe to exploit dependent peoples by extracting economic and fiscal profits from them" (Fieldhouse p. 380). In his view, "there was no guarantee that a colonial enterprise would produce high profits simply because it was in a colony" (Fieldhouse p. 389).

Patrick O'Brien, in a recent overview of discussions of British economic expansion and empire, has joined the consensus that economic expansion was not dependent on empire-building, and to some degree was even hindered by it. The lack of success of 19th-century charted companies such as the North Borneo Company, based in what is now the Malaysian state of Sabah, showed that the old mercantilist patterns of nationally based trade could not work (Fieldhouse p. 366). However, P.J. Cain and A.G. Hopkins have demonstrated that finance did play a major role as a motive in imperialism, even though they did not look at whether the motive matched the effects. In a study of selected British colonies (including Malaya) that goes on from Cain and Hopkins's work, Michael Havinden and David Meredith have shown that the British government did hope for great profits from small economic outlays. In the second quarter of the 20th century there was a marked turn away from free trade to empire-based trade systems, with hopes of increasing such national profits. By doing this, the imperial powers argued, they would bring development and improvement to the lives of their subjects. Such hopes were largely illusory, for "colonialism and development were largely contradictory": European "dreams (or myths)" of great profits were realized by some of the larger trading companies, but they produced "structural imbalances" in the colonies and uneven drains on government and private investment (Havinden and Meredith p. 317).

European nationalism, which took on a distinctive military flavor during the 1870s, was a greater impetus to imperialism than the profit motive, and helps to explain the strange mix of liberal and aristocratic ideologies in the age of high imperialism (see O'Brien). Ultimately, cultural attitudes, promoting "national greatness," were the political spurs to empire building, and can best explain continued public enthusiasm for the task. This does not mean, however, that culture explains all aspects of imperialism. Some colonies were very profitable in the new economy produced by the massive expansion of communication, infrastructure, and scales of capital investment between 1870 and the Great Depression at the end of the 1880s, and then up to the second Great Depression at the end of the 1920s. The colonies in Southeast Asia were sources of raw materials: Malaya's rubber, and to a lesser degree its tin, produced enormous returns for Britain (Havinden and Meredith p.103), as to a lesser extent did the products of Indochina and the East Indies for France and the Netherlands respectively. The new demand for oil proved lucrative in Burma and the Dutch East Indies, while Burma was also profitable because of a major increase in rice production created by the colonial government's opening up of new rice-growing areas through swamp reclamation in the Burma Delta (see Adas).

The main problems for most colonial governments were, first, that these profits did not go to the colony itself, nor even sometimes to the mother country, and, second, that these profitable colonies had to be made to subsidize the very unprofitable ones. Despite a speculative bubble in tin-mining in Laos, Cambodia and Laos alike represented expenses that the French administrators could only make less of a drain on Vietnamese income through onerous taxation regimes (see Stuart-Fox Chapter 2).

Taxation was the chief means of extracting resources from the colonies in order to make them self-sustaining. Even though the citizens of the metropolises usually paid the largest proportion of taxation for maintaining empires (see O'Brien), the proportional burden was often heavy, with disastrous results for the colonial economies. Head taxes were often the simplest solution – even the Thais imposed poll taxes – although the Dutch had a sophisticated set of ways of extracting money based on produce, which they called "land rent." After 1870, however, this had to be rethought after the "Cultivation System," which forced people to grow cash crops for a government trading monopoly, was abolished. By the early decades of the 20th century, the land rent system was being formed into a series of bureaucratic controls of land and villages in Java (see Breman).

In order to document populations, draw off forms of forced labor, and tax populations effectively, greater control had to be exercised further and further down the levels of society. The era of high imperialism saw a shift from colonialism based around port cities to the government of whole populations. Colonial regimes worked their way down from collaborating with regional rulers until village heads became instruments of knowledge and power over populations. In the process, villages were redefined as discrete and governable units, and the structure of rural society was altered (see Breman). This contributed to the creation of a peasantry who could be taxed and monitored (see Elson). In the process, landless peasants were displaced, to become either rural laborers for their better-off associates, or to move to cities and become the kernel of urban workforces.

Colonial Labor

Because the colonies in Southeast Asia were sources of raw materials, they needed their own regimes of labor. The story of the transformation of the peasantry is one part of the creation of the labor regimes; the other is the manipulation of systems of worker management. Rubber, sugar, tobacco, and other products needed to be extracted as cheaply as possible, but by a labor force that could be controlled. Groups of colonial workers who were neither enslaved nor free to choose when and where they worked were the end-product of the new labor systems. These groups were also ethnically diverse, the labor regimes being the engines of large-scale migrations that have continued into the postcolonial period.

Problems in obtaining workers manifested themselves in the early days of European colonization. Local populations were usually mobilized by a series of ties that ranged in form from slavery to patronage, and settlers found that they had either to enter into these systems, or to import workers from elsewhere. In the Philippines, the Spanish promoted the establishment of large estates owned by private landlords or the Catholic Church, with a semi-feudal patronage of local people as a ready workforce. In other colonies there was a need for workers who were outside local networks of loyalty and obligation. In the 17th century, in cities such as Batavia (now Jakarta), canals and other infrastructure were created by encouraging Chinese migration to provide a labor force of "coolies" (see Blussé). More sophisticated systems of providing "coolie" labor developed by putting the workers in the debt of the labor suppliers. Thus were created the indenture systems by which laborers from poor areas were induced to sign up for periods of overseas work for which they initially had to pay back the cost of their passage.

The networks for importing Chinese labor were based mainly around the areas of southern China most affected by wars and natural disasters, where dire poverty pushed different language groups out into the broader world. These networks, operating through clan chains called *kongsi*, were highly developed in Southeast Asia by the middle of the 19th century, providing tin-miners in southern Burma, southern Siam/Thailand, northern Malaya, and Borneo (see Burns). Such Chinese networks often shaded into "secret societies," and while they were useful for the colonial authorities they were also regarded with great suspicion. The age of high imperialism saw hundreds of thousands of Chinese imported to become dockworkers in the major cities of Indonesia or rickshaw-pullers in the thriving entrepot of Singapore (see Warren 1986). Such labor forces were exclusively male, creating a ready market for prostitution. In the late 19th century, thousands of women were imported from China and Japan, especially into centers such as Singapore (see Warren 1993).

The British also made use of their colonial networks and turned South India into the major source of indentured labor for their colonies in Southeast Asia, the Pacific, and the West Indies. In total, as many as four million Indians migrated for work in the colonial period, producing an Indian population in the Malayan colonies of Britain that numbered more than 500,000 in the 1930s (Marshall pp. 286–87). As well as becoming plantation laborers, tin-miners, and coolies in rice mills and docks, in both Malaya and Burma, Indians also became moneylenders and supplied the core of the colonial army in Burma, giving rise to ethnic conflicts in the 1920 and 1930s (see Adas).

Colonial labor regimes also involved movements within colonies. During the era of high imperialism, large numbers of laborers were moved from northern to southern Vietnam to work in the plantations of the Mekong Delta, while impoverished Javanese signed what became infamous as the "coolie contract" to work on the tobacco and rubber plantations of Sumatra (see Stoler 1985). The French moved other Vietnamese into Laos, to work in coal mines or tin mines, or to join the civil service, while the opening of the Burma Delta involved new intermingling among the ethnic groups in Burma.

Indenture meant that workers had to sign contracts that they often did not understand, and that specified wages and conditions, including accommodation, food, and medical treatment. In most cases, the understanding was that workers would be repatriated at the end of their period of work: indeed, of the four million Indians who went to Malaya under the indenture system, perhaps as many as three million returned (Marshall p. 286). Some indentured workers did not even survive the journey to their place of work, such were the conditions on board the ships. Life on most plantations was hard, workers having to endure 12- to 14-hour shifts, with inadequate food, and beatings or other forms of harsh punishment. Floggings and even executions were meted out on some plantations to enforce the absolute power of the planters, who worked through Asian overseers. One Dutch labor inspector in North Borneo was so appalled by the workers' secret evidence that he "tried to cut his own throat and had to be taken down to the coast" (Allen p. 51). Europeans tended, conveniently, to blame the poor conditions and brutality on Chinese, Javanese or Vietnamese recruiters and overseers (Allen p. 170). During the colonial period, the plantation economies were at the mercy of world commodity prices, Southeast Asia being severely affected by the Great Depressions at the end of the 1890s and the 1920s. Large estates were a triumph of control of labor and a way to ensure that the products went to European markets at European prices, so they persisted until the end of the age of empire. Yet, ironically, postcolonial smallholdings in local hands, worked by those who have a choice in the matter, are more productive than the imperial plantations were (see Sairin).

Plantation areas were a type of frontier society, where planters were hard men living a hard life (Allen p. 91). The smaller-scale planters and plantation managers came from all over the world to make their fortunes, and regarded themselves as pioneers carving out a space in the jungles. (The ethos and pathos of the lives of these smaller-scale planters is captured in the short stories of W. Somerset Maugham.) Their attitudes were carried over when they worked on the larger estates of Pacific Dunlop in Malaya or Michelin in southern Vietnam: Europeans saw themselves as "cut off," their social world being restricted to whites-only clubs.

Imperial Government

The harshness of the plantations revealed one of the underlying problems of empire: that very small numbers of Europeans had to find ways to rule over large numbers of "natives." On the plantations, this had to be done in a way that maximized productivity and enforced submission. In government, policies of indirect rule achieved the necessary effects. The proportions were shockingly small. In 1900, there were only 91,000 Europeans in the Dutch East Indies, out of a population of 28 million. The colonial military consisted of nearly 40,000 men at that time (Porter p. 114). Among the Europeans, civil servants numbered in the hundreds. Many of those classified as "Europeans" were not Dutch. In Indochina in the same period, there were only around 9,000 French inhabitants (Stoler 1997 p. 227 note 2). In these circumstances, a mystique of European status had to be rigidly policed to maintain power.

Indirect rule worked in different ways in different regimes. In the majority of cases, it was presented as a continuation of "traditional rule." Thus, the royal families of Vietnam, Laos, Cambodia, and several of the Malay states were maintained in hybrid forms of their offices under French and British control. In the Dutch East Indies, various local aristocracies were maintained and converted into implements of Dutch rule. This conversion reached its highest form in Java, where the offices of the *priyayi* or aristocracy became the apex of a native civil service hierarchy that paralleled the positions of the Dutch civil service. At each level, Dutch officials were regarded as "older brothers" of their native equivalents. Although the Burman royalty had been abolished with the fall of King Thebaw in 1885, Burma was divided into two areas. The first was governed directly, while in the second traditional aristocracies, especially the Shan lords or princes, were maintained. Only in the Philippines was there no surviving form of local aristocracy, although the larger-scale *mestizo* landowners formed a client class for the Spanish and then for the US rulers.

As well as meaning that the burden of administration was passed over, indirect rule created the appearance of maintaining indigenous traditions. European administrators and a succession of travelers to Southeast Asia contributed to the development of romantic ideas of "the East," in which sultans and rajahs stood out as the most exotic "natives" of all. One renewed, or perhaps invented, tradition that administrators were keen to continue was that of forced labor for the lords. Arguing that such obligations were owed to the state, regimes throughout the region forced villagers to build roads, railroads, bridges, and other signs of development without any payment. At the same time, most colonial rulers denigrated their subjects as "lazy natives"; hence the need to import workers from other parts of Asia.

Throughout the latter part of the 19th century, colonies went through processes of administrative reform that created modern bureaucracies with specialized departments and hierarchies, and roles regularized on a scale of salaries. The field officers who had most contact with their subjects were given professional training in management, languages, and traditional law. The Dutch in particular believed in having highly trained "Indologists," as they called them, and the University of Leiden became the center where they studied (see Fasseur).

Studies of Asian languages had been important to colonialism before the era of high imperialism, although most often this had meant interest in ancient languages, above all Sanskrit, which was not in great use in day-to-day life in Southeast Asia. In the 19th century, language policies evolved as specialist administrators not only studied languages but were instrumental in promoting prestigious dialects as the preferred forms. Thus, the Dutch promoted the Javanese of the court of Surakarta, and both Dutch and British administrators sought an exemplary form of the Malay language, eventually choosing the dialect of the Riau archipelago (which took in Johor). While French, US, British and Portuguese officials promoted the language of their respective metropolis among colonial subjects, the Dutch chose to promote a version of Malay as the lingua franca of their East Indies, and not to allow the study of their own language. Speaking Dutch became the privilege of a small native elite, and separated the Dutch themselves from the wider society (see Maier).

The new forms of administration inaugurated styles of governmentality or intensifications of efforts to know about and manage all aspects of life. As departments of agriculture were set up, scientific forms of agriculture were introduced to increase production. New monitoring of health began with programs of smallpox vaccination and surveillance of prostitution. The fight against typhus and malaria involved attention to daily life and dwellings in villages. Likewise, forest services had been established even before the 19th century in Burma, Malaya, and Java, but their size and powers were increased in an effort to manage forestry resources. The increased powers resulted in government monopolies over forestry that had positive ecological effects but were greatly resented by local populations as intrusions upon traditional rights.

New, modern and "rational" forms of the state came into being throughout the colonial empires, with mixed effects. The different imperial administrations developed in close relationship to each other. For example, US officials in the Philippines learned from the experiences of the British in Malaya, just as plantation owners compared management systems across colonial boundaries. Steamships, railroads, telegraphs, electric lighting, and new types of industrial machinery, followed by radio, motor cars, and aircraft in the 20th century, matched the new governmentality. Despite the variations between colonies, imperialism did have similar and sometimes uniform effects across the region, as the collective agent of European modernity.

Finally, while Siam/Thailand was not colonized, it was subject to European legal and trade regimes thanks to the constant pressure on its borders from the competing powers of Britain and France. The ruling class realized that their survival depended on adapting European modernity, so European advisers were brought in to provide guidance on everything from the building of railroads and canals to the education system, while royal princes were sent to Europe to study and observe the military, political and administrative systems of Germany, France, and Britain. Using this knowledge, they restructured their country into a state with government departments, administrative districts, and internationally recognized boundaries.

Imperial Culture and Domination

There were cultural undertones dominating imperial thinking in each of the European powers. The French were generally convinced that their empire was undertaken as a *mission civilisatrice*, a "civilizing mission." Similarly, Dutch colonial policy was dominated by forms of paternalism that reached their height in 1901, when the government arranged for Queen Wilhelmina to proclaim an "Ethical Policy," under which the Dutch paid back the debt of resources from the Indies in the form of development, including education. British rule was paternalistic as well, but with markedly contrasting attitudes toward Burmese, Malays, Chinese, and Indians. Public opinion in the United States was ambivalent about its colony of the Philippines, given the importance of anti-British and anti-Spanish colonialism in its ideology, and this led to the grafting of a halfhearted form of democracy on the US model on top of the Spanish system, which had been organized along regional lines. Portugal had long been left behind in the imperial competition in Europe, and as a poorer European power it put little into its colony in East Timor, preferring to rule through a *mestizo* Portuguese-Timorese elite.

Portugal was also distinctive among the colonial powers for its more relaxed attitudes to "race" or ethnicity. For most of the other powers, "racial" difference became a key point of status for the ruling minorities. Even the admired local aristocracies were not considered the equals of the colonials, and in particular there was much anxiety about the idea of western women forming relationships with "native" men. The opening up of the Suez Canal had meant a rapid increase in the number of Europeans coming to Asia, in particular in the number of women, thus beginning to correct the huge gender imbalance among the colonists. These women were seen as the "mothers of empire," the bearers of a new kind of respectability. It was no longer considered permissible for European men to have local concubines, or children by them, and the status of the children of "mixed" relationships was severely reduced.

The management of ethnic difference in colonial societies has been termed "pluralism," a euphemism for the separation and separate government of different groups. In the European empires, "race" became a criterion for imposing different systems of law, and different economic, social and political roles, on different groups. "Pluralism" was a term coined by John Furnivall, a British colonial official turned critic who compared Dutch management of their East Indies with British government in Burma, favoring the Dutch. "Segmentation" is a more accurate term for the "divide and rule" policies that separated out different groups for discrimination in different colonies (see van Doorn). The economic role that the Dutch assigned to the Chinese in their East Indies, as opium and tax farmers as well as traders, has been one of the causes of the postcolonial ethnic conflict between Javanese and Chinese. In Burma, it was Indians who received privileged roles in the army and the civil service, and were thus the target of Burmese rioters. In Malaya, different colonial policies and attitudes pitted Malays against Chinese. As was the pattern elsewhere in colonial regimes, certain minorities became the crack troops of empire: the Karen in Burma and the Ambonese in the Indies formed elite forces in the colonial army, again to the chagrin of other ethnic groups. Yet even those who believed that the "civilizing mission" might eventually "raise" their subjects to the level of Europe thought that this was a process that would take centuries. They conceived of empire as a permanent institution, and were deeply shocked at its abrupt ending in the 1940s.

The laws of empire, based on "race," status, and European constructions of local traditions, were always being made to manipulate the complexities of colonial societies. As much as legal anxiety was focused on "race," sexuality kept crossing the barriers, and contradictions manifested themselves. Impoverished Europeans represented one such contradiction, and had to be banned from the colonies in order to maintain white status. For some, like the former colonial policeman George Orwell, the contradictions of maintaining the fiction of empire through domination were too much: they were to be captured in his novel *Burmese Days* (1934). However, many colonial administrators and educators became champions of local societies, sometimes crossing the boundaries to feed notions of independence and freedom among the subjects of empire. Many of the members of the new bureaucracies expressed their antipathy to modern European life, and carried that antipathy over into attempts to limit capitalism in the colonies. Some of the more sensational conflicts arose between planters and administrators, but the planters and large companies could appeal to powerful lobbyists in the metropolis to silence their critics in the civil service, or to cover up "scandals" and "affairs" that threatened their power (see Stoler 1992).

Such "affairs" revealed the many gaps in the systems of colonial rule. The Europeans' dependence on the "natives" gave power to the intermediaries in the system. In Java, regents or Javanese rulers with a good grasp of the system could play off colonial administrators against each other. The stereotype of the "inscrutable Oriental" proved to be a convenient device in the complex politics of the colony (see Onghokham). European officials rarely if ever entered villages, so that the cracks in the system opened up opportunities for village headmen to accumulate power, and gave rise to criminal networks of hitmen, bandits, and magicians (see Schulte Nordholt). These could hire themselves out to village headmen or regents, or operate labor-hire services that included provision of prostitution and gambling in the growing colonial cities (see Ingleson).

In most colonial systems, the administrators accepted that they were dependent on village or regional officials, and turned a blind eye to the ways in which taxation was collected or censuses were conducted. Officials who carried out their work too efficiently invited trouble. In Burma, taxation inspectors were often murdered or robbed. In Cambodia in 1925, when one colonial resident, Félix-Louis Bardez, became concerned about "corruption" in the collecting of taxes and went to inspect one of the villages for himself, he was beaten to death by the villagers, who acted collectively to reinforce their separateness from the administration (see Chandler).

Law and order presented constant problems in the colonies. Some of the British saw Burma as almost ungovernable, a site of constant "banditry." Every colony had uprisings against its rulers, often in the form of messianic or millenarian movements that attempted to bring about an abrupt return to native rule. The Hsaya San uprising in Burma in 1930 is one major example of an appeal for a return to Burmese kingship, which could only be put down by importing an extra 3,000 Indian troops, swelling to 12,000 the military numbers needed to suppress it (see Adas). Sometimes these revolts merged with more modern forms of revolution, as US forces found when trying to take over the Philippines. The major anti-imperial conflicts occurred around the time of the Great Depression between the world wars. There were Communist uprisings both in Indonesia, in 1926–27, and in Indochina, in 1930–31, focused in both cases around plantation areas.

Indigenous responses to imperialism ranged from total rejection to forms of accommodation. The aristocracies who could operate within the politics of empire often thrived, while avoidance was the most effective survival mechanism for those not born to status and advantage. Armed conflict against imperial power became one manifestation of what the Indonesian nationalist intellectual Sutan Sjahrir characterized as the psychopathic nature of empire: "on the one hand the sadists and those who suffer from megalomania, on the other hand the souls that are distorted by inferiority complexes" (as cited in Maier, p. 39). Such distorted relationships meant that modernity was localized in unique forms in the colonies. The children of the new middle classes among empire's subjects, whether they were the beneficiaries of the Burma Delta's rice boom or members of the "new *priyayi*" of the Dutch East Indies, whose upward mobility came through joining the civil service, represented this ambivalent localization. They benefited from western education, studied Marx and Lenin, enjoyed the prestige of being able to wear shoes, coats, and ties, and were frequently lambasted by Europeans as *déraciné*, cut off from their own cultures and traditions. These enthusiasts for modernity became the nationalists who were to take over at the end of empire.

Further Reading

Adas, Michael, *The Burma Delta*, Madison, WI: University of Wisconsin Press, 1974

Adas demonstrates the relationship between the colonial economy, social upheaval, and the development of the anticolonial movement in Burma.

Allen, Charles (editor), *Tales from the South China Seas: Images of the British in Southeast Asia in the 20th Century*, London: Futura, 1983

Based on oral histories of the British in Malaya, Sarawak, and North Borneo (Sabah), this book vividly depicts the prevailing British attitude toward colonialism.

Blussé, L., *Strange Company: Chinese Settlers, Mestizo Women and the Dutch in VOC Batavia*, Dordrecht: Foris, 1986

This collection of articles on aspects of colonial culture in the 17th and 18th centuries demonstrates continuities with, and differences from, the later era of high imperialism.

Breman, Jan, *Control of Land and Labour in Colonial Java*, Dordrecht: Foris, 1983

A detailed study of the relationships between land and labor policies and the restructuring of village society in Java

Burns, P.L., "Capitalism and the Malay States," in Hamza Alavi et al. (editors), *Capitalism and Colonial Production*, London: Croom Helm, 1982

This paper concentrates on Chinese capitalism before and during the early stages of British imperialism in Malaya.

Cain, P.J., and A.G. Hopkins, *British Imperialism*, in two volumes (Volume 1, *Innovation and Expansion 1688–1914*; Volume 2, *Crisis and Deconstruction 1914–1990*), Harlow and New York: Longman, 1993

In one of the most important recent studies of the economic bases of imperialism, Cain and Hopkins reassess the recent literature.

Castle, Robert, Jim Hagan, and Andrew Wells, "Labour Relations in the Rubber Plantations of Indochina, 1910–1940," paper given at the Indian Labour History Conference, New Delhi, 1999

A study of French attitudes towards labor and labor management, detailing the plantation regimes in Vietnam

Chandler, David P., "The Assassination of Resident Bardez (1925): Premonition of Revolt in Colonial Cambodia," in *Journal of the Siam Society*, Volume 70, numbers 1 and 2, 1982

A case study of how rational approaches to administration conflicted with the workings of the French colonial system

Clancy-Smith, Julia, and Frances Gouda (editors), *Domesticating the Empire: Race, Gender, and Family Life in French and Dutch Colonialism*, Charlottesville, VA: University Press of Virginia, 1998

A collection of papers comparing French and Dutch attitudes toward gender and domestic life

Cooper, Frederick, and Ann Laura Stoler (editors), *Tensions of Empire: Colonial Cultures in a Bourgeois World*, Berkeley: University of California Press, 1997

This collection focuses on how imperialism was both shaped by the development of bourgeois culture and contributed to the development of that culture. The introduction provides an important overview of the recent literature on imperialism, particularly the debates of postcolonialism, and the book as a whole is a major example of cultural approaches to imperialism.

Cribb, Robert (editor), *The Late Colonial State in Indonesia: Political and Economic Foundations of the Netherlands Indies 1880–1942*. Leiden: KITLV Press, 1994

A collection of articles reassessing economic, military, administrative and social aspects of Dutch colonialism

Doorn, J.J.A. van, "A Divided Society: Segmentation and Mediation in Late Colonial Indonesia," in Gerrit Schutte and Heather Sutherland (editors), *Papers of the Dutch-Indonesian Historical Conference held at Lage Vuursche, The Netherlands, 23–27 June 1980*, Leiden: Bureau of Indonesian Studies, 1982

This article examines the problems of "racial" classification and social division in the Dutch East Indies. In particular, it looks at cases that crossed the boundaries, and the way in which Dutch people born in the archipelago attempted to create their own "Indies" society.

Elson, Robert E., *The End of the Peasantry in Southeast Asia: A Social and Economic History of Peasant Livelihood, 1800–1900s*, London: Macmillan, and New York: St Martin's Press, 1997

By examining the different economic, social and cultural issues that shaped the lives of peasants in Southeast Asia in the 19th and early 20th centuries, Elson demonstrates the close relationship between imperialism and the formation of the region's modern peasantry.

Fasseur, Cijs, *De Indologen: Ambtenaren voor de Oost, 1825–1950* [The Indologists: Civil Servants for the East, 1825–1950], Amsterdam: Bert Bakker, 1993

A detailed study both of the administrative structures that the Dutch established in the East Indies, and of the personnel and training of the administrators

Fieldhouse, D.K., *The Colonial Empires: A Comparative Survey from the Eighteenth Century*, second edition, London: Macmillan, 1982

Written from a perspective combining attention to politics and to economics, this is a definitive study of the nature of empire and the transformations of empire.

Furnivall, John S., *Colonial Policy and Practice: a Comparative Study of Burma and Netherlands India*, Cambridge: Cambridge University Press, 1948; New York: New York University Press, 1956

Furnivall's comparison between the two colonial systems is particularly critical of the British. This work provided the terms of much of the debate about colonialism, including the idea of "plural societies" (although Furnivall's term "Netherlands India," then used as an alternative for "Dutch East Indies," has passed entirely out of use).

Havinden, Michael, and David Meredith, *Colonialism and Development: Britain and its Tropical Colonies, 1850–1960*, London and New York: Routledge, 1993

Although this study does not include India or Burma, it demonstrates that the British empire did not provide "development" for its colonies, in the sense that it did not provide the basis for prosperous societies.

Hobson, J. A., *Imperialism: A Study*, third edition, London: Allen and Unwin, 1938; Ann Arbor: University of Michigan Press, 1965

One of the key works in the literature on imperialism and economics

Hong Lysa, *Thailand in the 19th Century*, Singapore: Institute of Southeast Asian Studies, 1984

Hong demonstrates the effects of imperialism on Siam/Thailand by outlining the economic and social consequences of the "free trade" agreements forced on the country.

Ingleson, John, *In Search of Justice: Workers and Unions in Colonial Java, 1908–1926*, Singapore, Oxford, and New York: Oxford University Press, 1986

This study of the development of urban proletariats under the Dutch looks at the relationship between the early labor unions and the nationalist movement.

Lenin, V.I., *Imperialism, the Highest Stage of Capitalism: A Popular Outline*, originally published in 1917, Moscow: Foreign Languages Publishing House, 1952

Lenin's thesis that imperialism was "the highest stage" of capitalist growth was developed primarily to explain the origins of World War I. It has been largely disproved, but the text remains influential because it has provided the terms for debating the relationship between imperialism and capitalism.

Maier, H.M.J., "From Heteroglossia to Polyglossia: The Creation of Malay and Dutch in the Indies," in *Indonesia*, number 56, October 1993

Through an assessment of language policies Maier looks at their role in separating Dutch from Malay in the psychology of imperialism.

Marshall, Peter James (editor), *The Cambridge Illustrated History of the British Empire*, Cambridge and New York: Cambridge University Press, 1996

This collection of articles by leading authors on imperialism is organized into two sections, the first surveying the major period of imperialism, the second providing essays on political, social, economic and cultural aspects of empire.

Murray, Martin J., *The Development of Capitalism in Colonial Indochina (1870–1940)*, Berkeley: University of California Press, 1980

Murray examines all aspects of capital investment, labor, and colonial policy in Indochina, within a Marxist framework.

O'Brien, Patrick, "Imperialism and the Rise and Decline of the British Economy, 1688–1989," in *New Left Review*, first series number 238, 1999

In this survey of the literature on the British economy and empire, O'Brien's arguments revolve around the contradictions of liberalism in imperial policy.

Onghokham, "The Inscrutable and the Paranoid: An Investigation into the Sources of the Brotodiningrat Affair," in Ruth T. McVey (editor), *Southeast Asian Transitions: Approaches through Social History*, New Haven, CT, and London: Yale University Press, 1978

A case study of the fraught relationship between Dutch and indigenous civil services and systems of rule in the late 19th century

Porter, Andrew, *European Imperialism, 1860–1914*, London: Macmillan, 1994

In this survey of the literature on imperialism, Porter classifies it into broad approaches to explaining imperialism that are either "metropolitan" or "peripheral." He further divides the "metropolitan" explanations into those that are politically based and those that concentrate on economy and society. The coverage is weakest on cultural approaches to imperialism, but this is a useful guide nonetheless.

Sairin, Sjafri, "The Appeal of Plantation Labour: Economic Imperatives and Cultural Considerations among Javanese Workers in North Sumatra," in *Sojourn*, Volume 11, number 1, 1996

A study of postcolonial plantation labor with critical implications for the colonial literature

Schulte Nordholt, Henk, "The Jago in the Shadow: Crime and 'Order' in the Colonial State in Java," in *Review of Indonesian and Malaysian Affairs*, Volume 25, number 1, 1991

In this examination of the "gaps" or "shadows" in the Dutch colonial system, the author demonstrates how imperfectly Dutch ideas of rational administration were realized on the village level, and how criminality developed within these "shadows."

Stoler, Ann Laura, *Capitalism and Confrontation in Sumatra's Plantation Belt, 1870–1979*, New Haven, CT, and London: Yale University Press, 1985

A major study of resistance and conflict in the rubber and tobacco plantations of Sumatra

Stoler, Ann Laura, "'In Cold Blood': Hierarchies of Credibility and the Politics of Colonial Narratives," in *Representations* number 37, Winter 1992

Stoler examines a colonial murder case and the ways in which it was represented, both in the reports that entered the official archives and in the restricted versions presented to the public.

Stoler, Ann Laura, "Sexual Affronts and Racial Frontiers: European Identities and the Cultural Politics of Exclusion in Colonial Southeast Asia," in Cooper and Stoler (editors), cited above

Stuart-Fox, Martin, *A History of Laos*, Cambridge, New York, and Melbourne: Cambridge University Press, 1997

The author places the history of French rule in Laos in the broader context of the country's history.

Taylor, Robert, *The State in Burma*, London: Hurst, and Honolulu: University of Hawaii Press, 1987

In this history of the idea of the state in Burma, the chapters on the colonial period demonstrate how relations between the state and the villages were restructured under the British.

Warren, James Francis, *Rickshaw Coolie: A People's History of Singapore, 1880–1940*, Singapore, Oxford, and New York: Oxford University Press, 1986

The majority of the rickshaw-pullers who are the subjects of this study were Chinese indentured workers. Warren outlines in depressing detail the circumstances of their brief and impoverished lives.

Warren, James Francis, *Ah Ku and Karayuki-San: Prostitution in Singapore, 1870–1940*, Singapore, Oxford, and New York: Oxford University Press, 1993

As in his study of the rickshaw-pullers, Warren draws on police and court records to reconstruct the social lives of the prostitutes of Singapore, most of whom were either Chinese (the Ah Ku) or Japanese (the Karayuki-San).

Wertheim, W.F., *Indonesian Society in Transition*, second edition, The Hague: Van Hoeve, 1959

One of the first critical studies of colonial and postcolonial society in Indonesia, Wertheim's work is particularly important for his attention to the class roles of the "new *priyayi*" who developed within the "native" civil service, and for the chapter on changing forms of labor.

Dr Adrian Vickers is Associate Professor of Southeast Asian History in the History and Politics Program at the University of Wollongong, New South Wales. He is one of the editors of the *Review of Indonesian and Malaysian Affairs*, and his many works on the history and culture of Indonesia include *Bali: A Paradise Created* (Harmondsworth, and New York: Penguin, 1989).

HISTORY AND CONTEXT

Chapter Three

World Economies: Southeast Asia since the 1950s

John Minns

In 1994, the respected economist Song Byung-nak predicted that by the year 2000 all the nations of Southeast Asia would have graduated to the status of "newly industrializing countries" (NICs) and that the center of the world economy would have shifted to East Asia by then (Song p. 218). Another estimate made at around the same time, by Anne Booth, suggested that by 2020 Indonesia would be the fifth largest economy in the world, after China, the United States, Japan, and India, while Thailand would be eighth, with Germany and South Korea in between (Booth p. 28.) A few analysts were suspicious of these forecasts, pointing, for example, to inefficient forms of production and low levels of labor productivity (see Krugman), but most were carried along with the prevailing euphoria and predicted waves of Asian "miracles." Barely half a decade later, with the experience of the crisis of 1997 intervening, these predictions now appear grossly and falsely optimistic.

Nevertheless, such optimism had some basis in fact. Some of the countries of Southeast Asia have per capita incomes among the highest outside the advanced industrial world, and Brunei, Indonesia, Malaysia, the Philippines, Singapore, and Thailand have all experienced significant economic growth since the mid-1960s (Oshima p. 203). Indeed, economic expansion in some of these countries has been faster than almost anywhere else, apart from the "miracle" economies of South Korea and Taiwan. Sometimes, then, optimism was based on simple but illegitimate extrapolations from growth that had already been achieved.

Sometimes, however, it was based on a lack of knowledge of the region, its great diversity, the stark differences in living standards within it, and the ways in which Southeast Asia as a whole, and its component parts, have been positioned within the world economy. To appreciate these questions we must begin with the structure and location of these economies as most of them became politically independent after World War II.

The Colonial Inheritance

The colonial powers oversaw economies in which primary production predominated. Although important local variations existed, it is broadly appropriate to view the late colonial economy as a triadic system. Long-distance, especially international trade, and the little manufacturing industry that existed, were largely in the hands of Europeans, along with plantation agriculture and mining. Most food for domestic consumption was produced by the native peoples, who also engaged in as well as small-scale handicrafts. Connecting these two economies was a class of merchants and moneylenders, often Chinese but occasionally Indian in background, who collected and traded produce, and provided small-scale credit to peasants (Wu and Wu p. 48). Trade and foreign investment connected colony to colonial power, rather than colony to colony: the tin and rubber of the Malay peninsula largely went to Britain in British ships and was sold there by British companies, and sugar went directly from the Philippines to the

United States. Intraregional trade was limited, since the first destination for each of the cash crops and minerals produced in the region was the country that had colonized it rather than any more diverse range of advanced, industrialized markets. Foreign investment came largely from the same colonial power.

These well-established patterns were disrupted by the advent of political independence, although to varying degrees. The French colonies in Indochina, in armed revolt against Paris, rapidly broke most of their economic ties with French firms upon declaring independence, or even beforehand. Many Dutch firms returned to Indonesia, although with reduced influence, after its war of independence, but most were then nationalized in 1957. British companies fared somewhat better in Malaya, and then Malaysia, until the early 1970s, when their interests in plantations and tin mining began to be acquired by the government. Burma cut its ties with British capital, and with the outside world as a whole, in a much more decisive way, even refusing to seek membership of the Commonwealth (Guyot et al. p. 190). More US companies survived and continued investment in and trade relations with the Philippines. Overall, however, few of the enterprises that had been dominant in the colonial period were able to make the transition to operating in the new environment after independence (Yoshihara p. 17).

Among the reasons for this discontinuity was the new geometry of global economic power after World War II. Britain, France, and the Netherlands, the great colonial powers of the region, had entered into long-term relative decline, and were displaced by the United States, a colossus that by 1945 accounted for half of the manufacturing output of the world and one third of its trade. First the United States and then, from the 1960s, Japan took over the dominant position in Southeast Asia's economies, occupying the top tier of their triadic systems.

Anticolonialism and postwar nationalism were propelled, at least in part, by economic motivations. As nationalism grew and succeeded in its political aim of independence, its economic aspects came to the fore. In Latin America in the 1950s and 1960s, theorists such as Raúl Prebisch were beginning to develop a radical critique of the exploitative relationship that had long existed between advanced and underdeveloped countries (see Prebisch). These theorists suggested that trading and other connections between rich and poor were the cause of underdevelopment, not a solution to it. Andre Gunder Frank suggested in the 1960s that underdevelopment was not a state of existence but a process, coining the phrase "the development of underdevelopment" (see Frank). Underdevelopment was something that had been done to poor countries by rich ones and it continued despite political independence. The potentially revolutionary implications of these ideas were made explicit in theories of dependency, its theorists often being known as *dependistas* in recognition of their origins in Latin America (see Harris for a short outline of the rise and decline of these theories).

However, not all those who were influenced by the ideas associated with dependency theory were revolutionary nationalists. Most leaders of newly independent states saw themselves as reforming nationalists, working to bring about a gradual upgrading of their countries' positions within the world system. Yet even they accepted that doing so required challenging the economic roles of primary producer and raw material supplier that the long colonial experience had assigned to them. To some degree at least, they demanded a break with colonial systems of production.

Industrialization and Urbanization

The key element in the economic transformation proposed by reformists and revolutionaries alike was to be a shift away from supplying primary products toward industrialization, which was widely seen as the hallmark of the wealthy former colonists. In some ways, especially given the context of the 1950s and 1960s, the adoption of such a goal represented a radical defiance of the colonial and neocolonial world order. Since then, industrialization and its likely partner, urbanization, have become and remained the aims of all the capitalist states of Southeast Asia.

A corollary of such ambitions was that, although agriculture occupied the vast mass of the workforce and dominated national production, it was considered to be secondary at best: its part in the economic strategies of the new governments was chiefly to subsidize manufacturing industry. Planners hoped that, where possible, agriculture could keep food prices low and in some cases governments intervened to force them lower, thus reducing the costs of urban industrial workforces. In other words, agriculture was to be bled to animate industry (Dixon pp. 149–50). For most of the 1950s, agricultural goods and minerals generally fetched good prices on the world market as a result of the commodity boom created by the Korean War, but by the end of the decade prices were falling rapidly relative to those of manufactured goods. Postwar capitalist production tended to use fewer inputs of raw materials (with the important exception of oil) and more inputs of capital-intensive technology in its products. The invention of numerous synthetic materials, given a boost by the necessities of wartime production, also undermined demand for natural ones. Finally, from the 1950s, capitalist production techniques were being applied more intensively to agriculture in the advanced countries, which were increasing production and reducing prices. The United States soon became the world's largest exporter of vegetable oils and by 1967 it was the world's leading exporter of rice (Fryer p. 14). As a result of all these trends, the terms of trade turned against Southeast Asia and the bias against agriculture was reinforced. With such political considerations reinforcing worldwide economic trends, the proportion of the workforce employed in agriculture fell, while the proportion employed in manufacturing rose almost everywhere.

For the dependency theorists, the major means by which industrialization was to be achieved was state intervention and planning. There were many models to follow. State intervention had been crucial to "late development" in Belgium, Germany, and Japan in the 19th century, and had become so once again in the reconstruction of the Japanese economy after US bombs had reduced much of that country to cinders in the latter stages of the Pacific War.

Perhaps the key model of state-led growth for poor countries in the 1950s and 1960s, even though it was rarely acknowledged, was the experience of the Soviet Union. It seemed to many that the transformation of this once overwhelmingly agricultural country into an industrial power, capable of outproducing West Germany and then of rivaling the United States, and all within a generation, had blazed a trail, while the weaknesses in that growth were yet to become apparent (see, for example, Myrdal pp. 726–27). However, state planning and a degree of state ownership had not been ruled out by the capitalist West in this period. Many social scientists and policy-makers in the United States, as well as in international institutions such as the World Bank, saw themselves as being engaged in developing a superior, democratic form of planning that was to underpin the mortal combat with Communist planning.

In any case, colonialism had left behind little in the way of a domestic bourgeoisie that might have provided the dynamism necessary to industrialize Southeast Asia societies. To the extent that such a class existed, it was often predominantly Chinese and therefore not of those ethnic groups which were central to the new nationalisms. Statist planning, on the other hand, offered a central role in the transformation of their countries to the nationalist intelligentsia and, sometimes, the armed forces, through their positions in the state apparatus. Thus it joined the material interests of these middle-class layers with an apparently high moral purpose: they could tell themselves and others that the advancement of their careers and the interests of the nation were intertwined. While the nationalist movements of the late colonial period had largely middle-class or elite leaderships, they needed to appeal to a mass audience by suggesting that national independence would create better lives for all. Hence, planning and economic intervention by the state became associated with the apparently radical rhetoric of domestic equity (Owen p. 470), and the "forced march" to industrialization, involving state control of external economic links and mobilization of domestic resources, could be presented as helping to redistribute wealth, not only between rich and poor countries, but also between the few rich and the many poor within each country.

Expanding the managerial activities of the state in the first two decades after independence was seen by some as part of a transition to socialism, combining greater equality (or at least paying lip service to it) with state management. It was not just those in the Communist parties of the region who understood state intervention in this way. At various times, Ne Win of Burma, Norodom Sihanouk of Cambodia, Lee Kuan Yew of Singapore, and Sukarno of Indonesia all claimed to be "socialists" of some kind. Even so, for the most part the expansion of state activity did not involve major conflicts with private capital. There were exceptions, such as the nationalization of Dutch companies in Indonesia in December 1957, and for all practical purposes the state monopolized markets and production in Burma after 1962, as well as in Vietnam, Laos, and Cambodia after the victories of their Communist parties over the course of 1975. Nevertheless, in general the states of Southeast Asia acquired enterprises or set them up to provide infrastructure and begin production primarily in industrial fields where private capital had not established itself. For example, Thailand experimented with state production monopolies under Phibun Songkhram, who was in power from 1948 to 1957, and although subsequent governments reduced the scale of state enterprise, they retained much of the state's role in the provision of infrastructure. Similarly, while Singapore invited foreign private capital to develop manufacturing industry after its separation from Malaysia in August 1965, the state also took a massive role for itself in providing public housing, training, and education, with the result that about one third of all investment in the city state came from the public sector. In Indonesia under Sukarno, the slogans of "Guided Democracy" and "Guided Economy," made it clear to all that the state was to be the guide in production as well as in politics.

State planning remained in vogue throughout the 1970s, even where full state ownership had never been widespread, as in the Philippines, or had been significantly reduced, as it was in Indonesia, under the "New Order" introduced after 1965, as well as in Thailand. Indonesia's first five-year development plan was implemented from 1969 onwards, notwithstanding the commitment of the Suharto government to private capitalist development. The New Economic Policy was instituted in Malaysia in 1971 under a similarly pro-capitalist government. In Thailand, despite the turn away from public enterprise after Phibun's fall from power, the National Economic Development Board was established in 1959 to maintain a degree of state planning, and key changes in the direction of the economy were introduced under the Third Five-year Plan from 1973. The desire to maintain state planning was also the reason for setting up the National Economic Development Authority in the Philippines in 1973. Even the tiny absolute monarchy of Brunei, which has long been completely reliant on oil exports, began implementing five-year plans as early as 1954, building a state apparatus that has since become the largest single employer in the country (Ali pp. 282, 287).

The nationalists of postwar Southeast Asia saw their key task as being to break away from the role of raw material supplier assigned to their countries by colonialism, in the belief that failure to do so was likely to lead to deepening poverty in a world dominated by industrial societies. Both nationalism and future prosperity demanded industrialization. Yet the human and physical resources required to begin the process were in short supply, and the lack of a substantial indigenous bourgeoisie or significant amounts of capital drove every country in the region to make some attempt at state planning and/or ownership between the 1950s and the 1980s. The location of the key nationalist leaderships within the state apparatus enhanced the argument for state economic tutelage. The attempt was made whether or not the government was committed to private capitalism, and whether or not a country was formally labeled "socialist" or "liberal democratic."

Import-substituting Industrialization

The strategy to achieve industrialization adopted, to varying degrees, by the region's planners was known as "import-substituting industrialization." Processing and manufacturing industries were to be developed to produce goods for the domestic market, and were to be protected from foreign competition by the state; over time, these industries would become strong enough to break out of the domestic market and begin exporting their products. At least in the early stages, domestic production was to center on satisfying mass consumer demand, for textiles and clothing, processed food, light and simple metal products, and other products that required relatively small amounts of capital investment and low levels of technology. Accordingly, states established a complex array of tariffs, controls on imports, and incentives and subsidies for manufacturers. The resulting higher prices of manufactured goods were considered to be a burden worth bearing if the outcome was an industrialized economy.

The Philippines, which had the best industrial infrastructure in the region at the end of World War II, seemed to offer a model of successful import substitution during the 1950s and early 1960s. Its imports of consumer goods fell from nearly 31% of total imports in 1948 to less than 5% in 1965 (Cho and Williams p. 230), while the growth of GNP per capita averaged 3.6% a year in the 1950s, the highest in Southeast Asia at the time (Oshima p. 75). Import substitution also appeared to work well in Malaysia, once it had brought production of all kinds back to the levels achieved before the war. Output from its manufacturing industries grew at a very respectable average annual rate of 17% between 1959 and 1968 (Cho and Williams pp. 236–37), and GNP per capita, which grew at a mere 1% a year in the 1950s, grew at 3.3% a year in the 1960s (Oshima p. 75). In the case of Singapore in the early 1960s, the exploitation of the small manufacturing sector developed under British rule to supply consumer markets elsewhere in Malaysia led to an apparently unequal division of the gains from import substitution, and contributed to Singapore's departure from the federation (Dixon pp. 157, 159). Thailand also pursued import substitution, with a heavy emphasis on supporting the private sector from 1961, when its first development plan began to be implemented (Jansen p. 15). Starting from a very low base as one of the poorest and most stagnant economies in the world in the 1950s, Thailand saw its GNP per capita increase by an annual average of 4.7% in the 1960s (Oshima p. 75). This growth was led by the building of infrastructure, but a significant import-substituting manufacturing sector had also begun to be developed (Warr pp. 29–30). Finally, Indonesia's Eight-year Plan, published in 1960, was strongly based on the goal of import substitution, in the hope that the country would become capable of producing all its own food, clothing and other basics in just three years (Hill p. 2).

Indigenism

Making up around 6% of the population of the region, the Chinese of Southeast Asia have played a crucial role in its economy for generations. Collecting local produce from farmers and marketing it, they were often the only group, apart from the colonists, with enough capital to be able to make loans to these same farmers. As a result, they were often accused of "taking their pound of flesh twice," first in trade and then in interest (Wu and Wu p. 49). Of course, most Chinese were not rich, but they were far more likely to operate businesses, especially retailing, than the people they lived among were. As a result, during the colonial period the Chinese occupied an intermediate position between the colonists, to whom they were useful commercial intermediaries, and the "indigenous" peoples.

Independence altered the situation. With their trading and moneylending activities under attack, and with many European firms having departed, the richer members of the Chinese communities attempted to shift their activities into manufacturing and mining (Wu and Wu p. 50), only to come up against the problem of economic nationalism, which was based, not on redistributing wealth and control among all residents within the boundaries of each state, but on favoring those ethnic groups whose elites had become politically dominant. Frank H. Golay and his co-authors are not alone in concluding that economic redistribution along ethnic lines came to dominate policy-making in these newly independent societies (Golay et al. pp. 7–8).

Such policies of "indigenism" attempted to transform the ethnic dimensions of the economies inherited from colonialism, and played an important part in economic restructuring, at least in Malaysia, Indonesia, Singapore, and Burma. Indigenism was a factor in the split between Singapore, where Chinese make up the majority, and Malaysia, where Malays are the largest group. The anti-Chinese riots of May 1969 prompted the Malaysian government to introduce the New Economic Policy, which explicitly sought to redistribute wealth along ethnic lines. Two main goals were proclaimed: the transfer of western-owned companies to Malays, so that they could control 70% of the economy by the year 1990; and the ownership by Malays and other "indigenous" peoples of at least 30% of the economy by the same date. Public enterprises were established to buy foreign businesses until the time came when they could be operated privately. The result was that state-owned companies came to dominate the domestic, import-substituting economy (Wong Tai Chee p. 106). Malaysia had been slower to make the shift toward a high level of state control than other countries in Southeast Asia. Now, driven by racism, the state was to play the leading role (Khoo p. 50). In Indonesia, discrimination against Chinese businesses was even more direct and began even earlier. In 1959, Chinese retailing in villages was simply declared illegal, and the systematic promotion and protection of the interests of "indigenous" (*pribumi*) businesses became, and has remained, a crucial government policy (Wu and Wu p. 61). In postcolonial Burma, meanwhile, many Indian-owned businesses were forced to leave the country and anti-Chinese prejudices were often encouraged by successive governments (Owen p. 477).

Chinese-owned businesses fared better in Thailand and the Philippines. Because Thailand was never formally colonized, the Thai elite had been able to retain its political power and had not felt so threatened by Chinese commercial activity (McVey p. 19). After World War II, Chinese businessmen in Thailand managed to consolidate their position by forming close links with the military and bureaucratic elites (Falkus p. 28). As for the Philippines, while there is often popular resentment against Chinese business, the landowning class had become tied up with Chinese merchants from an early period and felt less need to challenge them after independence (McVey p. 19).

Where "indigenism" resulted in action against Chinese or other minorities, it reinforced postcolonial tendencies toward state control of economic activity. In addition, however, because it discriminated against the most internationally or regionally oriented sections of capital, it propelled policy-makers even further in the direction of autarky than the concept of import substitution alone might have taken them. In the extreme case of Burma, near-total isolation ensued. The way in which the new states of Southeast Asia approached the world economy after World War II can thus be seen as a complex mixture of the economic heritage left by colonialism, the nationalist ideologies developed in response to that heritage, and the interests of the ethnic elites that predominated in the newly independent states.

Boom and Cold War

The world economy experienced boom conditions in the 1950s and 1960s: each year, the industrialized economies grew by around 5% and world trade increased by around 10%. Such sustained high rates of growth were unprecedented. The boom might have created the ideal circumstances for a poor commodity-producing region such as Southeast Asia to begin to transform itself, but the fruits of the boom were not evenly spread across the world system. While all the economies of Southeast Asia benefited from it in some way, in the long term the boom tended to marginalize them further. Sophisticated production expanded, but it was overwhelmingly based in the United States, western Europe, and Japan. The newer methods of manufacturing, employing high inputs of capital and technology, were expensive and beyond the reach of developing economies. What little manufacturing capacity existed in Southeast Asia became even more outdated and uncompetitive in world markets. Import substitution, with its emphasis on the domestic economy, was further entrenched as a result.

Further, the expansion of world trade was largely a result of increases in trade between the wealthy countries. Even in 1965, at the height of the boom, Southeast Asia still supplied the United States with only 3.7% of its imports and took just 2.8% of its exports. In other words, the whole of Southeast Asia was of less economic importance to the United States than Mexico was. Western Europe had even fewer economic connections with Southeast Asia: in the same year, the six member states of what was then the European Economic Community together took around 1.5% of the region's exports and accounted for roughly the same proportion of its imports (Fryer p. 11). Japan was beginning to pay more attention to the region as its own economy recovered and moved ahead, but overall Southeast Asia remained something of a backwater, of scant economic interest to the advanced countries.

However, while advanced capitalism cared little for the economies of the region, it was obsessed with its politics. The anticolonial struggles in Dutch Indonesia, British Malaya, and French Indochina gave way to clashes that had major ramifications for superpower conflict. Until 1975, the most important link between the countries of Indochina and the rest of the world was through war. Nor was conflict absent from the rest of the region: Communist revolts were sustained in the Philippines from 1946 to 1954, and in Malaya from 1948 until the "emergency" was formally ended in 1960, while the largest Communist Party outside the Soviet bloc and China made its presence felt in Indonesia until most of its members were massacred in 1965 and 1966.

As a site for Cold War contestation, Southeast Asia attracted massive Western military forces and military-related funds. Apart from Korea and Indochina, where US troops became directly embroiled, other countries effectively became "frontline" states. Thailand received vast amounts of military assistance, amounting to nearly 60% of the national defense budget, between 1950 and 1975, as well as US$1 billion in subsidies for US bases and Thai troops in Vietnam, and another US$650 million in economic aid (Owen p. 479). The economy of the Philippines also came to depend in part on direct US aid and the installations at Clark Air Base and Subic Bay Naval Base. Both Malaysia and Singapore received substantial amounts of Cold War funding, and aid was crucial to Indonesia's stabilization program, implemented under the "New Order" regime between 1966 and 1968. In addition, most countries in Southeast Asia, as well as several beyond the region, benefited from contracts associated with the war in Vietnam.

The End of National Economic Independence

All three countries in Indochina were devastated throughout the 1950s, the 1960s, and the first half of the 1970s: their peoples were barely able to survive, let alone undertake a process of industrialization. Meanwhile, the real incomes of most of the other countries in Southeast Asia went on growing, but not nearly as rapidly as the nationalists demanded, the masses expected or the leaders had promised. The apparent success of import substitution reflected the extremely low industrial base from which the new states began, since the addition of small amounts of manufacturing to economies where there had been very little could produce spectacular growth rates, at least for a time. In addition, the first phase of import substitution was bound to be easier than later stages of industrialization because it concentrated on finding substitutes that could most readily be produced domestically, with basic technology and small capital investments. Further success required much more of both, making the maintenance of high growth rates ever more difficult.

In any case, the structure of manufacturing had not developed in line with the original hopes of the policy-makers. The production of consumer goods tended to dominate, while the production of capital goods had barely begun, which meant that expensive machinery had to be imported, along with the raw materials that each economy lacked. The nature of the import pile had changed, but balance of payments problems remained. There was little sign that any of the industries developed behind protective walls was capable of expansion into the world market. Thus, throughout the 1960s and 1970s exports from Southeast Asia were still overwhelmingly primary products, as they had been under colonialism, and each country's exports tended to be dominated by just two or three primary items, making them extremely vulnerable to changes in world prices. By 1973, timber, copper, and sugar still accounted for 51.2% of exports from the Philippines, rice, rubber, tin, and corn for 40.9% of Thailand's exports, rubber and tin for 46% of Malaysia's exports, and wood, rubber, oil, and petroleum products for 81% of the value of Indonesia's exports (Wu and Wu p. 21). The oil price "shocks" of 1974 and 1979 brought crises to the oil importers, Thailand and the Philippines, even as they assisted the economies of Indonesia, Brunei, and (to a degree) Malaysia.

Signs that import substitution was not succeeding, and pressure to switch away from it, were already apparent in the early 1960s. By the end of that decade, most countries in Southeast Asia had begun to shift toward "export-oriented industrialization." Neither strategy was ever implemented in pure form, and elements of each coexisted in reality, but by the early 1970s the new focus on exporting was predominant in the region and had become closely linked to promoting openness to foreign capital investment.

Singapore, with its tiny domestic market, was the first to shift to an export-oriented strategy, starting soon after its attainment of full independence in 1965. It also opened its doors to foreign companies, providing them with extremely generous tax concessions, state-provided infrastructure, land, and relatively cheap labor. Thailand had been among the most open of the region's economies since the late 1950s and had not gone so far down the road to import substitution (Falkus p. 15). In the Philippines, the problems associated with import substitution in a small and impoverished economy had begun to make themselves felt by the 1960s, when the economy slowed down. However, pressures to expand exports were resisted by powerful domestic manufacturing interests and although economic growth picked up a little in the 1970s, to average 3% for most of the decade, the Philippines became one of the least successful economies in the region. Government and private debt accumulated alarmingly, especially as a result of the second oil shock, and income per capita contracted by 20% between 1980 and 1986 (Owen p. 489). Malaysia made a long-term attempt to develop heavy industry, including steel, shipbuilding, petrochemicals, and cars, from the 1970s onwards, much of it on the basis of state ownership. The country's balance of trade was somewhat cushioned by the richness of its resource base, but at the same time the need for manufacturing exports was beginning to make itself felt. Government policy began to provide more generous concessions for exporters from the early 1970s. In Indonesia, which pursued one of the most ambitious import substitution strategies in the region, the attempt to become self-sufficient in consumer goods collapsed within four years of its introduction in the Eight-year Plan of 1960. In fact, the economy failed to grow at all in real terms between 1961 and 1964 (Hill p. 2).

The New International Division of Labor

The new emphasis on export promotion was associated with a major shift in world capital investments that had a dramatic effect on Southeast Asia. In the 1950s and 1960s, foreign capital had mostly entered the region either to extract its raw materials or to use low technology to manufacture products for domestic consumption. That pattern began to change in the late 1960s, as some firms from advanced countries began to relocate parts of their more labor-intensive operations to the region. Japanese companies led the way in taking advantage of the cheap labor of Southeast Asia. These investments were different from earlier ones, in that they were designed to produce manufactured goods to be exported back to advanced countries. Becoming more established in the 1970s, by the late 1980s and early 1990s they had created massive manufacturing booms in several economies in the region. This rearrangement of world production, involving the partial transfer of labor-intensive production from the advanced industrial countries to some of the developing countries, became known as the "new international division of labor." An obvious manifestation of this trend was the proliferation of free trade zones and export-processing zones in Southeast Asia.

The spread of export-oriented manufacturing based on foreign capital investment was given a dramatic boost in the 1980s by an important restructuring of the world economy. The key element in these changes was the long-term growth of the Japanese economy. By the mid-1980s, as a result of its success in exporting, Japan had piled up massive trade surpluses and foreign currency reserves. The response from its major competitors in the United States and western Europe was to demand that Japan reduce its exports and begin to take more imports from them. They raised protectionist barriers to Japanese goods and demanded an adjustment of international currencies. The ensuing Plaza Accord of 1985 helped produce an appreciation of the yen by about 50% in the following two years. The high yen made exports from Japan less competitive but also made assets elsewhere extremely cheap for Japanese buyers. The obvious strategy for Japanese companies was to buy productive assets overseas, in many cases in Southeast Asia, and use them as an export platform. The strategy was reinforced because of another consequence of Japan's industrialization: its transformation into a high-wage economy. Since relatively low-skilled, labor-intensive production processes could no longer be undertaken competitively in Japan, it made even more sense to transfer them. A similar process unfolded in the United States and western Europe, though with less impact on Southeast Asia.

Japan had already relocated some of its production to the "tiger" economies of South Korea, Taiwan, Hong Kong, and Singapore in the 1970s. By the 1980s, however, labor costs in these countries were also too high and Japanese capital went looking for other destinations. At first, some of this investment was in textiles, but the bulk of the capital entering the region from Japan was invested in the production of electrical components, machinery, and appliances.

The resulting transformation of some of the economies in Southeast Asia was dramatic. By 1988, manufactures had replaced raw materials as Malaysia's largest source of foreign exchange, and by 1990 it was the third largest producer of semiconductors in the world, after Japan and the United States (Wong Tai Chee p. 111). The Asian "tigers" themselves launched a smaller wave of investment in Southeast Asia from the late 1980s, generally following the Japanese pattern. By 1988, total investments by Taiwanese firms in Malaysia were second only to those of Japanese firms; the third most important investor was Singapore, followed by the United States and Hong Kong (Wong Tai Chee p. 112).

The transformation was even more dramatic in Thailand. By 1985, there was general pessimism about the Thai economy, especially as growth had fallen to just 3.5% (Falkus p. 14), but during 1986, and for the first time, manufacturing outstripped agriculture in its importance to the country's GDP, stimulated by large-scale expansion of light, export-oriented industries and funded by foreign investment. By 1992, manufactured goods made up 77.8% of exports (Falkus p. 17), and extremely rapid growth continued for the rest of the decade. The long-term pegging of the baht to the US dollar was also a factor in this rapid growth, since it caused a significant depreciation of the currency against the yen, stimulating capital inflows and enhancing the export competitiveness of Thai-owned firms (Warr p. 34).

In the case of Indonesia, the steady decline in the price of oil, which, as we have seen, was a major source of export income, forced the government to liberalize its foreign investment guidelines in the late 1980s, a shift in policy that helped to speed the inflow of foreign capital (Djidin pp.16–17, Tzeng p. 375). Although foreign capital never became dominant in such labor-intensive manufacturing industries as textiles, clothing, and footwear, these and others began to take off, with the result that manufactured goods rose from just 2.3% of exports in 1980 to 47.5% in 1992 (Hill p. 164).

The Philippines too attracted some foreign capital, despite its bouts of political instability.

Nevertheless, the poor economic performance of the Philippines meant that one of its most important links to the rest of the world economy continued to take the form of exports of people. By 1993, around six million Filipinos were working overseas as "products" of an export trade that, for the most part, was sanctioned by the government, although around 30% of these migrant workers were undocumented and illegal (Gonzalez and Holmes p. 301).

By contrast, Burma clung consistently to its established policy of economic isolation. Following the events of 1988, when the military refused to recognize the overwhelming electoral victory of the National League for Democracy led by Aung San Suu Kyi, the military dictatorship, calling itself first the State Law and Order Restoration Council and then the State Peace and Development Council, has used high levels of repression and forced labor for infrastructure projects. Since the late 1980s, the regime has made attempts to reopen links with the rest of the world, and has sought foreign capital for ventures in tourism, resource exploitation, and manufacturing for the domestic market. There has been foreign investment in tourism in particular, but the level of military control and resistance to the dictatorship has meant that Burma has not developed the same kind of export-oriented manufacturing as other countries in Southeast Asia. By 1995, manufacturing made up only 9.1% of Burma's GDP (Wong, J., p. 345).

From the mid-1970s, Indochina, emerging from war with the United States and its allies, faced the task of repairing its devastated economies. Continuing conflicts between Vietnam and China, between Vietnam and Cambodia, and within Cambodia itself made the task even more difficult. By the mid-1980s, Vietnam was still heavily dependent on aid from the Soviet Union, but otherwise it remained in an isolation enforced by US sanctions. Declining living standards had already pushed the government to introduce liberal market reforms in the late 1970s; then the collapse of the Soviet bloc in 1989 forced it even further in this direction. The reform process known as *doi moi* opened the country to market competition and, increasingly, to foreign investment. By 1990, Vietnam had one of the most liberal investment codes in Asia (Cho and Williams p. 232) and in 1998 the government decided to allow foreign firms to operate alone, without Vietnamese joint venture partners (Nathan p. 342). Massive cuts have been made in the public sector: between 1988 and 1992, for example, around 1.5 million public sector jobs were abolished (Dollar p. 174). Laos and Cambodia, which are even poorer than Vietnam, have also begun to move in the same direction, introducing greater freedom for markets and autonomy for state enterprises.

"Miracle" Economies?

These changes in Southeast Asia, coming in the wake of the success of the four Asian "tigers," continue to challenge the notion that the world economy can be divided simply between rich exporters of manufactured goods and poor exporters of primary products. The new and complex division of labor within the world economy has undoubtedly changed the lives of millions in Southeast Asia, transforming them into industrial workers and urban dwellers. Yet the longer-term implications for the future of the world system remain a matter of dispute.

For some observers, the changes associated with this new international division of labor represent a triumph of the free market and have the potential to raise living standards and reduce inequality, both within each country in Southeast Asia, and between the region and the richer countries of the West. Others paint a less rosy picture, arguing that the new forms of industrialization introduced since the late 1980s are reliant on extreme exploitation of labor and loose or nonexistent government controls on pollution. The supporters of open markets reply that these are merely transitional phenomena, and that higher-value production will follow as investment continues, possibly involving subcontracting to firms based in the region, the training of the region's workers to higher levels, the continuing transfer of ever more advanced technology, and the upgrading of the manufacture goods produced in Southeast Asia.

However, it is still uncertain whether such upgrading, involving the eventual relocation of sophisticated production, can take place, in general or throughout the region. Some foreign firms clearly prefer to remain in enclaves, using cheap labor and government subsidies but importing most of their other inputs and establishing few if any other links to the host economy. Further, since the process of change has occurred in part because more established manufacturing centers have faced rising labor costs, there are obvious questions to be asked about what happens when wages rise in those countries that receive foreign investment. Capital is now highly mobile and is likely to move on in search of even cheaper labor, notably in China, which has already become a major competitor for labor-intensive, export-oriented manufacturing. In particular, if workers begin to organize themselves successfully in the new industries, as began to happen, for example, in the textile industry in South Korea in the 1970s, footloose manufacturing capital may move on in search of less assertive workers. It follows that at least some of the industrialization that has taken place in Southeast Asia may well be transient, and is unlikely to be sustained if profits are squeezed or wages rise.

Finally, in the face of the rhetoric about the "miracle" economies of the region, it is worth emphasizing that the new ways in which Southeast Asia is connected with the rest of the world economy do not benefit all the inhabitants of the region. There is no doubt that the average wealth of the region's people has significantly increased, whether measured over the whole period since 1945 or over the past decade and a half, the period that engendered the optimism mentioned at the beginning of this chapter. Yet averages do not tell the whole story. The jobless in the slums of Jakarta, the temporary workers in the sweated trades in Thailand, the forced laborers in Burma, and the children scouring the garbage tips of Metro Manila are not better off, and certainly do not feel better off. The wealth produced by the new industries and the new export markets does not necessarily trickle down at all.

Further Reading

Ali, Ameer, "Brunei Darussalam: An Oil Economy in Search of an Alternative Path," in *Asian Profile*, Volume 25, number 4, August 1997

Ali details Brunei's attempt to diversify its economy and reduce its dependence on oil.

Booth, Anne, "Southeast Asian Growth: Can the Momentum be Maintained?" in *Southeast Asian Affairs*, 1995

Booth discusses the basis for optimism about Southeast Asia before the crash of 1997 and outlines some of the problems the region may face in the long term, especially those associated with efforts to replicate the model of growth prevailing in Northeast Asia.

Cho, George, and Stephen Wyn Williams, "Trade, Aid and Regional Integration," in Denis Dwyer (editor), *Southeast Asian Development: Geographical Perspectives*, Harlow: Longman, and New York: Wiley, 1990

This paper looks at the region's trade from a geographical perspective, employing concepts derived from world systems theory.

Dixon, Chris, *Southeast Asia in the World Economy*, Cambridge, New York, and Melbourne: Cambridge University Press, 1991

Dixon offers several ways in which the region might be seen: as part of the periphery of the world economy, as part of an Asia-Pacific economy dominated by Japan, as part of an economy centered on the four Asian "tigers," and as a regional economy centered on Singapore. The book deals particularly well with questions of foreign direct investment and the "new international division of labor."

Djidin, Des Alwi, "The Political Economy of Indonesia's New Economic Policy," in *Journal of Contemporary Asia*, Volume 27, number 1, 1997

This article is an assessment of the major reforms introduced in Indonesia in the late 1980s in response to the impact of declining oil prices.

Dollar, David, "Economic Reform, Openness and Vietnam's Entry into ASEAN," in *ASEAN Economic Bulletin*, Volume 13, number 2, November 1996

Dollar assesses the progress of Vietnam's liberal reforms and the effect that the recent stalling of the reform process has had on foreign investment and economic growth.

Falkus, Malcolm, "Thai Industrialization: An Overview," in Medhi Krongkaew (editor), *Thailand's Industrialization and Its Consequences*, New York: St Martin's Press, and London: Macmillan, 1995

This comprehensive survey of the reasons for Thailand's rapid economic growth suggests that there is a need for caution in predicting continuing growth at these levels.

Frank, Andre Gunder, "The Development of Underdevelopment," in Andre Gunder Frank (editor), *Latin America: Underdevelopment or Revolution*, New York: Monthly Review Press, 1969

This key article in the school of dependency theory (first published in 1966) argues that underdevelopment is not a state of being but a process, and that the "development of underdevelopment" continues through market relations between metropolitan and satellite economies.

Fryer, Donald W., *Emerging Southeast Asia: A Study in Growth and Stagnation*, New York: McGraw-Hill, and London: George Philip, 1970

This overview of the economies of Southeast Asia suggests that two subregions were beginning to emerge at the time it was written: one outward looking and prosperous, the other inward-looking and relatively poor.

Golay, Frank H., Ralph Anspach, M. Ruth Pfanner, and Eliezer B. Ayal, *Underdevelopment and Economic Nationalism in Southeast Asia*, Ithaca, NY: Cornell University Press, 1969

Golay and his co-authors deal with the concept of economic nationalism in Southeast Asia and argue that "indigenism" is an important element motivating public policy in newly independent states in the postwar period.

Gonzalez, Joaquin, and Ronald Holmes, "The Philippine Labor Diaspora: Trends, Issues, and Policies," in *Southeast Asian Affairs*, 1995

In this overview of the history and current practice of labor export from the Philippines, the authors assess the management of the labor export program, its benefits, and its costs.

Guyot, James, "Burma in 1997: From Empire to ASEAN," in *Asian Survey*, Volume 38, number 2, February 1998

An account of Burma's entry into ASEAN and recent changes inside its military junta

Harris, Nigel, *The End of the Third World: Newly Industrializing Countries and the Decline of an Ideology*, London: I.B. Tauris, 1986; New York: Meredith, 1987

Harris argues that the concept of the "Third World" implied a form of nationalist economic and political thought, and suggests that the rapid rise of the Asian "tigers" in particular has undermined this ideology.

Higgott, Richard, and Richard Robison, *Southeast Asia: Essays in the Political Economy of Structural Change*, London, Boston, and Melbourne: Routledge and Kegan Paul, 1985

This book gives a broad overview of the transformation of Southeast Asia's economies, surveying major changes including industrialization and increased integration with world capital and commodity circuits.

Hill, Hal, *The Indonesian Economy*, second edition, Cambridge, New York, and Melbourne: Cambridge University Press, 2000

This major and comprehensive work on the Indonesian economy begins with the turn away from import substitution in the 1960s, and includes a section on the causes and effects of the crisis of 1997.

Jansen, Karel, "Thailand: The Next NIC?" in *Journal of Contemporary Asia*, Volume 21, number 1, 1991

Jansen suggests that the rapid growth of the Thai economy has been, in part, a consequence of favorable international circumstances but also of government policies emphasizing exports. He argues that gradual, rather than sudden, policy shifts are the most appropriate means of changing policy settings.

Khoo Kay Jin, "The Grand Vision: Mahathir and Modernization," in Joel S. Kahn and Francis Loh Kok Wah (editors), *Fragmented Vision: Culture and Politics in Contemporary Malaysia*, Sydney: Asian Studies Association of Australia in association with Allen and Unwin, 1992

An analysis of the internal politics of Malaysia's ruling party and its implications for government economic policy

Krugman, Paul, "The Myth of Asia's Miracle," in *Foreign Affairs*, Volume 73, number 6, 1994

Krugman argues that the Asian "miracle" is based on "perspiration, not inspiration," in other words on a simple mobilization of factors of production rather than on major increases in efficiency.

McVey, Ruth, "Materialization of the Southeast Asian Entrepreneur," in Ruth McVey (editor), *Southeast Asian Capitalists*, Ithaca, NY: Cornell University Southeast Asia Program, 1992

McVey addresses Chinese "pariah" capitalism in Southeast Asia, suggesting historical reasons for differences in the treatment of Chinese entrepreneurs from country to country in the region.

Myrdal, Gunnar, *Asian Drama: An Inquiry into the Poverty of Nations*, Volume 2, New York: Twentieth Century Fund, 1968

A massive and seminal work supporting state economic planning in developing countries

Nathan, Melina, "Vietnam: Is Globalization a Friend or a Foe?" in *Southeast Asian Affairs*, 1999

Nathan looks at the state of Vietnam's foreign relations, and the defense and economic implications, and also assesses the reform process, especially changes in state-owned enterprises and attitudes to market-oriented reform inside the ruling Communist Party.

Oshima, Harry T., *Economic Growth in Monsoon Asia: A Comparative Survey*, Tokyo: University of Tokyo Press, 1987

Oshima uses a comparative approach to view economic growth in Northeast as well as Southeast Asia, beginning with the precolonial period.

Owen, Norman, "Economic and Social Change," in Nicholas Tarling (editor), *The Cambridge History of Southeast Asia*, Volume 2, *Nineteenth and Twentieth Centuries*, Cambridge, New York, and Melbourne: Cambridge University Press, 1992

This is an excellent treatment of economic development in postwar Southeast Asia, covering changes in government policies and attitudes to development, the impact of nationalism on economic policy, and the region's changing role in the world economy.

Prebisch, Raúl, "Commercial Policy in the Underdeveloped Countries," in *American Economic Review*, 1959

Prebisch, the first of the "structuralist" theorists from Latin America, sets out his opposition to the argument that simply maximizing trade with industrial countries will lead to economic development in nonindustrial countries.

Song Byung-nak, *The Rise of the Korean Economy*, second edition, Hong Kong, Oxford, and New York: Oxford University Press, 1997

This neoliberal account of the "miracle" in South Korea predicts extremely rapid growth, along similar lines, in Southeast Asia.

Tzeng, Rueyling, "Foreign Direct Investment in Southeast Asia: Implications of Regional Economic Integration in the Western Society," in *Asian Profile*, Volume 25, number 5, October 1997

The author emphasizes the ways in which transformations of the world economy have influenced foreign direct investment in Southeast Asia.

Warr, Peter G., "The Thai Economy," in Peter G. Warr (editor), *The Thai Economy in Transition*, Cambridge, New York, and Melbourne: Cambridge University Press, 1993

This overview of economic growth since World War II includes a discussion of political changes in Thailand and relates these to changes in economic policy.

Wong, John, "Why Has Myanmar not Developed Like East Asia?" in *ASEAN Economic Bulletin*, Volume 13, number 3, March 1997

Wong suggests that it is bad policy, not lack of comparative advantage, that has cost Burma the chance to industrialize as rapidly as the Asian "tigers."

Wong Tai Chee, "Industrial Development, the New Economic Policy in Malaysia, and the International Division of Labor," in *ASEAN Economic Bulletin*, Volume 7, number 1, July 1990

This paper discusses ways in which state-owned heavy industry, local small industry, and multinational capital interact in Malaysia, and outlines the effects of policies discriminating against Chinese businesses.

Wu Yuan-li and Wu Chun-hsi, *Economic Development in Southeast Asia: The Chinese Dimension*, Stanford, CA: Hoover Institution Press, 1980

The authors provide a detailed account of the economic activities of the Chinese in Southeast Asia, and outline the reasons for discrimination against Chinese entrepreneurs.

Yoshihara Kunio, *The Rise of Ersatz Capitalism in South-East Asia*, Singapore, Oxford and New York: Oxford University Press, 1988

Yoshihara suggests that the predominant form of capitalism in the region is "ersatz" rather than dynamic, being based on a high level of speculative activity and rent-seeking.

Dr John Minns is a Lecturer in History and Politics and Coordinator of the Postgraduate Program in International Relations at the University of Wollongong, New South Wales.

Nation States

NATION STATES

Chapter Four
Thailand

Michael Kelly Connors

Playing alongside Marlon Brando in George H. Englund's film *The Ugly American* (1963), the aristocratic Kukrit Pramoj, later to be Prime Minister of Thailand (1975–76), charmed audiences around the world with his beautiful portrayal of the wicked, aggrieved, plotting leader of the fictitious country of Sarkhan. His poise and charming arrogance, his masterly manipulation, and his severe judgment thrilled cloistered western audiences, who mostly knew of "Asians" only as speakers of pidgin English possessing childlike simplicity or, conversely, as brutal, militaristic fiends.

The admiration and repulsion that Pramoj evoked in audiences might also be felt on learning of Thailand's contorted and contested history. At first glance, Thailand seems to be a charmed kingdom, having a peaceful and benign monarch who has presided over significant economic and political advances since June 1946. Closer inspection, however, reveals this image to be carefully crafted, designed to keep the lid on a history of conflict, civil strife, and massacres of the population by the military in 1976 and again in 1992.

Another story is that of Thailand's economic miracle, again a story with two paths. The first story, officially sanctioned, tells of sophisticated macroeconomic policy, an open economy, and uncanny entrepreneurial talent permitting an unmatched record of sustained growth over several decades (see, for example, World Bank). The second story is to be found under the bridges where itinerants spend their nights, and in shantytowns where collective bathing facilities are on public view. People embody statistics, living out skewed income distribution and poor provision of basic infrastructure. The environment suffers too: canals are black with industrial effluents and forests have been laid waste (see Bello et al.). For years, it seemed as if giddy rates of economic growth could excuse anything, from authoritarian government, to callous dam-building, to agricultural decline. Then came the currency crisis of 1997–98, when the Thai currency, the baht, lost more than half its value against the US dollar, leading to widespread bankruptcy and soaring unemployment. The ugly side of Thailand was there for all to see.

Political Development

Thailand's economy is articulated by both formal and informal systems of power, domestic and international. It underwent significant change, in the direction of capitalist market relations, from the mid-1850s, and began to experience significant growth a century later. Now, however, it has reached crisis point.

If there is any pattern to the narrative that forms Thailand's modern history, it embodies the expansion of politically engaged elite forces and the constant suppression of popular forces. This expansion of elite political space has meant a gradual and uneven transformation of the political system, from authoritarianism, undergirded by the practice of clientelism, toward a more rule-based political regime, characterized by the formal markers of liberal democracy. However, this regime continues to be challenged by the existence of "money politics," the infiltration of formal political institutions, such as the National Assembly and the political parties, by sectional moneyed interests that subvert these institutions to their own ends, making a mockery of the idea of a liberal, neutral state.

Another feature of the Thai system is that individual political fortunes are shifting and uncertain: no grand figure straddles Thai politics. Notably, however, King Bhumibon Adulyadej has transformed the monarchy from the cadaver he inherited into something of a dynamic parapolitical institution. The monarchy is now capable of shrewd political interventions whenever the compact among hostile elite social forces – capitalists, bureaucrats, and military – either breaks up or is challenged from below.

Building the State and the Nation

In line with this revival of the monarchy, it has become customary to begin discussions of modern Thailand by reference to the modernizing work of King Chulalongkorn (reigned 1868–1910). In the late 1880s, faced with the threat of the imperialist intrusion already suffered by Indochina and Burma, Chulalongkorn brought into being the modern Thai state, which, ironically, shared several features with the colonial states of Southeast Asia (Girling p. 46). Simultaneously, the monarchy sought to accelerate social change, by passing edicts abolishing slavery and the practice of forced labor, thereby consolidating the tripartite class structure of wage laborers, free peasants, and entrepreneurs that had already been developing within the international commercial economy of the time. These developments were crucially related to the Bowring Treaty of 1855, which effectively opened up Thailand's trade to British interests. Thailand came to be an exporter of primary goods, the most prominent being rice (see Chaiyan). With Thais dominating agricultural production, the roles played by the Chinese as tax farmers, entrepreneurs, and wage laborers in the new economy became central. Between the late 1880s and the early 1930s, more than one million Chinese entered Thailand (see Seksan).

While Chulalongkorn provided the basis for a modern bourgeois administrative state, his son and successor Vajiravudh (reigned 1910–25) provided the ideological content of the new state formation. By articulating an ideology of "Nation, Religion, and Monarchy," he provided successive regimes with the cement that was to bind the people to elite images of Thai society (see Murashima). Throughout Thailand's modern history, this triadic ideology has been propagated, producing "endless affirmations of the identity of the dynasty and the nation" (Anderson 1983 p. 95).

Mythologized in the country's classrooms, these nation-building achievements were secured at the cost of repressing regional differences and alternative identities. Official nationalism was not simply an attempt at evoking a sense of unity among diverse ethnic groups, but also served the purposes of alienating the increasingly strong Chinese capitalist class (Vella pp. 186–96), and of legitimizing the state as representative of the "national will." This should not be taken to mean that little resistance was offered. In Bangkok, in particular, there was a lively intellectual resistance to the pretensions of the absolute monarchy (see Yuangrat and Wedel). There were also provincial revolts against the dynasty in the North and Northeast, leading to a wave of repression and carnage (Keyes pp. 54–56).

In June 1932, the absolute monarchy was overthrown by the People's Party in a limited coup backed by businesspeople, the embryonic labor union movement, and some intellectuals. The party, composed largely of salaried commoner bureaucrats and military officers, castigated the monarchy for "planting rice on the backs of the people" and began to initiate constitutional reform. Forced to compromise, King Prachathipok (Vajiravudh's younger brother) took part in drafting a constitution allowing for universal suffrage and the gradual implementation of a representative parliamentary system. In 1935, the King abdicated, citing the failure of the People's Party to share power with other groups. Behind the abdication was the failure of the royalist camp to usurp the new regime, despite some boisterous attempts at counterrevolution (see Batson).

In a sense, Prachathipok was retrospectively vindicated. By the late 1930s, the military faction within the People's Party, led by General Phibun Songkhram (Phibun), had become dominant. Phibun's first term as Prime Minister, from 1938 to 1944, was characterized by the erosion of constitutional rule and by xenophobic nationalism. Phibun also collaborated with the Japanese during World War II, which, in part, led to his downfall and a short period of imprisonment.

Factions and Patronage, 1945–57

After the war had ended, it appeared for a time that a new constitutional alliance might be formed between royalists and the liberal faction within the People's Party, centered on Pridi Phanomyong. This alliance brought together figures from the Free Thai movement, which had operated as an underground resistance movement against Japanese occupation, with some royalists and provincial notables, supported by nascent business and labor groupings. Each element in the new alliance was afflicted with the bitter aftertaste of Phibun's centralist rule. The new regime drafted a highly progressive constitution, which banned civil servants and military officers from political activity, abolished the appointed half of the membership of the legislature, established an indirectly elected Senate, and legalized political parties. This potential basis for democracy gave way to a military coup in November 1947, an event widely seen as having initiated the entrenchment of the military as the leading political force in Thailand (Keyes p. 72).

The origins of the coup lay in splits within the postwar civilian regime, economic deterioration, popular antipathy to corruption, and military resentment of civilian interference in its affairs. Additionally, a number of returned royal exiles, among others, began to organize a royalist element in the legislature. This was to become the Democrat Party. After the young King Ananda Mahidol was found shot dead in June 1946, this royalist faction formed an alliance with the malcontented military factions (Pasuk and Baker pp. 267–68). Phibun was invited by the coup leaders to resume the premiership, in April 1948, but he ruled in a greatly diminished capacity, beholden to factional interests and rival cliques.

Initially cautious, the United States, which by then was strategically embroiled in the region, learned to embrace Phibun as a man willing to sing its anti-Communist tune. Phibun's position was pragmatically engineered to win military aid for Thailand. The rapprochement between the two countries also entailed repression of leftwing elements and Communist activities, as well as of Islamic elements in the South of Thailand (see Fineman).

As for the economy, it is possible to discern a nationalist strategy pursued by state agencies ever since 1932: the state implemented programs designed to stimulate a Thai entrepreneurial class distinct from the Chinese. In the late 1930s, Phibun imposed limits on Chinese immigration and occupational eligibility, and implemented policies to establish Thai industry and commerce under state control. Between 1932 and the mid-1950s, 90 state enterprises and public companies with substantial government holdings were launched in a series of attempts to "indigenize" the economy. Economic nationalism was skewed by different factions in the People's Party, which interpreted the policy differently, some establishing relationships and joint companies with Chinese entrepreneurs (Pasuk and Baker pp. 118–25). Despite these developments, most Thais continued to live in a rural subsistence economy, aided by what seemed like an endless land frontier, into which they could flee to escape the extending arm of the state. However, in Bangkok and in provincial centers these economic arrangements led to the emergence of a complex of capitalist-bureaucratic networks based on patronage and reward. It is not surprising that these networks flourished, and continue to do so even today in modified form, given the failure to establish a routinized, neutral administrative system, and a functioning rule of law. In such circumstances, patron-client ties have provided a measure of security and predictability in an otherwise potentially chaotic social field (Neher p. 8). By aligning with influential political figures, business groups have been able to protect their interests from arbitrary state vandalism (see Sungsidh).

Thus, Phibun's second premiership (1948–57) was characterized by the struggle for strategic advantage, material gain, and power (Kobkua pp. 192–212). Expanding patron-client networks grew around rival cliques, and each network spread across a range of activities, including the opium trade, ministerial finance, arms commissions, and directorships (Girling p. 110). In the mid-1950s, Phibun attempted to shore up his position by pursuing a populist strategy aimed at forging links with labor and peasant groups. As his rivals gained ground, Phibun also turned toward an attempt at invigorating democracy, promoting reforms of local government and reducing restrictions on the press, in the hope of winning "progressive" forces to his side (Keyes pp. 74–75). In the elections of February 1957, Phibun secured a victory based largely on fraud. The odious nature of this exercise, despite the resplendent democratic rhetoric, provided Phibun's chief rival, Marshal Sarit Thanarat, with his pretext to overthrow the government in September 1957.

The Saritian Restoration, 1957–73

After ruling by martial law decree for some months and then going overseas for medical treatment, Sarit assumed the premiership in February 1959, and began to recast political, ideological and economic relations in Thailand. He tore up the Constitution, continued to rule by decree, and fostered traditional notions of benevolent leadership to entrench his rule (see Thak). He strategically deployed the monarchy to support the regime, supporting the King's visits to the provinces, while the King himself commenced some much publicized ideological, developmental and philanthropic programs intended to restore the prestige of the throne. It is from this point that is possible to trace the rise of the King as a parapolitical institution in Thai politics. Sarit and his successors also mobilized Buddhist monks toward state ends, supporting programs of national integration (see Jackson). Sarit died in office in 1963, but power passed to his deputies, who ruled until 1973.

Under Sarit and his successors, western forces effectively penetrated the Thai state. Under the influence of the World Bank, economic technocrats were able to pursue a relatively liberal capitalist agenda (see Muscat). Thailand entered into a new phase of economic policy, entailing a degree of liberalization and involving a shift from state-led industrialization toward import substitution, a strategy that lasted into the early 1980s. This development was matched by a deepened relationship with the United States. Both the military and the civilian arms of the state were "advised" by the United States Operations Mission (later absorbed into the US Agency for International Development), notably on security and rural development programs to counteract the threat of rural insurgency (see Chairat). By some estimates, US military aid between 1951 and 1971 amounted to close to US$1 billion, or approximately half the defense budget for the period. This aid assisted in the massive expansion of the armed forces, from around 45,000 men in 1955 to 134,000 in 1961. The United States also played a key role in the formation of new governmental agencies such as the Community Development Department, and in providing educational and technical assistance to the expanding state (Chai-Anan et al. pp. 21–22; Connors 2000 pp. 114–50).

While battling on the fronts of security and rural development in the 1960s, the state was challenged in urban areas by a series of actors seeking an expanded political arena. During Sarit's years in power, it appeared that constitutionalism had been eroded, but by the late 1960s the political consequences of Thailand's economic growth were beginning to be felt. University students, products of an expanded tertiary education system and predominantly the children of the rich and middle-class Sino-Thais, were becoming discontented about the limited options facing them after graduation, and began to be politically assertive. Among middle-class and business elements, there was also a demand for an expansion of the political realm and a rejection of the Saritian formula for politics (Surin pp. 146–47). Between 1969 and 1971, another military Prime Minister, General Thanom Kittikachorn, experimented with a new constitution, but, finding the new legislature's scrutiny of his budget proposals inconvenient, he launched another coup, in November 1971, claiming that "the current world situation and the increasing threat to the nation's security require prompt action, which is not possible through due process of law under the present Constitution" (as quoted in Girling p. 115).

Progress and Reaction, 1973–91

In October 1973, hundreds of thousands of protesters rallied in Bangkok to demand the release of detained democracy activists. This movement snowballed into a confrontation with the armed forces, leading to hundreds of protesters being slaughtered. With the military split and the movement growing despite the repression, the King reputedly ordered Thanom and his deputies out of the country. This ushered in a period of mass struggle on an unprecedented scale.

Following the promulgation of a new constitution in 1974 and elections in January 1975, a coalition of conservative and liberal political parties, representative of business and bureaucratic-military interests, dominated the government led by Kukrit Pramoj, who had been not just a film actor but also a novelist, a historian, and an exponent of Thai classical dance. The coalition led by this witty but erratic polymath was ousted after the elections held in April 1976, and a similarly knife-edged coalition came to power. Both these governments were torn between a reforming impulse and the search for conservative entrenchment. Internationally, Kukrit's government attempted to redefine Thailand's relationship with the United States, taking into account the ending of the Vietnam War and the US rapprochement with China. Kukrit initiated the withdrawal of US troops and opened lines of communication with China, laying himself open to charges that he was a Communist sympathizer, particularly from sections of the military (see Morell and Chai-Anan). Both governments also had to contend with the rising tide of demands from the increasingly radical student-worker alliance, a protest movement of farmers, and a growing Communist insurgency.

While the state had banned unions in the late 1950s, by the late 1960s workers were becoming active once again. In 1972, some attempt was made to incorporate the underground union movement through legislative sanction, a process that ended in the passing of the Labor Relations Law in 1975. When the military regime fell in 1973, an avalanche of strikes and acts of working-class solidarity followed. Unionization and strike rates rose to unprecedented levels (see Hewison and Brown). During the "open" period of 1973–76, student protesters filled the streets of the capital and other cities almost every day. Joining them were reformist monks, who began to criticize the hierarchical structures, not just of their own religious organizations, but of society itself. Conversely, rightwing monks called for the elimination of the perceived Communist threat (Jackson pp. 77–88). Farmers also began to organize, raising demands for land reform and state measures to alleviate the impact of the encroaching commercial economy (see Turton).

The new and fragile parliamentary regime failed to contain these movements with its limited reforms, giving rise to a rightwing backlash sponsored by leading bureaucrats, elements in the palace, and rightwing politicians and capitalists. This conservative reaction was partly determined by the apparent Communist victories in Cambodia, Laos, and Vietnam, which drove large sections of the Thai establishment into a panic. More than 20 leaders of the Peasant Federation of Thailand were assassinated and the backlash culminated in rightwing forces launching an attack on Thammasat University, in which hundreds of unarmed demonstrators were killed. These events then served as the pretext for another military coup, in October 1976, and the installation of Thanin Kraivixien, a judge favored by the royal palace, as Prime Minister (see Anderson 1977). Thanin launched a program of inculcation into the ideology of "Nation, Religion, and Monarchy," reimposed censorship, and ordered the arrest and harassment of suspected dissidents. Several months into his term of office, Thanin, gloating about the seeming order that his ruthless regime had constructed, told foreign investors: "Strikes will not bother you anymore... I have a vision. It is of US dollars [and] Deutschmarks... all flying into Thailand – millions and millions of them for the great reunion..." (as quoted in *The Nation*, January 26, 1977).

The shocking nature of the repression caused thousands of students and intellectuals to flee to the jungle, where many joined the Communist Party of Thailand. Back in 1965, when the Communists launched their Maoist-inspired armed struggle after years of clandestine existence, they constituted a small and relatively unthreatening force. By the mid-1970s, however, it was influential over territory inhabited by around four million people, particularly in the poor, ethnically Lao Northeast, where a residual secessionist threat remained (Chai-Anan et al. pp. 63–64). After Thanin's repression, the Communist Party was the only genuine source of danger for the regime, yet within a few years it had begun to collapse, largely as a symptom of the broader conflict between China, on the one hand, and Vietnam and the Soviet bloc, on the other. By the early 1980s, many of those who had fled to the jungle had returned to Bangkok under a general amnesty.

The amnesty had been made possible because in October 1977 the military had launched a coup against Thanin, paradoxically initiating a controlled liberalization of politics. Under the initial leadership of General Kriangsak Chamanand, yet another constitution was promulgated in 1978, recognizing the emergence of "extrabureaucratic" players but maintaining the bureaucratic dominance of the upper house of the legislature (Prudhisan Jumbala pp. 89–117.) During the 1980s, a new dynamic interaction emerged, within the terms of this settlement, between the ruling bureaucratic and technocratic elites and the capitalist class, with the latter increasingly gaining political power through the legislature and the Cabinet, as well as through semicorporatist arrangements for formulating economic policy (see Anek).

From 1980 until 1988, General Prem Tinsulanonda presided over this semidemocratic system as an unelected Prime Minister appointed by the legislature. He survived the repeated collapse of governing coalitions, attempted coups in 1981 and 1985, and the burgeoning economic crisis, largely as a result of the monarchy's implicit support for him, as well as his own role in mediating the various social forces. Within this structure, a party system emerged that formally provided for nonbureaucratic and nonmilitary input into the regime, but also threw up new informal structures of political patronage and crony capitalism, otherwise known as "money politics" (see Ockey 1992).

Despite the rise of money politics, liberal forces, after a staggering start, won a series of constitutional battles, diminishing the powers of the military and the bureaucrats. These included reforms of the electoral system and the removal of the right of civil servants to serve as ministers. In 1988, an elected member of the House of Representatives, Chatichai Choonhaven, formed a coalition government in which, for the first time in many years, the main processes of decision-making were beyond the direct control of the military-bureaucratic complex. Instead, Chatichai's government, which found itself presiding over an economic boom, proved to be highly corrupt. Nevertheless, despite the political tensions that continued throughout the 1980s, the diversification of capitalist groups and the steady enlargement of the middle class led many to assume that Thailand had entered a new era, in which military intervention in politics was a thing of the past.

From Military Coup to Political Reform, 1991–2000

Within less than a decade, the military and the bureaucracy had been confronted with a steady decline in their political influence and a serious threat to their illegitimate commercial interests (see Hewison 1993). In February 1991, a military grouping calling itself the National Peacekeeping Council staged a coup aimed at restoring the military's deteriorating position. The rationale for the coup was the corrupt nature of what its proponents called the "parliamentary dictatorship." Many from the business sector embraced the new military-appointed government, led by a liberal-minded business manager, Anand Panyarachun.

While Anand's government lacked any popular accountability, it did see to the passing of numerous laws, including measures to liberalize the economy and the banning of labor unions from state-owned enterprises. For its part, the military began securing its future political position by intervening in the drafting of yet another constitution. Between February 1991 and May 1992, the political parties and the military-bureaucratic apparatus engaged in a complex struggle to define the constitutional order (see McCargo 1997 and Murray).

Elections were held in March 1992 under the terms of the new Constitution, which provided for a Senate selected by the National Peacekeeping Council. A newly formed conservative party, Sammkakhi Tham, which meshed the rentier interests of the military, the bureaucracy, and provincial capitalists, emerged victorious. In April, this party invited the leader of the coup, Suchinda Kraprayoon, to take office as Prime Minister. Subsequently, a broad-based opposition arose against Suchinda, forcing him from office in May 1992, although this was not until after his supporters in the military had shot dead scores of protestors. In September, new elections were held under the terms of constitutional amendments that made it necessary for the Prime Minister to be an elected member of the legislature. The relatively liberal Democrat Party won the largest number of seats and led the coalition government formed by Chuan Likphai.

The new government inherited a system dominated by political parties, including his own, that were beholden to financial interests. Thus, it seemed certain to most observers that Thailand was destined to experience another round of looting of state finances. Corruption scandals broke out frequently, the most notable being the controversy over the numbers of wealthy individuals connected to the Democrat Party who benefited from a land reform program officially intended to help landless farmers (see King).

In the face of chronic corruption, the military, deeply unpopular and still smarting from the events of 1992, was unable to step in. Instead, a new constitutional movement emerged, seeking reform of the political process. It aimed at establishing stable executive government by removing the informal influence of money in politics. Supported domestically by capitalist federations, banking associations, and reformist elements in the political system and the bureaucracy, the movement also won support from international interests, which had long been seeking a more stable political system (see Connors 1999 and McCargo 1998). Debate on the constitution raged during the first half of 1997, with the majority of the National Assembly seemingly opposed to a draft prepared by the independent Constitutional Drafting Assembly. In July 1997, however, the economy went into meltdown, as the Asian financial crisis got under way. Its political impact was immediate. Leading conservative forces previously opposed to the draft now argued against continuing active opposition to it, fearing that Thailand would be plunged into political chaos. Thus, despite opposition from vested interests, the new Constitution was enacted in September 1997.

This Constitution may be seen as a significant step toward the consolidation of an institutionally liberal regime. Its various provisions all point to a system of checks and balances, including specific measures against corruption and articles encouraging greater popular participation in the political system. In March 2000 the Constitution faced its greatest test yet. The unprecedented direct elections to the Senate resulted in widespread electoral fraud, particularly in the form of the buying and selling of votes. In this respect, the elections were no different than the many elections held since the 1970s for the House of Representatives. Unlike in those cases, however, there was now an Electoral Commission, armed with constitutional prerogatives, which proceeded to disqualify 78 of the 200 winners and to force new elections for their seats. In some provinces, the second-round results were also revoked by the Commission, and the elections went on into third and forth rounds.

While the Senate elections are a testament to the real and widespread desire to clean up politics, the reform process still faces the formidable opposition of moneyed politicians, for whom political parties are instruments to greater power and wealth. Many such politicians continue to shift unashamedly from party to party in their quest for continuing power and larger retainers. Nonetheless, one can speak of the late 1990s as inaugurating a new period in Thai history, one in which the institutionalization of political conflict within the framework of liberal democracy has become a realizable if still somewhat distant possibility.

Economic Transformation

Just as dramatic as the shift in political power during the 1980s and 1990s was the underlying factor that produced this shift: the growth of the Thai economy. From the 1950s, the economy began to enter a period of steady growth, averaging close to 8% a year in the 1960s, 7% a year in the 1970s, and 5.5% a year in the 1980s. From the late 1980s to the early 1990s, it experienced double-digit growth. During this whole period, manufacturing grew at roughly twice the rate that agriculture did. Agriculture as a proportion of GDP declined dramatically from 40% in 1960 to just over 10% in the 1990s, while manufacturing rose from 12.5% of GDP to more than 28% during the same period (see International Labor Organization).

Nevertheless, much of the economic growth was dependent on the expansion of agricultural production: through the transfer of resources from agriculture to manufacturing, domestic sources of capital could be mobilized. In 1950, farm holdings accounted for 15.4% of Thailand's surface area, but in 1991 they accounted for 41.6%. The same period saw a doubling in the area of paddy land, from 10% to 21.6%; a rise in the area given over to field crops, from 1.4% to 10.4%; and an expansion of forested land, from 1.4% of the total area to 6.3% (according to the Ministry of Agriculture, cited in Dixon 1999 p. 150). Although foreign direct investment and loans were also to play a part in Thailand's industrial growth, much of the necessary capital was transferred from the agricultural sector, which also provided the bulk of state revenues in the 1960s and 1970s, and accounted for the majority of export earnings until the mid-1980s (Dixon 1996 pp. 29–30). It is also commonly held that controls on the price and export of rice depressed its cost. This helped to keep labor costs relatively low, thus generating still more capital for industrial expansion and the growth of wage labor (Dixon 1996 pp. 29–40; but for the debate on this issue see Medhi pp. 53–56).

Meanwhile, promotional measures established under the policy of import substitution succeeded in establishing an industrial base by the 1970s (Dixon 1996 p. 30). Import substitution produced developing industrial conglomerates with significant financial partners; it also provided the basis for a score of families to dominate the protected financial sector. Blessed with access to foreign capital for development, these groupings were able to build up oligopolies in a number of areas of the economy (Hewison 1997 p. 103). Economic growth was also sustained by remittances earned by Thai expatriates working as laborers, as well as by income from tourism, which underwent a further increase in the 1980s.

However, in the late 1970s and the early 1980s the crises related to the huge increases in oil prices and to Thailand's own current account deficits led the government to seek aid from the IMF and the World Bank. This came in the form of a package of "structural adjustment": Thailand came under pressure to liberalize its markets, to abandon import substitution in favor of export-oriented industrialization, and to commit itself to revolutionizing agricultural production. None of these suggestions came as a novelty to many of the capitalist forces in Thailand. Textile and other industries that had grown up in the era of import substitution were already beginning to feel the strain of limited domestic markets and had begun to export, despite the policy bias against them. Nonetheless, there was residual resistance from stakeholders in the import substitution complex, so that for much of the 1980s Thai policy-makers lurched from position to position, made unstable compromises, and failed to implement the programs suggested by the World Bank.

By the 1980s, however, export-oriented industrialization had at least received official sanction: Thailand was to export its way out of the crisis. As it happened, it exported its way into another crisis. The shift to exports received added impetus with the flow of foreign direct investment from Japan and the newly industrialized economies after 1986, the result of the appreciation of the yen after the Plaza Accords of 1985. The influx of foreign direct investment, along with the continuing investment of domestic capital, produced an export boom.

In 1991, Anand's government liberalized the financial services sector, allowing the making of loans denominated in US dollars to Thai businesses at rates much lower than those available in the domestic market. Through this reform, Thailand sucked in billions of US dollars-worth of loans for which the ultimate collateral was, as Walden Bello has put it, the economy itself, which achieved the world's highest rates of growth from 1985 to 1995 (see Bello 1997). The new capital predominantly went into reckless speculation on the stock market and in real estate, rather than into productive investment. Growth began to stagger as the 1990s progressed, and as export markets contracted. In 1996 export growth ceased altogether. This decline highlighted the unrealistic pegging of the Thai currency, the baht, to the appreciating US dollar, which reduced the competitiveness of Thai exports. The government and the Bank of Thailand resisted calls for devaluation, but as capital began to flow out of Thailand in early 1997, the baht came under increasing pressure.

In June 1997, speculators moved into the markets and brought about the collapse of the baht. The devaluation that followed, along with the calling in of short-term dollar-denominated loans, led to the closure of more than 50 financial institutions. Liquidity dried up and the bewildered government turned to the IMF for help (see Bello et al.). This coincided with the constitutional debates discussed above. As the World Bank and the IMF began to publicize their familiar mantras of transparency, efficiency, and participation, otherwise known as "good governance," there was a sense, at least among some observers, that their prescriptions could be nicely dovetailed with the demands of the reform movement. The opening up of the Thai economic to further foreign influence and control – a condition for the receipt of IMF loans – now reinforced the liberal logic of political reform.

The Maintenance of Maldevelopment

Notwithstanding the economic crisis, Thailand appears to have undergone significant growth. However, well before the crash many commentators had questioned the sustainability of the country's growth rates, and pointed out its social and environmental consequences (see, for example, Suthy 1991). These critics pointed out the uneven nature of the growth in Thailand, which had resulted in an unequal spread of its benefits (Parnwall and Arghiros pp. 2–3). Uneven development is evident in many dimensions, most notably in the concentration of infrastructure and growth in Bangkok and provincial centers, at the expense of peri-urban and rural areas. In 1988, for example, the Bangkok Metropolitan Region accounted for 50% of GDP but only 10% of the population (Parnwall and Arghiros p. 13). Uneven development was also evident in income distribution. The proportion of incomes accruing to the richest 20% of the population rose from 49% to 57.5% between the mid-1970s and 1994, while the incomes of the poorest 20% dropped from 6% of the total to just under 4%; the richer 40% improved their position from roughly 70% of national income to 77% over the same period, while the proportion accruing to the poorer 60% dropped from 30% to 23% (Dixon 1999 p. 218).

These figures reflect the fact that Thailand has a dual economy, sharply divided between the industrial and modern services sector, on the one hand, and the underdeveloped agricultural sector on the other. There have been few linkages between the new export-oriented industries and the livelihoods of most people in the countryside, so that, while something like an industrial revolution has occurred in Thailand, it has occurred in decidedly insulated areas. Peter Bell has argued that it may therefore be better to think of Thailand's experience as one of "maldevelopment": the country has undergone a systematic process of growth that has produced structural inequalities and cultural fragmentation, and has had a negative impact on the environment and on the position of women (Bell p. 49). Political disempowerment made it difficult to forge effective resistance to maldevelopment, so that growth has been accompanied by the seemingly uncontrollable spread of AIDS, the destruction of forests and wildlife, the implicit state support of the sex industry as a source of earnings, and the indirect suppression of labor unions.

The question that now confronts progressive forces in Thailand is how best to address these problems. There are several influential lines of argument here. The first, which may be called the elite developmental model, attempts to address the problems of uneven development and to further the integration of the Thai economy into the global marketplace. In order to achieve these twin goals, it is argued, the growth strategy should be continued, in the hope that its "trickle down" effect can be complemented with well-placed infrastructure developments, as well as the decentralization of investment and production. Such a program would also entail further commodification of agriculture, with the consequence that there would be a substantial decline in the proportion of the population (now around 54%) that earn most or some of their livelihood from agriculture. This would require a type of industrialization that could absorb the displaced population, although so far industrialization has failed to do this. Further, resources could be put into upgrading the skills of the industrial workforce, in the hope that Thailand can make the transition from being a producer of industrial goods with low added value to producing goods with high added value. Such ideas have been the mainstay of the largely ineffective development plans concocted by the National Economic and Social Development Board, as well as those put forward by the World Bank.

In opposition to this elite developmental model, there has been a growing movement based on the idea of "localism" (see Hewison 2000). Merging together a range of ideas that have been developed over several decades among proponents of "alternative" agriculture, this current of thought proposes that the Thai economy can be reinvigorated by stimulating local markets and strengthening community organizations in rural areas. Although much of the rhetoric associated with this line of argument, and deployed by several nongovernmental organizations, might be taken to be anticapitalist, localism is about nurturing a capitalist ethic in local settings. The central demand of localism is for the recognition of the importance of agriculture in the economy and the need for a balanced form of development that supports it rather than letting market forces raze it to the ground (Connors 2000 pp. 336–72).

Ideas such as these, however, need political and organizational means of expression. As yet, Thailand's power structure remains firmly in the hands of large-scale capitalists and powerful bureaucrats bent on implementing the mainstream neoliberal strategy for growth. Their determination to continue as before did not wane even in the first few months after the crisis of mid-1997. At that time, many in the middle and upper classes of Thailand undertook what was presented as an act of national contrition, pouring their expensive gold bracelets, watches, US dollars, and other luxury items into a Thais Help Thais Fund managed by a charismatic monk named Phra Prayom Kulyano. Many rich Thais began to appear in magazines, speaking of their failings, their greed, and their will to start afresh. Yet for such people and for others in the middle classes, rural poverty and the lifestyle of the farmers continue to represent little more than a theme park, which at times is mocked in television soap operas, at other times valorized as a source of wisdom and profundity. Both forms of representation indicate an inability to recognize the reality of the class divisions that have been a constant characteristic of Thailand's modern history. In this light, rhetorical support for local development and self-sufficiency, whether from governments or from corporations, appears to be aimed largely at placating restive farmers and potential activists until the next round of uneven economic growth can get going.

Democracy, Kingship, and the New Constitutionalism

The future of Thailand's emerging liberal democracy is likely to be more problematic than the continuance of its liberal economic policies. Politicians, bureaucrats, and capitalists nurtured in the ways of money politics are still resisting the new liberalism, presenting reformers with endless battles that generally end in compromise, so that at best the system is only partially reformed. Further, while one aspect of political reform is more responsive and accountable government, a seemingly contradictory aspect is the constitutional article guaranteeing support for capitalism as the dominant economic system.

This hints at a broader picture of the reformation of the Thai state, linked to what Stephen Gill has called the "new constitutionalism." Gill's basic argument is that, as economic globalization proceeds, there is an emergent constitutionalism that aims at allowing the "dominant economic forces to be increasingly insulated from democratic rule and popular accountability" (Gill p. 23). Thailand's movement for political reform clearly embodies a concern for establishing the conditions of executive efficiency and a desire to remove the partisan influence of sectional interest groups from the executive policy process. The new constitutionalism finds expression in autonomous regulatory institutions, such as the international trade regime, which police the global economy and force open national economies. To the extent that Thai democracy has shifted toward liberal forms of democracy, it is also embracing aspects of this new constitutionalism. Disciplined by the IMF and the World Bank, Thai governments are now expected to govern under international scrutiny ("transparency"). As governments work to appease the dominant interests of capital, the political space for programs of popular reform is steadily diminished. It remains to be seen how the competing forces that have driven Thailand's political transition will evolve, but clearly the democratic elements of the emergent liberal regime are under pressure.

Any discussion of Thai politics and its future cannot ignore the pivotal role of the monarch. Indeed, some observers argue that political reform has been at least partly motivated by the desire to establish new mechanisms of stable rule and authority before the death of the present King, who is already the longest-reigning monarch in the world (see McCargo 1998). While it is true that the King has played a mostly conservative role in Thai politics, intervening to restore order and effectively arguing for gradual change and compromise (see Hewison 1999), many Thais have become accustomed to linking the well-being of the body politic to his wise counsel. This view is partly determined by elite distaste for the masses of workers and peasants that make up Thailand's population. Each time the democratic system has been opened up, so their argument goes, the people have failed to act as responsible citizens. Lured by demagogues or paid off by money politicians, they seem incapable of developing a democratic mentality. By such rationalizations is Thailand's graduated democracy justified. Given the failure to establish a fully functioning democratic polity, the role of the King is paramount: he intervenes when all else fails. As a symbol of national unity and good citizenship, the monarchy is held to be indispensable, and it has been deployed to produce and reaffirm the fundamental unity of all Thai people – a unity that simply does not exist (Connors 2000 pp. 209–50).

In these circumstances, much will depend on who succeeds the King. It may be his unpopular son, Crown Prince Vajiralongkorn; it may be his philanthropically engaged and popular daughter, Princess Sirindhorn. The question is further complicated by the issue of the monarchy's immense wealth, which is said to have been badly affected by the crisis and has been implicated in questionable business deals. Without the ironclad legitimacy and the wall of censorship that surrounds the present King, the monarchy may well be subjected to public criticism and inspection, diminishing its usefulness as a resource for political stability and order.

Thailand is also notable for the relative absence of mass organizations among workers and peasants, as well as for the pained efforts of generations of activists to form such organizations, with only limited success (see Ji). Rather than look to the somewhat spurious notion of a monolithic and unchanging "political culture" to explain this, it is more appropriate to cite the legislative restrictions and repressive measures that have been used to suppress emergent class formations, most notably in 1976, when the monarchy played a crucial role. If new formations emerge out of the present economic crisis, the absence of the present monarch could be an important factor in providing space for their radicalization. Indeed, when the crisis broke out in the summer of 1997 there were some signs of mass radical opposition, but the King mobilized much of his legitimacy to argue for reflection, self-reliance, and austerity. Incredibly, many "progressive" activists spread the same message. In this context, those workers who dared to strike were portrayed as selfish. By such ideological means has potential opposition to the project of neoliberalism been demobilized.

Conclusion

The broad picture of Thailand's economic and political development outlined in this chapter may appear to be bleak and even ungenerous, yet similar critiques will be found in Thailand itself, both in published sources and in conversation. Thailand never was a "land of smiles"; it has always been a country with clear class divisions. The instruments of national rule have always been used against the mass of the population, aiming variously at domesticating, educating, and disciplining them. The fact that struggle from below has constantly broken out over the decades is overlooked, yet it forcefully suggests that, just as the elite have failed to build a just and balanced economic and political order, so they have also failed to exhaust the sense of justice among the people. Therein lies the real hope for the people of Thailand.

Further Reading

Anderson, Benedict, "Withdrawal Symptoms: Social and Cultural Aspects of the October 6 Coup," in *Bulletin of Concerned Asian Scholars*, 1977

This is a classic article, widely read, which retains all of its passion and intelligence many years on.

Anderson, Benedict, *Imagined Communities: Reflections on the Origin and Spread of Nationalism*, London, Verso, 1983

A classic study of nationalism that includes a discussion of Siam/Thailand

Anek Laothamatas, *Business Associations and the New Political Economy of Thailand: From Bureaucratic Polity to Liberal Corporatism*, Boulder, CO: Westview Press, and Singapore: Institute of Southeast Asian Studies, 1992

Hailed as a breakthrough in Thai studies for its challenge to the orthodox interpretation of Thailand as a "bureaucratic polity," Anek's book usefully recounts the complex relationship of business and state officials, and the impact of the former on policy outcomes.

Batson, Benjamin, *The End of the Absolute Monarchy in Siam*, Singapore, Oxford, and New York: Oxford University Press, 1984

This is the standard text on the revolution in Thailand in 1932. It is rich in historical detail, with generous citations from historical documents.

Bell, Peter, "Development or Maldevelopment? The Contradictions of Thailand's Economic Growth," in Michael Parnwell (editor), *Uneven Development in Thailand*, Aldershot and Brookfield, VT: Avebury, 1996

Bell argues forcefully for a recasting of the way in which Thailand's "development" should be seen.

Bello, Walden, "Fast Track Capitalism," in *The Nation*, April 12, 1997

Bello, Walden, Shea Cunningham, and Li Kheng Poh, *A Siamese Tragedy: Development and Disintegration in Modern Thailand*, London: Zed Books, Oakland, CA: Food First, and Bangkok: White Lotus, 1998

An incisive, provocative and comprehensive discussion of Thailand's developmental experience

Chai-Anan Samudavinija, Kusama Snitwongse, and Suchit Bunbongkarn, *From Armed Suppression to Political Offensive: Attitudinal Transformation of Thai Military Officers since 1976*, Bangkok: Chulalongkorn University Institute of Security and International Studies, 1990

A good overview of political and ideological currents in the Thai military

Chairat Charoensin-O-Larn, *Understanding Postwar Reformism in Thailand*, Bangkok: Editions Duang Kamol, 1988

A critical look at rural development strategies in Thailand and their relationship to counterinsurgency measures

Chaiyan Rajchagool, *The Rise and Fall of the Thai Absolute Monarchy: Foundations of the Modern Thai State from Feudalism to Peripheral Capitalism*, Bangkok: White Lotus, 1994

This historical study provides a theoretical account of Thailand's emergence as a peripheral capitalist state constructed largely by the monarchy.

Connors, Michael Kelly, "Political Reform and the State in Thailand," in *Journal of Contemporary Asia*, Volume 29, number 4, 1999

A critical overview of the social forces behind Thailand's latest Constitution

Connors, Michael Kelly, *Subjecting Citizens: Democracy, National Ideology, and the Doctrine of Political Development in Thailand*, PhD dissertation, University of Melbourne, 2000

A study of the strategic use of democratic thought and nationalism for elite purposes

Dixon 1996: Dixon, Chris, "Thailand's Rapid Economic Growth: Causes, Sustainability and Lessons," in Michael Parnwell (editor), cited above under Bell

A concise and unusually readable study of the factors contributing to growth

Dixon 1999: Dixon, Chris, *The Thai Economy: Uneven Development and Internationalisation*, London and New York: Routledge, 1999

Dixon has done a wonderful job of making economics accessible, deploying a clear and lucid prose style to present a cogent case for rethinking the Thai economy.

Fineman, Daniel, *A Special Relationship: The United States and Military Government in Thailand, 1947–1958*, Honolulu: University of Hawaii Press, 1997

A well-researched and readable account of the evolution of Thai-US relations

Gill, Stephen, "New Constitutionalism, Democratization, and the Global Political Economy," in *Pacific Review*, Volume 10, number 1, 1998

Gill suggests that we rethink the notion of democratization through its relationship to economic globalization, in order to gain a better perception of its disciplinary aspects.

Girling, John, *Thailand: Society and Politics*, Ithaca, NY: Cornell University Press, 1981

Once a standard text on Thailand, this book is still well worth reading. Girling cuts through much of the mystique that surrounds Thai politics and provides a credible radical interpretation.

Hewison, Kevin, "Of Regimes, States, and Pluralities: Thai Politics enters the 1990s," in Kevin Hewison, Richard Robison, and Garry Rodan (editors), *Southeast Asia in the 1990s: Authoritarianism, Democracy and Capitalism*, St Leonards, NSW: Allen and Unwin, 1993

Hewison takes a multifaceted look at the forces pressing for change in Thailand.

Hewison, Kevin, "Thailand: Capitalist Development and the State," in Garry Rodan, Kevin Hewison, and Richard Robison (editors), *The Political Economy of Southeast Asia: An Introduction*, Melbourne, Oxford, and New York: Oxford University Press, 1997

An analysis of the Thai economy by one of the leading political economists studying the country

Hewison, Kevin, "The Monarchy and Democratization," in Kevin Hewison (editor), *Political Change in Thailand: Democracy and Participation*, London and New York: Routledge, 1997

This bold critique of the Thai King's political thinking is one of the few such pieces available.

Hewison, Kevin, "Resisting Globalization: A study of Localism in Thailand," in *Pacific Review*, Volume 13, number 2, 2000

A partly polemical piece opposing the "romantic, antiurban" ideas that Hewison attributes to various individuals, nongovernmental organizations, and social movements in Thailand

Hewison, Kevin, and Andrew Brown, "Labor and Unions in an Industrializing Thailand," in *Journal of Contemporary Asia*, Volume 24, number 4, 1994

The authors consider the changing historical context of the labor movement in Thailand, linking this to patterns of class struggle and state strategy.

International Labor Organization, *Macroeconomic Policies and Poverty: the Thai Experience*, available at www.ilo.org/public/english/region/asro/bangkok/paper/thaiexp.htm, no date given

Jackson, Peter, *Buddhism, Legitimation, and Conflict: The Political Functions of Urban Thai Buddhism*, Singapore: Institute of Southeast Asian Studies, 1989

This major study of state intervention in the Buddhist order, and its functionality for state legitimacy, usefully discusses the rise of dissident groups, linking changing beliefs to social and economic change.

Ji Giles Ungpakorn, *Thailand: Class Struggle in an Era of Economic Crisis*, Hong Kong: Asia Monitor Resource Centre, and Bangkok: Workers' Democracy Book Club, 1999

A refreshingly iconoclastic study of the role of workers in Thailand.

Keyes, Charles, *Thailand: Buddhist Kingdom as Modern Nation-state*, Boulder, CO: Westview Press, 1987; Bangkok: Editions Duang Kamon, 1989

A masterful survey of Thai history up to the 1980s

King, Daniel E., "Thailand in 1995: Open Society, Dynamic Economy, Troubled Politics," in *Asian Survey*, Volume 36, number 2, 1996

A useful survey of Thai politics during the years following the protests of 1992

Kobkua Suwannathat-Pian, *Thailand's Durable Premier: Phibun Through Three Decades, 1932–1957*, Kuala Lumpur, Oxford, and New York: Oxford University Press, 1995

It may be true that royalists have demonized Phibun's role in Thai history, but making him out to have been a saint, as this book attempts to do, is hardly the appropriate response. Nevertheless, the book provides insight into the period and into Phibun's motives.

McCargo, Duncan, *Chamlong Srimuang and the New Thai Politics*, London: Hurst, and New York: St Martin's Press, 1997

This book provides an overview of the philosophy and career of an enigmatic political adventurer. Although Chamlong's fortunes have faded, the book is still valuable for its incisive analysis of the democracy movement of 1991–92, as well as its extremely useful discussion of political Buddhism, parties, and the military.

McCargo, Duncan, "Alternative Meanings of Political Reform in Contemporary Thailand," in *The Copenhagen Journal of Asian Studies*, Volume 13, 1998

This article provides an overview of the competing perspectives and motives behind Thailand's political reform.

Medhi Krongkaew, "Contributions of Agriculture to Industrialisation," in Medhi Krongkaew (editor), *Thailand's Industrialisation and its Consequences*, London: Macmillan, and New York: St Martin's Press, 1995

This chapter gives a taste of the debate on the contested position of agriculture in Thailand's economic growth.

Morell, David, and Chai-Anan Samudavanija, *Political Conflict in Thailand: Reform, Reaction, Revolution*, Cambridge, MA: Oelgeschlager, Gunn, and Hain, 1981

Somewhat conservative in their interpretation of Thai politics in the 1970s, the authors succeed in bringing key events, personalities, and trends into focus.

Murashima Eiji, "The Origin of Modern Official State Ideology in Thailand," in *Journal of Southeast Asian Studies*, Volume 19, number 1, 1988

This heavily cited piece provides a good overview of the formation of modern nationalism in Thailand.

Murray, David, *Angels and Devils: Thai Politics from February 1991 to September 1992 – A Struggle for Democracy?*, Bangkok: White Orchid, 1996

A good account of Thai politics at a crucial turning point, complemented by full-spread cartoons

Muscat, Robert J., *The Fifth Tiger: A Study of Thai Development Policy*, Helsinki: United Nations University Press, and Armonk, NY: M.E. Sharpe, 1994

This is a partisan text on the side of elite development policies, which provides a thoughtful discussion of the intersection of policy and elite processes, and a readable account of Thailand's growth.

Neher, Clark, *Politics and Culture in Thailand*, Ann Arbor, MI: Center for Political Studies, Institute for Social Research, University of Michigan, 1987

A thoughtful discussion of political culture in Thailand, with an argument for the use of clientelism as an analytical concept for understanding Thai politics

Ockey, James, "Business Leaders, Gangsters, and Civilian Rule in Thailand," PhD dissertation, Cornell University, 1992

An excellent full-length study of the transition in Thai politics in the 1980s

Parnwell, Michael, and Daniel Arghiros, "Introduction: Uneven Development in Thailand," in Michael Parnwell (editor), cited above under Bell

A convenient analytical summary of just how spatially differentiated Thailand's development has been

Pasuk Phongpaichit and Chris Baker, *Thailand: Economy and Politics*, Kuala Lumpur, Oxford, and New York: Oxford University Press, 1995

This is a masterly synthesis of hundreds of monographs, research papers, and articles, including the authors' own work. The book fills a massive gap in Thai studies by providing one of the most comprehensive, fully referenced and genuinely analytical histories of modern Thailand. Written from the perspective of political economists, it has already become the standard reference work.

Prudhisan Jumbala, *Nation Building and Democratization in Thailand*, Bangkok: Chulalongkorn University Social Research Institute, 1992

A useful survey of constitutions and political developments from the 1930s to the 1980s

Seksan Prasertkul, *The Transformation of the Thai State and Economic Change 1855–1945*, PhD dissertation, Cornell University, 1989

One of Thailand's student politicians from the 1970s, Seksan undertakes a Marxist analysis of capitalist development in Thailand. The result is full of rich research and analysis.

Sungsidh Piriyarangsan, *Thai Bureaucratic Capitalism 1932–1960*, Bangkok: Chulalongkorn University Social Research Institute, 1982

A radical interpretation of Thai political economy up to the 1960s

Surin Maisrikrod, "The Making of Thai Democracy: A Study of Political Alliances Among the State, the Capitalists, and the Middle Class," in Anek Laothamatas (editor), *Democratization in Southeast and East Asia*, Singapore: Institute of Southeast Asian Studies, and New York: St Martin's Press, 1997

A thoughtful overview of the interaction of state, capitalist and middle-class forces

Suthy Prasartset, *Democratic Alternatives to Maldevelopment: The Case of Thailand*, Yokohama: International Peace Research Institute Meigaku, 1991

A highly prescient paper, worth reading not just for its diagnosis of maldevelopment, but also for the strategies proposed against it

Thak Chaloemtiarana, *Thailand: The Politics of Despotic Paternalism*, Bangkok: Thammasat University Thai Khadi Institute, 1979

This critical and well-argued text has become the standard reference work on the Sarit period.

Turton, Andrew, "The Current Situation in the Thai Countryside," in *Journal of Contemporary Asia*, Volume 8, number 1, 1978

A bold survey of rural movements and state repression in the 1970s

Vella, Walter, *Chaiyo! King Vajiravudh and the Development of Thai Nationalism*, Honolulu: University of Hawaii Press, 1978

An enjoyable discussion of Vajiravudh's troubled reign

Wedel, Yuangrat and Paul, *Radical Thought, Thai Mind: The Development of Revolutionary Ideas in Thailand*, Bangkok: Assumption Business Administration College, 1987

This book should be compulsory reading for all those who have accepted the myth that Thai political culture breeds apathy and indifference. For those who have not, this is a lively discussion of currents within Thai radicalism.

World Bank, *The East Asian Miracle: Economic Growth and Public Policy*, New York and Oxford: Oxford University Press, 1993

The authoritative text on the "miracle," published just a few years before the crash

---

Dr Michael Kelly Connors is a Lecturer in the Institute for Politics and International Studies at the University of Leeds in England.

NATION STATES

Chapter Five
Malaysia

Fiona Yap

In just a little under 60 years, the territories that now comprise the Federation of Malaysian States have survived occupation by the Japanese (1942–45), battled against a Communist insurgency (1948–60), obtained self-rule from the British (1957 onwards), and struggled with merger and identity (1963 and 1965), while facing down the Indonesian government's policy of *Konfrontasi* ("Confrontation," 1963–66). Since 1965, the 13 states of Malaysia have faced interethnic riots (May 1969), addressed episodic Communist guerilla activity, dealt with an increase in religious extremism and potential fanaticism (1978–80), and survived factionalism and leadership altercations within the dominant party in the ruling coalition (1985–88).

Alongside all these political challenges, the country has also had to confront the issue of economic development. In the 1950s and 1960s, as it became increasingly apparent that Malaysia could not continue to rely on primary exports as the source of development, the country embarked on import-substituting industrialization to propel economic performance. By the late 1960s, however, it became clear that this strategy had become exhausted and that the "easy growth" stage was over. Malaysia turned to export-oriented industrialization, accompanied by deficit spending to "pump prime" its economy. The strategy worked sufficiently well in the 1970s, as nonbank financing and balance-of-payments surpluses enabled the government to sustain public sector fiscal deficits without incurring significant external borrowing. However, by the beginning of the 1980s the initiation of a phase of heavy industrialization, coupled with balance-of-trade deficits, contributed to the "twin deficits" problem. As a result, between 1980 and 1982 Malaysia's external debt rose to more than RM$24.3 billion, or 41% of GDP. In 1985, Malaysia's economy plummeted further and went into a recession.

Just when it seemed that Malaysia's politics and economy had both bottomed out, there was a transformation in both. By 1989, it became clear that Mahathir Mohamad, the Prime Minister and the leader of the dominant party, the United Malays National Organization (UMNO), had not only survived leadership and political challenges but had emerged from them stronger than ever (see Means). Similarly, Malaysia's economy was not only rejuvenated but its performance was strengthened, to a point where the country was widely regarded as one of the high-performance economies known as the "Asian Tigers" or "Dragons." As Khoo Boo Teik has put it, the country that seemed unable to do much right now seemed unlikely to do anything wrong (see Khoo Boo Teik). In the face of double-digit growth between 1990 and 1995, few questioned the economic fortunes of the country or, for that matter, those of any of the other Tigers or Dragons.

Yet the political and economic fate of Malaysia turned out to be far from settled. A snapshot of political and economic conditions between 1997 and 1998 yields a picture almost diametrically opposed to what would have appeared in one taken, say, at the beginning of 1995. In particular, the onslaught of the Asian financial crisis in 1997 raised questions once considered resolved, notably about the underlying integrity and soundness of all the economies of East Asia. In this regard,

Malaysia's refusal to abide by the guidelines of the IMF and its subsequent "heretical" use of capital controls as a solution to the crisis increased skepticism and suspicion. To compound matters, Malaysia's politics was also shaken by the dismissal and subsequent prosecution of the Deputy Prime Minister, Anwar Ibrahim. Many observers, accordingly, have added political scapegoating and persecution to their list of concerns about the country, which already included the cronyism and patronage politics that are seen as clouding Malaysia's economic horizon (see Aziz, Crouch, Gomez and Jomo, Nesadurai, and Rhodes and Higgott).

The Strength of the Executive

It is easy, and indeed almost habitual, to credit Malaysia's economic performance to its political leaders, and particularly to the present Prime Minister, Mahathir Mohamad. Among the policies pursued since Mahathir came to power in 1981 have been Malaysia's efforts at heavy industrialization (1980), the "Look East" policy (1981–82), "Malaysia Incorporated" (1985), privatization (1983, and then a *Master Plan* in 1991), the "Second Outline Perspective" (1990), and "Vision 2020" (1991, 1997). However, Mahathir's long-lasting leadership probably makes it easy to forget that his predecessors were neither weak nor indecisive.

It is noteworthy that in nearly 50 years of independence, Malaysia has had only four Prime Ministers. Two of the four, Tengku Abdul Rahman (1957–70) and Mahathir, have enjoyed tenures exceeding 10 years. Tun Abdul Razak (1970–76) and Tun Hussein Onn (1976–81) were in office for shorter periods, but their tenures were cut short primarily because of illness: in fact, Tun Razak died in office. Of the four, only Tun Hussein Onn, who succeeded Tun Razak following the latter's death on January 14, 1976, suffered from having a weak political base. Despite this, he managed to remain in office until 1981 and left a legacy of tough measures against corruption (see Milne and Mauzy). Tun Hussein Onn's period in office makes it clear that, however much of the credit, or blame, for Malaysia's political or economic conditions may be attributed to individual leaders, the strength that has facilitated their effective decision-making has been fostered by the political system itself. In other words, it would not be surprising if Mahathir's successors are also strong policy-makers. Thus, it is relevant to examine how the political system imbues its leaders with such strength.

Malaysia has a parliamentary political system that operates with a strong principle of federal supremacy. Federal legislation takes precedence over state legislation and the federal government may enact any legislation affecting the states if necessitated by federal concerns such as foreign relations or security. Each state has a unicameral legislature, procedurally modeled on the House of Representatives (Dewan Rakyat), the lower house of the federal Parliament (Parlimen). Each state also has (in nine cases) a hereditary sultan or raja, or (in the other four states) a federally appointed governor. These positions are primarily ceremonial, although the rulers and governors do meet a few times each year in a Conference of Rulers, to discuss issues affecting their position and privileges, and, in every fifth year, to choose the paramount ruler, the Yang Di-Pertuan Agong (usually abbreviated to Agong). In theory, the Agong's seal of consent is required for any parliamentary bill to become law; however, a constitutional amendment enacted in 1984 effectively annulled that necessity by enabling all bills to become law automatically 60 days after presentation to the Agong (see Shinn).

The federal bicameral legislature, then, is the primary legislative authority, with the House of Representatives wielding more authority than the Senate (Dewan Negara), since the Senate has a power of delay but no veto over bills. The House of Representatives is elected for five-year terms, although it may be dissolved before the end of the term, as is typical in parliamentary systems. Elections must be held within 60 days of dissolution. Electoral victories are determined primarily on the basis of plurality voting ("first past the post") rather than proportionality.

The executive, comprising the Prime Minister and his Cabinet, remains in office with the support and confidence of the majority coalition in Parliament. The Cabinet operates on the basis of ministerial responsibility and meets once a week. Its decisions are implemented under the coordination of the Secretary to the Cabinet, who is also head of the Prime Minister's Department. The Prime Minister's Department in turn supervises the policies and actions of the ministries, and is responsible for administrative and economic research and planning.

This outline of the political system indicates the potential sources of the strength of the executive. First and foremost, the Prime Minister benefits from the generally solid support that majoritarian parliamentary systems typically vest in the executive. It is no coincidence that some of the world's most authoritative elected leaders, such as Margaret Thatcher in the United Kingdom or Indira Gandhi in India, have wielded power in parliamentary systems with plurality voting. The possibility that such authoritativeness can veer into authoritarianism was realized, to take just one example, when Indira Gandhi successfully imposed a state of emergency and suspended elections in the world's largest democracy between 1975 and 1977.

The executive in Malaysia is also strengthened by the preeminence of the Prime Minister's Department in policy-making. For instance, the Economic Planning Unit (EPU) in the Prime Minister's Department has the primary responsibility of monitoring and evaluating projects set up under the New Economic Policy (NEP) of 1971 to achieve the goal of interethnic redistribution. The significance and continued influence of the NEP will be discussed below; suffice to say here that it was launched primarily to address Malay grievances at their small share of Malaysia's economic prosperity and potential loss of political strength. That a unit in the Prime Minister's Department should have primary responsibility for a national policy that has been instrumental in expanding the role of government in the economy indicates the path and process by which the Prime Minister can exert his strength and overcome potential political challenges.

In 1987–88, the strength of Malaysia's executive was further enhanced, in this case at the expense of the judiciary. Malaysia's Constitution grants the courts the power of judicial review, and the importance of an independent judiciary was reiterated in the *Rukunegara*, a set of principles for preserving harmonious interethnic relations published by the government on August 31, 1970:

> The rule of law is ensured by the existence of an independent judiciary with powers to pronounce on the constitutionality and legality or otherwise of executive acts
> (as cited in Milne and Mauzy, p. 46).

The courts, then, have been the final recourse for attempts to restrain executive power. In general, the courts have not refrained from using their power of judicial review to rule against the executive. Perhaps as a result, following a series of judicial rulings against the government in 1986–87, including a decision that declared the very existence of UMNO illegal (see below), the government moved to amend the Constitution in order to confer judicial powers through parliamentary statute instead of through the Constitution. By this amendment, the judiciary was stripped of its powers of judicial review. The judiciary was also subordinated to the executive through the Attorney General, who gained control over judicial assignments and transfers, and the power to decide the cases that would be heard and which courts would hear them (see Milne and Mauzy, and Khoo Boo Teik). The severity of these blows against judicial power is evident from the suspension of Tun Salleh Abbas, Lord President of the Supreme Court, in May 1988, the suspension of five other Supreme Court Justices in July, and the subsequent dismissal of the Lord President (see Milne and Mauzy).

There is, then, an absence of formal vetoes against the executive in Malaysia's political system. This allows the executive to wield substantial leverage on decision-making and the policy process once in power.

The Significance of UMNO

Nevertheless, some constraints on the executive may be exercised at the party level. In particular, the sometimes fierce contestation in the elections of party officials within UMNO, every three years, has prompted the observation that democracy can flourish within this forum (see Khoo Khay Jin, Means, Jesudason). At a minimum, the need for a power base within UMNO in order to be elected as a party official demands that party loyalists be accommodated and critics coopted.

UMNO has been the dominant party in the two coalitions, the Alliance (1957–74) and then the National Front (Barisan Nasional, or BN), that have formed all governments since 1957. UMNO officials typically assume positions in the government administration, with the President and Deputy President of UMNO consistently occupying the positions of Prime Minister and Deputy Prime Minister. However, three additional factors suggest that the significance of UMNO goes beyond these basics, and that UMNO's dominance and influence on policy-making are likely to continue well into the 21st century. They are, first, UMNO's ability to foster orderly competition rather than factionalization within the party; its policy of multiethnic accommodation, while recognizing special rights for Malays and other "Bumiputras" (members of ethnic groups perceived as "indigenous," in contrast to the Chinese and Indian minorities); and the ability of its leaders to seek innovative solutions that preserve this accommodation.

The most contentious case of intraparty fighting in UMNO occurred between 1985 and 1988, and led to a split in the party ranks following the party elections in April 1987 and the subsequent judicial ruling (mentioned above) that declared UMNO illegal as a party after the validity of the elections had been contested in court. The subsequent registration, in February 1988, of New UMNO (UMNO Baru) reestablished its legality as a party. For the most part, however, party contests since the inception of UMNO in March 1946 have not been characterized by such drama. Tun Hussein Onn, whose political base, as we have seen, was the weakest of any of the four Prime Ministers, encountered the most challenges to his rule, yet in the party elections in 1978 he defeated his only opponent, Sulaiman Palestin, with 898 votes to 250 (see Khoo Boo Teik).

One of the ways in which the party manages to encourage orderly competition rather than factionalism is through its control of benefits and patronage. As the dominant party, UMNO is singularly well-placed to distribute the spoils of office, and to coopt opposition and criticism (see Means). For example, in response to party dissatisfaction with Tengku Abdul Rahman's stalwart stand on ethnic accommodation, his successor Tun Abdul Razak brought Mahathir Mohamad and Musa Hitam back into UMNO's ranks. Both were well-known for their stand on pursuing Malay special rights and had been expelled from UMNO for violating party discipline by vigorously espousing these beliefs. Similarly, in 1982 Mahathir Mohamad brought Anwar Ibrahim, then the leader of the Malaysian Islamic Youth Movement (ABIM) and a critic of UMNO's secular, pragmatic policies for improving the lives of Malays, into UMNO, and gave him a position in government, thereby depriving the rival Malaysian Islam Party (Parti Islam SeMalaysia, or PAS) of potential support from the segment of opinion that Anwar Ibrahim represented (see Means, and Milne and Mauzy). It has been suggested that the fracture in 1987 occurred primarily because the distribution of patronage and spoils diminished as a result of the recession of 1985–86 (see Khoo Kay Jin).

In any case, UMNO's ability to foster orderly competition stands in stark contrast to the factionalism suffered by some of its coalition partners, such as the Malaysian Chinese Association (MCA) or the Malaysian Indian Congress (MIC). The rifts in the MCA and the MIC have cost both parties electoral support from their key constituencies, which has, in turn, reduced their ability to influence government policies.

UMNO's policy of multiethnic accommodation is best evidenced in two incidents that will be discussed in greater detail below. First, it was through its policy of multiethnic compromise, encapsulated in the "Bargain" of 1957, that Tengku Abdul Rahman and other UMNO leaders obtained independence and recognition of the special position of Malays in the Federation. Second, it was in the same spirit of multiethnic accommodation that Tun Abdul Razak enlarged the ruling coalition and replaced the Alliance with the National Front in 1974. Although some in the Malay community continue to feel that UMNO could do more for the special rights of Bumiputras, it is important to note that both incidents succeeded in forging agreements, and buttressing political support for UMNO and its coalition partners.

In short, by advocating the policy of multiethnic accommodation as well as respect for special Bumiputra rights, UMNO stands to enjoy broader support than the main opposition parties, the PAS or the Democratic Action Party (DAP). At various elections, these rivals have corroded support for the National Front, most notably in 1969 (the PAS), 1986 (the DAP), and 1999 (both), yet both parties ultimately suffer from their inability to attract support beyond their core constituents, or provide a credible cohesive front against the ruling coalition. The PAS, with its party platform of establishing an Islamic state, can offer a viable challenge to UMNO but struggles to win non-Malay supporters. The DAP has its political roots in demands for racial equality in a "Malaysian Malaysia," as opposed to the system of endorsing special Bumiputra rights, which predictably makes it unattractive to Malay and other Bumiputra voters. The difficulty of reconciling such divergent party objectives and policies is evident from the fact that these opposition parties could not unite under a single banner until 1999, when they established the Alternative Front (Barisan Alternatif, or BA). It appears that they were spurred to unite primarily as a result of the dismissal and subsequent prosecution of the former Deputy Prime Minister, Anwar Ibrahim, rather than because of any lasting resolution of their conflicting party objectives. Perhaps as a result, the outcome of the 1999 elections was mixed. On the one hand, the National Front doubled the number of seats it holds in Parliament. On the other hand, two leaders of the DAP lost their seats. It remains to be seen if the election results reflected protest votes against UMNO in connection with the treatment of Anwar, or a real shift in support to the opposition. In any case, it is clear that, in order to mount a viable challenge to the ruling coalition, both the PAS and the DAP need to maintain a cohesive front, based on common objectives and policies.

Until they can do so, UMNO's policy of multiethnic accommodation will continue to ensure that it enjoys broader support than either of these opposition parties does.

The ability of UMNO's leaders to forge innovative solutions and to preserve the policy of multiethnic has also helped to seal the party's position as the center of substantive power supporting the government. For instance, Tengku Abdul Rahman, the first Prime Minister of the Federation, who is widely regarded as the founding father of independent Malaysia, stared down the threat of military retaliation from Indonesia to conclude a Federation agreement predicated on broad-based support for ethnic accommodation, and for coexistence between the traditional monarchy and the new representative government (see Milne and Mauzy, and Means). The Tengku persisted in this approach even when it cost him political support following the riots of May 1969.

Tun Abdul Razak, who succeeded the Tengku following the latter's retirement in September 1970, was the chief architect of the New Economic Policy (NEP), whose objective of interethnic redistribution to diffuse tensions following the riots greatly increased the scope of government activity. To his credit, alongside the effort to accommodate Malay grievances, Tun Razak also sought to broaden the political base of the government. To this effect, he replaced the governing coalition of the Alliance, comprising UMNO, the MCA, and the MIC, with the enlarged National Front, inviting any party willing to work with UMNO into the coalition. Between the inception of the idea of the National Front, in August 1972, and its formal registration, in July 1974, the Front incorporated several parties that had been rivals of the Alliance, including the PAS and the Malaysian People's Movement (Gerakan Rakyat Malaysia, usually known as Gerakan), which had competed with the MCA for Chinese support.

Perhaps UMNO's most innovative endeavor to preserve multiethnic accommodation with emphasis on Bumiputra rights is represented by Malaysia's use of capital controls in response to the Asian financial crisis of 1997–98. The Malaysian government had earlier accepted and begun implementation of an austerity package of fiscal and monetary contraction recommended by Anwar Ibrahim, then Deputy Prime Minister, in December 1997. Although the package presaged a sharp contraction in the economy, market analysts and economists welcomed it as putting the country on the road to recovery. However, the fallout from the austerity package was wider than expected. In addition to contracting government spending through deferred projects, some of which were Mahathir's favorites, the economic contraction also produced financial difficulties for many of Mahathir's political allies and UMNO-linked conglomerates. The government's subsequent bailout of three prominent politically connected companies, including the Konsortium Perkapalan Berhad (KPB, or Shipping Consortium Limited), in which Mahathir's eldest son, Mirzan, has a 51% stake, did nothing to restore market confidence (see Nesadurai). The austerity package came to be seen as a smokescreen rather than a committed effort to introduce corporate and financial reforms. It also increased discontent, both within UMNO and in the electorate, as businesses continued to fail in a market that showed few signs of recovery. In the face of mounting dissatisfaction, a reduced ability to distribute patronage to help cushion the economic impact, and indications in the first quarter of 1998 that the country had slid into recession, Mahathir changed course. By June 1998, the government had replaced the austerity package with one that provided economic stimulation to the tune of RM$66 billion. In September, capital controls were introduced (see Nesadurai). As unpopular as this approach was among foreign investors, it allowed the government to pursue and reconcile domestic political and economic priorities against global market pressures and constraints. In the process, the government successfully protected its key constituencies, including the Malay corporate elites that had risen to prominence through the affirmative action measures of the NEP, from the perils of economic adjustment. Once again, UMNO emerged relatively unscathed (see Nesadurai).

The results of the parliamentary elections of 1999 suggest that the opposition parties may provide a viable challenge to the ruling coalition if they can reconcile their conflicting party objectives and present a united front. Until then, however, UMNO's leadership will no doubt continue to navigate the country through its political and economic development. And, as long as UMNO is at the helm, Malaysia will continue to witness patronage, cronyism, and policies that favor the economic interests of UMNO's political and corporate constituencies, as the party uses these privileges to win converts or quell critics and preserve its support.

Interethnic Relations

The discussion of UMNO brings to the fore another important factor that influences Malaysia's political and economic landscape: the relations among its major ethnic groups. In the 1950s, the difficulties in recognizing and accommodating ethnic rights and privileges came close to preventing the establishment of the Federation. In the 1960s, interethnic tensions caused riots that led to a transformation in the role of the state. Even now, ethnicity forms the invisible divide between the main opposition parties, the PAS and the DAP, and continues to hinder their coalescence into a viable challenger to the National Front.

Ethnic tension in Malaysia exists primarily between the Malays and other Bumiputras, who comprise around 60% of the population and dominate the political sphere, and the Chinese, who make up around 30% of the population and dominate the economy. For the most part, the two groups coexist peacefully, but suspicion and fear, notably in relation to the tenuous boundary between their two spheres of activity, have caused tensions to flare and continue to be potential sources of political unrest. As recently as August 18, 2000, UMNO's Youth organization organized a rally to protest against what they perceived as an attempt by the Malaysian Chinese Organizations Election Appeals Committee (Suqiu) to question Bumiputra rights and privileges.

Some of these rights are encoded in the Constitution, forged under the "Bargain" of 1957. The Bargain was a set of explicit compromises, as set out in the Constitution, aimed at recognizing the special position of the Malays in exchange for Malay concessions on citizenship, as well as some tacit compromises regarding the protection of non-Malay economic interests. In exchange for the according of citizenship *jus soli* to every person born in the Federation after Independence (Merdeka) was achieved on August 31, 1957, the Constitution accepts Islam as the state religion, Malay as the national language, and the Malay monarchs as heads of state. In addition, the Constitution provides for the Yang Dipertuan Agong to "safeguard the special position of the Malays" by reserving positions in the federal civil service, through scholarships or other educational or training privileges, and by granting permits and licenses (see Hirschman). However, it is worth noting that, even as the Constitution identifies a national language and religion, it also sets out a policy of religious and linguistic freedoms. As testimonial of that earnestness, official publications are printed in both Malay and English. Implicitly, also, the Bargain meant that the Chinese would continue to enjoy the economic opportunities that had been open to them under British rule (see Milne and Mauzy).

The compromises set out in the Bargain were contingent on each ethnic group respecting the boundaries that had been established. Unfortunately, as political and economic conditions worsened in the 1960s, the bonds began to fray. Despite continued growth, the disparities in unemployment rates and income between the wealthy and the poor increased. Although the deteriorating conditions affected all Malaysians, the economic inequalities came to be perceived along ethnic lines. Malays in the countryside felt that the government was not doing enough for farmers, while urban Malays felt the economic gap most strongly in comparison to urban Chinese professionals and businesspeople. The Chinese, in turn, felt that their economic opportunities were threatened by the increasing economic encroachment of government, in the form of regulation and new public enterprises (see Gomez and Jomo, Jesudason, Means, and Milne and Mauzy).

The growing dissatisfaction of both ethnic groups with their representative parties, the UMNO and the MCA, and the willingness of the opposition parties to appeal for support along ethnic lines, led the Alliance coalition to lose its two-thirds majority in the parliamentary elections in May 1969. The biggest losses were sustained by the MCA, which lost more than half of its seats to opposition Chinese parties. Although the Alliance retained the majority of seats in Parliament (66 of 103 seats), the Chinese opposition parties held victory parades in the capital, Kuala Lumpur, to celebrate their successful erosion of MCA support and, hence, MCA's endorsement of special Bumiputra rights. Malays responded with their own pro-government demonstrations. Armed with weapons, many assembled to listen to Malay politicians recount the "insults" they had sustained and the challenge allegedly being mounted against Malay special rights. In this highly charged atmosphere, armed Malays started looting and burning Chinese shops and houses in areas where the Chinese and Malays lived in close proximity. Groups of Chinese launched counterattacks. The mob violence overwhelmed the police, and so the army was called in and a state of emergency was imposed. Severe rioting, looting, and arson continued for two days, despite the heavy military presence and the declaration of a curfew. Property damage left 6,000 people homeless and social calm was completely restored only after two months (see Gomez and Jomo, Jesudason, Means, and Milne and Mauzy).

The riots exposed the tensions underlying interethnic relations and the frailty of the government's efforts to maintain the balance set out under the Bargain. Following the confrontations of 1969, the government enlarged the governing coalition (see above); adopted a constitutional amendment to outlaw criticism and questioning of Malay rights and status, and prohibit any act, speech, or publication that produced "feelings of ill will and enmity between different races" (see Means); and substantially increased its own economic presence in order to redress the perceived imbalance. Among the measures adopted was the New Economic Policy (NEP), which embodies a set of objectives to improve the economic welfare of the Bumiputras through ethnic preferences in employment, education, corporate ownership, and business activities (see Nesadurai). To reassure the Chinese and the Indians that the government did not intend to improve Bumiputra welfare at their expense, the government took pains to reiterate that – in the words of the *Second Malaysia Plan* of 1971, which introduced the NEP:

> The New Economic Policy is based upon a rapidly expanding economy that offers increasing opportunities for all Malaysians, as well as additional resources for development. Thus, in the implementation of the Policy, the Government will ensure that no particular group will experience any loss or feel any sense of deprivation.

Despite these reassurances, it was inevitable that the growth of the public sector and regulations would constrict opportunities for non-Bumiputras. Statutes such as the Industrial Coordination Act of 1975, requiring licenses for all manufacturing activities, which could be revoked if the operations did not fulfill government requirements regarding Malay employment, were particularly daunting. Nevertheless, while many may argue with the overall effectiveness of government's efforts on improving economic welfare, there was one clear, if unintended, effect: it provided an impetus to forming "Ali Baba" collaborations between Malays and Chinese. Specifically, Malays fronted Chinese businesses to obtain licenses for them in return for payments (see Gomez and Jomo, Jesudason, Means, and Milne and Mauzy). Although these "Ali Baba" partnerships fall far short of the government's intended effect, they have at least provided the rudiments for interaction between the main ethnic groups. In some cases, the Malay directors have evolved from merely acting as fronts to playing real corporate roles (see Milne and Mauzy). Chinese businesses have also forged links with Malay politicians and power centers (see Gomez and Jomo, and Milne and Mauzy). However, the Indian minority, accounting for around 8% of Malaysians, has not fared so well.

Economic growth has also helped to smooth ethnic tensions between the main groups (see Milne and Mauzy). Since 1990, the government has also made several economic concessions to the non-Bumiputras, most notably by shifting the focus away from interethnic redistribution in the adoption of the National Development Policy (1991–2020), privatization of state businesses, and deregulation of the economy (see Gomez and Jomo, and the *Second Outline Perspective Plan*). On the political front, the government has also made explicit overtures to court Chinese voters. In July 1999, Mahathir stepped up the charm in a public speech, thanking the nation's wealthy ethnic Chinese community for standing behind him (as reported in *The Straits Times*, July 18, 1999). Whether these political and economic accommodations will persist depends largely on Malaysia's economic performance and, perhaps, on the extent to which the specter of the riots of 1969 recedes.

The New Economic Policy (NEP)

The significance of the NEP is twofold. First, the NEP led to a dramatic increase in the size of the state machine in order to implement the policy and address the concerns of particularistic interests, such as the UMNO-linked corporations and the Malay businesses, that the policy created. As a result, the influence of the NEP continues to be felt through the links and businesses that it nurtured, even though the government pronounced the successful attainment of the NEP's objectives in 1990 and even announced a new National Development Policy (NDP) in succession to it. Second, the NEP is a crucial manifestation of the interconnectedness between the economy and politics that continues to govern developments in Malaysia. It was the result of political instability and has in turn influenced politics, most recently in the fallout from the financial crisis of 1997–98 (see above).

The NEP in fact refers to a set of objectives, rather than specific policies, promulgated in 1971. The NEP aimed to eradicate poverty, restructure society, and rebalance the economy, in the hope of avoiding any repetition of the riots of 1969. As set out in the *Second Malaysia Plan*,

> The first prong is to reduce and eventually eradicate poverty, by raising income levels and increasing employment opportunities

for all Malaysians, irrespective of race. The second prong aims at accelerating the process of restructuring Malaysian society to correct economic imbalance, so as to reduce and eventually eliminate the identification of race with economic function.

The aims of the NEP were, then, ostensibly "irrespective of race"; yet the NEP target of indigenous corporate ownership of 30% by 1990 made clear who the intended beneficiaries of the government's efforts were.

To achieve the targets, the government essentially set itself up as trustee for the Malays in particular and the Bumiputras more generally. For instance, state enterprises such as the National Corporation (Pernas), the National Equity Corporation (Permodalan Nasional Berhad, or PNB), and the State Development Corporations (SEDCs) were established to accumulate corporate assets directly. In some cases, this has entailed buying shares in established companies for subsequent acquisition by Malays. Other state organs, such as the Council of Trust for the Indigenous People (Majlis Amanah Rakyat, or MARA) and Bank Bumiputra, have provided financial help to promote Malay entrepreneurship. To safeguard these acquisitions and businesses, regulatory committees were set up, including the Foreign Investment Committee (FIC), which monitors foreign acquisitions of Malaysian companies and has pressured large corporations into restructuring their equity, and the Capital Issues Committee (CIC), which has worked with the FIC to set the price of shares issued to Malays by Chinese and foreign companies. In addition, regulatory measures such as the Industrial Coordination Act were introduced to demand licensing of manufacturing activities and to monitor the activities of those medium-sized companies that escaped the purview of the FIC or the CIC.

As a result of the government's significant involvement in the economy, the NEP managed to achieve some of the objectives. The incidence of poverty has decreased across all ethnic groups. For the country as a whole, poverty levels decreased from 49.3% of the population in 1970 to 15% in 1990. The identification of occupation with ethnicity has also been reduced (see Gomez and Jomo). Finally, the Malay share of equity in publicly listed companies increased from 2.4% in 1970 to 20.3% in 1990 (according to the *Sixth Malaysia Plan*, 1991).

These accomplishments came at a price. Although wealth generally increased, the distribution of wealth across the nation remains uneven, for West Malaysia (or Peninsular Malaysia), which was Malaya until 1963, is still doing much better than East Malaysia (the states of Sabah and Sarawak, which are on the island of Borneo). Small farmers and rural workers continue to suffer economically and account for a majority of the nation's "hardcore poor," with limited access to basic services and amenities (see the *Sixth Malaysia Plan*, 1991). Further, the attempt within the terms of the NEP to regulate and control the economy, and the concurrent lack of administrative machinery to coordinate these tasks, left unlimited discretionary powers to government-owned or subsidized enterprises. Consequently, corruption, financial scandals, misuse of public funds, and administrative ineptitude have plagued some of these institutions, including Bank Bumiputra, much touted as the flagship of the NEP (see Aziz). Not surprisingly, as 1990, the year targeted for the conclusion of the NEP, approached, speculation was rife as to whether the NEP would be extended.

In 1991, Mahathir introduced the National Development Policy (NDP) to succeed the NEP. Where the NEP had emphasized interethnic redistribution, the NDP focuses on a "Malaysian nation" (*bangsa Malaysia*) transcending ethnic relations (see Gomez and Jomo). The NDP also shifted away from emphasizing Malay shareholdings in corporations and businesses to promoting "meaningful participation," the strengthening of Malay management, and the retention of existing businesses and corporations (see the *Second Outline Perspective Plan*, 1991).

While many have welcomed the ostensible shift in the government's policies, it is unclear if the shift is practicable. The government's responses to the crisis of 1997–98 revealed that the NEP had developed and nurtured a group of Malay corporate elites and businesses, in addition to state-bureaucratic organs and government-linked corporations that are unlikely to bear the burdens of economic contraction. As a result, although the NEP is no longer the government's guiding policy, it nonetheless continues to shape its economic responses (see Gomez and Jomo, and Nesadurai). The crisis suggests that only continued economic development can afford the government the luxury of taking a backseat in steering the economy.

Conclusion

Despite predictions by many market analysts that Malaysia would take some time to recover from the crisis, the economy took an upswing in 1998 and in 1999 GDP grew by 5.8%, sufficient for Mahathir to claim victory over the crisis just before the parliamentary elections in December. Yet not all are convinced that Malaysia's economic performance is sustainable. In particular, it is disputable whether growth is being achieved through real economic development or mostly through the state's largesse.

What, then, can we expect of Malaysia as it enters the 21st century? As far as the political system is concerned, there is every reason to expect that a strong executive, capable of deciding the country's course of development, will remain in place, although UMNO may well place constraints on the executive through challenges for political leadership. If the PAS and the DAP can maintain a united front, they may pose a serious challenge to UMNO, but it is more likely that UMNO, with some input from its coalition partners, will continue to dominate policy-making. The directions it takes will be guided principally by its need to go on making interethnic accommodations and by the legacy of government involvement that the NEP has left behind.

As for the economy, there is one particular factor that justifies taking an optimistic view: Malaysia has the rudiments for ensuring efficient economic performance, especially in its institutional infrastructure. The Economic Planning Unit (EPU) in the Prime Minister's Department has been the primary administrative unit for several projects, including the general administration of the NEP and the privatization plans. Similarly, the central bank displayed a remarkable depth of economic understanding during the recent crisis and, until Mahathir intervened, it was willing to pursue a strict monetary policy in order to cool the overheated economy. If Malaysia can supplement this infrastructure with a decrease in the government's involvement in the economy and an increase in the administrative machinery to achieve better oversight of existing enterprises, the prospects for continued growth, accompanied by and underpinning political stability, are strong.

Further Reading

Aziz, Tengku Abdul, "Malaysia Incorporated: Ethics on Trial," in *Australian Journal of Public Administration*, Volume 58, number 4, 1999

> Aziz discusses some of the financial scandals in Malaysia, attributing them mainly to the increase in the government's economic activities since 1971, and the lack of administrative machinery to monitor its activities and make them accountable.

Crouch, Harold, "Industrialization and Political Change," in Harold Brookfield (editor), *Transformation with Industrialization in Peninsular Malaysia*, Kuala Lumpur, Oxford, and New York: Oxford University Press, 1994

> Crouch discusses the complex influence of changes in the economy on political developments, focusing on West Malaysia.

Gomez, Edmund, and Kwanme Jomo, *Malaysia's Political Economy: Politics, Patronage, and Profits*, second edition, Cambridge and New York: Cambridge University Press, 1999

> The authors examine Malaysia from an economic perspective, paying particular attention to the objectives and effects of major economic policies since the New Economic Policy was introduced.

Hirschman, Charles, "The Society and its Environment," in Frederica Bunge (editor), *Malaysia: A Country Study*, Washington, DC: Foreign Area Studies, 1984

> The chapter describes interethnic relations in Malaysia and the compromises, set out in the Constitution, that are intended to balance these relations.

Jesudason, James, *Ethnicity and the Economy: The State, Chinese Business, and Multinationals in Malaysia*, Singapore, Oxford, and New York: Oxford University Press, 1989

Jesudason provides an in-depth discussion of the evolution of Malaysia's economic policies from *laissez faire* to government dominance of the economy, tracing the varying effects of these policies on the economic participation of the Chinese and foreign businesses.

Khoo Boo Teik, *Paradoxes of Mahathirism: An Intellectual Biography of Mahathir Mohamad*, Kuala Lumpur, Oxford, and New York: Oxford University Press, 1995

This book chronicles Mahathir's political career, and provides rich detail of the events surrounding his decisions and responses.

Khoo Kay Jin, "The Grand Vision: Mahathir and Modernization," in Joel Kahn and Francis Loh (editors), *Fragmented Vision: Culture and Politics in Contemporary Malaysia*, Honolulu: University of Hawaii Press, 1992

The author details the fracture in UMNO in 1985–88 and proposes that, in addition to the patronage issues that may have motivated factionalism, the fracture also revealed policy differences among the factions.

Means, Gordon, *Malaysian Politics: The Second Generation*, Singapore, Oxford, and New York: Oxford University Press, 1991

The author analyzes the three administrations following that of Tengku Abdul Rahman, and details developments within UMNO during Mahathir's period in office.

Milne, Robert S., and Diane Mauzy, *Malaysian Politics under Mahathir*, London and New York: Routledge, 1999

Milne and Mauzy examine the effects of Mahathir's tenure on political and economic development in Malaysia, with special focus on ethnic relations, foreign relations, human rights, and the succession.

Nesadurai, Helen, "In Defense of National Economic Autonomy? Malaysia's Response to the Financial Crisis," in *The Pacific Review*, Volume 13, number 1, 2000

This article provides a richly detailed chronicle of the changes in Malaysia's responses to the Asian financial crisis, showing how the changes reveal the continuing influence of the New Economic Policy.

Privatization Masterplan, Kuala Lumpur: National Printing Department, 1991

This official publication contains the government's description of the privatization process and the early stages of its implementation.

Rhodes, Martin, and Richard Higgott, "Introduction: The Asian Crisis and the Myth of Capitalist 'convergence'," in *The Pacific Review*, Volume 13, number 1, 2000

Rhodes and Higgott set out the theoretical objections to the prevailing notion that the Asian crisis will lead to a convergence of the economies of the region on a US model, in which the market dictates.

Second Malaysia Plan, 1971–1975, Kuala Lumpur: Malaysia Government Press, 1971

This is the official formulation of the objectives of the New Economic Policy, and includes the targets for Bumiputra participation in education, industry, commerce, and the bureaucracy.

Second Outline Perspective Plan, 1991–2000, Kuala Lumpur: National Printing Department, 1991

This government publication describes the change in emphasis from promoting Bumiputra rights to forging a new Malaysian unity based on ethnic diversity.

Shinn, Rinn-Sup, "Government and Politics," in Bunge, cited above under Hirschman

This chapter describes the political system of Malaysia, detailing the main institutions, and the checks and balances among them.

The Straits Times. Various issues. Singapore.

This long-established English-language newspaper provides accounts of developments in Malaysia as well as in its country of publication.

Dr Fiona Yap is an Assistant Professor in the Department of Political Science at the University of Kansas, Lawrence.

NATION STATES

Chapter Six
Singapore

Michael Haas

Few would have guessed in 1900 that a sleepy, swampy island, with an economy centered on fishing, smuggling, and the entertainment of sailors from passing ships, would have the fourth highest income per capita among the countries of the world by 2000. That Singapore has risen meteorically is undeniable. How did such a transformation occur? Does Singapore, which has been an independent state since 1965, provide a model for poorer countries? Can this island republic expect to enjoy further prosperity in 2100?

Economic Transformation

In 1819, the ambitious Sir Stamford Raffles landed at what is now Singapore, and persuaded the Sultan of Riau and Johor to cede the island to his employers, the British East India Company. In 1824, Britain assumed sovereignty over Singapore, and in 1826 it was absorbed into the Straits Settlements, along with Malacca and Penang, as part of British India. In 1867, however, the Straits Settlements became a crown colony governed directly from London.

After World War II, Britain granted independence to most of its colonies, but hesitated to do so in Southeast Asia because there was a Communist insurgency in Malaya and a strong Communist Party in Singapore. Nevertheless, Malaya received full independence in August 1957, and in June 1959 Singapore was granted self-governing status, for all matters except defense and foreign affairs, under the prime ministership of Lee Kuan Yew. Next, the formation of Malaysia in September 1963 involved the federation of the 11 states of Malaya with Singapore, as well as Sabah and Sarawak. This was not to last. Lee Kuan Yew's strident rhetoric on behalf of the economic development of Malaysia annoyed leaders in Kuala Lumpur, disputes arose over government spending and representation of the states in the federal Parliament, and Singapore left the federation in August 1965.

Lee, who had lobbied long and hard for Singapore's inclusion in Malaysia so that his country would be part of a larger economic market, then feared that the city state faced economic collapse. Yet Lee, who served as Prime Minister until November 1990, was wrong in fearing the worst: between 1965 and 2000, Singapore's economy grew faster than any other in Southeast Asia. Despite downturns in 1964, 1985, and 1996–97, the island republic's income per capita has surpassed that of most European countries, including that of its former colonial master, the United Kingdom.

Today, the most significant industries in this country of around 3.5 million people are shipbuilding, oil refining, electronics, banking, textiles, food, rubber, lumber processing, transportation, and tourism. Among the products manufactured are automotive components, computers, electronic instrumentation, machine tools, medical instruments, and precision engineering devices. Only 1.6% of the land is arable and the share of agriculture in the labor force is less than 1%; manufacturing and other industries account for 35%, and services for 64% (*FEER*, pp. 13 and 15). Since labor demand and real wages have grown rapidly, there is little unemployment. Poverty has been largely eliminated, and about 80% of residents own their own homes or apartments. Singapore's inflation rate is below the world average, despite the shortage of labor.

The standard indicators of quality of life also place Singapore among the leading countries in the world. Life expectancy, for example, has risen from 68 years in 1965 to 76 years in 2000. Improvements in schooling have resulted in sharp increases in literacy, and Singapore now has a highly skilled, technologically sophisticated labor force. Infant mortality has also fallen sharply, to around one half the average rate in the United States, although the birth rate is only slightly higher than the death rate.

Factors in Singapore's Success

Against this background, it is understandable that in 1993 the World Bank described Singapore as a "miracle economy," but opinions differ on how the "miracle" is to be explained.

Most observers start with Singapore's favorable geographic circumstances. Halfway between China and India, and blessed with a calm harbor, Singapore was already an important entrepot for international trade in 1900. A century later, the value of its exports was equivalent to 170% of its GDP (*FEER* p. 13), largely because many cargo-laden ships dock for a while at the port of Singapore before proceeding on their journeys to other destinations. During the colonial era, Singapore became a major refueling station and stopover point for ships in the British Royal Navy. As a result, the port became a major location for shipbuilding and repairing.

After World War II, Singapore's trade and investment became integrated with the world capitalist system, and its geographic location once again assisted its development. The shortest route for oil tankers from the Middle East to Japan and South Korea is through the Singapore Straits. Singapore is also an obvious stopover location and transshipment point for jet airplane traffic, since there are long distances between the region's central trading cities. Goods manufactured in Singapore were once sold primarily to adjacent countries, but they can now reach destinations in Europe and North America overnight. Thus, Singapore has continued to benefit as the world has become increasingly prosperous.

However, geographic location cannot provide the whole explanation for Singapore's economic success. Another major factor has been its population, which is highly educated, in excellent health, and connected to modern sources of information. In other words, Singapore has acquired the ideal labor force that most countries seek as a precondition to economic success. Singapore's economy has been able as a result to diversify beyond merely serving as an entrepot, and has shifted from low-paid, labor-intensive production to highly lucrative capital-intensive production of high-technology goods.

Specific government policies have promoted the transformation of Singapore from a sleepy, malaria-infested port town, dependent on unskilled labor by illiterate workers, to a bustling, clean, modern city with an educationally advanced workforce. Some credit for the change should be accorded to Britain, which made Singapore a jewel in the empire, providing paved roads, safe drinking water, the best schools in the region, and a fine university. It was also the British who first recruited hard-working and entrepreneurial Chinese to the island, in order to provide a middle-class infrastructure for urban life. Independent Singapore adopted the British conception of the island as a progressive urban location exploiting the rural Malayan hinterland, but went beyond the somewhat aristocratic approach of the British.

First, the government, committed to a parliamentary socialist ideology in the early years, expanded educational opportunities. A multilingual educational system, with English as a second language, steered students into technical and vocational studies. Parallel educational streams in Chinese, English, Malay, and Tamil languages assured the population that cultural diversity would be respected, although most parents pragmatically enrolled their children in English-language schools. Later, the government discouraged education along separatist communal lines. The greatest economic benefit was derived from enhancing quality at the lower levels of education, in order to produce workers with relatively high levels of literacy, well prepared to follow and interpret complicated instructions. At the higher levels of the educational system, the government provides scholarships so that the brightest are able to reach their intellectual potential through college education, both at home and abroad. Large proportions of students enroll in professional schools, such as colleges of business administration. Upon completion of college education, graduates are then required to use their newly honed skills as government employees for at least five years, assuring that the latest innovations are adopted by the civil service.

Second, in 1960 the government established the Housing and Development Board, giving it extensive powers of land condemnation and housing construction. Soon after independence, a downtown district of decaying shops was transformed into a cityscape of modern skyscrapers and shopping malls. New towns with government-financed housing and recreational facilities were built around the island, and most residents were able to buy their own apartments at reasonable prices. Drainage projects and sanitation improvements gave the island the appearance of a set of immaculate gardens. The government initially also provided affordable health care for all, so that chronic health conditions could be treated early and thus would not develop into acute disabilities. The disabled and infirm, however, were left to family and private care. As the population became more prosperous and could afford to pay for better health care by private physicians, the quality of the medical profession rose as well but government benefits were cut.

Third, Singapore's increasingly restrictive immigration policy has been biased in favor of workers in the most productive periods of their lifespans. Unskilled immigrants handle the more menial jobs, leaving employment in the most highly skilled occupations to Singaporeans or well-trained migrants, resulting in a class structure with large gaps in wages. Singapore has had a labor shortage as the economy has boomed, so immigration has been an important factor in accounting for increasing economic prosperity. In 1969, for example, after anti-Chinese riots broke out in Malaysia, a flood of well-trained Chinese migrated to Singapore. The labor force is also highly mobile: individuals expect to move through a hierarchy of jobs, acquiring new skills through education and on-the-job training, and thus to achieve upward mobility in a "meritocratic" environment.

Fourth, just as government policies promoted the development of a sophisticated and flexible labor force, so Singapore's leaders eventually adopted other policies to provide an economic infrastructure that many observers believe is second to none throughout the world. At first, the ruling People's Action Party (PAP), which outmaneuvered pro-Communist political parties on the road to power, used socialist rhetoric, and the prospect of government control of the economy frightened the business community and deterred foreign investment. While it was integrated with Malaya and later Malaysia, Singapore followed an import-substitution approach (see Chapter 3), but the result was a disaster. These policies were reversed when the Minister of Finance, Goh Keng Swee, embarked on a pragmatic program of industrial development to be financed by foreign capital. Singapore adopted export-oriented industrialization at a time when most developing countries, including Malaysia, were still committed to import substitution. In August 1961, an Economic Development Board was established and work began on Jurong Industrial Estate, a tax haven that sought foreign investment for new industries. One crucial problem in supporting the growth process in developing countries, the transfer of technology, was solved by welcoming multinational corporations into the city state. The high technology demanded by these corporations spurred the growth of domestic high-technology industries.

The Economic Development Board is now one of 70 statutory bodies that implement the government's economic policies. Some promote economic growth by developing infrastructure, ranging from bridges and piers to roads and telephone lines, and including the world's most modern airport at Changi, as well as soliciting foreign investment to finance construction. Other boards supervise broadcasting, finance, housing, industrial training, shipping, and urban redevelopment. These boards, manned by experts of exceptional technical knowledge, channel the foreign capital that accounts for around 80% of all capital in manufacturing (see Neher).

In addition, from 1968, when the PAP won all the seats in the parliamentary elections, economic development plans were implemented, including the formation of new "government-linked corporations" (GLCs) and government-financed joint ventures with the private sector. The GLCs, approximately 500 in number, provide the government with the opportunity to intervene in the market, although they are not counted within the government budget. Managed by able corporate executives, GLCs have been highly efficient and profitable. When efficiency declines, the government restructures or privatizes GLCs.

It should be emphasized that policies in specific industries developed step by step. Initially, high priority was given to transportation. The government-owned Keppel Corporation organized Singapore's port so that thousands of private companies could trade with the rest of the world. With the rising importance of air freight, the government also invested heavily in airport facilities. Changi Airport has acquired the same central role in facilitating air freight as Singapore's harbor has played in sea freight. Singapore Airlines benefited from the exploding demand for cargo and passenger transport in Asia.

New laws introduced in the 1960s permitted longer working hours, regulated wages, and restricted bonuses and overtime, but required employers to place large matching contributions in the Central Provident Fund. This institution, which was founded in 1954, before the PAP came to power, and is managed by the Government of Singapore Investment Corporation, now absorbs around 40% of salaries (see Lingle and Wickman). There was no provision for unemployment insurance, but the government's policy of full employment was largely successful. Welfare eligibility criteria became more stringent, so that the welfare state effectively ceased to exist in Singapore.

In June 1978, Parliament adopted laws to facilitate the internationalization of the economy. All restrictions on capital flows, in any currency, were abolished, as were capital gains taxes, inheritance taxes, and wealth taxes, but only for foreign corporations. Companies operating in Singapore were allowed 100% foreign ownership and foreign corporations received major tax reliefs. The government promised that no foreign company would be nationalized under any pretext. This change in economic policy proved eminently successful in attracting foreign investment. Before 1978, Singapore's participation in the international economy was mostly in the area of trade. The upswing that began in 1978 was a major turning point; thenceforward, foreign investment became the leading factor in the progress of Singapore's economy. In this respect, Singapore's economic liberalization was ahead of its time. By contrast to countries where local businesses have successfully lobbied to keep foreign competition out of the economy, the government of Singapore has given preference to large corporations over small firms: arguably, more higher-paying jobs have been created as a result.

Culture and Ethnicity

A largely unspoken but widely accepted explanation for Singapore's progress is that it is partly or even mainly due to the predominant ethnic group, the Chinese, who account for around 77% of the population. Singapore's leaders have criticized other ethnic groups, giving credence to the inference that they believe in a Chinese cultural advantage. Chinese trading groups have indeed prospered in Indonesia, Malaysia, the Philippines, Thailand, and Vietnam. Since most Chinese trading houses prefer to base their business activity in a Chinese social environment, the most successful Chinese trading families took up residence in Hong Kong and Singapore. The attraction of these cities was especially enhanced by China's political disintegration and subsequent transformation into a Communist country in the 1950s.

Although many South Indians occupy high positions in government, J.B. Jeyaretnam, an articulate South Indian opposition politician, has been slandered by government officials in demeaning racial terms. Malays, meanwhile, have been considered second-class citizens, as evidenced by a remark made by the Defense Minister, Lee Hsien Loong (Lee Kuan Yew's son), in 1985, to the effect that they would never be sent into the front line of battle in the event of war with Indonesia or Malaysia. The proposal to set up a government bureau to promote marriages among college graduates has been deconstructed by many observers to mean that Chinese, who dominate college enrollment, should have more children, whereas Malays should be sterilized (see Devan and Heng).

In 1989, Parliament adopted a quota system for the numbers of people from minorities allowed to live in public housing, inhabited by around 86% of the country's residents, in order to prevent the development of ethnic residential enclaves. Later that year, after a seven-year experiment, required courses in religion were abolished in favor of civics courses, and the government began to promote a curriculum and national ideology stressing shared core values, often known as "Asian values." In 1990, after some deliberation, Parliament passed the Maintenance of Religious Harmony Act, which proscribes religious organizations from engaging in politics. These and other measures suggest strongly that many PAP officials are acting on their private belief that minorities hold back progress because they are more communally oriented and therefore less productive.

Alternative Explanations

Most observers of Singapore's economic success stress the role of innovative government policies. At least two alternative explanations have been offered by those who dispute that view. According to some economists, the prosperity of the countries of East Asia is inextricably interlinked in the manner of geese flying in formation, with Japan in front (see, for example, James). As Japan became an economic superpower, manufacturing was outsourced to neighboring South Korea and Taiwan, and Japan developed more high-technology industries. As South Korea and Taiwan boomed, they too diversified into high-technology industries, and manufacturing at a lower level of technology moved to Southeast Asia, including Singapore. The economic successes of Japan, South Korea, and Taiwan resulted in more trade passing through the port of Singapore. Likewise, when manufacturing plants moved to Southeast Asia, Singapore capitalized on the increased trade flow.

A second alternative explanation has been developed by Christopher Lingle and Kurt Wickman. They note that few options were available to Singapore, given the natural resources and demographic profile of the city state, and argue that Singapore merely copied the policies of Hong Kong, another enclave with a small urban population that developed as a trade entrepot next to a large hinterland. They go on to explore the specific features of a city state economy.

First, in a city state individuals can more readily find jobs that match their productivity, and the costs of switching jobs are relatively low. High job mobility means that adjustments to the labor market, in response to changes in productivity and demand for skills, become continuous. The consequent repeated upgrading of workers' skills, along with continuing high levels of labor mobility, allows new technology to be easily integrated into the economy.

Second, both a flexible labor market and rapid economic growth allow individuals to choose their own income path. A rigid labor market, characterized by restrictive job definitions imposed by centralized trade unions, would facilitate collective bargaining for wages, but, especially when combined with state regulation and wage legislation, would tend to obstruct the matching of jobseekers with jobs. Leaders of a city state in which wages rise with productivity are less likely to be pressured to develop an advanced income distribution policy, and so this source of social tension is eliminated.

Third, a city state will tend to attract middle class citizens from the regional economy. A concentration of individuals earning relatively high incomes promotes such middle-class values as hard work, honesty in fulfilling contracts, life-long learning, and the habit of saving. Once known as the "Protestant ethic," these values are crucial for economic growth in any society and are acquired in the home even before children go to government-run schools.

Fourth, the industrial structure of a city state will best thrive under conditions of maximum openness in international economic relations. According to Lingle and Wickman, the Singapore government had no choice but to maintain a free port and to seek a common trade policy for as much of Asia as they could have stable relations with. The demand and scope for such policies were already present through the economic structure of the city states and the networks among overseas Chinese throughout Asia.

Instances of Failure

Whatever the reasons for Singapore's relative success, there have also been some failures. When the government has intervened directly in the economy, outside the narrow areas described above, the results have been mixed. The "new wage" policy of 1980–81, under which wages were generally raised, was based on the assumption that higher wages would force the economy to shift into more productive activities at a faster pace. By 1985–86, however, Singapore's economy was hit by a deep recession; economic growth fell almost to zero, and private investment declined by about 50%. In 1988, after the government withdrew the high-wage policy, the economy returned to its former path of steady high growth.

The downturn of 1985–86 proved Singapore's vulnerability to the health of the world economy, which a small state cannot control. Thereafter, Singapore quietly began to invest in neighboring countries in order to promote prosperity in the region. When the Asian financial crisis erupted in the late 1990s, many Singapore investments proved worthless, and the economy went into crisis once again.

Further, although the Singapore government is generally considered to be free from corruption, examples of crony capitalism have emerged in recent years. Lee Kuan Yew's youngest son is the chief executive officer of Singapore Telecom, his wife is the majority shareholder in a bus and taxi firm that receives government contracts, and Lee and Lee, a law firm in which Lee Kuan Yew's brother and his wife are partners, handles all housing contracts for government-owned apartments. In 1996, controversy raged after the public disclosure that a housing developer had offered apartments worth S$700,000 to both Lee Kuan Yew and his son Lee Hsien Loong, the minister mentioned above, at 5 to 12% below the standard prices. Lee Kuan Yew's brother was a director in the corporation selling the properties, and notification of the sale was delayed beyond the time limit required by the regulations of the Stock Exchange of Singapore, which therefore criticized the sale.

In addition, several Singapore firms are fronts for smuggling operations, for which the country has been famous for centuries. A US State Department report has revealed that the government's Singapore Investment Corporation played a role in financing drug smuggling from Burma (see Bernstein and Kean). Later information has identified the wife of Lee Hsien Loong, who is now Deputy Prime Minister, as the head of Singapore Technologies Group, which has sold electronic eavesdropping equipment to the Burmese government. Singapore now risks a trade embargo because of this involvement in the drug trade.

Problems on the Horizon

Singapore's leaders frequently admit that the economy is fragile, and they are right to do so. The future of their city state is uncertain.

To begin with, there are geographic limits to Singapore. Even including around 60 islets located between Indonesia and Malaysia, the national territory, at just 240 square miles, is smaller than, for example, Corfu, Guam, the Isle of Man, Isle Royale in Lake Superior, or O`ahu in Hawaii. In order to overcome these territorial limits, Singapore has begun to explore the possibility of locating Singapore-owned factories on a nearby Indonesian island. In 1990, the idea expanded into a proposal for adjacent parts of Indonesia (the Riau archipelago) and Malaysia (the state of Johor) to form a special economic zone, known as "Sijori." Additional factories and warehouses, where rental costs are below those of affluent Singapore, could then enable the island republic to share its economic growth with its neighbors (see Lee, Parsonage, and Vatikiotis). Sijori took off in 1992, when Indonesia allowed 100% Singapore ownership of factories on Batam Island. Corporations based in Singapore then relocated industrial plants to Batam, employing Indonesians at lower wages, although commuters from Singapore still provide much of the trained workforce.

A more serious problem lies ahead. Thailand has been considering the possibility of building a canal across the Kra Isthmus that would save billions of dollars in fuel by shortening trade routes from Europe to East Asia. Japan faces severe financial problems, but it could still finance a large portion of the project, which would decimate Singapore's economy. Similarly, since airplanes now carry enough fuel to overfly Singapore, air traffic may bypass Changi Airport and head for Bangkok as a much larger volume of business develops in Thailand, the principal hub for developments in Indochina.

Singapore has no significant natural resources, so minerals, oil, and water are imported. Indonesian oil could be withheld, as could Malaysian water. Both countries could easily blockade Singapore, as the Straits of Malacca are relatively narrow. Pirate ships ply the straits even now, making the route unsafe. Since Singapore also lacks any natural barriers to invasion, Indonesia or Malaysia could quickly overwhelm Singapore militarily, just as Japan did during World War II. Singapore's military defenses are probably inadequate, and only 4.7% of GDP is spent on defense (*FEER* p. 13).

Singapore is therefore crucially dependent on maintaining friendly relations with its larger neighbors. Nevertheless, government leaders repeatedly infuriate both countries with undiplomatic statements. In 1996, for example, Lee Kuan Yew, who wields significant influence even after leaving the prime ministership, intimated that an economically faltering Singapore might one day rejoin an economically resurgent Malaysia, provided that Kuala Lumpur abolished special treatment for its indigenous peoples and accepted meritocracy. After Malaysia filed a diplomatic protest, on the grounds that this obvious criticism represented interference in its internal affairs, other politicians in Singapore made variations on Lee's remarks. Two years later, Lee Kuan Yew referred to Johore Bahru as a place "notorious for shootings, muggings and car-jackings"; after yet another diplomatic protest by Kuala Lumpur, the government-controlled press in Singapore continued the undiplomatic discourse.

Another potential source of problems is the steady flow of emigration from Singapore. Faced with government controls over political discourse, dissidents tend to leave the country; on a larger scale, thousands depart in search of better-paying jobs in industrial democracies, having absorbed the government's exhortations to increase their incomes by shifting jobs. An estimated 5% of the population leaves Singapore each year (see Neher). Nor is this the only difficulty affecting the country's labor market. Although there is 92% literacy (at least according to official figures), around 20% of the population are literate only in Chinese. Only 10% of the labor force has had any post-secondary education. Further, even now, after the shift into high-technology industries, Singapore workers are only half as productive as Japanese and it has been estimated that only 1.75% of GDP growth can be attributed to increased productivity (see Chua). In addition, 26% of the labor force is imported from other countries (according to a report in *Asiaweek*, September 15, 2000). Many of these migrant workers are Filipino housemaids and Thai construction workers. Finally, around 10% of the population lives at or below the poverty line. It is clear that, without the incomes of the managers of expatriate firms and their employees, Singapore's GDP would be so much lower that it would rank as a developing country, not a developed one.

Social alienation is undoubtedly rife throughout Singapore. In many families, rising living costs have forced both parents to work, so that they are unable to manage their families at home, divorce rates are increasing, and their children are free to roam the streets. Singapore is not in fact the safest country in Asia, despite the widespread belief that it is. The tough approach to crime, which includes

1,000 canings each year, does not appear to be working, as rates of crime are higher than in Asian countries where penalties are less severe. In 1993, Singapore had a murder rate of 59.4 per 100,000 people, compared with 57.5 in Australia and 42.2 in Malaysia. Increasing juvenile delinquency prompted the government in 1997 to ban "tea dancing" among persons aged less than 16. Suicide has been common in housing projects; attempted teenage suicide has also risen in recent years. Drug rehabilitation cases went up by 50% between 1990 and 1995.

Finally, the economy itself is not as competitive as it once was. As mentioned above, Singapore took over technologies formerly adopted by South Korea and Taiwan, which then progressed to higher levels of industrial production; but, as some of the "flying geese" theorists have emphasized, so too did Malaysia and Thailand, which have larger populations of equally disciplined but significantly cheaper workers. The hinterland economy around Singapore is developing rapidly and competing directly with the island republic in manufacturing. Manufacturing labor costs grew by 10.8% a year from 1985 to 1995, while unit labor costs rose to US$7.28 an hour (according to a report in *Asiaweek*, July 7, 1995). In comparison, the average hourly wage in Malaysia in 1995 was US$1.59, in Indonesia 30 cents, and in both China and India 25 cents.

Since not all wage differences are compensated by higher productivity in Singapore, investment has tended to move toward these countries in recent years. For example, large parts of the strategic computer assembly industry have moved from Singapore to Penang in Malaysia, and the present glut of microchips on the world market has contributed to a contraction of the growth potential in Singapore's economy. Singapore may be squeezed out of the world market, unable to increase worker productivity fast enough to outflank other countries in the region, which will be able to manufacture the same goods at lower cost.

While economic progress in East Asia may well continue to spill over into Singapore, the largest source of trade remains the United States, which accounts for 20% of the total. Nevertheless, on several occasions Singapore has jeopardized its trade relations with the United States. In 1987, Washington threatened to withdraw tariff concessions granted under the generalized system of preferences (GSP) until Singapore agreed to stop the infringement of copyright by companies making and selling pirated computer programs and music tapes. In 1988, Singapore lost GSP status anyway, after a joint congressional resolution expressed concern over the decline of human rights in the island republic.

Authoritarian Politics

The government of Singapore has never made any secret of its contempt for democracy, which policy-makers regard as an impediment to economic growth. In most countries, the disruption inevitable in eras of rapid economic change has been the seedbed for strikes, government instability, and such anomic behavior as alcoholism and suicide. In Singapore, where Lee Kuan Yew and the PAP long ago accepted the argument that such side effects of development offer a justification for authoritarian rule (see, for example, Huntington), the autonomous civil society in which disruption might have originated has been stifled.

One of the earliest authoritarian measures came in 1961, when the government deregistered the Communist-dominated Singapore Trades Union Congress. It was replaced by the National Trades Union Congress, a body that accepted the government's policy of limiting labor disputes so as to encourage business. In February 1963, the Internal Security Council launched "Operation Cold Store," detaining more than 100 people, including 24 members of the opposition Barisan Socialist Party. The mass arrests led to protest riots; the unrest, in turn, served as a pretext to arrest more dissidents, thus crushing the party. Since then, no opposition party has ever developed a mass base in Singapore. In 1964, the government deregistered the powerful dockworkers' union. In 1965, Lee sued an opposition party candidate for libel and won, a practice that continues to bankrupt opposition politicians to the present. In 1985, a prominent opposition parliamentarian, J.B. Jeyaretnam, the leader of the Workers Party, was charged with making false declarations of his party's accounts; the following year he was convicted, fined, imprisoned, disqualified from sitting in Parliament for five years, and disbarred. In 1989, after he won an appeal on his disbarment to the Judicial Committee of the Privy Council, the British court that still serves as the highest court of appeal for many Commonwealth countries, the government abolished its jurisdiction over Singapore.

The government has also intervened in the domestic judicial system. In 1963, fearing that political opponents might gain sympathy from juries, Lee insisted on abolishing trial by jury except for capital offenses. The latter exception was dropped in 1969. Since all judges are appointed by the PAP, subject to review by a commission controlled by the same party, it is not surprising that rulings in political trials have never displeased the government (see Seow).

The media, another potential source of criticism or opposition, have been curbed ever since 1966, when the government took its first step toward suppressing independent sources of information by issuing the Essential Information (Control of Publications and Safeguarding of Information) Regulations. These have been interpreted to mean that news leaks from government sources constitute sedition. In 1971, charges of sedition were brought against a Singapore Chinese newspaper for editorials lamenting the declining state of Chinese-language education, and a reporter for the *Far Eastern Economic Review* was the first of many over the years to be accused of interfering in the internal affairs of the country (in this case, by reporting news about Singapore obtained from an employee of Amnesty International) and then ordered to leave the country. In 1974, Parliament amended the Newspaper and Printing Presses Act to require public offerings of shares and government approval of all foreign investment in newspapers. In 1977, a further amendment to the law enabled the government to take over the management of all domestic publications by 1983, and two foreign correspondents were arrested for alleged slanting of the news in favor of "Communists." They were released only after they had been coerced into falsely confessing pro-Communist leanings. In 1986, the Newspaper and Printing Presses Act was amended yet again, this time to authorize government ministers to blacklist any international publication in the *Government Gazette* for alleged interference in Singapore's domestic politics. A ceiling on circulation could be imposed if a news story or editorial comment was thought to discredit the government and the publication refused to honor the "right of reply," that is, to afford space in the publication for sometimes lengthy and argumentative statements in rebuttal by a government official. The aim of "gazetting" was to reduce the revenues of the publications, thus making them less likely to risk offending the government again. In 1988, after several American-owned publications were gazetted, the Newspaper and Printing Presses Act was amended to allow for the reproduction of gazetted publications for sale at cost. In 1990, all foreign publications circulating in Singapore were required to have a permit for sale and to post a bond of S$294,000 in case of future problems. Thus, the government has muzzled the press on the pretext of ensuring stable rule (see Davies).

In reaction to these media restrictions, the *Asian Wall Street Journal* has pulled out of the Singapore market, while *The Economist* and the *US News and World Report* have permanently closed their Singapore offices. Singapore has thus foreclosed on any opportunity it may once have had to become the media center for Southeast Asia.

The Singapore government has also intruded upon freedom of religion. In 1987, for example, the Christian Conference of Asia, a Protestant organization, was shut down, and in 1995 the government seized 50 Bibles as "undesirable publications" in connection with the arrest of members of Jehovah's Witnesses. Religious organizations of all sorts have taken note.

In 1997, Prime Minister Goh Chok Tong interpreted the latest electoral victory of the PAP to mean that Singaporeans "rejected a western-style liberal democracy and freedoms, putting individual rights over that of society" (as quoted by Szep). During the election campaign, he had warned that constituencies that voted for the opposition would be the last to receive needed government-financed improvements and might become "slums." Censorship was extended into a new area of technology when opposition parties were prohibited from disseminating their messages through videotapes.

The connection between these repressive actions and the economy is that they provide assurance to foreign corporations that the government will keep the lid firmly on any potential unrest. Votes against the PAP or speeches objecting to the repression are thus seen as posing direct dangers to the economy. Foreign corporations have largely ratified the government's view by continuing to do business in Singapore without making even the slightest objection to the lack of democracy.

The result is that Singapore now stands out as having one of the most authoritarian governments in the world, obsessed with the evils of chewing gum, littering, and failing to flush public toilets, and interpreting political opposition in the context of the long struggle waged two generations ago against Communist insurgencies. Singapore thus seems anachronistic: ironically, what was once a factor in attracting foreign investors may increasingly work against Singapore's interests as the country is more widely seen to be an unattractive place to do business. Yet because it lacks input from sources other than PAP, in which the rank and file defer to the leadership, Singapore may now be unable to adjust to changes in the world economy.

Meanwhile, discontent has been made manifest in election returns, even though elections are rigged through various devices, such as multimember districting, and opposition is discouraged by arrests and libel suits. Support for the PAP peaked at 84% in 1968, and in recent years it has been only a few percentage points above 60%. In the presidential election of 1993, the PAP's candidate won only 58% of the vote, despite the fact that the opposing candidate had no previous political experience.

Conclusion

Singapore has achieved economic success by capitalizing on its geographic location, developing its labor force, adopting sound economic policies, enforcing political stability, and receiving the spillover benefits of being part of a region that has enjoyed the highest rates of economic growth in the world for much of the postwar era. Nevertheless, the upward trend may not continue. In the 21st century, the Kra Canal may be built, increasing numbers of educated Singaporeans may leave the country, other governments will duplicate the more successful of Singapore's economic policies, authoritarian rule may flounder, and the region and the world may experience slower economic growth. If the leaders of Singapore continue to have contempt for democracy, responsible corporations will move away, and the boom years will end. The driving forces behind Singapore's continued success will be the youngest and best educated individuals in the private sector. What will stand in the way of future economic progress in Singapore will be a government with oldfashioned ideas about how to control the lives of its citizens.

Further Reading

Bernstein, Dennis, and Leslie Kean, "Singapore's Blood Money," in *The Nation*, October 20, 1997

Chua Beng-Huat, *Communitarian Ideology and Democracy in Singapore*. London and New York: Routledge, 1995

   Chua makes an interesting effort to identify the ideological presuppositions of the People's Action Party of Singapore, although misdefining "communitarian" as placing the state above the individual.

Davies, Derek, "The Press," in Haas (editor), cited below

Devan, Janadas, and Geraldine Heng, "State Fatherhood: The Politics of Nationalism, Sexuality, and Race in Singapore," in Andrew Parker et al. (editors), *Nationalisms and Sexualities*, London and New York: Routledge, 1992

FEER: Far Eastern Economic Review (editors), *Asia 2001 Yearbook*, Hong Kong: Far Eastern Economic Review, 2000

Haas, Michael (editor), *The Singapore Puzzle*, Westport, CT, and London: Praeger, 1999

   This collection of papers examines the reasons why Singapore has become more authoritarian while achieving greater prosperity. Most of the papers are by outspoken political opponents of the regime.

Huntington, Samuel P., *Political Order in Changing Societies*, New Haven, CT, and London: Yale University Press, 1968

James, William E. (editor), *Asian Development: Economic Success and Policy Lessons*, Madison: University of Wisconsin Press, 1989

The papers in this collection, which covers a wide range of countries and policy options, include some important expositions of the "flying geese" theory of development in East Asia.

Lee Tsao Yuan (editor), *Growth Triangle: The Johor-Singapore-Riau Experience*, Singapore: Institute of Southeast Asian Studies, 1991

A useful but already partly outdated collection focusing on the opportunities and problems arising in the development of this special zone

Lingle, Christopher, *Singapore's Authoritarian Capitalism: Asian Values, Free Market Illusions, and Political Dependency*, Fairfax, VA: Locke Institute, 1996

This books offers a brilliant analysis of the flaws in Singapore's economic policies.

Lingle, Christopher, and Kurt Wickman (1999). "The Economy," in Haas (editor), cited above

Neher, Clark, "The Case for Singapore," in Haas (editor), cited above

Parsonage, James, "Southeast Asia's 'Growth Triangle': A Subregional Response to Global Transformation," in *International Journal of Urban and Regional Research*, number 16, June 1992

Seow, Francis T., "The Judiciary," in Haas (editor), cited above

Szep, Jason, "Singapore Libel Case Sends Chill Throughout Asia," in *Reuter*, May 29, 1997

Vatikiotis, Michael, "Chip off the Block: Doubts Plague Singapore-Centered 'Growth Triangle'," in *Far Eastern Economic Review*, January 7, 1993

World Bank, *The East Asian Miracle: Economic Growth and Public Policy*, New York and Oxford: Oxford University Press, 1993

This report received extensive publicity because it presented a generally flattering picture of the rapid economic progress of East Asia, but it crucially failed to discern the problems that led to the liquidity crisis in the region only a few years later.

Dr Michael Haas teaches Political Science at California State University, Fullerton.

NATION STATES

Chapter Seven
Brunei

Geoffrey C. Gunn

Brunei, which had a population of only around 324,000 in 1998, possesses vast reserves of oil and natural gas. These have enabled it to lever its economy into providing an enviable standard of living for its population and enormous wealth for its ruling family, led by Sultan Hassanal Bolkiah, along with a network of close nonroyal collaborators. Accordingly, the country has been identified by analysts, variously, as conforming to the economic model of a "rentier state" (see Colclough or Gunn), or as suffering from "Dutch disease effects" (see Cleary and Wong).

The notion that states based on external sources of income are substantially different from states based on domestic taxation was first proposed with reference to a number of oil-exporting countries in the Middle East. In this connection, Hazem Beblawi and Giacomo Luciani have drawn distinctions between a rentier state, a "rentier economy," and a "rentier mentality" (Beblawi and Luciani p. 52). Rentier economies become problematic, on this theory, when a rentier mentality breaks the causal linkages between work and risk-taking, on the one hand, and the prevailing pattern of rewards, on the other. "Dutch disease effects" are much less fully theorized, being derived from studies of the economic impact of hydrocarbon revenues on the Netherlands in the 1960s, which is said to have taken the form of a bloated public sector and nonproductive social expenditure.

In any case, however they are theorized, Brunei and the states that resemble it face a common set of problems. Sooner or later, they must decide how to expand the skills base of the population, stimulate the private sector, diversify the economy (especially by expanding "downstream" activities), provide transparency in national accounting, and rein in the unbridled pursuit of ostentatious consumerism. As will be shown here, Brunei has not made much progress on addressing any of these issues.

The Political Economy of a Rentier State

A British protectorate from 1888, Brunei was guided by a British High Commissioner (who usually also served as Governor of Sarawak) from 1906 until September 1959, when the present Sultan's father, Omar Ali Saifuddin III, assumed executive power under a written Constitution. However, Brunei, officially known as Negara Brunei Darussalam, did not achieve full independence until January 1984. It was the British, and specifically the Shell group of companies, that from 1913 pioneered the search for oil in Brunei. Exploitation of the territory's hydrocarbon resources began in 1929, and went on expanding to the point where, at the time independence was achieved, Brunei was one of the richest states in the world.

Traditionally a nation of self-sufficient peasants and fishermen, living in a favorable ecological niche, Brunei has experienced a major shift away from agriculture during the last three decades. The rubber plantation economy that had been started before World War II had faded away by the 1960s, and the overall trend has been toward the loss of self-sufficiency in food, including even fish. For example, only 1% of the country's demand for rice can be met from domestic production. The one saving grace offered by the exploitation of oil has been to spare the country from the ravages of tropical hardwood logging such as has occurred, with very damaging effects, in the neighboring Malaysian states of Sabah and Sarawak.

Material improvements, including the development of basic infrastructure, public housing, educational facilities, and medical services, began to be implemented under the series of five-year National Development Plans that began in 1953. These have continued up to the present with the Seventh National Development Plan (1996–2000), under which expanded allocations have matched a more sophisticated macroeconomic environment. This expansion has been financed, needless to say, by the modernization of the oil industry, the discovery of new fields, including reserves of natural gas, and the windfall benefits of the large increases in oil prices in 1973–74 and 1979. Today, Brunei is the third largest oil producer in Southeast Asia, with an average output of 163,000 barrels a day (see Brunei Government).

The steady dependence on the hydrocarbons industry, especially exports of crude oil and liquefied petroleum gas (LPG), to the neglect of traditional agriculture and of other industries, has made Brunei extremely vulnerable in the global marketplace. Oil and gas account for more than 96% of the value of Brunei's exports, which suggests that there is an urgent need to match the rhetoric of diversification with action. Yet, while rentier states typically accumulate reserves sufficient to protect their economies against the vagaries of the global economy, Brunei's brief experience as an independent state has not been characterized either by prudent management of reserves or by accountability for its successes and failures.

The one major interruption to the steady growth of the hydrocarbons industry and the accompanying isolation of Brunei from international events came with the Japanese invasion, which lasted from December 1941 to June 1945, and led to the destruction of the oil wells as well as of agricultural systems. The other major challenge to the Sultanate was less economically disruptive but has had a lasting impact. In December 1962, Sheikh Ahmad Azahari, whose Brunei People's Party (Parti Rakyat Brunei) had won a majority in the country's only legislative elections in March, led a brief and unsuccessful armed rebellion, proclaiming republican goals and receiving support from Indonesia. The rising was suppressed by the British forces still stationed in the country, and the experiment in limited constitutional monarchy that had begun in 1959 was suspended. Brunei was placed under a state of emergency that has continued in force ever since. The present Sultan, who came to power after his father's abdication in October 1967, has maintained a broadly pro-British policy, but has also deeply accommodated Islamic and Malay traditions. As a result, governance in Brunei conforms closely to the pattern of dynastic monarchy that still prevails in much of the Middle East.

While independent Brunei has gone as far as creating a cabinet system of government, it has shown no signs of reviving its shortlived legislature or tolerating open opposition. While political parties have surfaced, they exist in token forms only. Rather than offering a political opening, as for example in Kuwait, Brunei has wrapped itself in the cloak of monoculturalism defined as Malay Islam *Beraja* ("royalty"), an ideological system that privileges those recognized as the Sultan's subjects – around 66% of the population – but excludes many others. Thus, the majority of Chinese, who comprise around 15% of the population, are stateless permanent residents rather than citizens. Meeting citizenship tests in Brunei is no small matter, as citizenship confers substantial rewards in what has been dubbed the "Shellfare" system. The state has been able to expand its social services to improve the livelihoods of larger and larger numbers of citizens, through better housing, health care, and education, as well as access to privileged positions in the civil service. No personal income tax is levied in Brunei. Simply stated, Brunei's affluence, as measured by consumption patterns and disposable income, makes the country the envy of its neighbors. In Brunei, as in other economies in Southeast Asia until the crisis of 1997–98, large numbers of people have been prepared to accept authoritarian rule as long as their rising economic aspirations are met. The establishment in Brunei does not need to emphasize economic success as a source of legitimacy, as some other authoritarian regimes in the region sought to do until they were swept away in the crisis: it is taken for granted.

Having taken control of defense and international relations when full independence was achieved, the new state entered the community of nations with membership of the UN, ASEAN, the Organization of the Islamic Conference, and other international organizations. It is notable, however, that Brunei is not a member of the Organization of Petroleum-exporting Countries (OPEC), and, as a relatively small producer country, is not constrained by OPEC's production quotas. In recent years, Brunei has raised the level of its oil output in accordance with the level of availability, even though in most years falling oil prices have tended to erode increases in revenues.

Where the global trend has been for producer countries to nationalize their oil industries, as in Indonesia (with Pertamina) or Malaysia (with Petronas), a distinctive feature of the industry in Brunei is its private character. Historically, the division of profits between Shell and the state in Brunei was to the advantage of the former, but in the modern period the terms have been altered to benefit the state and the royal family. Natural gas production is of more recent origin, starting with an LPG plant that was the largest of its kind in the world when it was opened in 1973 (Cleary and Wong pp. 44–46). Today, Brunei is the fourth largest producer of LPG in the world. The oil industry nowadays is dominated by four companies in the Brunei Shell group, in each of which the state holds 50% of the shares. The dominant company is Brunei Shell Petroleum, which is responsible for exploration and production as well as refining. Second in importance is Brunei LPG, which is owned jointly by the state of Brunei, Shell, and the Mitsubishi Corporation of Japan. Another company, Brunei Coldgas, buys the LPG, and a fourth, Brunei Shell Tanker, transports it to Japan, where it is used by the Tokyo Electric Power Company, the Tokyo Gas Company, and the Osaka Gas Company. The only outside player in this business is Jasra Elf, a joint venture between Jasra International Petroleum, which is owned by the royal family, and the French company Elf Aquitaine. Jasra Elf has made some important offshore discoveries, not only extending Brunei's known reserves but also breaking the long-established monopoly of Brunei Shell. In 1994, South Korea became an additional customer for LPG, representing an important step toward diversifying Brunei's markets. Additionally, certain technological innovations have recently been introduced, at some cost, in an effort to rejuvenate aging offshore oil and gas fields, such as the Ampa-Fairly Field.

Meanwhile, attempts by the state to stimulate the rest of the private sector and reverse the "dependence mentality" of the middle classes have been desultory. Numerous schemes have been routinely proposed or attempted, such as developing financial services, promoting foreign investment in new start-ups, stimulating the fisheries industry, or promoting ecological tourism. The most recent and grandiose scheme to diversify the economy was unveiled by the Brunei Economic Council in April 2000, with the aim of developing such downstream activities as petrochemicals, gas-based industries, refining, and bunkering. For the first time, the formation of a national oil company was publicly suggested (see Stephen 2000a). Even so, this plan is premised upon rationalization within the oil industry, including improvements in the efficiency of exploration and maximizing production, within the limits of a "flexible" conservation policy, as well as on increasing competitiveness within the industry.

In discussing the "diversification dilemma" in Brunei, Mark Cleary and Shuang Yann Wong have listed the obstacles to diversification as including deficiencies in "labor, capital, resources and management skills" in a "political and cultural system that is often highly rigid, conservative and traditionalist" (Cleary and Wong p. 123). Indeed, discussion of Brunei's national wealth would not be complete without reference to the privileged and secretive royal family, which can be said to have its own separate economy (Leake pp. 84–85). The Sultan and his relatives are perhaps the richest family in the world, although there is no easy way to distinguish the family's fortunes from those of the state. As supreme executive and sovereign, the Sultan has the power to dispose of all state assets as he sees fit. The Sultan may at one time have had control over between US$40 billion and US$80 billion, a figure equal to Brunei's reserves, and his three brothers have long had significant assets of their own. The Sultan's possessions range from luxury hotels in London, Singapore, and Bali to cattle stations in Australia, as well as jewelry and art collections. At home, the royal family's presence looms large, in the shape of sprawling palaces and domains that tend to expand as the family itself expands. One of the Sultan's brothers, Prince Mohammed, has a high business profile through his owning a controlling interest in QAF Holdings, a company registered in Singapore that has interests in a range of ventures from supermarkets to newspapers (the *Borneo Bulletin*), including joint ventures with the military regime in Burma. His other brothers, Prince Sufri and Prince Jefri, both have private investment companies. The royal family's construction activities generate considerable business for contractors and suppliers in Brunei. It is notable that the royal family's economic activities inside Brunei are not reflected in the national accounts and fall outside the scope of the surveys conducted by the government's Economic Planning Unit. The separation of the family economy from the Brunei economy also extends to the provision of separate electricity supplies and telecommunications, among other services (Australian High Commission pp. 24–25).

Brunei and the Asian Crisis

Brunei was largely shielded from the Asian crisis that started in 1997 by its small size and its irrelevance as a destination for foreign investment. Nevertheless, the fall in world oil prices dragged the economy down to its most critical state since the country became independent. During 1998, Brunei saw declines in the prices of crude oil and LPG of 37% and 25% respectively, as compared to 1998. With oil output constant at around 15,000 to 16,000 barrels a day, this meant that the contribution of the oil industry to GDP continued to decline, with the government and non-oil industries taking up the slack. As a consequence of revenue shortfalls and economic malaise, the growth of real GDP in 1998 was only around 1%, compared to 3.5% in 1996 and 4% in 1997. The budget balance in 1998 registered an estimated 6% of GDP, which led the government to corporatize some government agencies and privatize some projects. At the same time, efforts were stepped up to increase alternative sources of revenues to supplement declining hydrocarbon revenues (see APEC Secretariat).

In fact, the country had had its own domestic economic problems for some time. Unemployment had been rising since 1988, and the budget deficit had started growing in that year too. In 1992, for example, government expenditure had been increased by 10.8%, mainly in order to pay for the lavish ceremonies commemorating the silver jubilee of the Sultan's accession to the throne. However, the practice of issuing midyear supplementary supply orders, including additional money for the royal purse, as in 1993, makes the published figures for government expenditure unreliable (Australian High Commission p. 11). Essentially, deficits are financed by drawing upon foreign investment income, which is itself a closely guarded secret.

On June 27, 1998, the Sultan declared "a state of emergency in the whole state," invoking clauses in the Constitution of 1959 that allow the Sultan to make financial provisions "as may be necessary during the period of the emergency, including provision for the public service." Government allocations draw from two funds, the Consolidated Fund, which covers the operating costs of the administration, and the Development Fund, which finances projects under the National Development Plans. Using the emergency clauses in the Constitution, the Sultan issued a series of Emergency (Supplementary) Supply orders for 1999 and 2000, leading to special appropriations out of the Consolidated Fund.

This was not enough to render Brunei completely immune to regional economic trends. The Brunei dollar, which is pegged to the Singapore dollar, lost around 14% of its value against the US dollar during the crisis. Additionally, certain of Brunei's non-oil exports, such as textiles, were affected by the crisis in other ASEAN countries. While the section of the population that has US dollar accounts and other assets was shielded by the decline in the value of the Brunei dollar, expatriate workers, many private sector contractors, and marginal elements in the economy suffered. The construction industry in particular had already been ailing owing to the unfolding of the Amedeo scandal, which was linked to the collapse of the largest domestic business conglomerate (see below).

In September 1998, the Sultan created a special Economic Council, with members drawn from the civil service, government economic agencies, and the private sector. The Council determined, not surprisingly, that the prevalent trends indicated an unsustainable economy and, in particular, that the growth in national income would not be able to match the growth in population. Since 1994, the government's budget deficit had averaged B$1 billion each year, equivalent to an average year's GDP. This shortfall had been funded by drawing upon Brunei's reserves.

The Economic Council's report, issued in January 2000, announced that an "action plan" for economic recovery had been devised and was to be implemented with some urgency. This involved a set of measures to inject liquidity into the economy, especially through assistance to small and medium-sized enterprises, but there was also a great deal of rhetoric about "fast track" payments by the government to private sector contractors, investment in information technology and in people, the establishment of think tanks, the expansion of the private sector, including Malay-owned companies, and, last but not least, maximum transparency. It was widely expected that the Council's recommendations would be incorporated into the next major economic blueprint for Brunei. Of course, however, such rhetoric does not fit well with a public culture that has traditionally inhibited dynamic breakthroughs. The continuing phenomenon of royal favoritism and extravagance illustrates the obstacles that still remain.

The Brunei Investment Agency, Amedeo, and Prince Jefri

The Brunei Investment Agency (BIA), established within the Ministry of Finance in August 1983, was intended to manage the Sultanate's reserves, taking on a role performed up till independence by the British Crown Agents. As I have argued elsewhere (Gunn p. 129), the role of the BIA in the external recycling of oil rent conforms broadly to the model outlined by Mahmoud Abdul-Fadil in 1987, through the forging of a triple alliance between the state, the new business elite, and international capital (see Abdul-Fadil p. 86).

The way in which this triple alliance operated in the case of Brunei has long been the subject of speculation among journalists reporting on financial affairs (see Bartholomew). The BIA's *modus operandi* has seldom entered public discourse, but in 1991 its managing director, Dato Abdul Rahman Karim, who was also Permanent Secretary of the Ministry of Finance, explained that the BIA handled only 40% of the Sultanate's foreign reserves, which he estimated at US$27 billion. The remainder was divided among eight foreign banking and investment institutions, while between 50% and 60% of the BIA's funds was placed in bonds, with the balance in shares (Gunn p. 120).

The picture would remain rather incomplete, however, without further discussion of the nexus linking the BIA with the state, the business elite, and international capital. While the BIA's assets were estimated to have risen to US$60 billion by the end of the decade, its operations suddenly came to the attention of the media, both domestic and international, in the middle of 1998, when the British press reported on the financial problems facing the Sultanate's largest conglomerate, Amedeo, then under the control of the Sultan's youngest brother, Prince Jefri, who was head of the BIA. (He had also been Minister of Finance until February 1997, but his dismissal from that post had not attracted much attention.) In June 1998, an investigation into the misuse of the BIA's funds in the Sultanate was entrusted to a "finance task force," headed by senior government officials with international advisers. In July, however, Amedeo collapsed, leaving debts estimated at US$16 billion, thus depleting the BIA's assets by the same amount. It is probably no coincidence that in August 1998 the Sultan nominated his oldest son, Al-Muktahee Billah, as Crown Prince, presumably in an attempt to quell rumors that the dynasty as a whole might be in trouble.

Prince Jefri, dismissed from all his official positions, fled the country, just as the Sultan assumed the position of Finance Minister. In September 1998, the government announced that Amedeo and 26 other companies had been placed under investigation and taken over by the government on suspicion of misappropriating funds. These companies included an amusement park, an international school, a hotel, and firms connected with fisheries, insurance, and telecommunications (see *Straits Times* September 22, 1998). Jefri briefly returned to Brunei in October 1998, having sold off some of his holdings, including the Asprey Group, the couturier Romasz Starweski, and the jeweler Hamilton and Inches, but these gestures did not mollify the Sultan, and Jefri returned to exile in July 1999. In the same month, Anderson Consulting confirmed that Amedeo was unable to pay debts estimated at US$3.7 billion, most of which were owed to the BIA. Amedeo was subsequently wound up, leaving creditors unpaid (*Far Eastern Economic Review* p. 96).

The industry most badly affected by the collapse of Amedeo was construction. For more than four years, Amedeo had been involved in a number of grandiose projects in the Sultanate, each involving investments of many millions of US dollars and each exemplifying what Rex Mortimer has called (in connection with Suharto's Indonesia) "showcase capitalism" (see Mortimer). They included Prince Jefri's private offices, a private mosque in the grounds of those offices, and the Jerudong Hotel. According to Ignatius Stephen, a journalist working for the *Borneo Bulletin*, the vast majority of the projects were for the benefit of the Prince and his family. Of US$6.2 billion injected into Amedeo by the BIA and other creditors, only 10% could be regarded as having been spent on infrastructure improvements, including the Berakas power station, a telecommunications tower, and an international school. Such projects, built by foreign workers on temporary contracts, did not create employment for citizens of Brunei or improve their skills. The construction boom fueled by Amedeo, and now represented by numerous abandoned projects in Brunei, created only the illusion of economic progress (Stephen 2000a p. 26). With the bursting of the speculative bubble, the void left by the collapse of Amedeo has cast a pall over business confidence in the Sultanate. In 1999, the government invested around US$208 million in development projects in an attempt to aid the construction industry (*Far Eastern Economic Review* p. 87).

In March 2000, the Sultan took steps against his youngest brother that were sensational by any standards, launching an unprecedented lawsuit against him, his son Prince Hamid, and 71 others, on the grounds that they had squandered more than B$40 billion in state funds. In addition to the failed projects described above, Jefri had reportedly spent US$2.75 billion over ten years, buying himself 2,000 cars, 17 aircraft, including a private Airbus A310, several yachts, large quantities of jewelry, and more than a dozen homes. The coverage of this brazen display of extravagance overshadowed reports of other controversies in which Jefri had been involved, such as a lawsuit for US$288 million brought against him in London by business partners who claimed that he had cheated them (see Ford). It appeared that the Sultan had chosen a public and judicial method of gaining restitution only after private negotiations had broken down. In fact, as Prince Jefri himself revealed in a press release (see Stephen 2000b), negotiations between lawyers representing the two parties had been going on since May 1999.

The trial began with the High Court of Brunei imposing a freeze on the assets of Prince Jefri and the other defendants around the world. It became still more sensational, and scurrilous, when Prince Jefri's counsel produced an affidavit alleging that Pehin Isa, the Sultan's Home Minister and special adviser, had received hundreds of thousands of dollars from Prince Jefri over several years (see Stephen 2000b). This allegation was not widely believed, but it did serve to raise questions as to public probity in general. Meanwhile, it seemed clear that the Sultan was still in control of events: it was no accident that the heavily censored media of Brunei, including the English-language daily, the *Borneo Bulletin*, were given carte blanche to report the case, especially the damaging allegations about Prince Jefri's financial improprieties.

Just at a point when Prince Jefri and Prince Hakim had appealed to the Judicial Committee of the Privy Council in London, the highest court in the Commonwealth, against the High Court's order that they disclose the sources of their funds and their living expenses, the Sultan and his estranged brother made an out-of-court settlement. The Minister of Education, Pehin Abdul Aziz Awang Umar, who headed the task force investigating the disappearance of the BIA's funds, announced that Prince Jefri had agreed to return all the assets that he had taken from the BIA, including hotels, land, and other items in Brunei and overseas. The exact terms of the settlement were not revealed, however, and the details of a separate settlement arranged for Prince Hakim have also been kept secret.

It is notable that this contest, played out partly in private, partly in public, was framed within two different discourses. At times, it appeared to be a falling out between brothers, ending in a reconciliation in line with the providence of Allah and the blessings of His Majesty. At other times, it was presented as a purely legal affair between the state and the BIA, as plaintiffs, and Prince Jefri and others as defendants. It is clear that the private settlement was the optimum outcome for the Sultan, because it kept the details of the case within the family and ensured that the boundaries of the royal economy were not publicly breached. The court's victory was portrayed as the Sultan's victory and, implicitly, that of his *rakyat*, his loyal subjects. While Prince Jefri claimed that the case had been brought under pressure from Islamic conservatives, the issue was barely explored in public, so that the boundary between Jefri's public life and his private life, however widely details of it were disclosed in foreign media, was also preserved intact within Brunei. There would be no further public disclosures. Jefri promptly returned the Airbus A310 to the Sultanate as an indication that there would be more illegally acquired assets to come, but, in a system almost entirely lacking in any kind of accountability, there was still no accounting for the missing billions. It remained very much open to question – although no longer in the media in Brunei itself – whether the reconciliation between the members of the ruling family had helped to promote transparency or truly served the public interest.

Conclusion

The recovery in the price of crude oil that began in the latter half of 1999, taking it from US$10 a barrel to more than US$30 by the middle of 2000, brought a reprieve to Brunei. While the resulting windfall, if it is sustained, will no doubt go far to replenish the missing billions, it will also tend to increase the pressures on Brunei to open up to improvements in management and reforms in the economy. "Growth without development, wealth without employment" (Cleary and Wong p. 139) is still an accurate summation of the country's problems, however much they are masked by Brunei's impressive achievements. If international agencies were in a position to offer advice to Brunei, then they would undoubtedly urge better governance first and foremost. Corruption, collusion, and nepotism eventually contributed to the collapse of President Suharto's "New Order" in Indonesia, and have aroused opposition and calls for reform throughout the region. Economic recovery in Brunei has its limits and its oil reserves are finite. In any case, while economic activity received a boost from rising oil prices, the deficit for 1999 was still US$592 million, as compared to US$353 million for 1998 (see *Far Eastern Economic Review*). The hallmark of the ruling dynasty has been extravagance on a grand scale, but its survival in its present form may well now call for self-effacement and restraint.

In the end, then, it is the vulnerability of the economy that has obliged the authorities in Brunei, at least those among them with technocratic inclinations, to look again at some of the ideas and activities that have been taken for granted for so long, and to seek to diversify the economy away from oil and natural gas, in order to reduce exposure to externalities. In other words, the challenges to "economic legitimacy" and dynastic rule that have been thrown up by the regional crisis appear to demand creative and even radical solutions. Yet almost as soon as the Sultan sought to rein in a wayward relative through a sensational civil action, the established limits to transparency came under threat, and the private space of the royal family and the royal economy were quickly secured. There is no doubt that a state that can allow its national reserves to be spent without any accountability is deeply troubled. However, no fundamental breach of the rentier economy of Brunei appears in sight while the dynastic model remains embedded, and no productionist revolution is likely to appear on the horizon while the rentier mentality holds sway.

Further Reading

Abdul-Fadil, Mahmoud, "The Macro-behaviour of Oil-rentier States in the Arab Region," in Hazem Beblawi and Giacomo Luciani (editors), cited below

APEC Secretariat, "Brunei Darussalam: Overall Economic Performance," at www.apecsec.org.sg/member/brune_report.html, June 2, 2000

Australian High Commission, *Country Economic Brief: Brunei Darussalam*, Bandar Seria Begawan, Brunei: Australian High Commission, February 1994

Bartholomew, James, *The Richest Man in the World: The Sultan of Brunei*, London and New York: Viking, 1989

"Gazetted" (banned) in Brunei, this book offers a racy picture of the royal family and their accumulated wealth.

Beblawi, Hazem, and Giacomo Luciani (editors), *The Rentier State*, London and New York: Croom Helm, 1987

This collection of papers, including the one by Mahmoud Abdul-Fadil cited above, comprises a definitive text on oil-rentier states in the Middle East.

Brunei Darussalam Economic Council, *Report*, at www.brudirect.com/BruneiInfo/info/BDECR_PressReleas.htm, 26 April 2000

Brunei Government, "The Economy," at www.brunei.gov.bn/about_brunei/economy

Cleary, Mark, and Shuang Yann Wong, *Oil, Economic Development and Diversification in Brunei Darussalam*, London: Macmillan, and New York: St Martin's Press, 1994

This is the best and most comprehensive analysis of the oil industry and the dilemmas of development in Brunei, written by two geographers with firsthand experience of the country.

Colclough, Christopher, "Brunei: Development Problems of a Resource Rich State," in *Euro-Asia Business Review*, Volume 4, number 4, 1985

A pioneering essay on Brunei's oil-rent economy by a political economist.

Far Eastern Economic Review (editors), *Asia 2000 Yearbook*, Hong Kong: *Far Eastern Economic Review*, 1999

In this, the 42nd edition of an authoritative annual publication, the focus is largely on the impact of the Asian crisis on the economies of the region.

Ford, Maggie, "Brunei: Profligate Prince Jefri," in *Newsweek*, April 10, 2000

Gunn, Geoffrey C., "Rentier Capitalism in Negara Brunei Darussalam," in Kevin Hewison, Richard Robison, and Garry Rodan (editors), *Southeast Asia in the 1990s: Authoritarianism, Democracy and Capitalism*, St Leonards, NSW: Allen and Unwin, 1993

This is a discussion of the opportunities and problems facing Brunei before the Asian crisis, in a textbook on the economies of Southeast Asia edited by three Australian political economists.

Leake, David, *Brunei: The Modern Southeast-Asian Islamic Sultanate*, Jefferson, NC: McFarland, 1989; Kuala Lumpur: Forum, 1990

A concise and useful text covering many aspects of the country, including the royal family and its business affairs

Mortimer, Rex (editor), *Showcase State: The Illusion of Indonesia's Accelerated Modernisation*, Sydney and London: Angus and Robertson, 1973

This book, probably the first academic collection to diagnose problems in the Indonesian "New Order," is still of some relevance in assessing the difficulties in this development model and in those that resemble it.

Stephen 2000a: Stephen, Ignatius, "Jefri-controlled Projects Lie Abandoned," in *The Sunday Times* (Singapore), April 2, 2000

Stephen 2000b: Stephen, Ignatius, "Prince Jefri, Brunei Settle $40 Billion Case," at www.brudirect.com/DailyInfo/News/Archive/May2000/130500/nite01.htm, May 13, 2000

Dr Geoffrey C. Gunn is Professor of International Relations in the Faculty of Economics at the University of Nagasaki, Japan.

NATION STATES

Chapter Eight
The Philippines

Raul Pertierra

The Philippines sees itself, and is seen by its Asian neighbors, as different. Being a predominantly Catholic country, with a long history of institutions adapted from the West, it confounds orientalist expectations of Asia. English is widely spoken and its people's hospitality to foreigners is well-known. Manila and other major cities have most of the amenities available in the West, for those who can afford them. In contrast to other cities in Southeast Asia, such as Singapore, Bangkok or Kuala Lumpur, Philippine cities appear not to have retained an exotic strangeness: everything, except the poverty, appears familiar. Western pop culture is pervasive and even its indigenous versions seem readily accessible. As a result, many westerners seeking "otherness" dismiss the Philippines as a damaged and imitative culture. This familiarity is advantageous when establishing sexual or domestic ties: unsurprisingly, the Philippines is a major source of "mail order brides" and domestic workers for other countries. Outsiders generally perceive Filipino women as docile and eager to please, yet within the Philippines women are generally portrayed as strong, dominating, and highly protective of their families. What are seen as docility and eagerness from the outside are seen as flexibility and responsibility from the inside. The Filipino practice of acknowledging status differences and a preference for masking hostilities facilitates relationships with foreigners. This practice, however, reveals the Filipinos' basic Asian orientation. Thus, contrary to initial expectations, the Philippines is as exotic or, alternatively, as accessible as other Southeast Asian countries. How did this contradiction come about?

The Historical Background
Colonization by Spain

When Spain colonized the Philippines in the 16th century, it found a land roughly the size of what is now the United Kingdom, with chiefs ruling kinship communities, and both slavery and warfare common practices. Larger riverine communities were ruled by hereditary chiefs with links to other chiefdoms, including communities in China, Japan, and Indonesia. Even bigger Islamicized polities existed in Sulu, whose sultans exercised control over large areas of the southern Philippines and Borneo. The Spanish were interested in controlling the lucrative spice trade, but were frustrated in their aims by the Portuguese and, later, the Dutch. While the Philippines had long-established ties with the major centers of Asia, their influence on local life was relatively unmarked. Hindu and Buddhist ideas were present, but only Islam exercised a significant influence in precolonial times. The Philippines marked the boundary of Indic and Sinitic civilizations without, however, acquiring splendid monuments.

While Spain initially viewed the Philippines as a possible staging area for the religious conversion of China and Japan, it soon had to be content with converting only the local peoples. Missions were established in Japan, Taiwan, and China but the resources needed for their maintenance proved inadequate. By the 17th century, Spain had begun its inevitable decline. The Spanish then embarked on a massive transformation of the Philippines into a loyal Catholic colony. Given their increasingly limited resources, they largely succeeded in this project. Filipinos responded positively, after an initial resistance, once they realized that Spanish influence could be contained and often even subverted. Vincent Rafael gives an amusing example of this subversion: to the delight of the friars, Filipinos enthusiastically accepted the Catholic practice of confession, but it soon became evident that the natives only confessed other people's sins rather than their own (see Rafael). Today, similarly, public officials rarely, if ever, acknowledge culpability, but they are very willing to accuse others.

The successful hispanicization of the Philippines has to be placed in its context in order to understand its particular adaptations. The country was too far away from Spain for it to exercise effective control. For this reason, the colony was initially administered from Peru and later from Mexico. Many of the hispanicizing practices came via Mexico, having been previously adapted to conform to the needs of that colony. Only during the 19th century did Spain exercise effective and direct control over the Philippines. To encourage development, the country opened its ports to outside trade. This trade quickly came under the control of European and, later, of US merchants. Filipinos began to see themselves as having different interests from their Spanish colonizers, despite sharing important features such as religion and language. Spanish had become the unifying language of the elite, but local dialects prevailed among the peasants and other rural folk.

The Independence Movement

By the second half of the 19th century, Filipinos had begun to exercise greater control over local affairs. A class of educated Filipinos, often exposed to the liberal doctrines prevailing in Europe, started to organize. At first they asked for local representation in Spain's legislature, but they quickly changed their demands to complete independence. Spain rejected these demands and embarked on a policy of suppression that only increased the feelings of discontent. The execution of three priests, Fathers Gomez, Burgos, and Zamora, in 1872, on suspicion of having encouraged a rebellion, marked the hardening of Filipino attitudes towards independence.

Jose Rizal, who has become the Philippines' foremost national hero, was traumatized by the execution of these priests and felt compelled to dedicate his energies to his country's liberation. Rizal came from a wealthy Chinese-Filipino family who encouraged both his scholarship and his politics. After receiving his education in the Philippines, he traveled extensively throughout Europe, took part in scientific symposiums, and published two novels (*Noli Mi Tangere* and *El Filibusterismo*) in which he exposed the cruelties of Spanish rule. They had an enormous influence on literate Filipinos. Rizal himself did not favor armed rebellion but instead advocated education and a gradual evolution of Philippine nationalism.

Nevertheless, when a rebellion was initiated by Andres Bonifacio in 1896, Rizal was seen as the main instigator and, after a hasty trial, he was publicly executed on December 30, 1896. News of his martyrdom quickly spread throughout the country. Even today, certain religious cults in the Philippines venerate Rizal as a reincarnation of Jesus Christ (see Ileto). The Philippine government and the elite prefer to depict him as an outstanding intellectual who shaped a nationalist consciousness. These different perspectives indicate something of the subaltern and dominant forces acting in contemporary Philippine society.

The rebellion prospered and it seemed that the Philippines would achieve independence after nearly four centuries of colonization. However, while the Filipinos were defeating the Spanish on the battlefield, the US government was making its own plans following its victory in the Spanish-American War early in 1898. The United States was rapidly transforming itself into a major world power but it lacked overseas colonies, then a mark of imperial might. On April 24, 1898, US warships sank the Spanish fleet in Manila Bay. The Filipinos were elated and redoubled their efforts to defeat the Spanish. They managed to drive the remaining Spanish forces into the fortified Intramuros, the original site of colonization, and awaited their surrender. Unknown to them, the Spanish government was simultaneously negotiating with the US government, and in December 1898 it ceded the Philippines to the United States. The Filipino general Emilio Aguinaldo declared independence on January 23, 1899, but US forces engaged in a brutal campaign of repression over the following three years, prefiguring events in the Vietnam War by using terrorism against civilians and establishing closely policed hamlets (see Wolff).

Apolinario Mabini, the leading theorist in Aguinaldo's provisional government, argued (in a letter written to a fellow revolutionary, Felipe Buencamino, in July 1899), that to capitulate to the US forces "would reinforce the belief of others that Filipinos lack culture . . . or that they are an uncivilized country" (as cited in Diokno, p. 6). This was precisely the US justification for invading the Philippines: it was argued (see, for example, Worcester), that Filipinos had at most a blunted moral sense and could not be reasonably expected to govern themselves according to acceptable democratic standards. This view prevailed despite opposing voices (see, for example, Blount), who did their best to expose US duplicity. In 1899, Senator Carl Schurz, a noted member of the Anti-Imperialist League, described the Philippine-American war as criminal aggression against the people of the Philippines (Sullivan p. 81).

Colonization by the United States

Once the United States had decided to occupy the Philippines permanently, rather than simply liberate it from its Spanish masters, administrators set out to organize the country's civil government. Educated Filipinos quickly became part of this new bureaucracy and in little over a decade were starting to voice demands for greater political autonomy. While these demands were becoming more acceptable to the US authorities, the hinterlands remained as a major justification for a colonial presence. Dean Worcester, who founded the Bureau of Non-Christian Tribes in 1901, was largely responsible for classifying Filipinos by religious affiliation as well as into ethnological categories. As a consequence, Filipinos were seen as consisting of hispanicized Catholic lowlanders, non-Christian highlanders, and Moros (Moslems). Each required different colonial policies to satisfy their particular needs, and Worcester's own usage of the term "Filipino" in fact referred only to hispanicized lowlanders. Worcester's classification bedevils Filipino society even today: Moslems and cultural minorities are still commonly seen as outside national society.

The classification of Filipinos into religious categories partly followed the earlier Spanish practice, but it was premised on very different ends. While the Spanish considered religious conversion to be one of their major purposes, US officials, coming from a more secular background, saw their new subjects as requiring tutelage in the art of modern democratic government. To begin this task, they first had to classify Filipinos into appropriate stages of cultural and political development. Following conventional evolutionary models of the time, Filipinos were classified, in order of ascending civilization, into Negritos, Indonesians (Moros and people from the highlands), and lowland Malayans (see Sullivan). The first group was seen as vestigial and a remnant of a bygone past. The second, while adhering to objectionable practices such as headhunting and slavery, nevertheless had redeeming features, of which a developed love of freedom, expressed in a primitive democracy, was the most laudable. The last group was outwardly the most developed, following centuries of colonial rule, but extensive intermarriage with other Asians and Europeans, which had created a *mestizo* ("mixed") people, had made them less "authentic." This last group was, however, the most vocal and successful in convincing the United States that they merited political autonomy. During the early period of US colonization, ethnographers mapped the hinterlands, identifying cultural differences that were then solidified into political units called tribes, whose members henceforth began to view themselves as significantly different from one another. In the context of traditional village hostilities and conflicts, this new system of classification transformed past differences into ethnic divisions. This process of ethnicization is largely responsible for the chronic factionalism characterizing contemporary Filipino society (see Pertierra 1988).

The project of modernity gives central prominence to the control of nature and society. Spaces threatening such control usually elicit normalizing practices, with their corresponding discourses. To justify US imperialism, the Philippines had to be portrayed as requiring intervention. Initially, the country was seen as suffering from the neglect and abuse of the Spanish. When the Philippine revolutionary government, having expressed its gratitude for US assistance, began to exert control in the countryside, the United States embarked on a war of conquest in which peoples of the highlands and the Moros, who had until then been largely autonomous, were seen as potential allies against the Catholic lowlanders. The United States initially courted their support and encouraged their sense of difference from lowland Christian Filipinos. US officials were convinced that they could eliminate such practices as headhunting or slavery more effectively than any Filipino government, while at the same time protecting these minorities from the rapacity of *mestizo* officialdom. They succeeded in eliminating these offensive customs and practices, but in the process hardened the divisions among Filipinos. The peoples in the Cordillera are still victims of these divisions and the Moros continue their struggle for a separate nation.

The US "civilizing mission" had many benefits for Filipinos, including those in the hinterlands. Public health, education, and civil administration were all significantly improved. The structures for a modern economy were established and US companies seemed keen to become major investors in it. Great expectations were held for the colony of a nation whose imperial projects were meant to be exceptional. Unfortunately, the US colonial project never attained the unrealistic expectations of its planners. There were too many contradictions for it to succeed. Good intentions at times masked duplicity or self-interest. In most cases, the colonizers supported elite interests, thus ensuring the continuing basis of the exploitative relationships that were to be challenged by the Huk rebellion (see below). All considered, the civilizing mission offered, at best, mixed blessings.

The Republic and its Discontents

The Philippines obtained its independence on July 2, 1946, after, as an anonymous wit put it, 300 years of living in a convent and 50 years in Hollywood. Manila had been almost completely destroyed and the countryside devastated during the four years of Japanese occupation, and these were not propitious times for a young nation to declare its sovereignty. Curiously, independence is celebrated on January 23, following Aguinaldo's declaration in 1899. By this sleight of hand, many Filipinos deny the significance of the US colonial period, just as they have forgotten their Spanish heritage. Thus, history is put to the service of a strategic present.

Insurgencies

Soon after achieving independence, the Philippines became mired in insurgency. Socialist doctrines had permeated the countryside, resulting in the formation of various quasi-Marxist organizations. Their members had initially participated in elections but quickly realized that Philippine democracy had no room for socialist programs. The Huk rebellion that resulted reached its peak in 1950, when its forces threatened to enter Manila (see Kerkvliet). The Philippine government was in grave danger and requested US assistance. A rising politician, Ramon Magsaysay, managed to convince the peasants to give up their rebellion in exchange for land and promised reforms. The government slowly seized the military initiative and the Huks were defeated, although they were not completely suppressed until the late 1950s. The promised reforms were slow in coming and peasant discontent grew. Another decade passed before this discontent found a new leadership.

In 1968, Jose Maria Sison, an academic and student leader at the University of the Philippines who had been active in the long-established Communist Party of the Philippines (PKP), established a Maoist Communist Party and its military arm, the New People's Army (NPA). Starting from a small student-based movement, the NPA quickly grew and, like the

Huks before them, soon came to threaten the central government. Sison's party was ideologically more sophisticated than the old PKP, drawing into its ranks leading intellectuals, artists, and socially concerned Filipinos. In the following two decades, the NPA expanded its mass base to include members of the disaffected middle class. It controlled vital sections of the labor movement and enjoyed the support of many overseas organizations.

In addition to the new Communist threat, the Moslems in Mindanao organized themselves and were demanding autonomy. The steady encroachment on Moslem lands by Christian settlers aroused strong hostility between the two groups. In 1972, Nur Misuari, who was also a professor at the University of the Philippines, founded the Moro National Liberation Front (MNLF). The Philippine army found itself fighting major rebellions on two fronts. This situation was used by President Ferdinand Marcos as his main pretext for declaring martial law on September 11, 1972.

Martial Law

Marcos had dominated Philippine politics since the early 1960s. Through cunning, guile, and force, he managed to become the only President to win two successive terms, in 1965 and in 1969, both campaigns being marked by an increase in coercion and corruption. The Philippine Constitution, following its US model (as amended after Franklin D. Roosevelt's presidency), allowed for a maximum of two presidential terms, but in the declaration of martial law Marcos found a way to extend his presidency indefinitely. Freedom of the press, an established Philippine tradition, was abolished. Congress and other political institutions were abolished or continued only with Marcos's permission. A general purge of Marcos's perceived enemies was initiated, starting with the left and the Moslem separatists but going on to include a host of political conservatives. People with opposed political agendas found themselves sharing the same prisons. This eventually resulted in the formation of a broad coalition, culminating in the "EDSA revolution" of 1986 (see below).

Because of the considerable turmoil of the early 1970s, many Filipinos were willing to give Marcos a chance to reorder Philippine society. The first few years of martial law appeared benign, but, because of the oil price rises of 1974 and 1979, and other economic difficulties, no significant improvements were achieved. In time, martial law revealed itself for what it was: simply an excuse to perpetuate Marcos's power. The economy deteriorated rapidly, prompting the regime to resort to the confiscation of private property and eventually the transfer of state funds into the private accounts of the Marcoses.

Meanwhile, large sections of society were becoming politicized, extending from the left and the Moslems through many elements of the middle class and, most importantly, the Catholic Church and other Christian denominations. Following a phase of critical cooperation during the early days of Marcos's rule, the Catholic Church eventually turned to collaboration with other opposition groups, including the left (see Youngblood), while radical Christians formed revolutionary organizations such as the Social Democrats (whose founder, Father Romeo Intengan, is now superior general of the Philippine Jesuits). Women belonging to religious associations were particularly active, abandoning their exclusive schools to politicize the poor. Journalists became bolder and began printing stories critical of the Marcoses. One such story concerned Marcos's claims to have won medals with a series of military exploits. Al McCoy, a respected historian, discovered that most of these claims were fraudulent. Stories openly critical of Marcos, his wife Imelda, other members of his family, and their cronies circulated freely in Manila and the provinces.

On August 21, 1983, ex-Senator Benigno "Ninoy" Aquino, Marcos's most prominent opponent, was assassinated on his return from a visit to the United States for medical treatment. This event was filmed by scores of international journalists and it shocked the nation. Despite the violent nature of Philippine politics, no one of national importance had been so brazenly killed. There had been other political assassinations, such as the murder in 1949 of Aurora Quezon, the widow of the wartime

President Manuel Quezon, but she had been ambushed by Huk rebels who later acknowledged their error. Ninoy, on the other hand, was met at the airport by Marcos's elite security force. It was officially but incredibly claimed that a lone gunman had managed to shoot Aquino, and indeed the military escort had immediately killed one Rolando Galman, who was then accused of having been the gunman. Such a blatant lie outraged even some of Marcos's supporters. No one mourned Galman, a minor professional hitman, but Ninoy belonged to one of the country's most eminent political families. Marcos responded by setting up a tribunal to investigate the killing, but by then most people had already made up their minds.

Ninoy's funeral marked a turning point for the opposition to Marcos. Thousands of mourners shouted out their hatred for his regime. The procession lasted several hours and an estimated two million people participated. The subservient press barely reported the event, but scores of opposition publications and several radio stations covered it. The traditions associated with death evoke very powerful emotions among Filipinos, and the disrespect shown to Ninoy recalled the martyrdom of Rizal a century before. Almost overnight, Ninoy had become the new Rizal.

Since Aquino had come from an elite background and had supported relatively conservative policies, the left was totally unprepared for this latest turn of events. Some advocated riding on the back of Ninoy's martyrdom, but others objected strongly to such a purely strategic move. The left split between those who supported the massive outpouring of grief and those who clung to the established, largely Maoist line, insisting that the battle would be won in the countryside and not in the cities. Ninoy's supporters came mainly from the urban sector and especially from the middle class.

The EDSA Revolution and its Aftermath

Growing pressure from the United States and other western powers forced Marcos into calling presidential and legislative elections in 1986. The opposition chose Corazon "Cory" Aquino, Ninoy's widow, to challenge Marcos. The role of the aggrieved widow, seeking revenge for her husband's death, is a common theme in popular culture, and Cory's inexperience, despite her politically privileged background, worked in her favor. She quickly came to represent the whole of civil society against the evils of the Marcos regime. For once, Marcos was put on the defensive and was unable to deal with this new form of opposition. He used his old patronage system and employed coercion whenever possible, but he was unable to resist massive civil disobedience.

The excitement of the elections, with their claims and counterclaims of victory, was abruptly ended by the discovery of a coup against Marcos instigated by his close associates. Events moved quickly and attracted world attention. The Defense Secretary, Juan Ponce Enrile, and General Narciso Ramos, Marcos's most senior collaborators during the period of martial law, sensed the collapse of the *ancien régime*. Accompanied by a group of young military officers, the conspirators sought refuge in a large army camp located on the main highway of Metro Manila, Epifanio de los Santos Avenue, which is generally known as EDSA. Marcos's forces could have easily overwhelmed them if it had not been for the timely intervention of Jaime, Cardinal Sin, the head of the Catholic Church in the Philippines. He broadcast an appeal over the Catholic radio station for Manilans to show their support for the rebels by surrounding their camp. More than one million people responded, effectively preventing Marcos from using his superior force.

The events of the following three days, February 22 to 25, 1986, were closely watched by the world media. People held their breath as they watched the tense confrontation between unarmed civilians and the military. The police and army had been known to kill scores of protesting civilians in the past (and the next major police massacre of civilians occurred less than a year later, when Corazon Aquino was in power), but the events on EDSA were different. The demonstration had been called by the Cardinal, and scores of priests, seminarians, and female religious participated in it. While members of religious orders had participated in earlier protests, the EDSA revolution, as it came to be known, represented the first time that the Church hierarchy had given its clear endorsement. Just as importantly, thousands of relatively wealthy Filipinos were also present. Pictures of nuns and students from elite colleges offering flowers or sandwiches to both sides signaled a new phase in civil disobedience.

In the end, Marcos capitulated and was flown to the United States, even though a small but significant number of Filipinos still supported him. Naturally, there was elation in Cory's camp and stories of miracles were frequently recounted. A large statue of the Virgin Mary now stands on one of EDSA's main intersections. However, the next couple of years saw a series of coups against Cory's government, bringing with them economic hardship. Colonel Greg Honasan, a hero of the EDSA events, was one of the main instigators of the coups. He was jailed briefly but is now a Senator. Nor did the Marcoses' exile last very long. After Marcos himself had died, in Hawaii, his family returned and some of them now occupy high political offices. Imelda Marcos became a member of Congress and is planning to campaign to become Mayor of Manila. Many of Marcos's allies are now back in power and there has still been no accounting for the vast wealth accumulated during his regime.

However, things are not the same. Cory Aquino retains her moral authority, even if people admit that not much was achieved during her term in office (1986–92). Scores of Filipinos remember the EDSA events as a high point in the nation's history. The events have taken on mythical proportions and, like other myths, EDSA sets the parameters within which explanations are accepted, rather than providing the explanations themselves. Hence, most Filipinos readily agree that something important happened at EDSA but, when pressed, are unable to provide good examples. Often they point out its supposed beneficial effects in other parts of the world, such as the liberation of central and eastern Europe or the unification of Germany.

Perhaps most importantly, there has been a reorganization of the country's basic institutions. As mentioned above, the Philippines inherited a Constitution based on the US model, but in 1973, during the era of martial law, a new Constitution was written, changing Congress into Parliament and expanding the franchise to include the young. In 1986, a third Constitution was promulgated, marking a partial return to the earlier arrangements but including changes introduced during Marcos's years in power. Marcos had politicized the military, but Cory and her successors reasserted civilian control. Presidents Fidel Ramos (1992–98) and Joseph Estrada (elected in 1992) have both tried to amend the Constitution, but they have met strong opposition from the Church and prominent activists, including Cory Aquino.

Cory Aquino was the first Filipino politician who willingly left office after the constitutional term had ended. A provision in the present Constitution limits the presidency to one term and other offices to three, although most politicians get around this by nominating their wives or children to their positions. Ninoy Aquino's family have also profited from his fame: a brother, a sister, and a brother in law have been elected to Congress.

In many ways, then, Philippine politics has returned to the practices of the days before Marcos. The traditional elite, including many of Marcos's cronies, is back in power, and the Communist rebellion appears to have gained new ground, after disastrous experiences in the 1990s. The issue of Moslem autonomy is still far from resolved, and has recently entered a new and dangerous phase. Finally, while the Philippines was not as seriously affected as its neighbors by the Asian crisis of 1997–98, its recovery is proving to be much slower.

Fantasy, Patronage, and Emigration

In this context, the election to the presidency of Joseph Estrada has introduced a new variable. Estrada, or "Erap," as he is familiarly known, was elected with a large majority. While he served his apprenticeship as the mayor of a small municipality, his fame stems principally from his career in action films, in which he often played either a poor, uneducated man striving to retain his dignity against superior forces, or a tough guy with a heart of gold. Poor Filipinos readily identify with him, but many among the educated middle class consider him an embarrassment to the country. Other film stars have followed in his footsteps, indicating that positive media exposure has become a major element in political success. Television commentators, journalists, and sporting heroes are also treated like film stars. Politicians are quickly adjusting their policies and personalities to fit this new requirement, and political success increasingly draws on a common world of fantasy. Imelda Marcos has consciously employed this strategy, claiming that the poor vicariously enjoy her displays of wealth. Imelda criticized Cory Aquino for pretending to be an ordinary housewife and for not displaying her own enormous wealth. However, there are important exceptions to this trend, indicating that media manipulation and fantasy are not the only factors in politics. Politicians with integrity are still accorded strong support.

Social scientists have generally described Philippine politics as a system of patronage. Cross-class ties are more important than class loyalties, because patrons ensure that their supporters are rewarded. Patronage is also closely tied to the politics of kinship, and the Philippine state is often said to be a huge system of patronage, where loyalty rather than competence are rewarded. However, while this model generally applies to many examples of local politics, it is less applicable at the provincial and national levels. While patronage is also exercised at these levels, other important factors intervene. The Communist insurgency, Moslem separatism, and the popularity of many national politicians cannot be reduced to patronage. In the past, there were also movements based on class, regional or ethnic interests. Nor can President Estrada's popularity be explained by referring to ties of patronage, even if these operate among his close supporters. Instead, patronage should be seen as another aspect of kinship and other local ties. Filipinos prefer to operate within groups whose members are known to one another. This preference generally works in local and other communities but not in urban situations. Filipinos, with some initial reluctance, adjust their behavior to these new situations.

Particularly interesting compromises are achieved when Filipinos decide to pursue work opportunities abroad. The first recorded major movement of workers abroad was in 1905, when Filipinos went to Hawaii and other parts of the United States in search of work. Once established abroad, Filipinos often sponsored their relatives and friends to join them. Soon, Filipino communities were to be found in Hawaii, California, and other parts of the United States where agricultural workers had settled. In the 1970s, following the political and economic crises under Marcos, large numbers of Filipinos once again explored opportunities abroad (see Pertierra 1992). They went to the Middle East, worked on foreign ships or found employment in Japan, Hong Kong, and Singapore. They also sought migration opportunities in Australia, Canada, and the United States. The first wave of Filipinos to go abroad tended to have professional or other specialized skills, and were mostly men, but women have since joined this emigration. They now work as nurses in Saudi Arabia, domestic helpers in Hong Kong, factory workers in Taiwan, and entertainers in Japan. Filipinos are now also found in significant numbers in Italy, Spain, and other European countries. Initially, overseas Filipinos consisted of migrants seeking permanent settlement abroad, but most now work on short-term contracts and plan to return home.

Estimates vary, but it is believed that between six and eight million Filipinos work abroad. Their remittances are the largest single source of foreign currency and ensure the viability of the Philippine economy. Most of these Filipinos abroad are adults with tertiary qualifications, and their absence from their home communities has created a dependence on the international economy, as well as a new global consciousness. These developments express some of the contradictions of the "new world order."

Philippine Antinomies

The reasons why the Philippines considers itself, and is seen by its Asian neighbors, as different should now be more apparent. A central problem for Philippine nationalism has been to imagine itself independently of its colonizers. This is difficult because nationalist sentiment is itself a product of the colonial experience. The consequent dilemma is shared by other nations in Southeast Asia such as Malaysia and Indonesia, which are also products of their respective colonial experiences. In their cases, however, people can always point out the existence of important precolonial centers and expound a notion of continuity based on the presence of indigenous inhabitants (*bumiputras* in Malaysia, *bumiputeras* in Indonesia). The Philippine equivalent is the *indio*, a term that was introduced by the Spanish and has often been used abusively. The generation of 19th-century expatriate Filipinos such as Jose Rizal proudly called themselves "Indios Bravos" to distinguish themselves from Philippine-born Spaniards, who were then referred to as "Filipinos." US officials did not make the same distinction because a mixed-race population of Filipino-Americans did not exist. The term "Filipino" was henceforth used to describe all the peoples of the Philippines.

The United States described its colonial project as fundamentally different from Spain's: the Spanish had kept the natives in superstitious poverty, while US administrators introduced modern and liberal arts of government. The colonial past was replaced by the modern present. Filipinos also accepted this discontinuity and saw their present as a break from the Spanish past. This disjunction is reflected culturally and linguistically, causing problems for any attempt to imagine the nation.

In the 19th century, Spanish was the language used by the Filipino elite to imagine an otherwise diverse nation, but less than 20% of the population spoke it, the majority relying on regional languages. Rizal and other intellectuals wrote in Spanish, a language no longer accessible to Filipinos today. They therefore have to read Rizal's novels, considered to be the classic expression of Filipino consciousness, in translation. These translations were initially in English but more recently they have been in Filipino, the national language. Benedict Anderson has argued that these translations often negate Rizal's aim of concretizing his social critique; instead, the translations manage only to distance the past from the present (see Anderson). This in turn transforms the contemporary Philippines into a society without history.

The disconnection of the Philippines from its past is best expressed in its national language policy. Filipino, a language still in the making, is intended to be the medium for nationalist expression. It is meant to combine various languages, using Tagalog, Manila's lingua franca, as its base. Spanish has been almost completely forgotten, but English retains its importance in education, business, and public affairs. Yet most Filipinos are not at ease in English, preferring instead to mix it with Tagalog ("Taglish") or other regional languages. Domestically and informally, English is rarely used.

The Republic of the Philippines has often been described as a weak state, unable to enforce its dictates on its citizens. Even under martial law, Marcos was unable to control vital sectors of society: even his attempts to control the research agenda of Philippine social science failed and instead he provoked academics and other intellectuals to resist his regime. However, if the Philippine state is weak, the class that controls it is very strong. The Filipino elite has managed to survive the Spanish and American colonial regimes to continue into the present. The EDSA revolution failed because its main instigators had no intention of changing society's basic structures. While the Philippines has experienced fundamental changes throughout its history, including changes in its own self-perception, other things have remained the same. It is a society of contradictions, full of antinomies.

Further Reading

Anderson, Benedict, "Hard to Imagine: A Puzzle in the History of the Philippines," in *Review of Indonesian and Malaysian Affairs*, Volume 28, number 1, 1994

> Anderson applies the influential ideas he has developed at length in *Imagined Communities* (London and New York: Verso, 1983) to the creation of nationalism in the Philippines.

Blount, James H., *The American Occupation of the Philippines*, New York and London: Putnam, 1913; reprinted New York: Oriole, 1973

Blount was one of the strongest contemporary critics of the early stages of US colonialism in the Philippines. His book expounds a counter-view to the prevailing ideas of Dean Worcester and others.

Diokno, Maris, *Perspectives on Peace during the Philippine-American War 1899–1902*, London: University of London School of Oriental and African Studies, 1994

Ileto, Reynaldo, *Pasyon and Revolution: Popular Movements in the Philippines, 1840–1910*, Quezon City: Ateneo de Manila University Press, 1979

This is a classic account of the rebellions in the Philippines in the closing decades of the Spanish colonial period. Ileto approaches history from below to develop a "subaltern" account of colonialism.

Kerkvliet, Benedict, *The Huk Rebellion, A Study of Peasant Revolt in the Philippines*, Berkeley: University of California Press, 1977

Kerkvliet's is the best available account of the rebellion in Central Luzon during the years immediately following World War II and the attainment of independence.

Pertierra, Raul, *Religion, Politics, and Rationality in a Philippine Municipality*, Honolulu: University of Hawaii Press, and Quezon City: Ateneo de Manila University Press, 1988

A standard anthropological study of religion and politics in a community in Northern Luzon

Pertierra, Raul (editor), *Remittances and Returnees*, Quezon City: New Day, 1992

A study of the effects of overseas labor migration on home communities in the Ilocos region, one of the main sources of such migration in Philippines

Rafael, Vincente, *Contracting Colonialism*, Ithaca, NY: Cornell University Press, 1988

A groundbreaking study of the effects of religious and colonial conversion in the Philippines during the 16th and 17th centuries

Sullivan, Rodney, *Exemplar of Americanism*, Ann Arbor: University of Michigan Press, 1991

The best study of Dean Worcester, the scholar and technocrat who played a very influential role during the US colonial period

Wolff, Leon, *Little Brown Brother*, New York: Doubleday, and London: Longman, 1961

An account of the brutalities of the US military conquest of the Philippines from 1898 to 1902

Worcester, Dean, *The Philippines Past and Present*, New York: Macmillan, and London: Mills and Boon, 1914

An authoritative account of the US colonial mission as seen by one of the leading lights in its implementation

Youngblood, Robert, *Marcos Against the Church*, Ithaca, NY: Cornell University Press, 1990

A competent account of the complex and ambivalent relationship between the Catholic Church and the man who dominated the Philippines for 21 years

---

Dr Raul Pertierra is an Associate Professor in the School of Sociology at the University of New South Wales in Sydney, and a Visiting Professor in the Department of Sociology and Anthropology at Ateneo de Manila University in the Philippines. In addition to those cited above, his publications include Philippine Localities and Global Perspectives *(Quezon City: Ateneo de Manila University Press, 1995) and* Explorations in Social Theory and Philippine Ethnography *(Quezon City: University of the Philippines Press, 1997).*

NATION STATES

Chapter Nine
Indonesia

William Case

Indonesia, with its vast population and great territorial reach, has long been regarded as the heavyweight among the nations of Southeast Asia. By 1949, its military forces had waged and won a war of independence against its colonial overlord, the Netherlands; six years later, one of its provincial capitals, Bandung, was the birthplace of the Nonaligned Movement. More recently, and until the crisis of 1997–98, its economy became as large as those of all the other ASEAN member states combined, an achievement that seemed to many to confirm the country's preeminence in the region. Today, however, in the wake of the crisis, Indonesia's standing has been seriously eroded. To be sure, it has done much to develop its political institutions, allowing new civil liberties, giving freedom to political parties, and holding competitive elections. However, in contrast to most other countries in the region, it has so far done little to regain its earlier economic prowess. In addition, ethnic and religious violence, secessionist movements, organized crime, and vigilantism have risen sharply in many parts of the archipelago. Meanwhile, the armed forces, far from restoring order and ensuring national unity, appear today to be behaving in ways that only deepen the unrest. In these circumstances, many observers have begun to speculate openly about the possible break-up of Indonesia.

Many of the factors that threaten Indonesia's economic recovery and the consolidation of its new democracy are deeply rooted in the country's history, its economic structures, and, perhaps, its culture. Others are more political in nature, involving the strategies, intrigues, and resistances mounted by entrenched interests, sometimes under the cover of apparently sophisticated ideologies. Together, these various factors offer formidable impediments to Indonesia's resuming its economic and political development, even as the other countries in Southeast Asia make new progress, however tenuous it may be. Nonetheless, as we hope to show here, while Indonesia's economic prospects remain uncertain, its democracy will probably survive.

Ethnicities and Economic Development

The area under Indonesian jurisdiction represents the fortuitous outcome of Dutch colonial conquest, rather than any coincidence of solidary ethnic identity and state apparatus. Indeed, anthropologists have suggested that around 200 distinct ethnic groups can be identified on the 13,000 islands that make up the archipelago. They are committed to variants of five major religions and speak around 300 different languages or dialects.

Dutch and Japanese Occupation

The Dutch politicized these divisions, privileging first selected religious groups, later selected ethnic communities. For example, small minorities of Christian Ambonese and Minahassans joined Javanese in policing the vast empire; Javanese aristocrats, drawn from a social *aliran* (stream) known as the *priyayi*, were recruited as middle-level administrators (see Geertz); Javanese workers, drawn from a lower stratum of *abangan*, toiled on plantations;

and devout (*santri*) Moslems operated as small-scale landowners and petty traders.

What stands out from the record, however, is the Dutch deployment of ethnic Chinese in ancillary sectors of the colonial economy. While Dutch-owned companies occupied the commanding heights, running plantations, mines, and oilfields, the Chinese were recruited into low-level processing and commodities dealing, short-distance transport, urban retail distribution, moneylending, and the professions. In this way, the Chinese, who constituted a relatively small collectivity in the Indonesian setting, were generally enabled to become much wealthier than members of the communities conceptualized as "indigenous." They were differentiated culturally too, for they were required by the Dutch to reside in specially demarcated "Chinatowns" and to adhere to dress codes that the Dutch interpreted as Chinese (Anderson 1987 pp. 3–4). Accordingly, although they gained economic resources, the Chinese lacked social power, and were thereby prevented from ever challenging the Dutch politically. On the contrary, their ethnicity effectively served as a buffer for the Dutch against "indigenous" resentments (Habir pp. 168–69). As early as the 18th century, Jakarta's administrative precursor, Batavia, erupted, not in anticolonial rebellion, but in anti-Chinese rioting.

More fundamentally still, many Indonesians grew suspicious of the capitalist economic relations that perpetuated colonial inequalities and gave considerable privileges to people whom they regarded as foreign interlopers. These sentiments gained organizational force through religious revivalism and new ideologies. Leveling forms of Islam appeared early in the 20th century, explicitly targeting Chinese businessmen who had made inroads into the batik trade; the Communist Party (Partai Komunis Indonesia, or PKI), founded in 1920, attracted support among rice growers and plantation workers; and a charismatic engineering student named Sukarno gained prominence toward the end of the 1920s with his amalgam of Islam, socialism, and nationalist appeals for independence.

All these movements were suppressed by the Dutch during the 1930s, but they were given new life and made more militant by the Japanese forces that occupied Indonesia during World War II. In their attempt to galvanize Indonesians against the West, the Japanese portrayed themselves as the "light of Asia" and some of their ideologues even sought to steer Islamic loyalties away from Mecca and toward their own Emperor. As the Allies moved nearer, however, they readied Indonesia for independence, organizing youth groups into paramilitary formations as early as 1943. Finally, at the end of the war, Indonesia's nationalist leaders adopted a Constitution, parts of which intimated their deep distrust of market capitalism. Indeed, the Constitution's authors were much influenced by the notions of "organicism" and "integralism" to which they had been exposed during the 1930s, for they seemed likely to be useful for integrating Indonesia's competing social *aliran* and classes into a more manageable, even harmonious body. Hence, in a much-cited passage, Article 33 of the 1945 Constitution committed the state to public ownership, economic cooperatives, social prosperity, and family principles. So strong were these tendencies in favor of state control and against market capitalism that some analysts have interpreted Indonesia's first constitution as not so much socialist as fascist in tone (Elson p. 7).

The Sukarno Era

With the defeat of the Japanese and the return of the Dutch, Indonesia's leaders were soon distracted from their ideological concerns, and had instead to focus on armed struggle. Further, after winning independence, in December 1949, they introduced a new provisional Constitution, then a second one, in 1950, which unexpectedly committed Indonesia to relatively free markets and democratic politics. At first, these policy commitments were buoyed by the surge in commodity exports that accompanied the Korean War, but with the end of that conflict commodity earnings quickly declined. They were also eroded by the government's overvaluing of the Indonesian currency, the rupiah, a measure that effectively subsidized Java's importers and consumers at the expense of the commodity exporters in the outer islands. This apparent bias reinforced perceptions of increasing Javanese dominance, and many of the outer islands erupted in rebellion during the mid-1950s.

Sukarno, Indonesia's most prominent nationalist leader, had been named the country's President when independence was declared in 1945. However, he had long been confined to a ceremonial role, largely owing to official embarrassment over his wartime collaboration with the Japanese, the turbulence of the independence struggle that followed, and the rise of Prime Ministers with power bases in the legislature. In 1957, however, confronted by the rebellions in the outer islands, Sukarno began to cooperate more closely with the military in order to gain a new ascendancy. To this end, he declared martial law in March that year, enabling the military to suppress the rebellions and restore political power to the center. Sukarno also began to institute a wideranging though inefficient scheme of state ownership and licensing, a program of import substitution and redistribution that he labeled the "Guided Economy." Moreover, he fortified his statist interventions with radical, even aggressive foreign policies, proposing an "axis" linking Indonesia to the Communist countries of East Asia, and thereby greatly antagonized the United States. At the same time, the military bolstered its own business enterprises, which it had begun to acquire during the war for independence, and which it greatly expanded after the nationalization of most Dutch-owned enterprises in December 1957. The military quickly stripped these enterprises of their assets, leaving management to languish and the national economy to decline.

The new partnership between Sukarno and the military soon grew tense, however. Some elements in the armed forces, especially in the navy, the air force, and the presidential guard, shared Sukarno's statist visions, but many army commanders, while remaining ambivalent toward market capitalism, became increasingly wary of their President's recklessness. As suspicions deepened, Sukarno attempted to hold the army in check by turning to the PKI, having recognized its mobilizing appeal among landless agricultural workers. Indeed, he canvassed the formation of a "fifth force," to be formed by arming the peasantry, alongside the four existing military services. Meanwhile, the pace of economic deterioration quickened: shortages of food and clothing were generating annual inflation rates of 600% by the mid-1960s.

In this context of worsening political tensions and economic hardship, an apparent coup was attempted in October 1965. Sukarnoist elements in the military, apparently colluding with leading figures in the PKI, struck preemptively at the army, murdering six commanders (Crouch pp. 97–134). Major General Suharto then struck back. During the months that followed, soldiers worked alongside Moslem youth groups to wipe out the PKI and its vast social base, launching a momentous bloodletting in which upwards of half a million people were killed (see Cribb). Many times that number were then imprisoned for long periods, often on remote islands, while members of their families were punished through blacklisting. In addition, Suharto began deftly to marginalize Sukarno. He gained control over national security in March 1966, became Acting President in March 1967, and assumed the presidency itself in March 1968. Sukarno, placed under virtual house arrest, died two years later.

Suharto in Command

With the PKI and its following now emasculated, Suharto turned his attention to the economy, recognizing that economic development could underpin his rule. With barely an elementary education, Suharto possessed no formal training in economics, but he took advice from neoclassical economists at the University of Indonesia. Because some of them had been trained at the Berkeley campus of the University of California, they came to be known as the "Berkeley Mafia." By rolling back many of the statist policies associated with Sukarno's "Guided Economy," they soon attracted foreign investment and developmental assistance. The average annual growth rates of 6–7% that began in this period persisted for nearly three decades.

Despite this initial burst of reform, doubts about market capitalism persisted, especially among university students who could see for themselves that socioeconomic inequalities were widening. In particular, once Indonesia opened the door to foreign capital, Japanese investors appeared on the scene, as part of their first major foray overseas after World War II. Japanese companies moved quickly into the textiles industry, for example, pushing indigenous producers aside and recruiting ethnic Chinese as joint venture partners. University students were galvanized into protest, particularly during a visit to Jakarta by the Japanese Prime Minister, Tanaka Kakuei, in January 1974. These protests, which coincided with growing rivalries among the elite, quickly swelled into riots that have become known by the term *Malari*, an acronym for the phrase "disaster of January 15" (Bresnan pp. 135–63).

The Malari riots presented Suharto with his first great test as President. He responded astutely, by mixing coercive measures with concessions, in particular by making pledges to curb foreign investment in ways that would create opportunities for indigenous businesspeople. Later in 1974, the Organization of Petroleum-exporting Countries (OPEC) instigated a dramatic fourfold increase in oil prices, soon bestowing upon Indonesia, a medium-sized oil producer, some unexpected windfall revenues. In these circumstances, the national petroleum company, Pertamina, was able to promote some indigenous businesspeople, providing them with managerial positions, contracts, and credit. A scaffolding of supportive state enterprises was also erected. On the other hand, in order to blunt the rise of "indigenous" interests, Suharto tapped a network of Chinese businesspeople even more intensively. Hence, alongside the many new state enterprises that were created, Suharto encouraged new Chinese-owned conglomerates, using them as the main agents through which he set out to fulfill his developmentalist visions.

Meanwhile, Suharto also unveiled new populist programs, offering state cooperatives, village development schemes, subsidies for staple goods, and various charity programs identified personally with the President, thus enabling him to dispense benefits in broadly patrimonialist ways. Suharto revived the old slogans of organicism and integralism, then braced these doctrines with a renewed emphasis on Pancasila, the code of ideological principles that had first been devised by Sukarno to promote consensus among the diverse social *aliran*.

These populist programs and collectivist appeals barely concealed the harsh social realities. Large numbers of workers were conscripted into industrial estates and factories to generate the export earnings upon which their meager benefits depended. In order to keep labor costs low, workers were absorbed into the new corporatism through a single official labor union, the All Indonesia Workers Union, or SPSI, founded in January 1985. These arrangements were then legitimated ideologically through the concept of "Pancasila labor relations." Industrialization never became so extensive that the workforce could be fully absorbed, which undercut efforts to organize autonomously. Just to make sure that the SPSI had no rivals, the military worked assiduously in the background to enforce labor discipline.

In sum, although Suharto avoided disparaging market capitalism to the extent that Sukarno had, and although he possessed far more resources with which to realize his industrializing visions, his statist interventions were in some ways reminiscent of the "Guided Economy" organized by his predecessor. Further, his manipulation of ethnicity in business-government relations, undergirded by his "security approach" to labor, harked back to the practices of the Dutch. This amalgam of state-centered capitalism and tightly controlled social forces was then portrayed by Suharto's ideologues as the "Pancasila economy," an orientation said to be in keeping with the country's historical, structural and cultural legacies.

During the early 1980s, however, global petroleum prices suddenly declined. Suharto reacted initially with another quick burst of patrimonialist behavior. A unit called "Team 10," entrenched in the State Secretariat, gained authority over government contracts for supply and construction, and then allocated them almost exclusively to "indigenous" businesspeople (see Winters). It was during this period too that Suharto began to foster his children's business ventures by granting them monopoly import licenses and sole distributorships. His middle son, Bambang, his daughter, Tutut, and his youngest son, Tommy, came rapidly to establish some of the country's largest indigenously owned conglomerates, with interests in telecommunications, media, infrastructure, property development, and commodities. Suharto's support for his children's business expansion was regarded by many observers as his greatest weakness, since it undermined his standing among professional elements in the military and the bureaucracy, as well as among the more dynamic elements in the business community.

Suharto remained sufficiently pragmatic to search for an alternative engine for growth once he had grasped that petroleum prices would not soon recover. During the mid-1980s, he returned to the policies of deregulation and liberalization that he had implemented after first coming to power. He began by opening up the banking system and the Jakarta Stock Exchange to foreign investors, who now returned to Indonesia in force, and even acquired majority stakes in many firms. Suharto also undertook tariff reforms, using the APEC meeting in Jakarta in 1994 to speed up the ASEAN agenda for free trade by 2010. In this new setting, many small and medium-sized enterprises roared into life, exporting a range of low-technology, labor-intensive manufactures, especially clothing and footwear, furniture, office supplies, and household goods. Indonesia underwent a remarkable transition to a more deregulated and export-oriented economy that gave new impetus to its long record of growth.

However, behind the new deregulatory processes that enticed foreign investors and freed local producers, Suharto took care to protect the interests of his relatives and friends. Even his new campaign of privatization was undertaken in ways that ensured that state assets flowed, as equity and as contracts, into the hands of his favorites. Crude monopoly licenses on commodity inputs were supplanted by discounted shareholdings, state contracts, and cheap credit. During the 1990s, as Suharto gained greater confidence that growth would persist, he resumed making statist interventions into the economy. Indeed, he discovered new sources of justification, combining the established Pancasila emphasis on "social harmony" with a new focus on using the state to promote technological advance. In 1993, Suharto sidelined the technocrats in his cabinet, who had favored deregulation, and appointed "technologues," members of a faction committed to high-technology state enterprises and projects. This impulse was made manifest in the aviation and shipping industries associated with B.J. Habibie, Suharto's Minister of Research and Technology, as well as in the national car project led by Suharto's son Tommy. Consequently, as recently as 1995 state enterprises still comprised a larger proportion of Indonesia's economy than the private sector did (Habir p. 189).

Throughout his long tenure as President, Suharto's guiding principle was to avoid opening up to market capitalism wherever he could, and to deregulate only when he had to. It has been persuasively argued that this outlook had deep roots in Indonesia's historical and sociological bedrock. To be sure, Suharto cynically manipulated this principle in order to strengthen his own political ascendancy: while he consented to the promotion of some indigenous businesspeople, he made sure that they were few in number and remained dependent. He also maintained his favored network of ethnic Chinese businesspeople, leaders of a community that was still comparatively wealthy but remained socially disempowered by its "alien" ethnicity. Especially during periods of statist intervention, but also during the intervals of pragmatic retreat, Suharto continued to modulate patronage in ways that enriched his relatives and friends.

It was precisely the loss of such patronage resources during the crisis of 1997–98 that led Suharto's supporters to abandon him, quickly precipitating his departure. Stripped of his capacity to reward and placate, Suharto was left with a record of developmentalism, high-level patrimonialism, and mass-level populism that was now overshadowed by images of corrupt practices and coercion. In this situation, calculations changed swiftly among Suharto's close supporters and mass followings. It was the corrupt deployment of illiberal imperatives, rather than any avoidance of free markets, that finally brought Suharto down. Despite Suharto's removal, however, many of the currents of illiberalism that he once encouraged persist in Indonesia, weighing heavily on the country's prospects for economic reform and recovery.

Obstacles to Democratic Politics

Many analysts contend that, for a variety of reasons, democratic politics and free markets are mutually reinforcing. Most notably, the accountability imposed by competitive elections is said to limit the state's ability to pursue predatory behaviors and corrupt practices. However, at least in countries at relatively low levels of development, there is no clear correlation between political regime and market orientation. In the case of Indonesia, authoritarian rule, imposed first by the Dutch, and later by Sukarno and Suharto, doubtless impeded the development of a market economy. Nevertheless, it may be argued that democracy can only be consolidated if some market distortions persist, safeguarding the core interests of the weakened but still potent elites who were associated with Suharto and are still able to cause trouble.

The Colonial Legacy

Indonesia's colonial experience under the Dutch failed utterly to lay any groundwork for democracy. Under the "Ethical Policy" introduced by the Dutch government in 1901, a small legislative body, the Volksraad (People's Council) was established, but it mostly involved Dutch electorates who occasionally voted for a few "acceptable" Indonesians. Most secular nationalists refused to cooperate with the Volksraad. Accordingly, this body did little to expose indigenous leaders to democratic procedures, much less to inspire any deeper political culture of participatory behaviors balanced by restraint. Instead, as Benedict Anderson has pointed out, steep social hierarchies enforced by a stark sense of fatalism prevailed, at least among the Javanese (then as now Indonesia's largest single ethnic group), punctuated occasionally by violent political upsurges (Anderson 1990 pp. 17–77). As mentioned above, Java's social *aliran* were made vastly more complicated by the Dutch articulation of religious and ethnic identities, then rigidified by occupational structures and uneven distribution of wealth. Hence, Sukarno's attempts during the late 1920s to give a unified voice to different social aspirations – an array of Islamicist, socialist, and nationalist longings – failed to crystallize in any coherent democratic movement. Sukarno was arrested and exiled, and his new party was disbanded. In consequence, the Dutch colonial experience, in contrast to the "tutelary model" applied, for example, in many British possessions (see Weiner), generated a virulent brand of pluralism that proved to be an uncongenial setting for democratic politics.

As we have seen, Indonesia's accession to independence from the Dutch involved a prolonged campaign of bargaining and armed struggle during the late 1940s. At one point, Sukarno and other nationalist leaders allowed themselves to be captured and imprisoned, and the struggle was carried on by the military, enabling officers to claim afterwards that the infirmity of civilian politicians and the practical need to mount a "people's defense" merited giving the armed forces a permanent place in the country's political life. Hence, the military's sociopolitical and security roles, enshrined first as the "middle way" and later, in 1957–58, as *dwifungsi* ("dual function"), involved its sharing fully in bureaucratic appointments, legislative activity, and commercial opportunities. The military also maintained its territorial command structure, embroidered with business links, across the archipelago and down to the village level, creating what must surely be one of the greatest obstacles to democratic politics. Finally, when the Dutch agreed to depart, they attempted to dispense some democratic advice, notably on the benefits of federalist power-sharing. Because Indonesia's nationalist leaders had grown suspicious of Dutch motives, however, they chose instead to bind their new country into a tight unitary administration. In this way, they concentrated resources at the center, stifled provincial autonomy, and deepened resentment in many of the outer islands.

From Parliamentarianism to "Guided Democracy"

Having few other models to choose from, and wishing at first to impress the United States, Indonesia's independence leaders put aside the Constitution of 1945 in order to set up a democracy. Some analysts of democratic transitions have argued that an early record of democracy, even when broken, favors redemocratization later, the benefits of electoral accountability having been made plain (Huntington 1986 p. 85; Huntington 1991 p. 42). However, Indonesia's democratic governments during the 1950s were immobilized by relentless horsetrading and produced few good policy outcomes. They were strikingly unstable, with six different coalitions rotating swiftly through office. When the long-delayed legislative elections were finally held, in September 1955, the results merely reflected once again the diverse aspirations of the *aliran*, dispersing power nearly equally among the four major parties. Indeed, to the extent that there was any change in the distribution of seats, the Partai Nasional Indonesia (Indonesian National Party, or PNI), which was associated most closely with the Javanese *priyayi*, gained a slight edge over Masyumi, a *santri* Moslem party that had ties to the outer islands, thereby exacerbating regionalist discontents. The PKI also won more seats, animating the *abangan* and stoking class tensions.

It was in these circumstances that the rebellions in the outer islands took place. Most of them were led by regional military commanders who objected to Jakarta's centralized control. Sukarno responded by declaring martial law, abruptly suspending many of the country's democratic procedures, and in July 1959, again on the advice of the leading military commanders, he reinstated the Constitution of 1945, sidelined the political parties, and terminated elections. In seeking to legitimize these measures, Sukarno asserted that the parties had amplified the differences among Indonesia's competing social forces, and elevated some interests over others through a starkly competitive formula of "50% plus one." The party system thus partly gave way to new corporatist arrangements based on state-sanctioned "functional groups." Elections were replaced by the appointment of broadly inclusive cabinets that were portrayed as consultative and as subordinate to Sukarno, firmly installed as President for Life. The "Guided Economy" found a close parallel in what Sukarno conceptualized as "Guided Democracy."

However, the military had gained a new unity after suppressing restive commanders in the outer islands and had benefited from greater budgetary allocations under martial law, the campaign to seize West New Guinea from the Dutch, and the subsequent period of "Confrontation" with Malaysia. It now began to challenge Sukarno's ascendancy. As we have seen, Sukarno responded by showing greater favor to the PKI, rewarding its leaders with bureaucratic positions in return for their mobilizing mass support for the government. The military responded in 1964 by forming Golkar (in full, the Sekretariat Bersama Golongan Karya, or Joint Secretariat of Functional Groups), through which it affiliated its own mass constituencies. A year later, the military utterly destroyed the PKI, paving the way for Suharto's rise to the presidency.

"Pancasila Democracy"

The first legislative elections since 1955 were held in July 1971. Beforehand, Suharto refurbished Golkar as his government's electoral vehicle, then distorted electoral procedures in its favor, so that Golkar captured nearly 63% of the votes cast. In January 1973, Suharto compelled nine opposition parties, representing most of Golkar's potential rivals, to reorganize themselves into two officially recognized groups, their internal decision-making and organizing activities being subjected to close control. Golkar replicated its victory, often with even greater margins over the other two parties, in all of Indonesia's subsequent legislative elections. Suharto also perpetuated his own presidency through a separate set of elections in an electoral college whose members he personally vetted (MacIntyre pp. 22–65).

Suharto called his regime "Pancasila Democracy", but in many ways it closely resembled Sukarno's "Guided Democracy."

For example, Suharto attempted to legitimize his authoritarian rule by contrasting it regularly with the apparent failings of the parliamentarian democracy that Indonesia had adopted after gaining independence. He also made use of corporatist techniques, elaborating functional groups, while celebrating the virtues of consultation and consensus. While Sukarno had proclaimed himself President for Life, Suharto emerged as *Bapak Pembangunan* ("Father of Development"), embracing Indonesia's disparate subnational identities through the principle of *kekeluaragaan* ("familism"). However, there were also important differences between these two authoritarian regimes. Where Sukarno had galvanized social forces through his own charismatic authority and revolutionary appeals, Suharto reassured these forces with calm and sober rhetoric, then dispersed them as a floating mass. Where Sukarno had confronted the army, Sukarno shrewdly coopted it, gradually establishing the paramountcy of his office even as his generals continued to do his bidding. Most importantly, while Sukarno's policies had brought hyperinflation and shortages of basic goods, Suharto brought a measure of prosperity. Until the final year of his presidency, steady economic growth uplifted an urban middle class, while sustaining enough development to gain the acquiescence of the masses in the cities and the countryside alike.

In short, the pace and contours of Suharto's industrializing strategies helped to perpetuate his authoritarian regime. Specifically, the country's industrializing bourgeoisie, which remained numerically small, politically dependent, and ethnically divided, relied largely on personalist understandings with Suharto, rather than any democratic procedures or the rule of law, to safeguard its dealings. The broader middle class, though more distant from Suharto and often impatient with inequitable and inefficient governance, nonetheless remained fearful of democratic participation by the far larger collectivities of workers and peasants. Hence, nongovernmental organizations, which in other national settings have helped to enliven civil society, mostly remained tame under Suharto. University students became politically active from time to time, but normally they focused on preparing to realize their ambitions. Workers, meanwhile, were weakened and divided by their uneven incorporation into the new manufacturing centers, as well as by the fragmented systems of agricultural production and land tenure.

Economic Crisis, Reform, and Democracy

Although the ground has shifted somewhat since the recent economic crisis, it has done so in ways that have permitted only gradual democratic progress. However, precisely because of this moderate pace, which coincides squarely with the avoidance of economic reforms, the prospects for consolidating Indonesia's new democracy have been somewhat improved.

Indonesia was profoundly affected by the regional economic crisis of 1997–98. At first, many analysts anticipated that Indonesia's strong leadership and its record of sustained success in macroeconomic management would enable it to cope effectively with the crisis, but it soon became the most seriously dislocated country in East Asia. During 1998, GNP shrank by 15%, a greater contraction than any country at Indonesia's level of development had experienced in the period since World War II (*Far Eastern Economic Review*, April 29, 1999, p. 23). The modern service and manufacturing industries were especially hard hit, buffeting the new urban middle class, while industrial workers were driven back into the informal economy or even into the rural hinterland from which most of them had migrated.

In these circumstances, large numbers of university students, their career chances now blighted by the crisis, began to mobilize against Suharto. When four student protesters were killed by security forces at the Trisakti University in Jakarta, urban mobs were inflamed, and their protests culminated in the enormous destruction wreaked by the Jakarta riots of May 1998. The homes and business premises of some of Suharto's closest supporters were looted and burned. Jakarta's Chinatown was the target of some of the worst violence: Chinese women were systematically assaulted and shopping complexes were razed to the ground. Suharto, having lost his patronage resources and his capacity to keep order, could no longer protect his supporters. On the contrary, he stood now as a lightning rod for mass grievances, further endangering the elite. In rapid succession, Golkar politicians, cabinet ministers, state enterprise managers, and leading businesspeople began to denounce Suharto publicly. Realizing that his generals had been discredited by their failure to prevent or curb the rioting, and stricken with indecision, Suharto agreed to step down.

Given that Indonesia's experience of economic crisis was widely attributed to cronyism, rather than to the actions of foreign investors, currency traders, and multilateral agencies (see Emmerson 1999a), one might have expected a corresponding upsurge in demands for market reforms. One might also have anticipated that, given the insolvency into which most business conglomerates had plunged, few of Suharto's erstwhile supporters would be able to resist such reformist pressures. Yet soon after Suharto had been sacrificed, popular ferment appeared to dissipate. Suharto's successor as President was his former Minister for Research and Technology, B.J. Habibie, who provided new scope for civil liberties and a timetable for elections, thus avoiding any meaningful bankruptcy proceedings or corruption investigations. In addition, when legislative elections were held in June 1999, Golkar won the second largest share of the votes cast. Megawati Sukarnoputri's success in winning a plurality of votes in these elections augured no better for market reforms, since her Indonesian Democracy Party of Struggle (PDI-P) was largely carried aloft by the nostalgia of the urban poor for the charismatic but vacuous "leftism" of her father, Sukarno.

The election campaign displayed some other continuities with earlier regimes. For example, Habibie did little to reassure foreign investors or ethnic Chinese businesspeople, thus discouraging the return of much-needed capital. Adi Sasono, Minister of Cooperatives and director of an Islamic think tank once associated with Habibie, the Center for International and Development Studies, helped to form a new political party, the People's Sovereignty Party, and then appealed for a "people's economy," in which state cooperatives would be strengthened and cheap credit allotted to small enterprises operated by Moslems. Other candidates canvassed more comprehensive schemes for crossethnic redistribution, modeled on the reverse discrimination of the New Economic Policy in neighboring Malaysia.

A presidential election was held in November 1999. Habibie was forced out of the running shortly beforehand, but this was due more to nationalist resentment over the loss of East Timor (see Chapter 10 of this volume) than to his refusal to reform the economy (Emmerson 1999b pp. 344–62). Although Megawati was also denied the presidency, she did not lose to an ardent proponent of free markets – for there was none – but to Abdurrahman Wahid, who was backed by Islamicist parties that had found unity, at the very last minute, in opposition to the prospect of a female President. Wahid's candidacy was also bolstered by breakaway elements in Golkar. Thus, as Marcus Mietzner has observed, Wahid's victory was not so much the result of democratic procedures as an exercise in frenzied powerbroking (Mietzner p. 56), a throwback to Indonesia's democratic experience during the 1950s.

Observers were unsure what economic policy Abdurrahman Wahid would adopt. On the face of it, he took a pragmatic approach, much as Suharto had done during the late 1960s and in the mid-1980s. Wahid even conducted numerous "road shows" in foreign countries, seeking to lure foreign investors back to Indonesia, and cooperated with the IMF in order to secure new tranches of financial support. He recruited Kwik Kian Gie, an ethnic Chinese from the PDI-P and a critic of Suharto's corrupt practices, as his leading economic minister. Wahid's tolerant demeanor and his commitment to a modernist Islam finally began to ease the suspicions of the ethnic Chinese, and the rupiah started to strengthen during the first few months of 2000.

However, following these early gains, Wahid's decision-making has appeared to lose discipline. He has acted spontaneously and informally, refusing to consult his cabinet, which in any case has remained fragmented because it included representatives of numerous parties following the deal-making in November 1999 and the persistent logic of unreformed, if now decentralized, patronage. Wahid rashly sacked a number of his ministers, including the few technocrats committed to reform, and, when asked by the legislature to account for his actions, made undocumented allegations of corruption. At the same time, he sought funding from off-budget sources and unaccountable donors, including the state-owned distributor of foodstuffs, BULOG, and the Sultan of Brunei (see reports in the Indonesian weekly news magazine *Tempo*, May 22–28, 2000, pp. 18–22, and June 12–18, 2000, p. 23.). Wahid's aims appear to have been honorable, to the extent that he was seeking funds with which to ease long-simmering grievances in the province of Aceh, for example, rather than to enrich himself or his family. However, his sidestepping of formal procedures recalls the mode of administration once practiced by Suharto, although, unlike Suharto, he has to face hostile coverage from a significantly freer press.

Even in those instances where Wahid has attempted to reform the economy, he has met stout resistance, usually cloaked in familiar appeals for indigenous or Islamicist favor. As one example, Wahid approved the creation of the Indonesia Bank Restructuring Agency (IBRA), designed to revitalize the country's moribund banks, and then approved its plans to begin selling off some of the corporate assets that it had acquired. In early 2000, it put Astra International, a key corporate subsidiary geared to automobile assembly, out to tender. Two US investment firms, Gilbert Global Equity and Newbridge Capital, were attracted to Astra (see *Far Eastern Economic Review*, February 10, 2000), and were supported in their bids by Edwin Soeryadjaya, a member of a prominent Chinese business family who enjoys close ties to Wahid. However, Astra's new manager, Rini Soewandi, the daughter of a former Governor of the Bank of Indonesia, strongly opposed the sale. She also gained support from Islamicist parties, some of which had representatives in Wahid's cabinet, by citing the need to protect indigenous business interests. In addition, Fuad Bawazier, who had been Finance Minister in Suharto's last cabinet, joined in the argument, using the rhetoric of economic nationalism. After several months of strained negotiations, the US investors and their Chinese partner withdrew. Astra was later sold off to a Singaporean company, in IBRA's first large-scale divestment after two years of operation. The case remains as a clear example of the obstacles that still hamper market reforms in Indonesia.

Foreign investors have found their dealings no easier outside the more modern industries. In particular, their access to gasfields and mining sites has been unsettled by the Indonesian government's commitments to political and financial devolution, which have enabled authorities at the provincial level to retain a greater share of the natural resource revenues that once flowed unimpeded to Jakarta. Moreover, two of the provinces with the richest resource endowments, Aceh and West Papua (the former West New Guinea or Irian Jaya), now bristle with militant secessionist movements. Elsewhere, large numbers of illegal miners have emerged in many provinces, driven by desperate circumstances to seize control of mining areas, where they deploy primitive and hazardous techniques. The security forces have done little to curb these activities, raising suspicions that entrepreneurial local commanders are benefiting from at least part of the proceeds of these ventures. It was against this background that in mid-2000 Kwik Kian Gie warned foreign investors that it was too soon for them to contemplate returning to Indonesia. Shortly afterward, he resigned as Coordinating Minister for the Economy.

Some observers will argue, on normative or developmentalist grounds, that Indonesia ought to rebuff foreign investors: if it can resist the high tide of global investment and the concentrated patterns of ownership that result, its economy will doubtless take longer to recover, but at least it might then be in a position to attempt to ease the social disparities that have been worsened by the emasculation of the middle class and industrial labor. However, Indonesia's political record suggests that the chief beneficiaries of such a reversion to economic nationalism, whatever rhetoric of justice and compassion is deployed, are likely to be, first, members of the political elite's families, then crony business elements, leading bureaucrats, and the generals. The record also suggests that the mass of the population would acquiesce in this maldistribution, however resentfully, in return for small, if absolute gains, but at the risk of being dispersed or harshly repressed yet again. Even after the crisis of 1997–98, the protesters remained cohesive for long enough to bring Suharto down, but failed to cause any lasting changes in the intricate social order that surrounded him. One must, then, be cautious in arguing that a rejection of foreign investment, or, more precisely, of the market reforms required to attract it, would ensure any more equitable distribution of resources.

On the other hand, it may well be the case that the avoidance of reform will improve the prospects for Indonesia's new democracy. As Guillermo O'Donnell and Philippe C. Schmitter have suggested, on the basis of their study of democratic transitions in southern Europe and South America, the "vital interests" and the "corporate autonomies" of leading businesspeople and the military generally remain inviolable in the course of such transitions (O'Donnell and Schmitter p. 38). Accordingly, any transition that brings excessively far-reaching reforms, or proceeds at too brisk a pace, is likely to become vulnerable to an authoritarian backlash. Indeed, in the case of Indonesia, one can be sure that the business elite and the generals, who were only temporarily disorientated by the crisis and the departure of Suharto, retain the tools with which to make considerable trouble if they choose.

In July 2000, Indonesia's Defense Minister, Juwono Sudarsono, voiced his suspicions that such trouble had already begun. In an interview with *Tempo*, he stated that whenever Wahid's government had shown serious intention to commit Suharto to trial, ethnic and religious violence had suddenly erupted, especially in the Moluccas, stirring once more the specter of Indonesia's dissolution (*Tempo*, July 3–9, 2000, p. 26). The business elite and the generals apparently fear that, if Suharto is brought finally to justice, a barrier to investigations into their own activities would be removed. Hence, they collaborate in pooling their money and influence, effectively signaling that the price of market reform must be social unrest. On the other hand, the avoidance of reform appears to perpetuate political stability, and indeed, may even promote the consolidation of Indonesia's new democracy (see Case).

One may question the worth and the authenticity of a democracy created under such conditions, but, to the extent that civil liberties are now broadly respected, and relatively free and fair elections can take place, Indonesia's regime meets the minimum standards of many of the "low quality" democracies elsewhere in the world (see O'Donnell). Wahid has promised that even if Suharto is tried and convicted, he will be granted a pardon in exchange for surrendering some of his enormous wealth. Meanwhile, most of the leading businesspeople, despite being frequently vilified in the press, have escaped even cursory judicial scrutiny. Their insolvency has rarely been formalized as bankruptcy, enabling them to retain their corporate armatures and personal fortunes. Similarly, while the leading generals have had to accept reduced representation for the military in the cabinet and in the legislature, they have so far avoided state sanctions for their many documented atrocities in Aceh, East Timor, West Papua, and elsewhere. They have even gained a new immunity to prosecution for acts that were not crimes when they were committed through a constitutional amendment passed in August 2000. The armed forces may now be largely discredited, but they still operate their territorial command, a network that is justified as necessary for national security but also provides them with sufficient scope for manipulating levels of unrest across the archipelago.

Conclusion

Although Suharto has been removed as Indonesia's national leader, and appears likely to have to face trial despite his lawyers' pleas that he is very ill, many of his erstwhile supporters have remained at the peak of their business and military hierarchies. Indeed,

Suharto's opaque style of rule appears to have infected the outlooks of his two successors as President, even though both have ostensibly been committed to market reform. Despite the urgent need for such reform if Indonesia's economy is to recover, many historical, structural, and cultural obstacles persist. In particular, we can expect to see sustained resistance from the business and military elites that became embedded under Suharto, for, while they are tenaciously self-interested, they are adept at using the rhetoric of nationalist causes and collectivist traditions. In these circumstances, foreign investors are likely to refuse to return to Indonesia in significant numbers, while new kinds of entrepreneurs will be unable to begin operations, precluding the internet start-ups that are helping to energize South Korea, Hong Kong, Singapore, and, to a lesser extent, Malaysia.

Indonesia is unable to revitalize its old economy and there are few prospects for any transition to the new economy. Nevertheless, and somewhat paradoxically, it is in part because entrenched elite interests have been so minimally disturbed that Indonesia's new democracy will probably be consolidated.

Further Reading

Anderson, Benedict, "Introduction," in *Southeast Asian Tribal Groups and Ethnic Minorities: Prospects for the Eighties and Beyond*, Cambridge, MA: Cultural Survival Inc. and Harvard University Department of Anthropology, 1987

Anderson provides a concise examination of colonial strategies of "divide and rule" in Southeast Asia.

Anderson, Benedict, "The Idea of Power in Javanese Culture," in *Language and Power: Exploring Political Cultures in Indonesia*, Ithaca, NY: Cornell University Press, 1990

An intriguing study of Javanese attitudes to power and authority

Bresnan, John, *Managing Indonesia: The Modern Political Economy*, New York: Columbia University Press, 1993

Bresnan gives detailed yet concise narrative accounts of some of the major issues that arose during Suharto's presidency.

Case, William, "Revisiting Elites, Transitions, and Founding Elections: An Unexpected Caller from Indonesia," in *Democratization*, forthcoming

This paper argues that Indonesia's democratic transition constituted a "bottom-up" process of replacement, but that its gradual pace has improved prospects for consolidation.

Cribb, Robert (editor), *The Indonesian Killings, 1965–66*, Clayton, Victoria: Monash University Center of Southeast Asian Studies, 1990

An edited collection of papers examining various facets of the mass killings during 1965–66

Crouch, Harold, *The Army and Politics in Indonesia*, Ithaca, NY: Cornell University Press, 1978

This book has come to be regarded as the classic examination of political developments under Sukarno and the transition to Suharto's regime.

Elson, Robert, *The Tragedy of Modern Indonesian History*, inaugural professorial lecture at Griffith University, Brisbane, Queensland, 1998

This published lecture provides an excellent overview of many of the leadership failings that have bedeviled Indonesia's political development.

Emmerson, Donald K. (editor), *Indonesia Beyond Suharto: Polity, Economy, Society, Transition*, Armonk, NY: M.E. Sharpe, 1999

This collection of essays by leading Indonesianists addresses Suharto's rule and the transition that followed his resignation; it includes the two chapters cited as Emmerson 1999a and 1999b.

Emmerson 1999a: Emmerson, Donald K., "Exit and Aftermath: The Crisis of 1997–98," in Donald K. Emmerson (editor), cited above

Emmerson provides a closely detailed account of the ways in which various forces combined in ending Suharto's rule.

Emmerson 1999b: Emmerson, Donald K., "Voting and Violence: Indonesia and East Timor in 1999," in Emmerson (editor), cited above

Emmerson offers an on-the-spot description of polling processes in East Timor during the independence referendum in 1999.

Geertz, Clifford, *The Religion of Java*, Glencoe, IL: Free Press, 1960

This once highly influential assessment of deeply entrenched social identities on Java retains much of its interest today.

Habir, Ahmad D., "Conglomerates: All in the Family," in Emmerson (editor), cited above

Habir provides a novel insider's perspective on changing attitudes to capitalism in Indonesia.

Huntington, Samuel P., "Will More Countries Become Democratic?", in Roy C. Macridis and Bernard E. Brown (editors), *Comparative Politics: Notes and Readings*, sixth edition, Chicago, IL: Dorsey Press, 1986

This paper has long been regarded, at least in some quarters, as a seminal overview of the kinds of forces that create pressures for democracy in developing countries.

Huntington, Samuel P., *The Third Wave: Democratization in the Late Twentieth Century*, Norman: University of Oklahoma Press, 1991

Huntington gives a highly influential but not uncontested account of democracy's "third wave," carefully enumerating its preconditions, processes, and agents.

MacIntyre, Andrew, *Business and Politics in Indonesia*, St Leonards, NSW: Allen and Unwin, 1991

An excellent study of the political influence gained by business associations during the middle decade of Suharto's "New Order"

Manning, Chris, and Peter van Diermen (editors), *Indonesia in Transition: Social Aspects of Reformasi and Crisis*, London: Zed Books, and Singapore: Institute of Southeast Asian Studies, 2000

This is probably the most comprehensive of the several volumes that have already appeared focusing on the period after Suharto, assessing both the brief presidency of B.J. Habibie and the election of Abdurrahman Wahid; it includes the paper by Mietzner cited just below.

Mietzner, Marcus, "The 1999 General Session: Wahid, Megawati and the Fight for the Presidency," in Manning and van Diermen (editors), cited above

A detailed discussion of how Abdurrahman Wahid came to win Indonesia's presidential election in 1999

O'Donnell, Guillermo, and Philippe C. Schmitter, "Tentative Conclusions about Uncertain Democracies," in Guillermo O'Donnell, Philippe C. Schmitter, and Laurence Whitehead (editors), *Transitions from Authoritarian Rule: Prospects for Democracy*, Baltimore, MD: Johns Hopkins University Press, 1986

A contribution to one of the most important books in the enormous literature on democratic transitions that began to appear during the 1980s

O'Donnell, Guillermo, "Illusions About Consolidation," in *Counterpoints: Selected Essays on Authoritarianism and Democratization*, Notre Dame, IL: University of Notre Dame Press, 1999

A collection of essays from one of the most insightful observers of democratizing processes in Latin America and southern Europe

Weiner, Myron, "Empirical Democratic Theory," in Myron Weiner and Ergun Ozbudun (editors), *Competitive Elections in Developing Countries*, Durham, NC: Duke University Press, 1987

Weiner investigates the historical and structural conditions that make for stable and competitive elections in developing countries.

Winters, Jeffrey A., *Power in Motion: Capital Mobility and the Indonesian State*, Ithaca, NY: Cornell University Press, 1996

Winters offers an excellent assessment of the ways in which fluctuating petroleum prices helped to shape policy shifts in Indonesia during the era of the "New Order."

Dr William Case is a Senior Lecturer in the School of International Business at Griffith University, Nathan, Queensland.

NATION STATES

Chapter Ten
East Timor

Damien Kingsbury

On September 4, 1999, the wave of violence and destruction that had been building across East Timor over the preceding months broke in a fury. Within a few days, it had swept staff of the UN Assistance Mission to East Timor (UNAMET) out of the territory, more than 250,000 East Timorese into forced exile, and almost all the rest of the population of more than 800,000 from their homes into the mountains, to seek sanctuary. The burning of homes and the murders of key leaders that had been increasing up to this point gave way to a scorched earth campaign. More than 70% of all dwellings and business premises were burned, while almost all infrastructure, apart from roads and bridges, was either destroyed or stolen. When soldiers of the UN-sponsored International Force to East Timor (Interfet) arrived in the territory, toward the end of September, they found a scene of almost total destruction, with soldiers from the Indonesian army burning and looting what little remained intact. East Timor had always been among the poorest of the territories ruled by Indonesia, and had been a neglected economic and social backwater under the Portuguese before that. Now the plans for rebuilding East Timor as an independent state had been set back, effectively, to starting with a sheet that was blank but for the three quarters of a million people who had returned to find almost total devastation.

Background

In 1518, the Portuguese explorer Duarte Barbosa became one of the first outsiders to visit the island now known as Timor, which was then a collection of small, competing kingdoms. The ethnic groups inhabiting the southern part of the territory were predominantly Polynesian, those in the North were mostly Melanesian, and those in the central regions largely aboriginal. As a result of this ethnic variety and the rugged, difficult terrain, groups speaking around 16 separate languages, as well as around 35 other dialects, had developed in the area (see De Matos, and Leitao).

Following Barbosa's discovery, the Portuguese established an outpost of the Dominican order and trading posts on the island, which they called Timor Loroe Sa'e (Land of the Rising Sun). Portugal gradually extended its control over most of the island, but by the time that it had managed to subjugate most of the local chiefs, the Dutch had begun to move onto the island too. Under agreements signed following conflicts between the Dutch and the Portuguese, in 1859 and in 1913, Timor was partitioned along its present lines (Hiorth pp. 6–7). This was the state of the island when Japanese forces occupied Dutch West Timor in February 1942. In the ensuing conflict, many local people supported Australian commandos against the Japanese. Around 40,000 East Timorese were killed by the Japanese in retribution, and many more died later in the course of Allied bombing (Jardine pp. 21–22; see also Hastings).

Portuguese East Timor was quiescent in the postwar period, for the colony was the least developed of Portugal's overseas possessions. However, the cost of retaining its colonies burdened Portugal and, after the dictatorship of Marcello Caetano was overthrown on April 24, 1974, Portugal quickly moved to cast off its colonial dependencies. As a consequence, political parties began to be formed in the colony. Founded on May 20, 1974, the Timorese Social Democratic Association, soon renamed the Frente Revolucionária de Timor L'Este Independente (Revolutionary Front for an Independent East Timor, or Fretilin), was the largest and most popular party. Its main rival was the União Democratica de Timor (Timorese Democratic Union, or UDT), founded on May 11, 1974, and comprising small businesspeople, property owners and officials. The UDT initially supported continued integration with Portugal, but soon opted for independence. The Associacão Popular Democratica de Timor (Popular Democratic Association of Timor, or Apodeti) was founded soon after, on May 27. Other, minor parties also came into existence, including the Partido Trabalhista (Workers Party), the Associacão Popular Monarquia de Timor (Popular Monarchist Association of Timor, or APMT), and the Associacão Democratica Intergracão Timor-Leste Australia (Democratic Association for the Integration of East Timor into Australia, or Aditla). Of these, the Partido Trabalhista had perhaps a dozen members, the APMT was unpopular and quickly became Klibur Oan Timur Aswain (the Sons of the Mountain Warriors, or Kota), and Aditla collapsed after receiving a negative response from Australia. Kota and Apodeti supported integration with Indonesia, and were essentially products of Indonesia's military intelligence service (Dunn pp. 65–66, Hiorth pp. 21–23; see also Nicol pp. 35–147 and Jolliffe Ch. 2–5). Between them, these smaller parties probably accounted for less than 10% of popular support.

Fretilin and the UDT, meanwhile, had found that their aims were similar. They came together in an alliance in January 1975 to seek independence, ahead of an election for a popular assembly planned for October 1976. However, as Indonesian intelligence was conspiring against the more leftist Fretilin, the UDT was led to believe that Fretilin would soon launch a coup in order to obtain power for itself alone. To forestall this possibility, on August 11, 1975, the UDT, supported by police units, made a preemptive strike against Fretilin. However, Fretilin was supported by around 3,000 Timorese (Portuguese army) soldiers and around 7,000 members of the local militia (Dunn p. 38). Fighting was fierce but it was over within days, having cost around two to three thousand lives. The Portuguese administration fled to the nearby island of Atauro and then to Portugal, although the claim to East Timor was formally maintained. In late August and early September 1975, remnants of the UDT, Apodeti, Kota, and the Partido Trabalhista fled to West Timor – part of Indonesia ever since the Dutch left – from where, with the assistance of the Indonesian military, they began crossborder raids the following month (see Dunn Ch. 8 and 9).

The first raids, in October, were on the northern coastal border town of Batu Gade and then at Balibo. Troops belonging to Falintil, then the armed wing of Fretilin, retreated in the face of superior numbers (and, notoriously, five Australian journalists who stayed in Balibo to film the invasion were killed by the invading force the following morning). However, Falintil put up stiff resistance near Maliana, prompting Indonesia's military planners to reconsider their approach to occupying the territory. On November 24, 1975, Fretilin asked the UN to press for the withdrawal of Indonesian forces and four days later it declared East Timor an independent republic. On November 30, the UDT and Apodeti formally announced their support for integration into Indonesia.

Finally, on December 7, Indonesia launched its acknowledged invasion of East Timor, starting with a parachute landing on the capital, Díli, ahead of the arrival of more than 40,000 occupying troops. The slaughter of around 60,000 East Timorese by Indonesian troops that immediately followed presaged a wholesale campaign of violence and dislocation. In May 1976, East Timorese delegates handpicked by the Indonesian army voted for East Timor's incorporation into Indonesia. This decision was never accepted by the UN Security Council, which voted on November 28, 1976, July 17, 1977, and November 20, 1978, for East Timor to be allowed an act of self-determination.

Despite the overwhelming presence of Indonesian armed forces in East Timor, Falintil kept up an active military campaign throughout the later 1970s. In the face of continuing resistance, the Indonesian forces instituted a policy of relocating villagers believed to be supporting Falintil to "secured" sites. Relocation was later estimated to have contributed to around half of the deaths of more than 200,000 people, around one third of East Timor's population, by late 1981 (ACFOA p. 3). However, Indonesia's own official assessment, for what it is worth, put the drop in population from 626,546 in 1973 to 555,350 in 1980, a decline of 15%, not including expected population growth.

Under Indonesian Occupation, 1975–98

Indonesia moved rapidly to put its stamp on the territory. Senior military figures took over or set up businesses that traded directly with Singapore, bypassing Jakarta and hence the need for taxation or other forms of accountability. For example, the military held a monopoly on East Timor's exports of coffee, by far its most profitable trade. Businesses and property once owned by Portuguese citizens were also taken over by Indonesians.

At the same time, education was expanded and the number of secondary schools increased rapidly. In part, this policy of educational development was intended to inculcate in the East Timorese a sense of belonging to the greater Indonesian nation. Indonesia also built a number of health clinics, although ordinary East Timorese felt little benefit from them, especially in the earlier years of the occupation. The people that the clinics and, increasingly, the schools did serve were the 150,000 or so "transmigrants" from various provinces of Indonesia. A large number of administrative posts were created and most were given to incomers. Even the roads that Indonesia built in East Timor served mainly to transport troops and equipment from district to district. According to aid workers whom I spoke with in East Timor when I first visited the territory in 1994, many of the major bridges and water supply projects were built with foreign aid.

Throughout all of this, the people of East Timor were often regarded as less than human by their new Indonesian masters. Women were frequently raped and opponents of Indonesian rule were murdered with monotonous regularity. The vast majority of the people felt that they lived in a land occupied by an alien force (AETA pp. 10–17). Meanwhile, having been reduced to a small, desperate force of a few hundred people, Falintil was rebuilt from the late 1980s as a small but tough and highly disciplined military group, consisting of four semiautonomous units with a total strength of around 2,000 fighters (Jardine pp. 54–57 and Singh pp. 145–46).

As the senior ranks of Falintil were thinned out by continued Indonesian attacks, leadership passed in 1981 to a regional guerilla leader, José Alexandre Gusmão, whose *nom de guerre* was Kay Rala Xanana. Gusmão's first major initiative as leader, taken on April 26, 1986, was to establish the National Council of Maubere Resistance (CNRM) as a clandestine coalition of all East Timorese groups, notably Fretilin; the UDT, which by now was anti-Indonesian; and the student group Resistencia Nacional dos Estudantes de Timor L'Este (Renetil). Falintil now began working more openly, emphasizing political education and relying on the people to provide support bases for their activities.

The next major move was the formal separation, in 1987, of Falintil from Fretilin, reflecting the need for the armed resistance to represent all anti-Indonesian political groups. From this point onward, Gusmão led the independence movement to reorient itself away from open armed conflict toward a policy of promoting civil disobedience and attracting international attention. The recruitment of young, urban East Timorese into its ranks breathed new life into the independence campaign.

Despite the already long and bloody history of Indonesian atrocities in East Timor, the issue that effectively galvanized world opinion was the killing of protestors at the Santa Cruz Cemetery in Dili on November 12, 1991. The protesters were attending the funeral of a student who had been killed two days previously, but the event also represented an opportunity to protest against Indonesia's continued occupation (Kingsbury 1998 pp. 119–20). Parts of the massacre were recorded on videotape and broadcast throughout the world, arousing considerable international sympathy for East Timor. The Indonesian government succeeded in making its own image worse by initially claiming that only 19 had been killed, then raising the estimate to 50, while other accounts put the toll at 273 (Grant p. 40). The Santa Cruz massacre was far from being the bloodiest of Indonesia's attacks against East Timorese, but it came at a time when the issue of East Timor was widely thought to have been resolved.

It was from around the time of the massacre that the Catholic Church, its congregations swelled with pro-independence activists, was transformed from the bottom up. By 1995, Bishop Carlos Belo had become an outspoken advocate of the human rights of his people and an opponent of the Indonesian occupation. In 1996, Bishop Belo and Fretilin's international representative, José Ramos Horta, were jointly awarded the Nobel Peace Prize.

As a part of his policy of bringing the struggle from the countryside to the towns, Gusmão moved to Dili in 1991. However, this strategy came undone on November 20, 1992 when, betrayed by an informer, Gusmão was arrested. Plans were made to charge him with subversion, which could be punished by execution, but eventually he was accused of rebellion, illegal possession of weapons, and separatism. He was sentenced to life imprisonment but, following international pressure, in 1993 his sentence was commuted to 20 years in prison.

While Fretilin and the UDT had worked closely together since the mid-1980s, there had been some reluctance among their members to form a full coalition, because of their ideological differences, the animosities lingering from August 1975, and the UDT's association, however forced, with the Indonesian army for some years thereafter. However, it became clear to both organizations that they needed to present a cohesive front in order to secure international support. This was done through the establishment in 1997 of the Conselho Nacional de Resistencia Timorense (National Council of Timorese Resistance, or CNRT), although Fretilin and the UDT continued to exist as separate parties. Falintil shifted its political association to the CNRT, bringing some UDT members into its ranks in the process. Gusmão, still leader of both Falintil and Fretilin, was elected president of the CNRT.

The "Popular Consultation" and its Aftermath

Following the resignation of President Suharto of Indonesia in May 1998, his successor, B.J. Habibie, released most political prisoners, lifted many restrictions on the media, and made fitful moves toward cleaning up the most obvious cases of corruption and nepotism. The cost of maintaining the military and administrative presence in East Timor, estimated at US$50 million a year, was higher per head of population than for any province of Indonesia itself. According to the World Bank, around 85% of recurrent and capital expenditure in East Timor came directly from Jakarta, although "leakage" made the amounts actually spent on government activities "substantially lower" (World Bank p. 2). Partly as a consequence, and also because of international concern over East Timor, Habibie opened negotiations with Portugal, still widely recognized as responsible for the territory, on June 20, 1998. On January 27, 1999, Habibie announced that the people of East Timor would be given the chance to vote on whether they wanted absorption into Indonesia or independence.

Most parties followed Gusmão's lead in asking for more time before the "popular consultation" was held; so too did the UN, which was to supervise the ballot. However, Habibie insisted that the ballot be conducted within seven months. This was an acknowledgment that he might well no longer be President at any later date and that his successor might adopt a very different policy, but it was also a response to demands from the army, which was aware that its chances of destabilizing the territory were greater in the short term. The army also insisted that the agreement between Indonesia, Portugal, and the UN, signed on May 5, 1999, should preclude the stationing of any non-Indonesian troops in East Timor. This reflected the deep reluctance on the part of many senior army officers, including the commander in chief,

General Wiranto, to accept Habibie's decision on East Timor. The UN Assistance Mission to East Timor (UNAMET) arrived in the territory on June 11, 1999.

Thirteen pro-Indonesian militias, all backed by the army, had become active in East Timor soon after Habibie's announcement in January 1999 that the ballot was to be held. Wiranto's "Crisis Team on East Timor," headed by Major General Zacky Anwar Makarim, a former head of military intelligence, set up the Pasukan Perjuang Integrasi (Integration Struggle Soldiers, or PPI) as an umbrella group for the militias (see Kingsbury 2000), which were based on existing "home defense units" created as early as 1977. By 1999, it was clear that these units were no longer capable of containing the groundswell of opinion in favor of independence: hence the perceived need for new organizations. Increasing numbers of men were recruited from among incomers to East Timor, joining former and serving Indonesian soldiers in the new units, and being given standard army weapons and equipment. Thus, any distinctions among the militias, the army, and the police became blurred.

The day of the "popular consultation," August 30, 1999, was relatively peaceful and, in the face of widespread intimidation, 98.6% of registered voters took part. However, violence resumed that afternoon and increased in the days leading up to the announcement, on September 4, that 78.5% of those voting had favored independence. Within minutes of the announcement, gunshots began to echo around Dili. UNAMET, which had promised to stay after the ballot, fled on September 14. In the ensuing carnage, virtually the whole population of East Timor was displaced: around 250,000 fled or were forcibly shipped to West Timor, and the rest took to the hills. At least 1,000 people, and probably many more, were killed by the militias and the army, and at the time of writing tens of thousands of East Timorese remain unaccounted for.

Interfet, UNTAET, and the Politics of East Timor

On September 12, 1999, Indonesia agreed to allow a multinational force, authorized by the UN Security Council and known as the International Force for East Timor (Interfet), to enter the territory. The Interfet troops, from Australia, New Zealand, Thailand, Portugal, the Philippines, South Korea, Cambodia, Britain, the Irish Republic, the United States, Singapore, and Malaysia, arrived in East Timor in late September, and the militias dispersed into West Timor soon afterwards.

On October 19, 1999, the Indonesian People's Consultative Assembly formally recognized the result of the "popular consultation" and on October 25 the UN Security Council passed Resolution 1272, establishing the UN Transitional Administration in East Timor (UNTAET) as an integrated, multidimensional peacekeeping operation responsible for the administration of the territory during its transition to independence. The withdrawal of Indonesian soldiers that had been taking place since late September was concluded on October 29, 1999, and the transfer of military operations from Interfet to UNTAET was completed on February 28, 2000.

East Timorese who had fled to the mountains or across the border began to trickle back into the villages and towns, in most cases finding them completely destroyed. Infrastructure such as electricity supply, water and sewage systems, and telecommunications had also been made unusable or, where possible, removed. Shelter was scarce, food was even scarcer, and a major international relief operation coordinated by UNTAET was required to address the immediate needs of those returning to their homes. Increasing attention was also paid to arranging the voluntary repatriation of refugees remaining in West Timor and other parts of Indonesia.

In the period before full elections for a new government, which are scheduled for mid-2001, East Timor is being administered by UNTAET in collaboration with the CNRT, which has been recognized as the de facto government, through the National Consultative Council. Nevertheless, reflecting East Timor's earlier political development and its geographic diversity, the territory's domestic politics remain highly factionalized.

The major division is still between Fretilin and the UDT, but there are others. Fretilin has tended to work closely with the CNRT, while coping with its own internal divisions, not only between radicals and conservatives, but also between its younger activists and its older leaders. The latter division has been highlighted by controversy over the choice of an official language. It has been the policy of the CNRT's elders, including those of Fretilin, to adopt Portuguese, but this has been contested by younger members who speak Tetun Díli and Bahasa Indonesia, and who recognize the value of English as a language giving access to the wider world (see Aditjondro 2000). (Tetun Díli is a simplified version of the Tetun language, originally used by the Portuguese as a lingua franca, which is now spoken by about 60% of the population of East Timor.)

There has also been some division between CNRT-Fretilin and Falintil: many in Falintil believe that, although they sustained the armed opposition to Indonesia and have had considerable support outside Díli, they have been largely excluded from decision-making. There are also regional differences within both Fretilin and Falintil, notably between the eastern Falintil (Zone I), the central Falintil (Zone II), and the western Falintil (Zones III and IV). Within Fretilin and the CNRT, there have also been divisions between those who wanted to maintain the clandestine organization successfully built up under the Indonesian occupation, and those who wanted CNRT and Fretilin to be fully public. It seemed that Falintil would have to reinvent itself as an institution of state – a police service or a paramilitary force – rather than as the armed wing of a political organization.

There are also older and more subterranean distinctions between those who see themselves as ethnically "pure" *kampones* (peasants) – also known as *indijenas* or *rai-na'in* – and the *mestisu* (mixed Portuguese-Timorese) elites. Many in the older generation of Fretilin/CNRT leaders tended to be better educated and to come from the *mestisu* elite, which was traditionally favored by the Portuguese. A gap also existed between Fretilin and the UDT, and within Fretilin, over ideology and therefore over such issues as public ownership and land redistribution. A Fretilin congress in Baucau in May 2000 made significant progress toward repairing the ideological and regional splits within the party.

There are two small parties at the leftwing end of the political spectrum, the Timorese Socialist Party (PST) and the Council for the Popular Defense of the Proclamation of the Democratic Republic (CPD-RDTL). Both have younger memberships than Fretilin and are more radical in their views. They may play larger roles in the future.

Meanwhile, the National Union Front for Political Affairs, the Popular Timorese Party (PPT), and other groups that used to favor integration with Indonesia have been revived and reorganized. The PPT has expressed an interest in registering for the elections in 2001, from its base in Kupang, in West Timor, in what has been widely seen as a maneuver intended to disrupt the voting process.

Finally, a new center-right political grouping, the Social Democratic Party, was launched early in September 2000, intending to capture potential voters who were disenchanted with what one Vice President of the CNRT, Mario Carrascalao, has called the "revivalism of the past." The new party has received the blessing both of Xanana Gusmão and of East Timor's two bishops. It counts among its leading members another Vice President of the CNRT, the Nobel laureate José Ramos Horta.

The Economy

Despite Indonesian claims that the economy of East Timor was developed during the occupation, the territory remained one of the poorest regions in Southeast Asia. Since around 90% of the population under the Indonesian occupation lived in rural communities, engaged in small-scale agricultural activities, and used cash relatively rarely, official statistics did not accurately represent their standards of living. Nevertheless, GDP per capita was just US$431 in 1996 and in the same year around 30% of households existed below the poverty line, double the average in Indonesia itself (World Bank p. 2) It has been claimed that after the collapse of the Indonesian economy, in 1997–98, GDP per capita in East Timor fell as low as $138.

East Timor's capacity for economic development rests, in the short term, on foreign aid and exports. The budget for redevelopment and reconstruction during 2000–02, which the World Bank has estimated at a little over US$307 million (World Bank Annex 2), is being administered through the Trust Fund for East Timor (TFET), but is in addition to around US$200 million required for immediate needs and a UN humanitarian aid program dispensing a further US$200 million. Such major inflows of funds would ordinarily be expected to circulate in the domestic economy, but as with similar aid programs, in Cambodia in 1993 for example, a very large proportion has already been earmarked to be spent on external goods and services, salaries of foreign officials and workers, and services provided by foreign-owned companies.

In the longer term, development is likely to depend in part on the revival of economic activities in which some progress was made under the Portuguese and the Indonesians. In particular, the coffee industry had been the source of the territory's main export earnings since the days of Portuguese rule. It has been estimated that the industry's output was worth around US$40 million a year when the economy was fully functioning, and that around 25% of East Timor's population relied wholly or mainly on incomes from the industry. East Timor's coffee had a number of advantages on the international market. It was exclusively derived from the Arabica bean, which produces the finest-quality coffee; it was grown at higher altitudes in soil that remains dry for much of the year, and was therefore highly aromatic; and, being grown without fertilizers or pesticides, it could be marketed as authentically organic. This factor alone ensured that, on an open market, East Timorese Arabica could be sold at prices around twice those of other premium coffees, although in practice, when East Timorese coffee was sold by such Indonesian organizations as P.T. Denok, it was blended with inferior Javanese coffee, but was still sold as East Timorese. In addition, as in other coffee-producing areas, what the smallholding coffee-growers were paid and what the international wholesalers received was quite different, and most of the profits from the trade did not remain in East Timor.

There have been high hopes for the discovery of hydrocarbons in the Timor Gap, between Timor and Australia, and these were reflected in the signing of a treaty by Australia and Indonesia in December 1989, creating the "Timor Gap Zone of Cooperation." However, the area did not live up to expectations, and only around AU$2.5 million-worth of royalties were distributed to each side between 1989 and 1999. The 1989 treaty was revised in January 2000, when UNTAET, acting on behalf of East Timor, replaced Indonesia as Australia's partner. Soon afterward, Phillips Petroleum announced a liquid petroleum gas project worth US$1.4 billion, while Royal Dutch Shell and Woodside Petroleum also expressed interest in further development in the region. It was widely expected that Australia would renegotiate the boundaries laid down in the treaty, which favors Australia, in order to grant East Timor a larger share of potential royalties. Indeed, in May 2000 José Ramos Horta, in his capacity as the CNRT's foreign affairs spokesman, asked Australia to hand over either the field or the royalties from it wholly or mainly to East Timor. Australia's Foreign Affairs Minister, Alexander Downer, responded by saying that the future of the treaty would be subject to further discussions with the East Timorese authorities. If Australia accedes to East Timor's wishes, the new state could benefit from new projects that could bring in tens of millions of US dollars a year (see AFP and AP).

Beyond coffee, petroleum, and natural gas, East Timor also has commercially viable sources of marble in the eastern districts, which have been only partially exploited, and retains significant stands of sandalwood, which have only minor economic value. Another potentially significant source of export income to East Timor is fishing, particularly through the leasing of fishing rights in its territorial waters. Coastal fishing is a major source of protein for the population and there is some capacity for further development. However, there is also concern, which has been expressed within the CNRT and elsewhere, that it is crucial to avoid depleting East Timor's natural resources.

Tourism has also been suggested as a potentially significant source of foreign income for East Timor, and it has even been claimed that East Timor might once have developed a tourism industry as lucrative – but also, perhaps, as destructive – as Bali's. East Timor certainly has many fine beaches, and some rugged and spectacular mountains and valleys, as well as an airfield capable of handling international aircraft at Baucau. However, not only was all of East Timor's infrastructure ruined or removed by the retreating Indonesian forces in September 1999, but the territory continues to be ridden with diseases, including dengue fever and malaria. The eradication of these diseases is one of the highest priorities for UNTAET and the CNRT. It is possible that small-scale "adventure" tourism may begin to attract visitors, but tourism on any larger scale seems unlikely for some time to come.

As of April 2000, economic activity in East Timor was almost entirely related to the UN, aid agencies or small businesses contributing to UN programs, for example in construction work, and unemployment was running at around 80% of the potentially active population in urban areas (see CDPM 2000a). In the period leading up to the "popular consultation," UNAMET had employed 28,000 people, but it seems likely that fewer than 8,000 jobs will be made available by UNTAET during the whole of 2000, and by the beginning of May only 1,100 posts had been filled. UNTAET's view was that most applicants had few or no skills to offer and required further training.

However, other employment programs are being established in East Timor. These include a public employment program, funded by multilateral pledges of US$520 million made in Tokyo in December 1999; a Community Empowerment and Local Government Project, which is to receive US$21.5 million from the World Bank and is to be managed by democratically elected local councils; the Quick Impact Program, intended to rehabilitate public facilities; and a transport infrastructure project. The Portuguese government and the US Agency for International Development (USAID) have also begun to finance a small enterprise program with loans of between US$500 and US$50,000: of the 772 applications for funding received by May 2000, 672 were from East Timorese (see CDPM 2000a). In order to help businesses to conduct transactions and to allow foreigners to exchange currency, two Australian banks have opened branches in East Timor.

After the destruction of September 1999, the once relatively widespread telecommunications system of East Timor was reduced to an Australian Telstra mobile network in Díli and two-way radio or satellite telephones outside the capital. Telstra has stated that establishing a wider network, especially for land lines, would be prohibitively expensive, but the Portuguese Telecommunications Company has announced that it hopes to form a consortium with PT Telkom of Indonesia to develop services throughout East Timor (see IO).

Education

Under the Indonesian occupation, primary education was usually conducted by untrained teachers in buildings with almost no facilities. According to Unicef, 90% of the 900 or so primary and secondary schools in East Timor were destroyed in 1999, and, with virtually all teachers having returned to Indonesia, only 6% of those remaining are fully qualified (see CDPM 2000b). Primary education has been put back in place on an ad hoc basis, with considerable support from UNTAET and the Catholic Church. "School in a box" kits have been distributed throughout East Timor by UNTAET and by May 2000 an estimated 150,000 children were receiving some sort of primary education. Taking into account the proportion of the population that has not yet returned, this actually represented a higher proportion of school-age children than in 1998. Unicef has launched a national teacher training program under which 30 trainees will complete courses and then train others throughout East Timor's educational districts. Both teaching and teacher-training are being conducted in Tetun, Portuguese, and Indonesian.

Following the destruction of the University of East Timor, the Díli Polytechnic, the Teachers Training College, the Health Academy, and the School of Economics, and the effective expulsion of East Timorese students studying at other institutions throughout Indonesia, efforts began in late 1999 to establish a single National University of East Timor, incorporating all of the former higher education institutions. Its first phase is planned to begin by October 2000. There is also a major effort to find scholarships in third countries for students whose studies in Indonesia have been disrupted.

The Rule of Law

East Timor's legal system ceased to function with Indonesia's withdrawal, and has since been completely overwhelmed because of limited jail space, the lack of courts, and the large numbers of alleged criminals awaiting trial. Almost all these alleged criminals are facing charges relating to the destruction and killing that took place before the ballot on August 30, 1999. However, large numbers of these alleged offenders remain free until more jail space can be established and a legal system put in place to hear their cases.

UNTAET is undertaking training of magistrates to hear such cases, and a legal code is being established, initially on the basis of Indonesian law, which requires that either the alleged offender be caught in the act or there is judicial approval for the laying of charges (see Shaw). Eight judges have been appointed by UNTAET to the District Court in Dili, along with prosecutors and defenders. Policing is being undertaken by UN soldiers, but plans are in place to recruit and train police officers from among the East Timorese themselves. Meanwhile, organized crime and petty offences are both on the increase in East Timor, especially in the larger towns, reflecting, among other factors, the high levels of unemployment and the related frustration; the continuing clashes between supporters and opponents of independence; and a general economic and social desperation.

One major issue that the new courts face is the question of property rights. There were few records of land ownership before Indonesia left East Timor, and most of those have been destroyed. Further, much of the possession of property between 1975 and 1999 disregarded property ownership prior to 1975, for which there were also few records. The head of UNTAET, Sergio Vieira de Mello, has acknowledged that land and property ownership will need to be resolved as a precondition of significant foreign investment in the territory (see MNNA). Fretilin has abandoned its previous policy of enforcing land redistribution, but is still committed to considering government ownership or apportionment of land previously owned by Indonesians or Portuguese.

Conclusion

The reconstruction of East Timor after the almost total destruction of September 1999 is expected to take at least several years, and to depend to a large extent on the establishment of the institutions of a new state. The high level of international support seems likely to ensure that there will be at least a reasonable start to this process. A number of leading figures in the CNRT have stated that East Timor will not attempt to embark on industrialization or become an economically developed state, but instead will remain a predominantly agrarian state, only slowly diversifying its economic base. They recognize that East Timor has a limited capacity for economic development and prefer to focus on making development sustainable. This suggests that they have learned some lessons from the experiences of developing countries since the 1960s.

Apart from the potential for internal political instability, which can quickly undermine any economic gains, the greatest threat to East Timor's development comes from West Timor. Although crossborder militia activity has been limited as long as there is an international military presence, the maintenance of border security will become more problematic once the peacekeepers withdraw. Many observers assumed that, with Indonesia having formally accepted the independence of East Timor, militia activity emanating from West Timor would fade away. However, while many members of the militias have been disarmed and have given up the fight, between 1,000 and 2,000 fighters remain armed and active, and are supported by senior officers of the Indonesian army, who would very much like to regain access to East Timor's potential economic capacity, not only in coffee and hydrocarbons, but also as a site for direct trade relations with third states. These militia fighters and the generals who support them remain a source of potential problems as long as the generals retain their positions within the Indonesian army. This in turn depends on the provincial structure of the army and its involvement in business, both legal and illegal. The resolution of these issues is now on the political agenda in Indonesia itself.

Even more disturbingly, many, probably most, Indonesian people, whether they are members of the decisionmaking elites or not, still believe that East Timor has been stolen from them. The fact that East Timor's closest neighbor, and by far its largest, regards this tiny new state with an ambivalence bordering on hostility is one more alarming sign that the future of East Timor is far from settled.

Further Reading

ACFOA: *East Timor: Keeping the Flame of Freedom Alive*, Melbourne: Australian Council for Overseas Aid, 1991

Aditjondro, George, *An Official Language for a New East Timor*, Darwin: East Timor International Support Centre, 2000

AETA: *East Timor: Betrayed But Not Beaten*, Melbourne: Australia East Timor Association, 1983

AFP: "Ramos Horta Calls on Australia to Renounce Timor Gap Oil Treaty," Sydney: Agence France Presse, May 8, 2000

AP: "Australia May Renegotiate Timor Gap Treaty – Downer," Canberra: Associated Press, May 9, 2000

Budiardjo, Carmel, and Liem Soei Liong, *The War Against East Timor*, London: Zed Books, 1984

A detailed account of Indonesian military activity and the circumstances that surrounded it up to the early 1980s, with particular emphasis on human rights issues

CDPM 2000a: "East Timor's Unemployment (80%) the Fundamental Problem," Lisbon: Comissão para os Direitos do Povo Maubere, June 5, 2000

CDPM 2000b: "Education in East Timor," Lisbon: Comissão para os Direitos do Povo Maubere, June 5, 2000

De Matos, Artur, *Timor Portugues 1515–1769*, Lisbon: Universidade de Lisboa, 1974

Dunn, James, *Timor: A People Betrayed*, Sydney: ABC Books, 1996

This thorough account of East Timor up to the 1980s is probably the best source on the period surrounding Indonesia's invasion. It includes detailed coverage of relations between Fretilin and the UDT, as well as of Australia's role in encouraging the invasion.

Grant, Bruce, *Indonesia*, third edition, Melbourne: Melbourne University Press, 1996

Hastings, Peter, "Timor – Some Australian Attitudes, 1941–1950," in J. Cotton (editor), *East Timor and Australia*, Canberra: Australian Defence Studies Centre, 1999

Hiorth, Finngeir, *Timor: Past and Present*, Townesville: James Cook University, 1985

IO: "Portugal Telecommunication to Invest in Indonesia and East Timor," in *Indonesian Observer*, July 10, 2000

Jardine, Matthew, *East Timor: Genocide in Paradise*, Monroe, ME: Odonian/Common Courage Press, 1995

A reliable book on East Timor, with an emphasis on historical development and human rights issues

Jolliffe, Jill, *East Timor*, St Lucia: University of Queensland Press, 1978

A clear and detailed account of events leading up to Indonesia's invasion of East Timor, including coverage of events in the period immediately afterward

Kingsbury, Damien, *The Politics of Indonesia*, Melbourne, Oxford, and New York: Oxford University Press, 1998

Kingsbury, Damien, "The TNI and the Militias," in Damien Kingsbury (editor), *Guns and Ballot Boxes: East Timor's Vote for Independence*, Clayton: Monash Asia Institute, 2000

This account of the circumstances surrounding the "popular consultation" that led to East Timor's independence is based on information from observers, UNAMET workers, and others who were present at the time.

Leitao, Humberto, *Vinte e Oito Anos de Historia de Timor*, Lisbon: Agencia Geral do Ultramar, 1956

MNNA: "UNTAET Wants to Restore Timor Land Records," Putrajaya: Malaysian National News Agency, April 28, 2000

Nicol, Bill, *Timor: the Stillborn Nation*, Melbourne: Widescope International Publishers, 1978

Shaw, James, "Establishing Rule of Law in Dili," in the *Sydney Morning Herald*, April 20, 2000

Singh, Bilveer, *East Timor, Indonesia and the World: Myths and Realities*, Singapore: Singapore Institute for International Affairs, 1995

Surya Timor: "East Timor Elections Will Happen with or without Pro-integration Group," in *Surya Timor*, June 6, 2000

Taylor, John, *Indonesia's Forgotten War: The Hidden History of East Timor*, London: Zed Books, and Leichhardt, NSW: Pluto Press, 1991

World Bank, *Report of the Joint Assessment Mission to East Timor*, Washington, DC: World Bank, December 8, 1999

Dr Damien Kingsbury is the Executive Officer of the Monash Asia Institute at Monash University, Melbourne. He coordinated the Australian East Timor International Volunteer Program from July to September 1999, and now assists in education support and assessing self-government in East Timor.

NATION STATES

Chapter Eleven
Vietnam

Athar Hussain

The Geographic Setting

Wedged between the Annamite Mountains and the sea, Vietnam is a thin strip of territory over much of its length of almost 1,600 kilometers (around 990 miles) from North to South, narrowing to as little as 40 kilometers (around 25 miles), but broadening out at each end. The Vietnamese often describe their country as resembling a bamboo pole with a wide-brimmed basket at each end, a traditional and still common method of carrying loads.

Topographically, Vietnam can be divided into three regions, the North, the Center, and the South, which were respectively called Tonkin, Annam, and Cochin China during the French colonial period. Much of the country is mountainous and the population is concentrated in the principal agricultural regions at the two extremities: the Red River delta in the North and the Mekong River delta in the South. The North is the original heartland of Vietnamese civilization (see Karnow) and contains the present capital, Hanoi, which is situated in the midst of the Red River delta. The Center comprises the Annamite Mountains and the narrow coastal lowlands, and is the location of the former imperial capital, Hue. With relatively poor soils and uncertain rainfall, the Center has traditionally been, as it still is, the poorest region of the country. The Mekong delta and the rest of the South were settled by Vietnamese relatively late in the country's history, but they have long since become vital to the economy, not only because Vietnam's most fertile areas are in the South, but also because of the commercial activities developed in and around Saigon (now part of Ho Chi Minh City).

Vietnam shares around 1,200 kilometers (750 miles) of land borders with China to the North, and around 2,800 kilometers (1,740 miles) with Laos and Cambodia to the West. Its coastline, which is more than 3,000 kilometers (around 1,860 miles) long, forms the basis of its longstanding claim to sovereignty over the Paracel Islands in the South China Sea. This island group, known as Hoang Sa in Vietnamese, is also claimed by China, under the name Hsisha. Vietnam has remained particularly sensitive to the political situation in Laos and Cambodia, especially the latter. It has also had ambivalent and complex dealings with China, which asserted suzerainty over Vietnam until the late 19th century, and alternated between supportiveness and mistrust during the 20th century. It was only in December 1999 that Vietnam settled its dispute with China over their common border, which runs through mountainous, sparsely populated and little studied terrain. Today, Vietnam still has boundary disputes with Cambodia, and there is some potential for further conflict between Vietnam and other countries in East Asia, not only over the Paracel Islands, but also over another island group, the Spratlys, known as Truong Sa in Vietnamese and Nansha in Chinese. Some or all of the Spratlys are claimed by Vietnam, China, Brunei, Indonesia, Malaysia, the Philippines, and Taiwan.

The People

Vietnam had an estimated population of just over 78 million in 1998 (see World Bank), making it the second largest nation in Southeast Asia (after Indonesia). Relative to its population, the land area is small and the cultivable area even smaller. As a result, Vietnam has some of the world's highest population densities, and, with 76% of the population living in rural areas and mostly dependent on farming, this is one of the principal causes of its continuing poverty.

Ethnic Vietnamese, sharing a common language and identity, constitute around 88% of the total population, which makes for a relatively high degree of cultural homogeneity compared to most other countries in Asia. Ethnic cleavage and conflict are by no means absent from Vietnamese society, but they have not been prominent in shaping its history and institutions.

Among the remaining 12% or so of the population, overseas Chinese, who are known as Hoa in Vietnam, form the largest single ethnic minority. Most of the Hoa are descendants of migrants who came, not overland but by sea, from the southeastern provinces of China. Like the Chinese minorities in other countries in Southeast Asia, the Hoa have preserved their cultural identity, despite the fact that, in most cases, their families left China several generations ago and have had little or no contact with their places of origin. They are mostly urban residents, working as artisans, skilled personnel, merchants, and industrialists. They are concentrated in the South, primarily in and around Ho Chi Minh City, and especially in the district of Cholon, which was administratively separate from the city of Saigon until both were absorbed, with other communities, into Ho Chi Minh City in May 1975. Most members of the small Hoa community in the North live in coastal localities bordering China.

Like the overseas Chinese elsewhere in Southeast Asia, the Hoa, with their combination of cultural distinctiveness and entrepreneurial flair, have attracted resentment from the majority population. This resentment has traditionally been low-key in Vietnam, but it came to the surface following the reunification of the country, and turned into active hostility in the wake of the border skirmishes between China and Vietnam, which began in February 1979 and lasted well into the 1980s (see Bowman, and Karnow). A substantial number of Hoa fled the country in the 1980s, and some attracted worldwide attention as "boat people," but many have since returned following the liberalization of the economy after 1989.

Unlike the Hoa, the members of Vietnam's other ethnic minorities are almost entirely rural, eking out their livings from farming. They tend to be poorer than the ethnic Vietnamese. The Southwest, the region where the Khmer (Cambodian) and Vietnamese civilizations intersected and their empires frequently clashed, now contains a significant Khmer minority, just as there is still a substantial Vietnamese population in Cambodia. Numerous tribal peoples inhabit mountainous areas in the North and the Center.

The annual income per capita of the Vietnamese population was US$330 in 1998, which makes it one of the poorest countries in Asia, and substantially poorer than India or China (see IMF and World Bank). This low average goes together with wide disparities in income, not only between regions but between rural and urban areas, and some of the rural localities are desperately poor. The level of urbanization is relatively low, but similar to that in a number of other Asian economies, such as China. According to official figures, only 34% of the population are permanent residents of cities and towns, but the proportion of urban dwellers is higher if rural migrants, working and living in cities on a long-term basis, are also included. Nevertheless, the Vietnamese population is still predominantly rural and agriculture is the dominant influence on the living standards of the population. Driven by acute pressure on land and comparatively low incomes, emigration out of rural areas is rising, and rapid urbanization is a salient feature of Vietnam's social landscape, as it is in much of the rest of Asia.

Despite its relative poverty and the remoteness of some rural localities from the centers of population, Vietnam has a well-educated population by international standards, especially as compared to the general level in most low-income economies. At 94.1%, the official literacy rate is substantially higher than either India's or China's and is comparable to literacy rates in Asian countries that are much richer than Vietnam is. This high level of literacy reflects the emphasis that the Vietnamese leadership has consistently placed on education as a means of nation-building and ideological inculcation. Vietnam's achievement in providing basic education serves to underline the general point that illiteracy and poverty need not go together.

As elsewhere in Asia, the Vietnamese population is undergoing a demographic transition from a regime of high fertility to one of low fertility. The rate of population growth has fallen steadily since the mid-1970s and the fertility rate is down to little more than two children for each woman of childbearing age. However, because of the inertia effect of the traditional practice of high fertility, the population is still growing at a relatively high pace of around 1.8% each year. It has been estimated that it will take another 50 or so years to stabilize the population at around 120 million, or 50% higher than the present total (see World Bank). It seems clear, then, that the creation of jobs for an expanding labor force, and particularly for those leaving the land and entering the cities, will remain the single most pressing social and economic issue in Vietnam for some time to come.

Amid the upheavals of recent decades, Vietnam, like many other Asian countries, has retained a lively tradition of veneration for ancestors. The expression of deference to deceased relatives through various rituals pervades Vietnamese life, cutting across formal religious boundaries. However, the spiritual life of the Vietnamese is also shaped by four systems of conduct and ethics: Confucianism, Taoism, Buddhism, and Christianity. All originated as imports from other civilizations, but all have struck deep roots within Vietnamese society and have taken on local coloring. Confucianism and Taoism originated in China, and their lasting influences reflects the centuries of Vietnam engagement with China, intermixed with cultural osmosis and resistance to attempts by the Chinese (whether actual or perceived) to take control of a country that was part of the Chinese empire until late in the tenth century CE (see Karnow). In Vietnam, Confucianism and Taoism have tended to merge with Buddhism, forming the *Tam Giao* ("three teachings") that lie at the heart of popular religious beliefs and practices. The Christian minority, which includes both Catholics and Protestants, owes its origins to the Catholic missionaries who were among the first westerners to venture into Vietnam. One of the most illustrious of the early missionaries was the French Jesuit Alexandre de Rhodes (1591–1660), the creator of the modified Roman alphabet that has since completely displaced Chinese characters as a vehicle for written Vietnamese. Finally, and again like other Asian countries, Vietnam has also produced some indigenous syncretic religions. Among the most prominent of these has been Cao Dai, a sect founded in the 1920s that seeks to combine the best of eastern and western religions and philosophies. The sect now has more than two million followers and a colorful headquarters at Tay Ninh in the South.

History

Unlike the many Asian states that took on their present forms only during and after the colonial period, Vietnam has existed as a distinct political entity for more than 1,000 years. A major theme resounding through its long history and rich folklore is resistance to foreign domination, first against China but more recently against France, the United States, and China once again (see Karnow, and Tarling).

Vietnam's engagement with the West began with the arrival of merchants and missionaries. The Portuguese established a trading post at Hoi An (which they called Faifo), to the South of the modern city of Danang in central Vietnam, in 1535, and an Italian Jesuit missionary established the first Catholic church in 1615. French colonial control proceeded in stages – hence the formal division of the country into one full colony (Cochin China) and two protectorates (Annam and Tonkin) – and was fully established only in 1883 (see Bowman, and Tarling). It ended at Geneva 71 years later, with the full decolonization of Vietnam, Cambodia, and Laos, the components of what had once been the French Indochinese Union.

The proclamation of the present Socialist Republic of Vietnam in July 1976 marked the formal reunification of the two parts of the country, North and South, which had been in a state of war from 1955 to 1975. The Democratic Republic of Vietnam, widely known as "North Vietnam," had been proclaimed in Hanoi in September 1945, shortly after the Japanese Imperial Army, which had wrested direct control of Vietnam from its former French collaborators only six months before, surrendered to the British. After the defeat of Japan, the French tried to move back into the country and resume their previous colonial role, but they encountered powerful resistance from the Viet Nam Doc Lap Dong Minh (Vietnam Independence League, or Viet Minh). This liberation force had emerged and gathered strength during World War II.

The Viet Minh was led by Ho Chi Minh, who later came to be widely regarded as the father of modern Vietnam. "Ho Chi Minh" ("Ho who enlightens") is the best known of the numerous aliases of Nguyen Ai Quoc (1890–1969), who had spent many years in the United Kingdom, France, the Soviet Union, and China before arriving back in Vietnam in February 1941 (see Halberstam). The founding members of the Viet Minh also included many of those who would later lead first North Vietnam and then reunified Vietnam. Prominent among them were Pham Van Dong, who served as Prime Minister from 1955 to 1987, and the legendary military strategist Vo Nguyen Giap, a former schoolteacher who masterminded many of the Viet Minh's major battles against the French and later US forces (see Pike).

In May 1954, at the height of the Cold War that divided much of the world into two antagonistic blocs, the Viet Minh inflicted a massive defeat on the French at Dienbienphu, a military outpost in the remote Northwest of the country. The French rapidly withdrew from Vietnam, Laos, and Cambodia, and two months after the defeat accorded full recognition to the three new independent states at the Geneva Conference. The same gathering provisionally divided Vietnam into North and South by drawing a line through the middle, at the 17th parallel. This division was exactly the same as the one imposed by the United States, the United Kingdom, and the Soviet Union, at the Postdam Conference in August 1945, on the British and Chinese Nationalist forces then occupying Vietnam. In 1954, as nine years earlier, it represented an approximate reflection of the balance of forces on the ground at the time; but it also took advantage of the geographic circumstances outlined above, with the result that the Red River delta became the center of "North Vietnam" and the Mekong delta the center of "South Vietnam." At the same time, it was decided, and formally announced, that the country would be reunited under a government chosen by the population through "free and fair" elections, which were to take place by July 1956.

According to neutral observers, such elections would have been won by the Viet Minh under the leadership of Ho Chi Minh. The elections were never held, however, and the provisional line of demarcation running through central Vietnam became a battleline between two antagonistic regimes, each supported by outside powers (see Tarling, and Karnow). The Vietnamese leadership in Hanoi responded to the division, not by breaching the dividing line, but by outflanking it, establishing the "Ho Chi Minh Trail" through the mountainous border provinces of Laos and Cambodia. In this way, Vietnam reinforced the practice that it was to pursue throughout the sequence of wars that dominated its history from the 1940s to the 1980s, treating Laos and Cambodia as its strategic rear, and using them as sanctuaries and routes of communications between North and South.

In the conflict that soon followed the Geneva Conference, the South, known as the Republic of Vietnam after the overthrow of the figurehead Emperor Bao Dai in October 1955, was supported by the United States and, with varying degrees of enthusiasm, by its allies. The Democratic Republic of Vietnam was supported by China and the countries of the Soviet bloc, although relations were overshadowed for a time by the Vietnamese leadership's resentment of Chinese and Soviet pressure, first to accept the Geneva Accords, and then to seek compromise with the regime in the South. Despite this pressure, the government in Hanoi never accepted the division of the country and set out to achieve reunification, initially through propaganda, but soon through war (see Kolko 1986, and Karnow). Despite massive financial and military assistance from the United States, including the participation of more than 500,000 US troops, the regime in the South collapsed in the face of a combined force of guerrillas (the "Viet Cong") and regular troops from the North, aided and abetted by a substantial section of the southern population. The two halves were reunified after a civil war that claimed millions of lives, instead of through the ballot box.

The involvement of the United States in Vietnam, which was in fact a continuation of the war started by the French in November 1946, began in October 1955 with financial and military assistance to the regime based in Saigon and led by Ngo Dinh Diem, a staunch Catholic from an aristocratic family in central Vietnam and a fervent anti-Communist. Diem, who had previously worked with the French colonial administration for ten years, had been appointed Prime Minister of the "Empire of Vietnam" in June 1954 by Emperor Bao Dai, who in turn was dependent upon the French. Diem had soon cast the Emperor aside and set out to assume total control over South Vietnam, creating a large number of enemies in all sections of the society in the process. He was killed in November 1963, during a military coup that turned out to be the first in a long series of coups up to the disappearance of the Republic of Vietnam in April 1975.

Confronted with the difficult choice of either accepting the impending collapse of South Vietnam or averting the takeover of the South through direct military intervention, the US opted for intervention (see Kolko 1986, and Karnow). However, faced with mounting domestic opposition to the war, the United States withdrew from South Vietnam, beginning in January 1973. Over the next two years, the army of South Vietnam disintegrated. The forces of the Democratic Republic of Vietnam entered Saigon, the southern capital, in April 1975, bringing to an end the 20th century's longest-lasting armed conflict, which had cost the lives of at least three million Vietnamese and around 58,000 Americans.

The reunification of the country did not immediately bring full peace. It was, in effect, a takeover of the defeated South by the victorious North, involving a significant degree of coercion and a settling of accounts with all the millions who could be regarded as having been collaborators. It also involved the forced transformation of the economic structure of the South, in an attempt to bring it into line with the state-controlled economy of the North, even though the process of imposing uniformity impeded recovery from the devastation caused by the war (see Tarling). The arrival of peace was further delayed by Vietnam's invasion of Cambodia in December 1978 and January 1979, which was aimed at toppling Pol Pot's murderous regime and installing a government that would be friendly to Vietnam. In retaliation against what it denounced as Vietnam's attempt to dominate Cambodia (and Laos), China invaded the northern part of Vietnam one month later, launching an inconclusive border war in which both sides suffered heavy casualties. The war did not last long, but skirmishes along the border continued into the 1980s. Thus, the protracted sequence of wars that had begun in 1946 went on for 43 years in total, ending in September 1989, when almost all the Vietnamese forces remaining in Cambodia withdrew.

The legacy of this long sequence of wars runs wide and deep in Vietnam's society and economy. Most Vietnamese families have been affected by war in one way or other, and the country still suffers from the loss of resources that could have gone into economic development but were wasted on the battlefields instead. The exceptionally low living standards of the Vietnamese population are the main result, and it will take many years to improve them. The protracted wars have also had a stultifying effect on Vietnam's political structure. Aside from changes caused by death or illness, the leadership of the ruling Communist Party remained the same from the Geneva Conference of 1954 up to the party congress 30 years later, and the last remaining member of the first generation of leaders who witnessed the establishment of the Democratic Republic of Vietnam in 1945 withdrew from the political scene as recently as 1997 (see Kolko 1997, and Porter). Finally, the wars also forced Vietnam into international isolation. At loggerheads with almost all its neighbors in Asia, including China, it had diplomatic relations with only a few countries outside the Soviet bloc from 1975 until the early 1990s, and found political and economic allies only among countries that were thousands of miles away, although it was admitted to the UN in 1977. Normal diplomatic relations with the United States were established only in 1995, the same year that Vietnam joined ASEAN, which had been founded precisely to serve as a bulwark against Communist North Vietnam and China.

The Political System

Together with China, North Korea, Cuba, and Laos, Vietnam is one of only five surviving examples of a Communist polity. The Communist Party of Vietnam (Dong Cong San Vietnam) holds the monopoly of power, being designated in the Constitution as the "sole force leading the state and society" (see Kolko 1997, and Porter). The party permeates the government and the armed forces, and appointment to any notable position, whether civil or military, is contingent on holding a party post at an appropriate level. Corresponding to each position in the state there is a party position with a higher rank. Associations and organizations are permitted to operate outside the aegis of the party, but strictly on the condition that they accept the leading role of the party.

In principle, the party's role consists of setting broad policy guidelines, and undertaking the training and appointment of personnel in the party and the government. In practice, however, the lines of division between government and party are blurred, and the leading posts overlap. Hence, the structures of government are inextricably intertwined with the party's machinery, nationally, regionally, and locally, and on issues regarded as crucial the party intervenes in the details of policy-making and implementation, contrary to the formal division of labor between the two sets of institutions.

The internal structure of the Vietnamese Communist Party closely resembles that of the now defunct Communist Party of the Soviet Union or that of the still very much alive Chinese Communist Party. Formally, the Central Committee is the "supreme leading organ" of the party, but its actual power is limited, largely by the fact that it meets only twice a year. Effective power rests with the Political Bureau (Politburo), a small group of leading members of the Central Committee, some of whom also occupy important positions in the government. The Politburo is headed by the General Secretary, one of the three individuals at the apex of the system, alongside the Prime Minister and the Chairman of the State Council, a crucial body that acts on behalf of the National Assembly by supervising the work of ministers and the judiciary. As in other Communist polities, the Constitution of the Republic and the statutes of the party lay down formal checks and balances, and provide for systems of accountability within the party and also within the government. In particular, the Central Committee is supposed to devote much of its infrequent meetings to examining and ratifying the decisions taken by the Politburo in the interim, and then to make its own detailed report on the course of events to the party congress, which meets for a short period at intervals of four years. Given the infrequency, and the stage management, of meetings of both the Central Committee and the congress, political power is concentrated, by default if not by design, in smaller bodies that meet much more frequently and control communication within the party structure.

What, then, are the tangible effects of one-party rule? The party functions primarily as a machinery for the selection and promotion of functionaries of the state, with the result that a few thousand meticulously selected, trained and vetted party "cadres" occupy all the leading positions in the government, the armed forces, and "mass organizations" such as the labor unions or the Women's Association. Second, the monopoly of power vested in the party predictably causes the leadership to place a high value on loyalty, although reference to ideological commitment, if only as a badge of such loyalty, is no longer as crucial as it once was. Not only is there a wide divergence of views among the party's leaders, but these divergences have widened in recent years, and are now aired in public more frequently, and with fewer repercussions, than was permitted in the years of war and initial reunification. Nevertheless, dissent has to stay within the limits imposed by adherence to whatever is the "party line" laid down by the leadership from time to time. The increasing diversity of views and their exposure outside closed party meetings both represent positive developments, but at the same time they implicitly call into question the rationale for permitting only one political party to operate. Meanwhile, the detailed supervision of government policy by the Politburo of the party not only restricts the range of policy options, but also significantly slows down the decision-making process.

The party apparatus is closely paralleled, at all levels, by the executive, legislative and judicial branches of government, which, in form at least, closely resemble the governmental structures familiar in liberal democratic parliamentarian polities. The National Assembly, the highest authority under the Constitution, elects the members of the key government bodies, such as the Council of Ministers, which is headed by the Prime Minister, the State Council (mentioned above), the National Defense Council, and the Supreme People's Court. The members of the Assembly are elected through a complicated system of segmented electoral colleges, not through universal suffrage. The weight of these elections in the selection of Assembly members has increased since December 1986, when the leadership launched the program of *doi moi* ("new structure"), a series of economic reforms complemented by a loosening of political control (see Nugent, and Kolko 1997). As in other one-party states, however, elections in Vietnam are still far from satisfying international standards as being "free and fair." Since the launch of *doi moi*, the National Assembly has also become more assertive and has even served at times as a forum for lively debates on government policy, but its power is limited by the fact that it too meets only twice a year. In practice, then, the legislature and the judiciary remain subordinated to the executive, in which the People's Army of Vietnam (Quan Doi Nhan Dan Viet Nam), comprising land, sea and air forces, still constitutes a central component (see Pike). It was estimated in the late 1980s that one in three adult males belonged to a military organization (Porter pp. 82–83), and retired military personnel tend to dominate a number of government bodies.

The Economy since 1986

As has already been suggested, the year 1986 can be seen as the watershed in Vietnam's recent economic and political history, as it was the year in which the National Assembly formally approved the program of radical economic reforms and modest political changes known as *doi moi*. The initial program was in itself a modest one, but it set in train a series of reforms that have fundamentally transformed the structure of the Vietnamese economy (see IMF, Kolko 1997, Nugent, and World Bank for a variety of data and of perspectives on this transformation).

Up to the end of 1986, Vietnam was a command economy. Agriculture was collectivized: land was collectively owned and farmers were organized into production teams consisting of groups of households. The government had a monopoly over agricultural marketing, which it applied coercively and exploited as an instrument of taxation All industrial and commercial enterprises were publicly owned, and private ownership was restricted to houses and small household enterprises. The price system and markets were both largely absent. Farms and enterprises were told what to produce, from whom to purchase, and to whom to sell. Economic incentives to improve efficiency were minimal or, in many cases, entirely absent. The few markets that existed were either composed of exchanges of small daily items or black markets in US dollars and smuggled goods. Most international trade was with countries in the Soviet bloc and foreign direct investment was not permitted. A large majority of the population lived in abject poverty.

By the mid-1990s, however, much had changed beyond recognition. Agriculture had been completely decollectivized: land remained nominally in public ownership, but was being leased out to rural households, on 20-year renewable leases that could be bought, sold, or transferred through inheritance. Peasants were free to sell their produce to whomever they wished. The state sector had shrunk: the number of state industrial enterprises had fallen by half from 12,000 to around 6,000. Private enterprise was not merely tolerated but actively encouraged. The prices of most commodities were determined by the market rather than, as in the past, set by the government. So too were the exchange rates of the Vietnamese currency, the dong, against the US dollar and other currencies. The economy, previously closed, was wide open to foreign investment and international trade. Foreign investment was solicited through preferential concessions. Foreign trade was no longer monopolized by government-owned trade companies. The economy was predominantly steered by the decisions of a myriad of enterprises and households rather than by just the government commands. Finally, and most importantly, the living standards of most of the population had improved and the proportion of Vietnamese seeking to survive below the official poverty line had fallen sharply.

However, while this kind of sharp dichotomy between the economy before *doi moi* and the economy after *doi moi* has its uses, in highlighting the huge transformation that has taken place in Vietnam, it is a simplification of a complex historical process. The adoption of *doi moi* did not represent a sudden shift in policy, for it had been preceded by a step-by-step retreat from collectivization in agriculture, starting in 1979, about the same time as decollectivization started in China. Even then, market-oriented economic reforms did not gather full momentum until 1989, around three years after the announcement of *doi moi*, and 11 years on the process of economic reform is still far from finished. Vietnam has still some way to go before it become a well-functioning market economy. Pockets of command management remain, and many policy-makers are suspicious of private enterprise and markets. Nevertheless, as a form of structural transformation the Vietnamese *doi moi* has turned out to have an impact at least as radical and far-reaching as the impact of economic reforms in China, the countries that formerly constituted the Soviet Union, or the majority of states in central and eastern Europe.

The question then arises as to what factors led the Vietnamese leadership to discard the command economy that it had espoused for three decades in favor of a market economy, which it had firmly rejected for just as long. As might be expected, the change in ideology was gradual, and the leadership was deeply divided over the extent of the shift toward a market economy. Indeed, *doi moi* itself was initially intended as a package of reforms to improve the operation of the command economy itself, and the transition to a market economy that followed from *doi moi* was forced on the leadership by the circumstances, rather than brought about by any act of abstract choice on their part. It can be argued that the leadership had little option but to dismantle the stifling command economy, a view aptly expressed in the slogan "*doi moi* or death," which was adopted by the sixth congress of the Communist Party in 1986. The country was simply not capable either of producing enough rice, the staple food, to meet the demands of the population, or of exporting enough to pay for imports of food, fuel, and inputs for its small manufacturing sector. The population was suffing periodic bouts of hunger, food shortages were endemic, and as recently as 1987–88 famine prevailed in many rural areas in the North. The inefficiencies of the command economy were compounded by the heavy toll of Vietnam's seemingly interminable wars. In particular, the war in Cambodia was costly in terms of lives and resources, especially for an economy that had already been enfeebled by a long sequence of wars. Unlike the patriotic war to reunify the North and the South of the country, it was also an unpopular conflict. Faced with the worsening economic situation and social uncertainty, those who could do so were fleeing in flimsy boats to neighboring countries. On its own, the Vietnamese economy would have collapsed. It was able to survive only because of subsidies from the Soviet Union, which was itself undergoing its terminal economic crisis by the mid-1980s.

Thus, a combination of internal and external factors steered the *doi moi* program in the direction of a full-scale transition to a market economy. Among the domestic factors, the most important in the short term was the spiraling rise in the rate of inflation, which climbed to an astronomical 1,000% a year in 1987, before falling to 300% a year in 1988. There was widespread flight from the dong into the US dollar, while the balance of payments deficit (the excess of the value of imports over the value of exports) had reached an unsustainable level. The external factors included, first, the increasing unwillingness of the Soviet Union to provide credit and, second, the collapse of Communism in central and eastern Europe and, two years later, in the Soviet Union itself. In 1988–89, Vietnam embarked on a drastic stabilization program, which was intended to rein in inflation and reduce the balance of payments deficit. The program worked, but it also undermined the foundations of the command economy, paving the way for the transition to a market economy.

The stabilization program and the ensuing reforms were remarkably successful. The inflation rate fell sharply, while the output of rice increased substantially. Vietnam stopped importing rice in 1989 and in 1990 it became the world's third largest exporter of rice (after the United States and Thailand), achieving an astonishing turnaround given that parts of the country had been undergoing famine conditions only a few years earlier. Vietnam also managed to overcome the problems created by the breakup of the Soviet Union, its major trading partner, without severe disruption. By the middle of the 1990s, the country was well-integrated into the international economy and had become a favored destination for foreign investment. Marking a complete break from its past economic dependence on the Soviet Union and its satellites in Europe, Vietnam's main trading partners in the 1990s have all been Asian economies. Those Vietnamese who had earlier left the country in the hope of escaping from abject poverty and political persecution began returning in droves, either as tourists or, in increasing numbers, to make money. During the 1990s, the annual inflation rate averaged around 10%, while the annual rate of economic growth averaged around 9% between 1990 and 1997, the year when the financial crisis swept though Northeast and Southeast Asia. In both these respects, Vietnam now compares favorably with other low-income countries around the world, including China. Some observers have even begun to label Vietnam as one of the new generation of "dragon economies" in Asia.

There are striking similarities in the transitions to a market economy in Vietnam and China, but there are important differences as well. Unlike the former Communist economies of the Soviet Union or central and eastern Europe, China and Vietnam are agrarian economies, with large majorities of their populations wholly or mainly dependent on farming. In both economies, the process of reform began with agriculture and immediately yielded huge benefits. The decollectivization of agriculture in both countries has taken the form of the parceling out of land to rural households. In both countries, economic reforms have led to an immediate acceleration in the growth rate rather than, as in the countries of the former Soviet bloc, a sharp fall in incomes followed, eventually if at all, by positive growth. Both remain one-party states and profess to be "socialist." However, China was and is far more industrialized than Vietnam, and is endowed with a better infrastructure.

Future Prospects

Vietnam remains a very poor country, its income per capita being far below even the average for ASEAN countries. Even if it maintains the high growth rates achieved in the 1990s, it would take Vietnam 20 years or more to approach the living standards of the other comparatively low-income economies in the region, such as the Philippines. The economy remains heavily dependent on agriculture, where the possibilities of further development are severely limited by scarcity of land and the relatively small average size of landholdings. The manufacturing base remains strikingly small and underdeveloped, while the state-owned enterprises that form the core of the mining and manufacturing sectors suffer from the problems of low efficiency and heavy dependence on government subsidies that are familiar in many Communist and post-Communist states. The indigenous private sector has grown and is now flourishing, but it still consists largely of small-scale enterprises in trade and other services. Reflecting the cumulative effects of over 40 years of war, the economic infrastructure is poor and mostly outdated. For example, the main rail link between Ho Chin Minh City in the South and Hanoi in the North was constructed under the direction of the French in the 1930s. The country is poor in natural resources: the reserves of oil and natural gas reserves off the southern coast, which began to be exploited in the 1990s, have been a valuable source of foreign earnings and have served to cushion the adverse economic impact of the collapse of the Soviet Union, but the reserves discovered so far are limited and production has already begun to taper off. One major potential source of economic problems is the huge volume of foreign debt, a large part of which is owed to Russia as the principal successor state of the Soviet Union. With export earnings insufficient to cover the import bill, the Vietnamese economy remains precariously dependent on fickle flows of foreign capital to finance its balance of payments deficit. Vietnam's most valuable resource is its population, which is hard-working, mostly literate, and, at least for the near future, willing to accept much lower wages than workers in most of the other countries of Southeast Asia.

Economic inequality, although not extensive by international standards, is significant and has risen with the transition to a market economy. The southern half of the country is richer than the northern half, which is handicapped by its endowment of comparatively poor land. There are also substantial regional inequalities within both the North and the South, and the gap between urban and rural areas is not only wide but also widening: the average income in urban areas is twice the average rural income. Income inequality has risen sharply, especially in the cities. However, thanks to the rapid economic growth in the 1990s, especially in agriculture, poverty has fallen substantially. According to the World Bank, the proportion of the population living in poverty fell from 58% in 1992–93 to 37% in 1999, due largely to an increase in rural incomes. Poverty remains a largely rural phenomenon, with around 45% of the rural population living below the poverty line, and is associated with the small area of cultivable land per person and the low level of available technology. It is estimated that around 10–15% of the urban population also falls below the poverty line. Unemployment remains a serious problem in urban areas, and is likely to grow as the rate of migration from rural areas increases.

The impressive momentum of economic growth between 1990 and 1997 is threatened from two directions. First, the rate of growth slowed down sharply in 1998, as the Asian financial crisis began to take effect. Although Vietnam was not in the eye of the storm, it felt the impact indirectly. Foreign direct investment, most of which came from Asian economies, dropped by more than half, and the rate of growth in exports fell sharply, with a substantial knock-on effect on the whole economy. Second, the momentum of growth was already slowing in the closing years of the 1990s because of internal factors. The reform program has slackened in vigor and the impact of the earlier reforms on growth has been fading. The decision-making process remains slow, and a source of frustration for foreign investors and international organizations. Added to that, corruption is widespread. It follows from all this that substantial political reforms are needed in order to sustain further economic reforms and help Vietnam to return to a rapid rate of growth.

Further Reading

Bowman, John. S. (editor), *Columbia Chronologies of Asian History and Culture*, New York: Columbia University Press, 2000

   A useful collection of materials on the historical development of Vietnam, as also of many other countries to which it can be compared and contrasted

Halberstam, David, *Ho*, London: Barrie and Jenkins, and New York: Random House, 1971

   This short and readable biography of Ho Chi Minh remains of interest both because Ho's life was exceptionally interesting in itself, and because it encapsulates the largely favorable but also somewhat naive view of Ho adopted by some in the West at the time it was published.

IMF, reports and data available at www.imf.org

The IMF's formal involvement in Vietnam began with a detailed study and analysis of its economy, conducted in 1994. Since then, this international agency has built up an important database of information, and opinions, on the course that the country's economic reforms have taken.

Karnow, Stanley, *Vietnam: A History*, revised and updated edition, New York and London: Viking Penguin, 1991

This book was originally written to accompany *Vietnam: A Television History*, produced by WGBH of Boston, Massachusetts and first broadcast in 1978. Its main focus is on the Vietnam War and the involvement in it of the United States, but Karnow also takes care to set that conflict in a broader historical and geographic context, and provides a wealth of detail on many aspects of Vietnam's modern history.

Kolko, Gabriel, *Anatomy of a War: Vietnam, the United States, and the Modern Historical Experience*, New York: New Press, 1985, reissued 1994; as *Vietnam: Anatomy of a War*, London: Allen and Unwin, 1985

Professor Kolko has studied Vietnam for many years, and, as a Canadian inclined to the left, provides an unusual and refreshing perspective on the war that dominated and shaped Vietnam from the 1940s to the 1970s.

Kolko, Gabriel, *Vietnam: Anatomy of a Peace*, London and New York: Routledge, 1997

In this later work, Professor Kolko eloquently expresses and explains his largely negative view of the changes that have taken place in Vietnam since the mid-1980s, challenging the somewhat onesided optimism of the government and the international agencies that collaborate with it.

Nugent, Nicholas, *Vietnam: The Second Revolution*, Brighton: In Print, 1996

An interesting study of the *doi moi* process and its impact on Vietnam's economy and society

Pike, Douglas, *PAVN: People's Army of Vietnam*, Novato, CA: Presidio Press, 1986, and New York: Da Capo Press, 1991

Professor Pike, a former US military intelligence officer turned academic, paints a detailed and fascinating picture of the Vietnamese armed forces, and incidentally offers insights into US attitudes to Vietnam as of the mid-1980s.

Porter, Gareth, *Vietnam: The Politics of Bureaucratic Socialism*, Ithaca, NY: Cornell University Press, 1993

Porter gives a critical and informative analysis of the convoluted processes of decision-making in the Vietnamese one-party state.

Scalapino, Robert A., *The Politics of Development: Perspectives on 20th-century Asia*, Cambridge, MA: Harvard University Press, 1989

A historical survey of diverse experiences of political change and development in a number of broadly comparable countries, including Vietnam

Tarling, Nicholas (editor), *The Cambridge History of Southeast Asia*, Volume 2, *Nineteenth and Twentieth Centuries*, Cambridge, New York, and Melbourne: Cambridge University Press, 1992

A number of distinguished scholars have contributed essays on all the countries in the region, allowing Vietnam's unique development path to be understood in comparison and contrast to the colonial and postcolonial experiences of its neighbors.

World Bank, reports and data available at www.worldbank.org.vn

Like the IMF, the World Bank has been increasingly involved in advising and supporting the Vietnamese government on its program of economic reforms since the mid-1990s, and this dedicated website contains detailed social and economic indicators that highlight the impact of the reforms.

Dr Athar Hussein is Deputy Director of the Asia Research Centre at the London School of Economics and Political Science. He has been a regular visitor to Vietnam, and has also been a member of the International Committee for Economics Research and Teaching in Vietnam.

NATION STATES

Chapter Twelve
Cambodia

Sue Downie

Cambodia's "golden years" of independence, relative peace, and relative prosperity under the rule of Norodom Sihanouk ended when the war between the United States and Vietnam spilled across the border. US bombings, the destructive policies of the Khmer Rouge, the Vietnamese invasion, and the subsequent civil war all left Cambodia shattered, economically, politically, socially, and culturally. An enormous effort will be required to rebuild a wartorn country that was on the verge of being a failed state.

The international community's involvement in Cambodia began in the early 1980s and was then aimed at ending the civil war, which was seen as a continuing threat to regional stability. Cambodia's apparent relative insignificance belied its geopolitical importance. The United States, China, and the Soviet Union became interested in peace in Cambodia, a tiny, economically insignificant country, as part of the much larger process of ending the Sino-Soviet proxy war in Indochina, in which Cambodia had been a pawn. In 1991, the international community brokered a peace settlement that culminated in Cambodia's four factions signing the Paris Peace Agreement, under which the UN established its largest peacekeeping mission to date. The agreement authorized the UN to disarm the Cambodian factions and organize national elections. While many in the UN saw this as bringing democracy to Cambodia, in fact it was only the first step down a long road. One test of the UN's legacy came in the elections of 1998, which showed that the government had made very little progress in fostering political development, although the situation was considerably better than it had been during the Khmer Rouge period (1975–79). The government had demonstrated its reluctance to acceptance a political opposition, it had failed to separate the state from the ruling party or the executive from the judiciary, and it had displayed little or no concern for the rule of law or for human rights. Cambodia has made significant economic progress since 1991, boosted by massive humanitarian and development aid; social changes have also been considerable, some of them negative.

The Background

Cambodia, one of the smallest and poorest nations in Southeast Asia, is an almost circular country surrounded by elongated neighbors: Thailand to the West and North, Laos to the Northeast, and Vietnam to the East and Southeast, with the Gulf of Thailand to the Southwest. The center and Southeast of the country are flat, but there are mountain ranges along its western, northwestern and northeastern borders. Cambodia's most prominent geographical feature is the vast Tonle Sap, the largest freshwater lake in Southeast Asia, which has a catchment area covering one third of the national territory. It is linked by a river of the same name to the Mekong, the longest river in Southeast Asia. During the wet season, when the Mekong floods, the Tonle Sap river reverses its course and flows into the lake, which, being shallow, floods the neighboring lands and expands to up to six times its dry season area. The Cambodian capital, Phnom Penh, is built at the confluence of the Mekong and the Tonle Sap rivers, which join together for a short stretch (around one kilometer, or three fifths of a mile), then separate again and head towards southern Vietnam, where they become part of the "Nine Dragons," the tributaries of the Mekong Delta, then empty into the South China Sea.

An estimated 90% of the 11.6 million people of Cambodia are of ethnic Khmer origin (see EIU), the remainder comprising descendants of Chinese immigrants and more recent Vietnamese immigrants, Chams, Lao, and ethnic minorities in the Northeast. Khmers tend to be huskier, taller, and darker than their counterparts in Thailand, Vietnam, or Laos, and sometimes have curly hair and rounder eyes. Khmers rarely marry Vietnamese or Chams, but marriage between Khmers and Chinese is common. The dominant religion is Theravada (Hinayana) Buddhism, but there are also small numbers of Moslems and Christians. The official language is Khmer, while Vietnamese, Chinese, and Cham are spoken among their respective groups. French was popular among the elite during the colonial period, English began gaining currency from the late 1980s, and since the late 1990s both have been taught in schools.

Cambodia's Constitution, enacted in 1993, calls on its citizens to "preserve and defend... the prestige of Angkor civilization" (CLRDC p. 1), the national flag bears the outline of Angkor Wat, and most Cambodians dream of visiting the ancient temple complex near the northwestern town of Siem Reap. Over the generations, the temple, the complex, and the empire that built it have come to embody Cambodia, which can trace its roots to the year 800, when King Jayavarman II inaugurated a dynasty in what is now northwest Cambodia. The subsequent Hindu-Buddhist kingdom flourished, and at its peak it covered what are now northern Thailand, southern Laos, and southern Vietnam. Following the decline of the Angkor empire, the capital was moved to Phnom Penh; there were then four centuries of conflict with Vietnam and Thailand. Fearing that it might be swallowed by either of these neighbors, Cambodia was obliged to become a protectorate, as part of French Indochina, in 1863. Cambodia was occupied by the Japanese during World War II, but France then reclaimed its control over the country (see Chandler 2000).

In April 1941, when King Monivong died, the French had replaced him with his 18-year-old grandson Prince Norodom Sihanouk. The French assumed that he would be a malleable monarch, but instead the nationalistic young king launched a "royal crusade for independence," which was finally granted in November 1953. The Geneva Accords of July 1954 formally concluded France's rule over Cambodia, Vietnam and Laos, and ended what is often called the First Indochina War. Sihanouk reigned for 14 years, but then, unhappy with the limited powers of the monarchy, abdicated in March 1955 so that he could set up his own political movement, the Popular Socialist Community. The 1950s and 1960s were years of political turbulence, characterized by extravagance, corruption, and the suppression of any opposition. However, it was also a relatively happy, healthy, prosperous and peaceful time, Cambodia's "golden age" (see Chandler 1991 and Osborne).

A serious decline in Sihanouk's relations with the United States had already begun after Washington had refused to support his independence crusade. While he continued to accept military assistance from the United States, he increasingly aligned himself with such leaders as Sukarno of Indonesia, Josip Broz Tito of former Yugoslavia, Zhou Enlai of China, and Gamal Abdel Nasser of Egypt (see Norodom). In 1959, Sihanouk accused the US Central Intelligence Agency of supporting a rightwing plot to overthrow him; in November 1963, he renounced US economic and military aid, and closed US aid missions; and in May 1965 he severed diplomatic relations (although they were restarted in June 1969). As the Vietnam War escalated, the forces opposed to both the United States and its allies in South Vietnam established sanctuaries inside Cambodia. Sihanouk wanted Cambodia to remain neutral, but allowing the Vietnamese Communists access led to Cambodia being seen as assisting the enemy, and therefore as an enemy itself. There were raids by US bombers on border areas of Cambodia as early as 1965, but in March 1968 the United States began 14 months of sustained and secret air raids against

Communist bases inside the country. The bombings devastated the eastern provinces, disrupted production, and drove an estimated 1.5 million peasants, one third of the population, to refugee camps in and around Phnom Penh (see Chandler 1991).

It was customary in the 1940s, 1950s, and 1960s for the children of the Cambodian middle and upper classes to complete their studies in France. Some of these students joined the French Communist Party, including Saloth Sar (who later used the alias Pol Pot), Ieng Sary, and Khieu Samphan. All three returned to Phnom Penh in the 1950s and subsequently set up an underground Communist movement that Sihanouk called the "Khmers Rouges" (French for "Red Cambodians"). In 1961, Sihanouk's government launched an anti-left campaign, harassing, humiliating and imprisoning leftist radicals and intellectuals, and eventually driving the leaders of the Khmer Rouge to the jungles, where they set up political bases (see Chandler 1991 and Kiernan 1985). For the moment, they remained relatively ineffective and largely unknown.

On March 17, 1970, Sihanouk was on his way from Moscow to Beijing when he was overthrown in a bloodless coup instigated by one of his cousins, Sirik Matak, and the Prime Minister, General Lon Nol, whose new regime was immediately recognized by the United States. Sihanouk flew on to Beijing, where he established himself in exile and announced that he was joining the Khmer Rouge in their fight against Lon Nol and the "US imperialist forces." Despite massive US aid to Lon Nol's forces, the resistance gradually took control of the countryside, and on April 17, 1975, after five years of civil war, overran Phnom Penh, forcing the entire population to the countryside. Not all resistance fighters were members of the Khmer Rouge, and not all members of the Khmer Rouge would call themselves Communists, but it was the hardline faction that took charge.

Pol Pot, Ieng Sary, Nuon Chea, Son Sen, and Khieu Samphan emerged as the leaders of the new regime. They set out to turn Cambodia into one giant farm. With little economic or administrative experience or bureaucratic control, the Khmer Rouge implemented their agrarian philosophy. Factories and businesses were abandoned or destroyed. The cities, including the capital, were deserted, the occupants forced into the countryside to take part in massive labor gangs. Money, commerce, medicine, education, religion, and private ownership were banned. Many monks were murdered, and temples were turned into prisons and pigsties, while schools and hospitals became torture chambers or ammunition stores. In Phnom Penh, the National Bank was blown up and the Catholic Cathedral was torn apart brick by brick so that it could not be rebuilt. Often, children were separated from parents, and husbands from wives, and many single people were forced to take part in mass weddings. Huge communal kitchens became the norm. Virtually all the members of the traditional elite, and virtually all educated people, professionals, and businessmen were killed, including architects, artists, bankers, dancers, doctors, engineers, journalists, princes, teachers, and traders. Only 7,000 of the country's 22,000 teachers and 45 of its 450 doctors remained in Cambodia; of 1,600 agricultural planners, technicians, and policy-makers, only 200 remained, of whom only 10 were graduates; and there was only one licensed veterinarian in the country. The Khmer Rouge expelled all foreigners, closed the ports and airports, and sealed Cambodia's borders. Unknown to the outside world, at least one million people, an estimated one sixth of the population, were killed or died of torture, disease or starvation. Thousands were buried in mass graves, which became known as "the killing fields." Meanwhile, around 400,000 people fled to Vietnam and Thailand. (See, among many publications, Becker, Kiernan 1996, Mysliwiec, Ngor, Picq, and Ponchaud.)

The Khmer Rouge's reign of terror lasted for three years, nine months, and 20 days. It ended when the Vietnamese, accompanied by freshly retrained and equipped Cambodian soldiers, liberated Phnom Penh, on January 7, 1979, and established the People's Republic of Kampuchea three days later. Ever since late 1975, the Khmer Rouge had been attacking villages in southern Vietnam from across the border, with greater intensity from early 1977, and had forced almost half a million

Vietnamese to abandon their homes and about 100,000 hectares of farmland (Mysliwiec p. 9). Vietnam had sent troops into southeastern Cambodia in retaliation before launching a full-scale offensive. Within weeks, the Vietnamese had swept right across the country and reached the western border, forcing the remaining Khmer Rouge to seek sanctuary in Thailand.

At first, Sihanouk had served as the Khmer Rouge's reluctant "Head of State," having returned to Phnom Penh in September 1975. He remained there, under virtual house arrest, until the day before the Vietnamese entered the capital. He then fled to Beijing, which remained his base during the resistance years and has continued to be his second home.

Civil War and Peace Talks, 1979–91

To this point, Cambodia had three key players: Sihanouk, the Khmer Rouge, and China, which had been a friend of Sihanouk for many years, and which sponsored the Khmer Rouge before, during, and after its years in power. From January 1979, three additional players entered the scene: Vietnam, which continued occupying and administering the country; Thailand, which accepted refugees, and provided sanctuary and support to the Khmer Rouge and other military groups and civilians fleeing the Vietnamese; and the new Cambodian regime, the People's Republic of Kampuchea. Two more players then entered the Cambodian arena: ASEAN, which supported the parties opposing the new regime; and the UN, which attempted to broker a peace settlement. Cambodia was also a pawn in the superpower rivalry of the early 1980s: the United States joined China in supporting the resistance factions, while the Soviet Union was a patron of Vietnam. China invaded northern Vietnam in February 1979 in an attempt to "punish" the Vietnamese government for invading Cambodia (Chanda 1988 p. 352), starting a brief, ineffective but politically significant border war. China then began supplying the surviving Khmer Rouge on the Thai border, where they had regrouped.

Other refugees on the border formed themselves into several resistance groups, which eventually distilled down to two: those led by Sihanouk, and those led by one of Sihanouk's former Prime Ministers, Son Sann. Although these two factions were ideologically opposed to the Khmer Rouge, both Sihanouk's Armée Nationale Khmer Indépendante (Independent National Khmer Army, or ANKI) and Son Sann's Khmer People's National Liberation Front (KPNLF) fought alongside the Khmer Rouge against their common enemies, the Vietnamese and the forces of the new regime. The Cambodian web of support links was complex and changeable, but most of the time the Khmer Rouge were supplied by China, ANKI and the KPNLF by the US and ASEAN, while Thailand provided sanctuary and transport of weapons to both the Khmer Rouge and the non-Communists. Under pressure from the West and ASEAN, the exiled non-Communists agreed in June 1982 to form a coalition with the Khmer Rouge, known as the Coalition Government of Democratic Kampuchea, which was headed by Sihanouk and held Cambodia's seat in the UN General Assembly.

The People's Republic of Kampuchea was headed by former Khmer Rouge officers who had fled to Vietnam, including Heng Samrin, who was named President, Chea Sim, who became Interior Minister, and Hun Sen, the Foreign Minister. In one sense, the regime represented a relief for those who had survived the Khmer Rouge. However, as the regime had been installed by Vietnam, it was not recognized by any non-Communist country except India, so for 13 years Cambodians were denied bilateral assistance and all forms of development aid from western countries or ASEAN. Deprivation, poverty, and war continued, although with far less severity. The first task facing the new regime was the threat of famine. Initial reports suggested that two million people faced starvation, although this figure was later reduced. A massive appeal by international nongovernmental organizations raised US$100 million and staved off disaster.

Despite the international isolation and the continuing war, the regime also set out to rebuild social institutions and the physical infrastructure. Gradually, offices, farms, and factories that had been damaged or destroyed during the Khmer Rouge period were restored, schools, hospitals, and temples were reopened, roads and bridges were rebuilt, and ports and airports resumed operations. A free market economy and private ownership were reintroduced in 1989. At the same time, efforts were made by many individuals, countries, and organizations to end the fighting.

The resulting peace process comprised a series of negotiations, some close together, some years apart. During the 1980s, what was universally referred to as "the Cambodia problem" was a microcosm of world politics in what can now be seen as the closing stages of the Cold War. It was both a spillover from the Vietnam War, and the last battlefield for a proxy war between the Soviet Union on the one hand, and China and the United States on the other. At the same time, ASEAN was building itself as a non-Communist counterbalance to China. Thus, Thailand agreed to allow Chinese supplies to reach the Khmer Rouge if China agreed to stop supporting the Communist insurgency in Thailand.

The UN began the peace process by organizing a conference in New York in July 1981; Indonesia hosted three "informal meetings" in its capital, Jakarta; Australia laid the groundwork for the peacekeeping plan; and France hosted the first and second international conferences on Cambodia in Paris, in August 1989 and October 1991. Throughout the negotiations, the forces of the two governments of "Kampuchea" continued fighting, each side hoping that their battlefield victories would help them at the negotiating table. One of the key factors in reaching a peace settlement was the withdrawal of Vietnamese troops, which was finalized in September 1989, although the Khmer Rouge declared that some remained in the country and continued fighting the "Vietnamese puppets." Finally, on October 23, 1991 representatives of 17 other countries and of the four Cambodian parties, including the Khmer Rouge, gathered in Paris to sign the peace agreement (see UN). This authorized the UN to establish a transitional authority and a peacekeeping mission to oversee a ceasefire and elections. An unspoken aim of the agreement was to provide a mechanism to allow the Soviet Union, China, and the United States to extricate themselves from Cambodia.

The Transition, 1991–93

Seventeen days after the signing in Paris, the UN began deploying the Advance Mission in Cambodia (UNAMIC), which helped to maintain the ceasefire and to prepare for the arrival of the 22,000 personnel of the Transitional Authority in Cambodia (UNTAC). The first UNAMIC peacekeepers arrived in time to form a guard of honor for the return from exile, on November 14, 1991, of Sihanouk, who had hardly been seen in public in Cambodia for 21 years. He was given a hero's welcome. Following Sihanouk came members of ANKI, the KPNLF, and the Khmer Rouge, some of whom had been in exile of up to 18 years. The return of Sihanouk and the arrival of the UN also opened the way for the new four-party Supreme National Council to be established in Phnom Penh, and for non-Communist governments to establish missions in the capital for the first time since 1975.

UNAMIC handed over authority to UNTAC in mid-March 1992, when the UN Secretary General's Special Representative, the Japanese diplomat Akashi Yasushi, arrived in Phnom Penh ahead of the UN's 16,000 troops. UNTAC combined all aspects of the UN's new provisions for "comprehensive" peacekeeping activities: it was the largest and most intrusive peacekeeping mission the UN had ever undertaken. Never before had a mission been given the task of disarming warring factions, repatriating refugees, controlling a country's civil administration, organizing elections, and promoting awareness of human rights.

Citing a variety of reasons, the Khmer Rouge refused to disarm, to allow the UN into the areas that they controlled, or even to register for the elections. Nevertheless, UNTAC proceeded with repatriation, its programs on human rights and civic education, and the elections themselves, which were held from May 23 to May 28, 1993. Twenty parties registered to contest the 120 seats in the new National Assembly. The main contenders were the Cambodian People's Party (CPP), supporting the de facto government, which had renamed itself the State of Cambodia; the royalist party, the Front Unie Nationale pour Camboge Indépendante, Neutrale, Pacifique, et Coopérative (United National Front for an Independent, Neutral, Peaceful and Cooperative Cambodia, or FUNCINPEC), which was aligned with Sihanouk's ANKI and was led by his son Ranariddh; and Son Sann's new organization, the Buddhist Liberal Democratic Party (BLDP).

The election campaign was marred by political violence, perpetrated by the Khmer Rouge against CPP supporters and ethnic Vietnamese living in Cambodia, but also by agents of the CPP against supporters of FUNCINPEC and the BLDP. In the month before the election, for example, UNTAC recorded 106 politically motivated killings (see Akashi). In the week before polling, UNTAC was forced to reduce the number of polling stations by almost half, the CPP threatened to pull out of the election altogether, Son Sann attempted to have his deputy barred from standing as a candidate, and the Khmer Rouge threatened to attack polling stations. Despite all these problems, 90% of the 4.7 million registered voters cast their ballots, 45% voting for FUNCINPEC, which won 58 seats, 38% for the CPP, which won 51 seats, and 4% for the BLDP, which took 10 seats. (The remaining seat went to the Molinaka Party, which won 1.37% of the votes cast but ceased to function soon after the elections.)

The formation of a new government required the assent of at least two thirds of the members of the new Assembly. The CPP used military threats to force FUNCINPEC to form a coalition government in which the two parties were formally equal partners.

Four months after the election, on September 24, the Assembly adopted a new Constitution, making Cambodia a constitutional monarchy in which the King would reign but not rule. Three days later, Sihanouk was reinstated as King, and UNTAC's mandate came to an end. On October 29, the Royal Government of Cambodia was formed, with Ranariddh as First Prime Minister and Hun Sen as Second Prime Minister. Similarly, all the leading positions in the various ministries were divided between FUNCINPEC and the CPP, a mechanism that forced the two former enemies to work side by side, but also generated a cumbersome administration and friction between the two parties. As Hun Sen controlled the bulk of the armed forces, and threatened to use them, he was able to extract from FUNCINPEC more concessions than would be normal in a coalition government, and indeed more than have been made in the coalition government formed in 1998.

There can be little doubt that UNTAC generated many positive legacies. The most important was the international recognition of a single government, based in Phnom Penh, which facilitated the return of multilateral and bilateral aid, as well as foreign investment. These in turn generated domestic investment. In addition, the launching of political parties was significant in a country that had previously been denied any genuine pluralism. UNTAC also contributed to the increased awareness of human rights, and helped to promote the subsequent establishment of genuinely independent nongovernmental organizations and media outlets. Many thousands of Cambodians benefited from exposure to international practices, and enhanced their capacities in administration, diplomacy, technical areas, and trade.

The First Coalition Government, 1993–98

Cambodia was rapidly readmitted to the Mekong River Commission, after an absence of 20 years, and was able to resume its UN seat unchallenged. Nevertheless, the new coalition government faced many problems. The Khmer Rouge, which had not taken part in the election, became an enemy of the state. Corruption, impunity, cronyism, and nepotism were rampant. Reform of the military, the civil administration, and the financial system was desperately needed. The value of the national currency, the riel, fell steadily and inflation rose. The redevelopment of infrastructure needed massive injections of funds, as did the rural areas where 80-85% of the population lived. It became clear that many of those who entered government for the first time, from FUNCINPEC and the BLDP, did not have the experience or competence to run the country.

It also became clear that this new democratically elected government was not prepared to tolerate political opposition.

The Problem of the Khmer Rouge

Among the new government's many challenges, the most demanding was the problem of the Khmer Rouge. Its long-established, well-armed, and jungle-hardened army was still led by committed ideologues, some of whom had been fighting for up to 25 years. The Khmer Rouge had persuaded the international community to include it in the peace process, despite its genocidal past, and it had then defied UNTAC; now it continued to dominate domestic politics, and to stretch the government's limited financial, material and human resources. From 1994 to mid-1996, small units of Khmer Rouge guerillas attacked and burned villages and transport infrastructure, including road bridges and rail tracks. Its armed forces numbered about one tenth of the government's, yet it had the ability to consume resources, disrupt development programs, and scare investors away. While the government desperately needed funds for infrastructure and rural development, the Khmer Rouge profited from the gem and timber operations it controlled on the Thai-Cambodian border. In May 1995, it was estimated that the timber alone was providing the Khmer Rouge with US$10million a month (see *Economist*).

On the other side, the government forces were tired of fighting, were disheartened because they had expected the UN mission and the election to bring peace, and, significantly, were preoccupied with integrating the units of the three armies of the CPP, FUNCINPEC, and the BLDP. In July 1994, however, the National Assembly passed legislation outlawing the Khmer Rouge. The government offered amnesties to rank and file members, and within seven months more than 6,000 soldiers and militia defected. The most significant defections came after August 1996, when there was a split in the Khmer Rouge leadership: Ieng Sary and 2,500 of his supporters went over to the government. These defections created friction between FUNCINPEC and the CPP, as they competed with each other to attract Khmer Rouge dissidents into their respective armies, which remained separate despite formal integration.

In July 1997, FUNCINPEC, without the agreement of the CPP, imported weapons, moved troops, and was on the verge of accepting a significant faction of the Khmer Rouge that wanted to defect. Feeling threatened, the CPP retaliated on July 5 with tanks and troops. Fighting between CPP and FUNCINPEC forces in and around Phnom Penh lasted for only 36 hours, but the political and economic ramifications, including a postponement of Cambodia's entry into ASEAN, lasted for more than 36 months. Troops loyal to Ranariddh continued fighting Hun Sen's soldiers in the Northwest for 12 months, initially forcing 40,000 civilians to flee to Thailand.

After repeated defections, the leadership of the Khmer Rouge began disintegrating in August 1996. By June 1997, it was in turmoil. Pol Pot ordered the execution of his "Defense Minister," Son Sen, and the ailing leader himself was then sentenced to life imprisonment at the end of a show trail in the jungle. In April 1998, he was found dead in his bed.

Political and Social Issues

In a speech given in Bangkok in April 1995, the Information Minister Ieng Mouly presented a summary of the situation in Cambodia. He began by saying that "without doubt the greatest achievement is the fact that the two main parties, who were once opponents, are running the country in coalition." People were able to form associations, to say what they liked, and to go where they liked, for the first time in almost 20 years. Free media were flourishing, with more than 40 publications, four private radio stations, and three private television stations. Thailand no longer had Cambodians living in camps along the border; the National Assembly had passed 17 laws; 130 domestic nongovernmental organizations, including a dozen human rights organizations, were operational; Cambodia had signed several regional and international agreements of cooperation, and its representative was chairing the Mekong River Commission (see Ieng).

Certainly, the situation had improved and citizens were experiencing considerably more freedom. By the end of 1997, human rights awareness had increased, as had respect for the law; labor unions were active; several institutes had been established; and a number of political parties were beginning to prepare for the next legislative elections, to be held in 1998. However, several issues overshadowed the achievements: impunity, corruption, and the absence of the rule of law; intolerance of opposition; and reluctance either to separate party from state, or to reform the military, the administration, and the judiciary. In the mid-1990s, the government and the development agencies also faced severe challenges on the economy, health, and education. Annual income per capita was estimated at just US$243, and the area of land under rice cultivation in 1992–93 was 20% less than it had been in 1967; yet landmines were distributed over an estimated 30% of agricultural land and less than 2% of the country's roads were paved. Average life expectancy was just 52; infant mortality was estimated at 97 per 1,000 live births, and maternal mortality at five per 1,000 births; only 20% of the urban population and 12% of the rural population had access to safe drinking water; and about 50% of the population had no access to the public health care system. Almost 45% of the population were under 15 years of age, but the primary school enrolment rate was only 47%, and somewhere between 35% and 65% of the population were illiterate (all figures from EIU). Cambodia's problems were highlighted when the IMF, which had approved a three-year enhanced structural adjustment facility in 1994, withheld this financing in 1997, on the grounds that the government had failed to improve its management of the budget to a satisfactory standard (see EIU).

Cambodia also received unfavorable coverage in the international media because of the intolerance of the governing parties, particularly the CPP. Journalists and editors who criticized the government or exposed corruption were killed and jailed; politicians were expelled from their positions or parties, and some were exiled; nongovernmental organizations were regarded as antigovernmental; and political parties, both inside and outside the coalition, were deliberately fragmented. With few exceptions, the influence of the Assembly was limited to speeches of support, and generally legislation was rubberstamped. Except for the main three parties, virtually all the political parties that had contested the elections in 1993 collapsed or went into hibernation. By 1997, the only effective opposition was being provided by Sam Rainsy, who had been dismissed from his post as Minister of Finance in October 1994, then removed from FUNCINPEC and from the National Assembly. He had established his own political party and while he was addressing a rally outside the National Assembly in March 1997, he narrowly escaped injury in a grenade attack that killed at least 16 people and injured around 100 others. Although the names and descriptions of the alleged perpetrators were widely circulated, by mid-2000 none had been arrested.

The grenade attack and the fighting that erupted in July (see above) had a devastating effect on political expression and activity at a time when Cambodians were beginning to speak out. Many provincial party offices were closed and agents went into hiding; newspapers linked to Ranariddh and Rainsy were closed when their journalists fled to Thailand; and human rights activists either fled or scaled down their activities. It was only a few months before the elections in July 1998 that they began to reemerge, tentatively and with less vigor than before. The most significant long-term impact of the events of July 1997 was the splintering of FUNCINPEC into four parties, only one of which, led by Ranariddh, won any seats in 1998.

The Administration and the Economy

In 1994, the government claimed that the state's revenues had increased three fold, that the budget deficit had been eliminated, that the banking, customs and tax systems had all been overhauled, and that measures were being taken against corruption and smuggling. The World Bank had approved a loan of US$100 million to Cambodia and donor countries had agreed to forgive US$92 million in arrears on debts to the IMF. However, Phnom Penh was still plagued by inadequate electricity, water and telephone systems, and suburban streets became bogholes in the wet season. In the countryside, there was an increase in the number of houses with tiled roofs, often taken to be an indicator of rising disposable income.

By April 1995, the riel had remained stable for almost two years, Cambodia had received pledges of more than US$700 million from the international community, and more than US$2 billion had been committed to the country by investors. In his speech already quoted above, Ieng Mouly declared that "Phnom Penh sometimes resembles a giant building site, with construction and renovation going on all over the city" (see Ieng). He did not, however, mention the corruption that pervaded the administration, or the illegal deforestation that had drastically reduced Cambodia's forest cover and deprived the government of millions of dollars in revenues (see Global Witness). In March 1995, Keat Chhon, then the Minister of Finance, estimated that every year corruption was depriving the state of around US$100 million, or more than one third of internally generated revenues. Even experienced international aid agencies with elaborate control systems routinely budgeted for losses of between 20% and 30% (Curtis pp. 147, 146).

The three main obstacles to development were Cambodia's relative lack of funds, its shortages of skilled and experienced personnel, and its inadequate institutional framework. In many cases, laws and regulations simply did not exist; where they did exist, they were often left unimplemented or were easily flouted. The government failed to undertake large-scale legal reforms or to create an independent judiciary, but 40 laws were passed by the National Assembly during its first five years (see CLRDC), and international nongovernmental organizations undertook "capacity-building" projects in relation to the courts, the police, and prisons.

While the government struggled to raise domestic revenues, almost US$1.4 billion-worth of external assistance was disbursed between 1992 and 1995 (Curtis p. 78). Most of this aid was spent in Phnom Penh, despite the fact that at least four fifths of the population lived in rural areas and the government had repeatedly committed itself to making rural development its first priority. The gap between Phnom Penh and the rest of Cambodia remained enormous. In 1993–94, for example, the average monthly household expenditure in Phnom Penh was more than twice Cambodia's annual average income per capita; annual spending on health care in Phnom Penh was US$36 per capita, compared with US$2 elsewhere; and 67% of the city's residents used electricity, compared with only 3% in rural areas (Curtis pp. 80–81).

Agriculture, accounting for almost 80% of total employment, continued to underpin Cambodia's economy. However, services, construction, and manufacturing accounted for more than three quarters of the estimated growth in real GDP, which amounted to 6% between 1991 and 1996, and exports of clothing in particular increased by almost 200% from US$63 million in 1996 to US$180 million in 1997, making a considerable contribution to total exports, valued at US$534 million in the latter year (Hang pp. 66, 70).

The events of July 1997 had significant consequences for the economy: the Ministry of Commerce estimated that it would take three years to reverse their impact. The suspension of international aid was followed by a dramatic decline in the numbers of tourists visiting the country. The impact was compounded by the Asian economic crisis. Foreign currency deposits fell during 1997; private capital inflows declined from US$170 million in 1996 to US$150 million; and during 1998 capital flight was estimated at US$100 million. These trends put pressure on exchange rates, and forced up both interest rates and unemployment. State revenues fell from 9.2% of GDP in 1997 to 8.5% in 1998. In its budget for 1997–98, the government responded to the crisis by making cuts of between 10% and 25% in its spending on education, health, agriculture, and rural development, while allowing spending on defense and security to overrun by 4% (Hang pp. 63, 65, 67).

The Second Coalition Government, 1998 onwards

During the campaigning for the elections held in July 1998, the CPP demonstrated its intolerance of opposition through violence, intimidation, and coercion. After the elections, it displayed once again its unwillingness to embrace transparency and accountability.

The elections were organized and monitored by the National Election Committee, which was established only five months before the polls were to be held, while the Constitutional Council, which is required under the Constitution to mediate electoral disputes, was convened for the first time just one month before the election and almost five years after the Constitution had been promulgated. The CPP dominated both these bodies, as well as the state media. Despite the difficulties, 39 parties contested the elections. Only the three largest were officially declared to have won any representation in the National Assembly, which now has 122 seats: the CPP received 41% of the votes cast and took 64 seats, FUNCINPEC won 32% and 43 seats, and the Sam Rainsy Party won 14% and 15 seats. The National Election Committee and the Constitutional Council dismissed almost all the 850 complaints of electoral fraud and irregularities lodged by parties other than the CPP. This led to a boycott of the Assembly by some members of the opposition, who led protests outside the Assembly building for two weeks, until they were violently dispersed; at least three people were killed. Four months after the election, the CPP and FUNCINPEC reached agreement on a second coalition government, Hun Sen became sole Prime Minister, and Ranariddh agreed to take on the presidency of the National Assembly. The agreement between the two main parties left Sam Rainsy and his supporters to form a small and ineffective opposition in the Assembly.

The new government immediately introduced a value-added tax that was expected to raise US$6.6 billion in revenues in the fiscal year 1999–2000. The economic crisis continued: during 1998, real GDP showed no growth at all, inflation averaged 15%, the current account deficit was US$23 million and foreign debt stood at around US$2.15 billion. However, 130 foreign investment projects, worth a total of US$840 million, were approved in the first ten months of 1998, compared with a total of US$760 million in foreign investment for the whole of 1997 (see EIU).

The political tasks confronting the second coalition government included, as before, the continued strengthening of law and order, and the implementation of administrative, financial and legal reforms. The government also faced three new challenges: reconciling and/or bringing to trial the Khmer Rouge leaders responsible for the genocide of 1975–79; demobilizing 50,000 members of the armed forces; and holding elections for local government bodies, known as communes on the French model.

Following the defection of Ieng Sary, the execution of Son Sen, and the death of Pol Pot, the only Khmer Rouge leaders still at large were Khieu Samphan and Nuon Chea, who both defected in December 1998; the military commander Ta Mok, who was captured in March 1999; and the former chief executioner Kang Kek Ieu (alias Duch), who was arrested in May 1999. The demise of the Khmer Rouge leadership opened the way for the government, with the agreement of the UN, to put Ta Mok, Duch, and perhaps others on trial. Negotiations and preparations for the trial began in November 1998, and were continuing when this chapter was written.

Overview and Conclusion

From the 1860s to the 1990s, Cambodia was never anything more than a pawn in the hands of French colonialism, rival superpowers (the United States, China, and the Soviet Union) and their allies, and its stronger neighbors, Vietnam and, more recently, the member states of ASEAN. Throughout these years, it had been pulled from left to right and back, depending on the politics of its patron states. Following the intervention of UNTAC, which operated on a scale unprecedented in the history of the UN, Cambodia has experienced relatively little direct interference by foreign countries, although it has had to meet certain conditions both to receive the external aid on which it depends so heavily, and to gain entry into organizations such as the UN and ASEAN.

The 1990s also saw the beginning of the end of the Khmer Rouge as a political and military force, although former members of that organization continue to figure in the political, social and legal life of the country. The collapse of the Khmer Rouge can be traced to the success of the government's campaign to encourage defections, as well as the coincidental splits within the organization's leadership, but it also reflects the realization among the rank and file that the country was not in fact being overrun by the Vietnamese, and that prosperity and stability might be achievable.

The rebuilding of Cambodia has undoubtedly gone at a slower pace than many Cambodians, tired of war and anxious for a new start, had expected and hoped. The regime installed by the Vietnamese in 1979 made a significant contribution to rebuilding the country after the overthrow of the Khmer Rouge, even though economic, political and social advancement were hampered by continuing civil war, international isolation, and the government's restrictive policies. The second phase of rebuilding began when multilateral and bilateral aid were resumed, supporting a government that faced resistance from only one rival army rather than three. Most Cambodians, and many foreigners, expected UNTAC and the first elections to bring democracy, peace, and prosperity almost instantaneously. The difference between expectation and reality can be explained in part because foreigners generally failed to understand or appreciate the complexity of Cambodian politics. This failure was compounded by a lack of understanding of the democratic process, both among Cambodians, who were not familiar with the concept and tended to see voting itself as a means to peace, and among foreign observers, especially those from the West, who expected Cambodia to reach overnight a level of democratic stability that their respective countries had taken more than 200 years to attain.

By mid-2000, there was no longer any fighting taking place in Cambodia, for the first time in 22 years. The Sam Rainsy Party was functioning as an opposition in the National Assembly, but faced severe constraints, and it is clear that the government has yet to accept the full implications of open competition between political parties in a liberal democracy. FUNCINPEC has again joined the CPP in coalition, although its status has been reduced, and members of both parties are attempting reconciliation. However, arbitrary force, impunity and corruption remain pervasive, the judiciary is not independent, and administrative reform still has a long way to go.

Further Reading

Akashi Yasushi, speech opening a meeting of the Supreme National Council of Cambodia at the headquarters of UNTAC in Phnom Penh on May 11, 1993; author's record

Becker, Elizabeth, *When the War was Over: The Voices of Cambodia's Revolution and its People*, New York and London: Simon and Schuster, 1986; second edition, as *When the War was Over: Cambodia and the Khmer Rouge Revolution*, New York: Public Affairs, 1998

After a brief historical sketch of the Angkor and French colonial eras, Becker describes the Sihanouk, Lon Nol and Khmer Rouge periods in detail. The second edition of her book takes the narrative up to the death of Pol Pot, whom she once interviewed.

Chanda, Nayan, *Brother Enemy: The War After the War – A History of Indochina since the Fall of Saigon*, San Diego and London: Harcourt Brace, 1986; New York: Collier, 1988

This book by a veteran Asian journalist covers events in Indochina, and the region's relations with China and the United States, from 1975, when Communists took control in Vietnam, Cambodia, and Laos, to 1979, when the Vietnamese ousted the Khmer Rouge.

Chandler, David, *The Tragedy of Cambodian History: Politics, War, and Revolution since 1945*, New Haven, CT, and London: Yale University Press, 1991

Chandler, the single most influential western historian of Cambodia, provides a detailed account of events from the end of World War II to 1979, focusing on the politically turbulent period when Sihanouk suppressed opposition groups and aligned his country with China.

Chandler, David, *A History of Cambodia*, third edition, Boulder, CO: Westview Press, 2000

This is a highly readable account of Cambodia's history, from the era of "Indianization" through the period of the Angkor empire, French colonialism, and the Sihanouk era, to the brutal years of the Khmer Rouge regime.

Chandler, David, *The Land and People of Cambodia*, New York: HarperCollins, 1991

An overview of Cambodia's demography, climate, agriculture, traditions, social structures, and history from the Angkor period to the 1980s

CLRDC: Cambodian Legal Resources Development Center, *Laws of Cambodia 1993–1998*, Phnom Penh: CLRDC, April 1998

This collection of statutes passed by the National Assembly elected in 1993 includes the current Constitution.

Curtis, Grant, *Cambodia Reborn? The Transition to Democracy and Development*, Washington, DC: Brookings Institution, 1998

This analysis, by a development consultant who has worked on Cambodia for several years, addresses the long-term impact of the UN peacekeeping mission, and the subsequent efforts by the Cambodian government and development agencies to rebuild the country.

Economist: anonymous, "Cambodia's Wood-fired War," in *Economist*, June 17, 1995

EIU: Economist Intelligence Unit, *Country Profile: Cambodia*, London: Economist Intelligence Unit, 2000

This is one volume in an annual series that has proved invaluable as a source of up-to-date statistics, perceptive economic analysis, and summaries of recent political events.

Global Witness: report by the nongovernmental organization Global Witness, cited in *Phnom Penh Post*, Volume 4, number 6, March 24 to April 6, 1995

Hang Chuon Naron, "The Greater Mekong Subregion: The Impact of the Asian Crisis on Cambodia," in Kao Kim Hourn and Jeffrey A. Kaplan (editors), *The Greater Mekong Subregion and ASEAN: From Backwaters to Headwaters*, Phnom Penh: CICP, 2000

Ieng Mouly, speech to the Economic Forum in Bangkok, April 3, 1995

Kiernan, Ben, *How Pol Pot Came to Power: A History of Communism in Kampuchea, 1930–1975*, London and New York: Verso, 1985

Kiernan traces the growth and activities of the various Cambodian groups that have called themselves "Communist," including the Phnom Penh branch of the Indochinese Communist Party, the Cambodian United Issarak Front, and the Khmer Rouge.

Kiernan, Ben, *The Pol Pot Regime: Race, Power and Genocide in Cambodia under the Khmer Rouge, 1975–79*, New Haven, CT, and London: Yale University Press, 1996

This magisterial work on the era of the "killing fields" has been well worth waiting for. Kiernan, who has devoted his academic career to research on, and analysis of, the Khmer Rouge, combines interviews with survivors and detailed background information to present a convincing portrait of a genocidal regime.

Mysliwiec, Eva, *Punishing the Poor: The International Isolation of Kampuchea*, Oxford: Oxfam, 1988

This little book, packed with statistics, has been dismissed as propaganda in some quarters, but it has lasting value as a record of how the regime installed by the Vietnamese began rebuilding the country.

Ngor, Haing, *Surviving the Killing Fields: The Cambodian Odyssey of Haing S. Ngor*, London: Chatto and Windus, 1988

Haing won an Oscar for his role in Roland Joffe's film *The Killing Fields*, making this biography among the better known (and less likely to become unavailable) of more than a dozen books by Cambodians who survived the Khmer Rouge period.

Norodom Sihanouk, with Bernard Krisher, *Sihanouk Reminisces: World Leaders I Have Known*, Bangkok: Duang Kamol, 1990

By recalling his meetings with world leaders, Sihanouk exposes many of his own political and personal preferences.

Osborne, Milton, *Sihanouk: Prince of Light, Prince of Darkness*, St Leonards, NSW: Allen and Unwin, and Honolulu: University of Hawaii Press, 1994

Osborne, a former diplomat, presents Sihanouk as prince, king, chief of state, resistance leader, and king once again. Cambodia's recent history is so closely bound up with Sihanouk's life that this book amounts to a study of Cambodia from 1941, when he was crowned, until 1993, when he was restored to the throne.

Picq, Laurence, *Beyond the Horizon: Five Years with the Khmer Rouge*, translated by Patricia Norland, New York: St Martin's Press, 1989

Laurence Picq was the only westerner remaining in Phnom Penh after the Khmer Rouge took control, and her account of her life under that regime with her husband, a middle-ranking official, is therefore uniquely valuable.

Ponchaud, François, *Cambodia: Year Zero*, New York: Holt, Rinehart, and Winston, and London: Allen Lane, 1978

Ponchaud, a French Catholic missionary, had been living in Cambodia for ten years when the Khmer Rouge took control of Phnom Penh. After he was forced to leave the country, he gathered interviews and letters from refugees in Thailand and Vietnam, and compared them with the Khmer Rouge's radio broadcasts in order to build up a picture of life inside "Democratic Kampuchea."

UN, *Agreements on a Comprehensive Political Settlement of the Cambodia Conflict*, New York: UN Secretariat, 1991

Sue Downie is a doctoral student in the Department of Politics at Monash University in Melbourne. She worked in Cambodia for six years, first, from 1990 to 1994, as a correspondent for UPI, the BBC, and other organizations, and then, from 1994 to 1995, as an adviser to the government on media affairs. She returns regularly to the country as a consultant and researcher.

NATION STATES

Chapter Thirteen
Laos

Gerald W. Fry and Manynooch Nitnoi Faming

Among the five Communist countries that remain in the world, Laos, formally known as the Lao People's Democratic Republic, receives the least attention either from the media or from scholars. The journalist Stan Sesser has called it the "forgotten country" (see Sesser). Our purpose here is to discuss the future of the political and economic system of Laos, and its potential for long-term sustainability. Of special interest is the distinctive Lao approach to development, which we shall term "Laoism."

Background and Historical Context

The Lao People's Democratic Republic is a landlocked country with a population of 5.5 million, as of 2000, and a total area of 237,715 square kilometers. It is therefore around the same size as the United Kingdom, but has less than one tenth of its population. The terrain is largely covered with rugged mountains. Laos shares a border in the North with China and Burma, in the East with Vietnam, in the West with Thailand, and in the South with Cambodia. The country has great ethnic diversity, with 47 main ethnic groups and 149 subgroups, which are normally grouped into three basic categories: the Lao Lum, ethnic Lao living in the lowlands and valleys; the Lao Theung, non-Lao ethnic groups living on the slopes of the mountains; and the Lao Sung, non-Lao ethnic groups living at the highest elevations (see Chazée, and also Appendix 5 of this book). The country is basically agricultural, with a high proportion of subsistence farming, and 87% of the harvested area is devoted to rice production (UNDP 1999 p. 17).

Roughly 80% of the population is employed in agriculture. Unusually for Asia, Laos has an extremely low population density, at 57 people per square mile, as compared, for example, to 615 people per square mile in Vietnam or 313 people per square mile in Thailand. It is ranked by the UN Development Program as one of the poorest countries in the world, with income per capita of only US$280 in 1999. This was, however, a significant improvement on US$77 in 1966 or US$80 in 1981. The UN Development Program's Human Development Index ranks Laos 136th out of 174 countries, the lowest rank of any country in Southeast Asia with the exception of Cambodia, which is 140th (UNDP 1999 p. 8). On average, Lao children undergo less than three years of schooling and the quality of that schooling is highly uneven. Life expectancy at birth was only 53.81 years in 1998 (according to the World Bank), and it has been estimated that 43% of the population is illiterate (UNDP 1999 p. 6).

The major theme of Lao history has been the remarkable survival of a small landlocked nation surrounded by powerful and assertive neighbors, namely, China, Burma, Thailand, Cambodia, and Vietnam. The first Lao polity was Lan Xang, meaning "land of a million elephants," which was founded by King Fa Ngum in 1353. During the most glorious period of its history, the reign of King Souligna Vongsa (1637–94), Lan Xang controlled much of what is now northern and northeastern Thailand, north and northwestern Cambodia, and parts of Burma and Vietnam, in addition to the territory of modern Laos. During Souligna Vongsa's reign, Lan Xang was a major center of Buddhist learning and art for the whole region (see Stuart-Fox 1997). However, there was a succession crisis following the death of Souligna Vongsa, and the country was subdivided into three kingdoms, Luang Prabang, Vientiane, and Champasak. The country was subsequently colonized by Siam (as Thailand was then known) and, later, by the French (1893–1953) as part of Indochina.

The French generally viewed Laos as a hinterland of Vietnam, to be exploited for the benefit of the French and those sections of Vietnamese society that they favored. One result of French colonialism was that many Vietnamese migrated to Laos, particularly to cities along the Mekong River. During World War II, France lost control of Laos to Nationalist China, Thailand, and the Japanese. After the end of the war, however, the French reoccupied Laos, despite the resistance of a group known as the Lao Issara ("Lao Independence"). This group later broke up into factions, one of which became the leftist Neo Lao Hak Sat ("love the country movement"), the Lao Patriotic Front, and the predecessor of the currently ruling Lao People's Revolutionary Party (LPRP).

After receiving its independence from France in October 1953, Laos became a constitutional monarchy under King Sisavang Vong, who reigned from 1904 to 1959. Soon after achieving independence, Laos was drawn into the Cold War as its political factions became proxies for the superpowers. Thus, the dramatic battle of Vientiane in December 1960 pitted rightwing forces, backed by the United States and Thailand, against the "neutralist" forces of Captain Kong Le, which were backed by the Soviet Union. Civil war was nearly constant throughout the first two decades of independence, there were frequent coups by sections of the military, and all three attempts at coalition government (1957–59, 1962, and 1974) ended in collapse. A remarkable facet of the civil war was that one of the leftist leaders, Prince Souphanouvong, known as the "Red Prince," was a half-brother of Prince Souvanna Phouma, a "neutralist" who cooperated with the faction favoring the United States.

At this stage, US military involvement in Laos, which had been recognized as neutral under the Geneva Accords of July 1954, was secret, and was masked by CIA front organizations such as Air America (see Robbins 1979 and 2000). During eight years of secret war, US planes dropped around 1.60 million tons of bombs on Laos, as compared, for instance, to around 1.36 million tons dropped on Nazi Germany (Robbins 2000 p. 379). The economic development of Laos during this period represented what could be termed "hyperdependence," for the government in Vientiane depended almost entirely on financial support from the United States, and the economy was dominated by US and French business interests. The primary beneficiaries of foreign aid were members of the governing elite and military officials closely associated with the secret war against the Pathet Lao, as the Communist movement had come to be known (see Parker, and Robbins 1979 and 2000). "Pathet Lao" means "Lao nation," reflecting the strongly nationalistic orientation of the revolutionary struggle.

Various social and economic indicators at the time indicated that the vast majority of Lao people were suffering extreme poverty and deprivation. In 1966, as we have seen, income per capita was US$77, while the illiteracy rate was 85% and average life expectancy was only 30 years. The formal monetary economy was limited to imported products from France and Thailand, and small amounts of agricultural exports. The sale of elegant postage stamps to philatelists was the leading source of foreign exchange, apart from the massive US aid, which rarely reached the poorest in Laos, and did little to compensate for the decadence, chaos, and political intrigues that characterized the country (see Strong).

Consolidating the Revolution, 1975–79

With the declaration of the Lao People's Democratic Republic on December 2, 1975, the country achieved genuine unity and independence for the first time in centuries. The monarchy was abolished, after more than 600 years, and peace was restored after nearly 20 years of civil war. The new government received support from the Soviet Union,

Vietnam, China, Cuba, and other Communist countries, but had very limited relations with western states.

By the time the war was over, Laos had lost more than 50% of its population, including large numbers of schoolteachers, professors, administrators, technicians, mechanics, and other professionals. There were only around two million people left in the country, living in widely dispersed and often remote areas. The overwhelming majority were farmers, and most had little knowledge of how to read and write. One of the main aims of the new government was to reform the Lao language in order to facilitate literacy initiatives and, it was hoped, to reduce hierarchical thinking (see Phoumi). During the period of French influence, education had been grossly neglected, except for a small elite. Given the weakness of the wartorn economy and the poor infrastructure, the new regime faced great difficulties in seeking to provide an improved standard of living for its people, which had been one of the major promises of the revolution.

In these and other ways, the Lao revolution was more about securing cultural and political sovereignty than about transforming the economic system. Traditionally, Lao peasants had not suffered the rural inequities associated with land tenancy and plantation systems. Nevertheless, under the leadership of the revolutionary hero Kaysone Phomvihane the new Lao government implemented policies to restructure the economy and society, aiming to establish genuine socialism in a subsistence economy. Kaysone frankly admitted that Laos did not possess a unified national economy, but had a combination of what he called a central economy and local economies. Kaysone attempted to articulate a development strategy appropriate to the basic village nature of Lao society and the special historical background of Laos.

Accordingly, the government introduced economic planning of a special Lao type, developed from the "rice roots." For example, each province was to become self-sufficient in food. The strategy was designed to aggregate local plans and integrate them at the national level. In 1978, a plan for agriculture was introduced, emphasizing the formation of cooperatives and calling for the strengthening of the emerging state sector as a base for future socialist transformation. At this stage, Kaysone was still following the model established in the Soviet Union.

During 1978 and 1979, more than 3,000 agricultural cooperatives were established (Bourdet 1995 p. 165). Some private sector agriculture continued to exist, but it was closely regulated and its output was limited. The cooperatives did contribute to remarkable progress in economic transformation. Numerous localities where traditionally only one crop had been grown now developed new farming techniques, facilitated by technical assistance from other Communist countries. This led to a significant increase in agricultural productivity. However, farmers were not satisfied with the cooperatives because they felt that they received far too little from their hard work and that they were subsidizing those who worked less. As a result, there was opposition to plans for further collectivization.

The government also owned and operated around 300 factories, as well as thousands of small industry and handicraft establishments. However, with such a small population and such an underdeveloped economy, Laos lacked the savings and other resources that were required for nonagricultural production to take off. In these circumstances, the government turned to the Soviet Union, Vietnam, and China for assistance. Up to the late 1980s, the Soviet Union played a key role, providing nearly one half of the foreign aid reaching the country, frequently in the form of commodity shipments (Yerofeyev p. 311). Laos has major foreign debt obligations amounting to nearly US$2.4 billion, of which around US$1.34 billion is owed to Russia alone (calculations based on Europa Publications p. 650).

The New Economic Policy (NEP), 1979–86

The performance of the agricultural cooperatives proved to be disappointing, and in July 1979 the government abruptly abandoned the cooperative movement and started decollectivizing. Farmers were given the option work on their own farms or plots, or voluntarily stay with the cooperatives, while the government started to liberalize trade and to implement market-oriented reforms under what was called the New Economic Policy (NEP). Despite this considerable shift in policy, the government still emphasized the enhancement of the economic role of the villages. Villages were allowed more self-government as the number of cooperatives gradually declined. However, the relatively low levels of education and experience among provincial and local leaders often created problems for the central government, resulting in special complexities and inefficiencies.

The first stage of experimentation under the NEP ended in 1982–83, with the government facing rising inflation and inadequate coordination of activities between the various levels of administration. Once again, the government changed its policy, this time to focus on strengthening planning mechanisms and to renew the emphasis on general cooperatives. Nonetheless, the shift back to cooperatives did not result in significant improvements in the economy. At the same time, this failure, along with the accompanying political repression, caused the country to lose between 10% and 15% of its population between 1975 and 1985 (see Dommen).

The regime was still pursuing a foreign policy based on isolation from the West and close relations with the Soviet bloc, Vietnam, and China. As it became clear that Laos was increasingly modeling its economy and society on the Communist pattern, many educated Lao, as well as those who had been associated with the United States during the secret war, fled as refugees to Thailand or to the West. The harshness of the Khmer Rouge regime in neighboring Cambodia reinforced the fears of such individuals. As the result of their exodus, there is now a large Lao diaspora scattered around the globe, primarily in the United States, Canada, France, and Australia (see Mayoury). The fact that, in many cases, some members of urban bourgeois families stayed the course, while others fled, can be seen either as an example of the Lao Buddhist "middle way," or as a tribute to their enduring pragmatism. In any case, remittances from the Lao diaspora have since become an important source of foreign exchange. More recently, some skilled individuals from the diaspora have returned to Laos to help in the development of the country.

The Shift to the New Economic Mechanism (NEM), 1986–92

In 1985, even as the NEP was still officially being implemented, the ruling party openly acknowledged the need to slow the pace of the "transition to socialism," primarily because of the various economic problems that had arisen. The government was particularly concerned about two major issues. The first was the country's poor physical infrastructure. The years of warfare had left little communication infrastructure functioning to serve people living in the rural areas. There were only a few roads connecting the valleys and the main alluvial plains, and these roads were inaccessible during the rainy season, making communication between the capital and the provinces almost impossible. Air travel, which represented a major cost, was the main and often the only way to travel from one province to another. The second issue was that there were inadequate incentives for building the economy. The prerequisites for achieving the goal of a more advanced economy were the generation of a marketable surplus in agriculture and the achievement of sustained improvements in productivity, through a deeper division of labor and a greater accumulation of capital.

A new slogan, "produce as much as your capacity, consume as much as you desire" (as quoted in the newspaper *Vientiane Mai*, May 15, 1983), reflected the much more liberal orientation of the government, which also underpinned the new reform plan introduced at the fourth party congress in 1986. This new policy was known as *Jintanagan Mai*, which literally means "new thinking," although the English term officially adopted was "New Economic Mechanism" (NEM). The reform program called for an opening of the economy to foreign investment, and the use of prices and other market mechanisms rather than state planning. In these respects, it was clearly related both to the restructuring (*perestroika*) then being attempted in the Soviet Union and to the similar "new structure" (*doi moi*) program being formulated in Vietnam around the same time (Schultze p.182). Like the ruling parties in those countries, the Lao party had recognized that rigid planning would not result in needed improvements in the economic welfare of the people.

The first law related to foreign investment was promulgated on July 25, 1988 (Schultze p. 184; see also Sunshine), and at the fifth party congress, in 1991, the government set out the new national agenda, building on the first few years of the NEM. Laos now had to increase exports, promote tourism and rural development, entice its shifting cultivators into stationary jobs, reform its financial system, encourage more investment from abroad, and introduce extensive administrative and legal reforms, in order to make the economy and the investment climate more transparent.

Domestically, the NEM has provided the context for the Lao government to orchestrate a smooth but gradual transition from a centrally planned system to an emerging market system. The government has avoided taking draconian measures to reduce the size of the public sector, which would have had extremely adverse effects on those losing their sources of livelihood. Nevertheless, since 1989 the state has sold off a large proportion of its enterprises through various divestment procedures. Even the national telecommunications company is now a joint venture with foreign corporations (Stuart-Fox et al. p. 74). Such joint ventures are among the most remarkable results of the opening up of Laos to increased foreign direct investment, aid, and tourism. This reversal of the country's ingrained isolation from the West was one of the main features of the NEM, particularly at the beginning of the 1990s, when the government began to anticipate that the country could no longer depend on the Soviet Union, or its former allies in central and eastern Europe, for large-scale development assistance.

The reforms introduced under the NEM also included the liberalization of imports and the abandonment of attempts by the government to fix the exchange rate of the Lao currency, the kip, against the US dollar and other major currencies. The present form of the currency, known as the "new kip" or "National Bank kip," was introduced in 1979, after the "liberation kip," adopted in 1976, had been subjected to a major depreciation (Hopkins p. 187). In recent years, following liberalization, the exchange rate has once again undergone considerable fluctuation (see below).

Perhaps the most remarkable indication of the impact of the NEM is the fact that, on average between 1992 and 1996, Thailand accounted for 74% of all the foreign direct investment in Laos and received 30% of all exports from Laos (Okonjo-Iweala et al. p. 49). As a landlocked country, Laos is heavily dependent on both Vietnam and Thailand for access to the sea, yet as recently as late 1987 and early 1988 there was a border war between Laos and Thailand, which lasted for several months. Relations have significantly improved since then, although in March 1997 several high-ranking technocrats within the ruling party of Laos were said to be excessively "Thai-oriented" and were demoted. A major concern of the Lao leadership is what is seen as excessive cultural and linguistic influence from Thailand, particularly given the popularity of Thai television and radio in Lao border towns along the Mekong River (see Mayoury and Pheuiphanh). In this and other respects, the relationship between these two neighbors has analogies with the relationship between the United States and Canada.

The United States is the source of the second largest block of foreign investment in Laos. In 1995, President Bill Clinton approved a change in US policy to make Laos eligible to receive development assistance. However, Laos has not yet achieved "most favored nation" status in its dealings with the United States, and, with the tariff on US imports from Laos set at 40%, most Lao products are simply not competitive in the world's largest single market.

Laos also receives considerable economic aid and technical assistance from the World Bank and the Asian Development Bank (ADB), from Japan, the leading bilateral donor, and from Australia, France, Germany, Switzerland, and the Nordic countries. Laos is now on good terms with almost every other country in the world, including many that it formerly had no diplomatic relations with. In July 1997 it became a member of ASEAN, and in 1998 the government submitted an application to join the World Trade Organization (see Anderson).

Nowadays, physical infrastructure, including projects to build dams and telecommunications systems, as well as the upgrading of health and education services, accounts for the largest single proportion of investment in Laos, both by the government and by bilateral or multilateral donors such as Japan, the World Bank, the ADB, the UN Development Program, and member states of ASEAN. The ADB, for example, promised to provide US$270 million in concessionary loans for the period 1998–2000. The funds from these various sources combine to make Laos a country with one of the highest levels of aid per capita in the world. However, Laos has also been successful in attracting foreign direct investment from private sector sources.

Finally, it should be noted that the first stage of implementation of the NEM was accompanied by a modest degree of political liberalization. In March 1989, in the first nationwide elections since the revolution, two candidates were permitted to run for each seat in the legislature, following "vetting" by the ruling party, and 18% of those who won seats were not party members. In 1991, the hammer and sickle were removed from the state symbol of Laos, to be replaced by an image of the country's most revered Buddhist temple, the That Luang. The present Constitution, the first since the revolution, was also promulgated in the same year, to institutionalize the new reforms. Nevertheless, freedom of expression, particularly the advocacy of alternative political systems, remains strictly limited.

The Boom Years, 1992–97

The substantial progress made in economic liberalization and structural reforms during the 1980s, under the visionary leadership of Kaysone, established the basic foundations for the private sector activity that contributed to the economic success of Laos from the late 1980s to 1997 (Kalra and Sløk p. 20), although Kaysone himself did not live to see them, for he died in November 1992. Laos achieved an unprecedented level of economic growth in the period 1992–97, averaging 7% a year, and showed clear signs of becoming as dynamic as other economies in the Asia-Pacific region. During the ten years following Kaysone's introduction of the NEM, that is, 1986–96, the economy nearly tripled in size, while income per capita more than doubled. During this period, inflation was generally held below 10%, the exchange rate was relatively stable, and gross official reserves grew substantially.

At the beginning of the revolution, in 1975, Laos had little international trade. Gross international trade was equivalent to just 11.1% of GNP and exports were sufficient to cover only 12% of imports. During the phase of consolidation, trade was almost exclusively with other Communist countries. However, by the time that the NEM was introduced in 1986, the economy had become considerably more internationalized: there was an increase of 434% in international trade between 1976 and 1986. Exports then increased more than tenfold between 1986 and 1996. By 1998, international trade was equivalent to 72% of GDP and exports were running at a level sufficient to cover 67% of exports, the highest level achieved since before the revolution of 1975.

The internationalization of Laos is also evident in its opening up to tourism since 1986. Before the NEM was introduced, Laos had extremely few foreign visitors, and most of those who did reach the country came from the Soviet Union, East Germany, and other Communist countries. Nowadays, by contrast, tourism is being actively promoted and encouraged, although it also continues to be carefully monitored. In 1993, for example, the government set aside 17 national biodiversity conservation areas, covering more than 10% of the country's total land area, in order to preserve its impressive and often unique flora and fauna (Europa Publications p. 642). The recent decision by Unesco to designate Luang Prabang, the former royal capital, as a World Heritage site has also given greater visibility to Laos as a tourist destination: many consider Luang Prabang to be the best preserved traditional city in Southeast Asia. With economic and technical assistance from Thailand, the city's airport has been remodeled, and now permits direct flights to Hong Kong, Kunming in southwestern China, and destinations in

Burma and Thailand. In 1991, the total number of tourists visiting Laos was only 37,613; by 1995, that number had grown dramatically, to 403,000. Laos now receives more tourists per head of its population than most of its neighbors and nearly as many as Thailand.

The Asian Crisis and After, 1997–2000

On July 2, 1997, the Thai government announced that it would allow the exchange rates of its currency, the baht, to float. This triggered the Asian financial crisis, as the baht and the Bangkok stock market went into free fall, and "contagion" quickly spread to other countries such as Indonesia, South Korea, and Malaysia. Since Laos does not have a stock market and the kip is not an internationally traded currency, it appeared at first that, like China, Vietnam, and Cambodia, Laos might be immune, and for a short period the kip retained its stability. However, it too was affected, and its value fell by a greater margin than that of any other Asian currency, from NK920 to US$1.00 in September 1997 to NK9,350 to US$1.00 by November 1999. The exchange rate crisis had a devastating impact on the stability of the Lao economy, and the purchasing power of the Lao people, particularly those living in urban areas, receiving salaries paid in kip, and dependent on the import of Thai commodities, declined dramatically. At the same time, there were some windfalls for some Lao, particularly members of the socioeconomic elite associated with the dollarized part of the economy. Thus, the crisis had the effect of further exacerbating the increasingly visible signs of inequality in Laos.

The free fall of the kip can be explained by several factors, which were all in play simultaneously. Given the increasingly intertwined relations between the Thai and Lao economies, serious contagion was inevitable. Initially, with the significant drop in the value of the baht, the trade balance shifted even more in favor of Thailand, as goods imported from that country became much cheaper for Lao purchasers. For example, Thai exports to Laos during the first three months of 1998 were worth B4.4 billion, while Lao exports to Thailand over the same period were worth only B384 million. Soon, however, as industrial activity in Thailand underwent a dramatic contraction, Thai imports of Lao electricity decreased significantly and Thailand was forced to renegotiate the price paid to Laos for the electricity. At the same time, investment in Laos, not only from Thailand but also from Malaysia, South Korea, and other Asian countries, fell sharply. As a result, 237 foreign companies, most of them Thai-owned, ceased their operations in Laos between 1997 and 2000 (*FEER* p. 150). All these trends led to rising demand for foreign exchange at a time when the supply was falling, which in turn exerted strong downward pressure on the value of the kip. As it declined rapidly, reaching NK3,682 to US$1.00 by October 1998, there was a panic flight out of the kip, which further exacerbated the crisis. With negative real interest rates, there were virtually no incentives for Lao nationals to keep any savings they had in kip.

The Lao government's fiscal and monetary policies also tended to aggravate the currency crisis. The central bank was pursuing a loose monetary policy, and was unable to control the expansion of credit by the numerous public and private banks in Vientiane. Monetary growth in 1998 was extremely high, at 113.3%. The government was also embarking on a major agricultural infrastructure project aimed at a significant expansion of the area of irrigated land available for growing rice. The project required the purchase of an extremely large number of waterpumps, which absorbed a large proportion of the nation's foreign exchange reserves. However commendable this project might be, given the goal of promoting more even development, the timing was unfortunate to the extent that it directly contributed to macroeconomic instability. From a Keynesian or Marxian perspective, such pump-priming at a time of deepening recession could be interpreted as a welcome exercise in humanistic economics, but it clearly contradicted the normative philosophy of the international agencies, such as the IMF and World Bank, which were highly critical of the government's "lax" policies (see Okonjo-Iweala et al.).

Nevertheless, by late 1999 there were positive signs that the economy was beginning to recover from the crisis. The value of the kip appreciated considerably, reaching a peak of NK7,600 to US$1.00 in May 2000, while the rate of inflation decreased by around one third. The recovery was also supported by assistance received both from China and Vietnam in a concerted effort to stabilize the currency. While the international agencies and the leading bilateral donors remained frustrated with the fiscal and monetary policies of the Lao government, they have also proved to be willing to go on assisting Laos. In addition, the drive to promote tourism in "Visit Laos Year" (1999–2000) made an important contribution foreign exchange earnings and by April 1999 the industry had become the country's leading source of such revenue (*FEER* p. 151).

The political impact of the economic crisis led to a major change of leading personnel in August 1999, notably including the dismissal of both the governor of the central bank, Cheuang Somboukham, and the Deputy Prime Minister for Economic Affairs, Kamphoui Keoboualapha (Stuart-Fox et al. p. 75). For Kamphoui, a prominent technocrat already known for his "Thai orientation," this was a second dismissal in little more than two years. The crisis has also contributed to the tendency within the Lao leadership to turn toward Vietnam and China, and away from those donors and agencies that have been critical of government policy.

Goals for the 21st Century

Against this background, the most critical goal is clearly to try to restore the macroeconomic stability that characterized the boom years of the 1990s, described above. A second and related goal is to diversify and enhance the country's foreign exchange earnings. According to Hans U. Luther, a German development economist who specializes in analyzing Laos, the last year in which Laos had a trade surplus was 1928 (Luther p. 44). Despite the small scale of its economy, and its reluctance and inability to develop heavy industry, it is possible for Laos to achieve such diversification and enhancement if it can continue to develop its existing industries – chiefly mineral extraction, hydro-electricity, textiles and garments, and tourism – and begin to develop cash crops on a larger but sustainable scale, such as coffee, tea, timber, and new types of rice suitable for the Japanese and other markets. Laos certainly has excellent potential in these areas. For example, as Thailand continues its own recovery from the crisis, its demand for Lao electricity will rise once again; Nike is already beginning to get involved in garment production in Laos; and the drive to promote tourism has provided the country with improved infrastructure and marketing. Laos could also benefit from a more active encouragement of remittances from the Lao diaspora and a reinvigoration of its philatelic exports, for which China and France in particular may provide significant markets.

Laos also needs to restore the confidence of foreign investors in order to revive and then increase the flow of foreign direct investment, particularly in those industries that will best facilitate an increase in foreign exchange earnings. With economic recovery gathering pace in Asia, Thailand, Malaysia, and South Korea may well become important sources of finance once again, while China also represents a potential source of large-scale investment, particularly in tourism and other services.

Attracting investments from abroad would be made easier if the government's reform program could be accelerated. Many of the enterprises that remain in the hands of the state have not yet felt the effects of the NEM, and there is persistent weakness in the financial sector, which is burdened with nonperforming loans (Kalra and Sløk, p. 18). The non-labor costs of doing business in Laos are still relatively high, and the state's regulations are still relatively onerous and/or ambiguous, to such a degree as to deter some potential investors. The government has not yet learned how to facilitate private sector development rather than put obstacles in its way, although it is to be hoped that it will not go so far in the quest for investment that it abandons its crucial role in preventing development that damages the natural environmental or the national culture, for example through inappropriate logging or the spread of a commercialized sex industry.

While it seeks to promote economic growth, the government is also seeking to address the problems of uneven development related to the country's poor physical infrastructure (see Bourdet 1998). A major highway from North to South has recently been completed, and projects are under way for roads that will link Laos more conveniently with ports in Vietnam, but the commitment to secure more even development remains essential. The recent major expansion of the area of irrigated agricultural land and the achievement of self-sufficiency in rice production are both encouraging signs of such a commitment on the part of the ruling party and the government. Laos is suitably located to become the site of important land links between southern China and major ports in Southeast Asia; alternatively, Laos could gain little from such links, while suffering social and environmental degradation.

In relation to foreign investment and domestic infrastructure alike, Laos also needs to maintain its focus on developing the skills and talents of its people. This goal represents a major challenge in itself, given the country's remarkable ethnic diversity and the remoteness of many of its rural communities. In an economy that is increasingly oriented to private sector development, the government will need to do even more than in the past to enhance the quality of people's lives, to provide greater opportunities for women, to improve productivity and competitiveness, and thus continue to attract foreign investment.

Finally, Laos faces the problems and opportunities inherent in its relatively high fertility rate, which in 1998 stood at an average of 5.48 live births for each woman of childbearing age (according to the World Bank). In 1999 alone, the country's population grew by 2.44%, which meant that its real economic growth, at 5% of GDP, was equivalent to growth per capita at less than 2.5%. The high fertility rate places a tremendous burden on Lao women and partially explains their generally low level of access to opportunities for personal development and employment. However, because the density of population is so low, and there are continuing concerns about the country's security in the midst of much larger neighbors, the government has been unwilling to introduce serious measures for family planning and population control, preferring to maintain its limited policy of encouraging the greater spacing of births.

Conclusion and Prospects

The regime in Laos has had to face major challenges in 1999 and 2000. The free fall of the currency in the wake of the Asian crisis has adversely affected the urban populations and their purchasing power, and overall prosperity has declined. Yet in the 21st century, as in the closing years of the 20th, the future of the regime, and the economic system, depends fundamentally on the capacity of the government to fulfill the revolutionary ideals for which it claims to stand, and to implement the basic ideology articulated by its revolutionary hero Kaysone.

There are three main possibilities. The first is that the regime collapses, because the ruling party fails to live up to its proclaimed ideals, does not deliver prosperity to the people, and allows corruption and inequality to become excessive. The second is that the existing political and economic system continues in place, burdened with persistent problems related to the small scale of the Lao economy and its dependence on external financial support. The third possibility, and the most optimistic prediction, is that Laos may move toward being the Switzerland or Luxemburg of Southeast Asia, exploiting its solid resource base population and the revival of the Asian economies to sell increasing amounts of hydroelectric power to Thailand and other neighbors, and to attract extensive investment and aid that would enable it to enhance the development of its labor force and its export capacities. Such a development path would permit Laos to preserve its unique political and economic system, which (in our view) creatively blends capitalism, Buddhism, and Marxism.

Indeed, although Laos is not, of course, a significant global player either economically or politically, it represents, at least for some observers, a fascinating case study in key issues of political economy as the new decade and century unfold. Soon after taking power, the

Lao People's Revolutionary Party realized how integral Buddhism was to the way of life of the majority of Lao people and embraced Buddhism as part of the new state's official ideology. Laos can therefore be seen as both Buddhist and Marxist (see, for example, Stuart-Fox 1996). From the perspective of the Lao leadership, pure capitalism and pure Communism are both too extreme, and the Lao mix of Adam Smith, Karl Marx, and the Buddha represents an interesting and innovative attempt at political economy, which may be called "Laoism." The approach is well-illustrated in the following passage from an address given by Samane Vignaket, President of the Lao National Assembly, to the New York Conference of Presiding Officers of National Parliaments (as reported in *Vientiane Times*, September 12, 2000):

> "The lifestyle and moral civilization of any society or of any nation should not be regarded as a criterion or model for any nation. The direct or indirect imposition of form, social political system and ideology by one nation on another is unacceptable, and will only undermine the fundamental principles of international relations."

Integral to "Laoism" is the *selective* pursuit of development and internationalization, avoiding the potentially excessive costs of conventional transitions to modernity. As a latecomer to the development process, Laos can benefit from the mistakes of its neighbors, and seek more sustainable and even development. For example, private cars are not permitted access to the Friendship Bridge across the Mekong River to Thailand, which was completed in 1994; buildings from the French colonial era cannot be destroyed but must be restored; and there are strict limits on the heights of buildings in the capital. The government has also acted to prevent the importation of such elements of foreign popular cultures as karaoke, the sex industry, and the depiction of violence, crime, and sex on television. The Lao system emphasizes moral education and high-quality parenting, and the country is largely free from crime.

As it carefully and cautiously pursues its own path to development, Laos faces many complex challenges. Many observers are skeptical that any country can develop and internationalize with any significant degree of selectivity, on the grounds that all nation-states are now subject to the same powerful forces of globalization. With Communism on the decline throughout the world, the question therefore arises whether Laos might eventually be the last remaining Communist state. Two basic challenges face the current regime and largely determine its potential for survival. First is its ability to deliver economic prosperity to its people. Second is its ability to ensure reasonable equity and opportunity for all, particularly those living in remote areas and members of ethnic minorities. Finally, and perhaps most importantly, there is the question of the survival of "Laoism" itself, as an innovative ideology that may have increasing relevance and appeal to a world in which there is growing disillusionment with the excessive materialism, violence, and social disorder associated with industrialization and modernity.

Further Reading

Anderson, Kym, *Lao Economic Reform and WTO Accession: Implications for Agricultural and Rural Development*, Singapore: Institute of Southeast Asian Studies, 1999

This study assesses the potential impact on Lao agriculture of a movement toward freer trade.

Bourdet, Yves, "Rural Reforms and Agricultural Productivity in Laos," in *The Journal of Developing Areas*, number 29, January 1995

This is a systematic assessment of changes in the Lao rural economy associated with major reforms, written by an economist at the University of Lund, in Sweden, who has done extensive research on the Lao economy.

Bourdet, Yves, "The Dynamics of Regional Disparities in Laos: The Poor and the Rich," in *Asian Survey*, 1998

The focus of this article is the persisting problem of regional disparities and uneven development. Bourdet examines the effect of reform policies on provincial development.

Chazée, Laurent, *The Peoples of Laos: Rural and Ethnic Diversities*, Bangkok: White Lotus, 1999

Chazée provides a valuable and extensive overview of ethnic diversity in Laos, based on detailed data on various ethnic groups and subgroups.

Chi Do Pham (editor), *Economic Development in Lao PDR: Horizon 2000*, Vientiane: Bank of the Lao People's Democratic Republic, 1994

This anthology of diverse articles on the Lao economy in the early 1990s was compiled and edited by the former Resident Representative of the IMF.

Dommen, Arthur J., *Laos: Keystone of Indochina*, Boulder, CO: Westview Press, 1985

This remains a useful assessment of the early period after the Lao revolution.

Europa Publications, *Far East and Australasia 2000*, London: Europa Publications, 1999

This major annual reference work gives detailed assessments of each of the countries of the Asia-Pacific region, with sections on physical and social geography, history, and economy, as well as a statistical survey and a directory of the government.

FEER: *Far Eastern Economic Review* (editors), *Asia Yearbook 2001*, Hong Kong: *Far Eastern Economic Review*, 2000

The latest edition of an annual review of the economies of the region that emphasizes the connections between economic activity and politics

Hopkins, Susannah, "The Economy," in *Laos: A Country Study*, Washington, DC: Library of Congress, 1995

A useful survey chapter that forms part of a standard reference source for basic data on the country

Kalra, Sanjay, and Torsten Sløk, "Inflation and Growth in Transition: Are the Asian Economies Differerent?", Washington, DC: IMF, 1999

The authors use detailed economic data on China, Laos, Vietnam, and Mongolia to make some valuable comparisons and contrasts.

Kaysone Phomvihane, *Revolution in Laos*, Moscow: Progress Publishers, 1981

An English translation of Kaysone's basic work articulating Lao political ideology

Luther, Hans U., *Learning from the Asian Crisis*, Vientiane: Lao-German Economic Training and Advisory Project, February 1999

A respected development economist specializing in Laos surveys the impact of the recent crisis on the country's economic performance and prospects.

Mayoury Ngaosyvathn, *The Lao in Australia: Perspectives on Settlement Experiences*, Nathan, Queensland: Centre for the Study of Australian-Asian Relations, Faculty of Asian and International Studies, Griffith University, 1993

An excellent overview of the Lao diaspora in Australia, by one of Laos's leading scholars

Mayoury Ngaosyvathn and Pheuiphanh Ngaosyvathn, *Kith and Kin Politics: The Relationship between Laos and Thailand*, Wollongong, Australia: Journal of Contemporary Asia Publishers, 1994

This is the best and most comprehensive analysis of the complex and important relationship between Laos and Thailand, presented from a Lao perspective.

Okonjo-Iweala, Ngozi, et al., "Impact of Asia's Financial Crisis on Cambodia and the Lao PDR," in *Finance and Development*, Volume 36, number 3, 1999

An IMF perspective on how the Asian economic crisis has affected banking, investment inflows, and society at large

Parker, James E., *Codename Mule: Fighting the Secret War in Laos for the CIA*, Annapolis, MD: Naval Institute Press, 1995; Bangkok: White Lotus, 1997

A firsthand account by a participant in the CIA's secret war in Laos

Phoumi Vongvichit, *Waiyagon Lao* [Lao Grammar], second edition, Sam Neua: Central Education Department, 1991

This text, originally published in 1967, articulates the language policy of the Lao government, which is intended both to facilitate literacy training and to develop what the government calls "revolutionary consciousness."

Robbins, Christopher, *Air America*, New York: Putnam, 1979; as *The Invisible Air Force*, London: Pan, 1981; and *The Ravens: Pilots of the Secret War on Laos*, Bangkok: Asia Books, 2000

The first of these two books is an exposé of Air America's role as the airline of the CIA in the secret war in Laos; the second uses materials made public since 1979 to add further detail to the author's account of a military venture that is still remarkably little known.

Schultze, Michael, *Die Geschichte von Laos: Von den Anfängen bis zum Beginn der neunziger Jahre* [The History of Laos from its Origins until the Beginning of the 1990s], Hamburg: Institut für Asienkunde, 1994

A survey that provides an interesting German perspective on Lao history

Sesser, Stan, *The Lands of Charm and Cruelty: Travels in Southeast Asia*, New York: Knopf, 1993; London: Picador, 1994

One chapter in this book, "Laos: The Forgotten Country," provides a journalistic account of the country at the beginning of its transition to a more internationalized and market capitalist economy. It is less ethnocentric than many accounts of this type.

SPC-NSC: *National Human Development Report 1998*, Vientiane: State Planning Committee and National Statistics Center, 1998

An official publication providing a wide variety of statistical indicators related to social development and quality of life

Strong, Anna Louise, *Cash and Violence in Laos and Vietnam*, Peking: New World Press, and New York: Mainstream Publishers, 1962

A devastating critique of the political economy of Laos after "independence," written from a strong leftist perspective

Stuart-Fox, Martin, *A History of Laos*, Cambridge, New York, and Melbourne: Cambridge University Press, 1997

A comprehensive and balanced history of Laos, the best by a western scholar

Stuart-Fox, Martin, *Buddhist Kingdom, Marxist State*, Bangkok: White Lotus, 1996

In this book, the leading western historian of Laos provides valuable insight into the unique political economy of Laos.

Stuart-Fox, Martin, Tin Maung Maung Than, and Melina Nathan, "Indochina and Myanmar," in Daljit Singh and Nick J. Freeman (editors), *Regional Outlook Southeast Asia 2000–2001*, Singapore: Institute of Southeast Asian Studies, 2000

Sunshine, Russell B., *Managing Foreign Investment: Lessons from Laos*, Honolulu, HI: East-West Center, 1995

This valuable and informative case study, by a former adviser to the Lao government on international investment, documents how Laos attracted more than US$1 billion in international investment from 30 countries in the early 1990s.

UNDP 1998: UN Development Program, *Lao PDR's Development Partners: Profiles of Cooperation Programmes*, Vientiane: UN Development Program, 1998

UNDP 1999: UN Development Program, *Development Cooperation: Lao PDR. 1998 Report*, Vientiane: UN Development Program, March 1999

A comprehensive report, with detailed statistics and budgets, on the development assistance activities of major bilateral and international donors

Vientiane Times: "NA President Addresses New York Conference," in *Vientiane Times*, September 12, 2000

Yerofeyev, Alexander, "Foreign Economic Assistance to Lao PDR: Transition from Soviet Aid to other Bilateral and Multilateral Aid," in Chi Do Pham (editor), cited above

Yerofeyev, who once served in Laos as a diplomat, provides an excellent overview of the history of Soviet economic assistance to the country.

Dr Gerald W. Fry is Professor of International/Intercultural Education in the Department of Educational Policy and Administration at the University of Minnesota. He has been the team leader for major projects funded by the Asian Development Bank in both Laos and Thailand. *Manynooch Nitnoi Faming* is a doctoral candidate in sociology at the University of Hong Kong. She has worked with the UN Development Program in the Office of the Prime Minister of Laos.

NATION STATES

Chapter Fourteen
Burma

Clark D. Neher

The last decade of the 20th century saw the democratization and liberalization of dozens of nations around the world. Countries that had formerly been controlled by Communist governments became quasidemocracies; military dictatorships in Latin America and Africa gave way to more liberal regimes; and, in Asia, a number of civilian dictatorships were replaced by elected governments seeking to create open societies. The transition was not a smooth one in any of these cases, and it is not yet clear whether democratization will prevail in all of them. Burma presents one of the most obvious exceptions to this trend toward transformation. As the world moved toward the 21st century, not only did Burma – in contrast to all the other countries of Southeast Asia – remain closed, but its government actually became more despotic.

The anachronistic nature of modern Burma is reflected in the contrast between its oppressive political system, on the one hand, and, on the other, its open culture and its highly sophisticated people. With its abundant natural resources, Burma should be flourishing, like its neighbor Thailand. With its strategic location, between South Asia and the rest of Southeast Asia, Burma should be a center for international trade and diplomacy. Instead, it is the most isolated nation of any in Asia, with the exception of North Korea. With its traditionally strong educational infrastructure, Burma could be the most educated society in Asia, but the universities have been closed for over a decade and the school system in general has atrophied as a result of the lack of government support. The tragedy of Burma is that the nation could be much more than it is, and yet it remains one of the world's poorest societies.

At the same time, the fact that it is impossible to be certain just how poor the Burmese are is just one example of how the regime has prevented its own people, let alone the outside world, from receiving reliable information about the country. The Burmese authorities do not want accurate data made available, at least partly because much of it makes the leadership look ineffective. Thus, while the US Central Intelligence Agency estimates that GNP per capita (at purchasing power parity) is around US$1,200, the World Bank estimates GDP per capita at around US$300, while other authorities quote a figure for annual income per capita of around US$700, compared to US$3,000 in Thailand and US$30,000 in the United States. Similarly, while it is widely estimated that average life expectancy is less than 60 years, compared to 70 years in Thailand and 77 in the United States, it is possible that the true figure is even lower.

Political Culture

In any society, every facet of politics is affected by the prevailing pattern of values, beliefs, and attitudes. This pattern provides some order and meaning to the political process. In Burma in particular, the religious beliefs of its people are central to understanding how the political system works. Although Christianity, Islam, and other faiths have significant numbers of followers, Theravada Buddhism, the faith of the majority of the population, is at the heart of Burmese culture. This form of Buddhism teaches moderation in all areas of life, and the acceptance of, and acquiescence in, the place in life given to one by fate. The political passivity of most Burmese and the country's low level of economic development have often been explained in terms of the Buddhist ideas that desire and greed cause suffering, and that there is a unity of virtue and power. Those who have power are good and deserve power. Power justifies itself. The traditional Burmese deference to authority is congruent with Buddhism's teachings (see, for example, Maung Maung Gyi's book, cited below).

Burmese political culture has also often been said to display a lack of trust in human relations. This distrust has led to an incapacity to form the associational organizations that in other nations have become the core of "civil society," generally regarded as a prerequisite for modern democratic government. This distrust has been used by the present military dictatorship to turn Burmese against Burmese and to impose new political values on the people.

However, the fact that Burma is an ethnically diverse nation makes it difficult to generalize about patterns of attitudes, values, and beliefs. Burma has been threatened by civil war ever since it achieved independence from the British in 1948. Around one third of the population is made up of minority peoples, including the Shan, the Karen, the Rakhine (or Arakhanese), the Chin, the Kachin, and the Mon, as well as people of Chinese and Indian descent. These minorities often live separately from the majority Burmans; they speak different languages and have different traditions. It is this diversity that the military dictatorship has used to rationalize its oppression of the Burmese people. There is a divided system in the ethnic states, which lie mostly in northern and eastern Burma, consisting of territories under the control of the military and those that are dominated by insurgent forces demanding independence from Burma or autonomy within it.

In 1989, the military dictatorship changed the English name of the country from "Burma" to "Myanmar" and also changed the English names of the major cities: for example, the capital, generally known as Rangoon, has been renamed "Yangon." However, these changes have not been accepted by many in the international community, nor by opposition groups in Burma itself, who prefer to retain the older names. "Myanmar" comes from the name used for themselves by the majority people, the Burmans, who are not related ethnically to many of the inhabitants of the northern mountain areas mentioned above. The relative proportions of these minority groups within the total population are as uncertain as most data on contemporary Burma, partly because, for obvious reasons, it serves the regime's interests to promote figures for the minorities that underestimate their true numbers. However, assuming, as most international agencies do, that the total population is around 48 million, then it seems likely that Burma is home to around 32 million Burmans, and to at least four million Shan, three million Karen, 2.5 million Rakhine, one million Mon, and 600,000 Kachin, as well as up to one million people of Indian descent. The truth is that no one, even in Burma itself, knows the exact figures. In any case, given that the Burmans probably account for only around two thirds of the population, and that the name changes were made without consulting representatives of any of the country's diverse peoples, using the name "Burma" instead of "Myanmar" is one small way of showing solidarity with the forces that are working to democratize the nation.

Political History

Burma, like all the nations of Southeast Asia except for Thailand, was colonized by a western power. Britain absorbed Burma into its Indian Raj (empire) in 1886, subsequently exploited the economy, and treated the Burmese as second-class citizens in their own country. At the same time, the British strengthened the nation's infrastructure by building railroad and port facilities, and schools, and established a functioning bureaucracy. In 1937, Burma was formally separated from the Raj and made into a British colony with a degree of internal self-government, but this arrangement had little time to develop before World War II began and the Japanese invaded. Burmese nationalists, with support from the

Japanese, eventually ousted the British, and the country became fully independent, as the Union of Burma, in January 1948. Burma's nationalist leader Aung San was to be the first leader of independent Burma, but he was assassinated by rival nationalist rebels just months before he was to take office.

Replacing Aung San, the venerated U Nu became Burma's first Prime Minister. Although he was personally incorruptible, he was not an effective administrator. The most prominent general in the Burmese army, Ne Win, took over the country in 1958, claiming that it was necessary to restore order, and then turned the government back to U Nu in 1960. However, Ne Win had tasted power, liked it, and took over the country once again, by coup d'état, in March 1962, this time permanently.

Ne Win promised to modernize the nation and to prevent the contending hill peoples from attaining their goals of independence or autonomy. Ne Win's goals were not achieved, however. The economy went into a long-term tailspin when the government nationalized the private sector, drove out minority entrepreneurs, took over peasants' land, and centralized all economic decisions in the hands of generals who knew nothing about economics. The result was a depression that has continued up to the present. Ne Win outlawed all political parties except for the Burma Socialist Program Party (BSPP), which had been created by the army, arrested oppositionists, censored the press, exploited and oppressed the hill peoples, and denounced the concept of democracy, which Ne Win deemed inappropriate for Burmese culture.

The Burmese were humiliated by the declining conditions within their country, including the deterioration of the educational and health systems. Their frustration led to a massive uprising in August 1988. Farmers, students, bureaucrats, business leaders, monks, and laborers went onto the streets of the cities and villages, demonstrating against the government. The protesters had no real leadership or program except for the abstraction "democracy," although many were infuriated by the government's disastrous economic policies, including the voiding of the Burmese currency, the kyat, an act that impoverished millions of Burmese. Ne Win had nullified the standard denominations of the kyat and substituted new banknotes in denominations of 45 and 90 kyat simply because his lucky number was nine. Students had lost their tuition money, shopkeepers had had to close their businesses, and a black market had replaced the formal economy. The Burmese were also embarrassed that their nation had become one of the poorest in the world, while its neighbors were experiencing an economic boom.

The citizens' confidence in their attempts at protest was heightened by the example of the Philippines, where "people's power" had ousted the corrupt dictator Ferdinand Marcos in February 1986. This confidence did not last. Although Ne Win himself stepped down as chairman of the one political party, the BSPP, his successor was Sein Lwin, who was infamous as the person most responsible for killing demonstrators. He was too controversial even for the dictators, and was soon replaced by weaker leaders until September 18, 1988, when General Saw Maung, the army commander in chief, seized power and suppressed the demonstrations, killing an estimated 3,000 to 4,000 persons.

The Military Regime

Saw Maung abrogated the Constitution, and took all power for himself and his cronies, while relaunching the BSPP as a fitfully active propaganda machine, the National Unity Party. He then organized the State Law and Order Restoration Council (SLORC), a secretive group of military officers that has run the country ever since.

International outrage over these events grew to such an extent that SLORC agreed to schedule elections for a new People's Assembly in 1990. The military believed that they could win the elections because of their domination over the press and other media, their control over resources necessary for a campaign, and their perception that the opposition could not unite. However, in a stunning act, the Burmese overwhelmingly voted for the opposition and against the military, giving candidates of the National League for Democracy (NLD) 81% of the seats in the Assembly (on this and other events at this stage, see Bertil Lintner's authoritative text, cited below). This unexpected result so shocked the military that they declared the elections null and void, arrested or executed many oppositionists, and vowed that they would continue to run the government for the sake of "national salvation."

In November 1997, the generals changed the name of their regime from SLORC, which a public relations firm had explained sounded too much like a barbaric, oppressive organization, to the State Peace and Development Council (the SPDC). Unlike the old acronym, the new one could not easily be pronounced with a sneer. Several members of SLORC were purged and dropped from power, but the name change did not bring about reform.

The SPDC is a junta of 19 high-ranking military officers led by a ruling troika: Lieutenant General Khin Nyunt, the Director of Defense Services Intelligence; Senior General Than Shwe, military commander in chief as well as self-appointed Prime Minister and Defense Minister; and General Maung Aye, deputy military commander in chief and heir apparent to Than Shwe. Maung Aye is reported to be in charge of the economy. General Ne Win, 89 years old in 2000, lives in seclusion in Rangoon, but still is said to influence important decisions. The leadership continues to defer to Ne Win, who ruled Burma between 1962 and 1988, even though he is too old to have a formal position of power. Much of Khin Nyunt's formidable power comes from his reputation as a political protégé of Ne Win. Very little is known about how decisions are made among the troika or within the SPDC junta as a whole. All their meetings are secret and no one is allowed to report on how policy is determined.

The military had completely misread the desires of the people and had underestimated the rise of the most prominent oppositionist, Aung San Suu Kyi, leader of the NLD. She was the daughter of Burma's founding father, Aung San, the hero of the struggle for independence from Britain. She had taken his name as part of her own. Suu Kyi had arrived in Burma in 1988, during the demonstrations, after many years abroad, which she had spent mostly in Britain, where she lived with her English husband, Michael Aris, and their two sons. She had returned to Burma to take care of her ailing mother, a former Ambassador to India, and had unexpectedly become a symbol of Burmese aspirations for democracy. It soon became clear that she could mesmerize audiences when she spoke about military corruption and the people's desire for freedom. As her fame grew, Burmese women wore clothes and combed their hair in the same style as Suu Kyi's. Her appeal was so great that the generals arrested her on July 20, 1989, before the elections had been completed, and placed her under house arrest. She received the Nobel Peace Prize in December 1991 for her courageous stand against the military, and for her determination in working to attain democracy and human rights in Burma.

SLORC promised to move the nation toward democracy by writing a new Constitution. However, some ten years after the voided election, no such constitution has been written. When it became clear that the economy was in shambles, the generals liberalized the more restrictive economic regulations and opened the country to foreign investment. Shops were allowed to reopen under private ownership. The economy improved but the people were not satisfied because they were still not free. The generals became involved in huge business corporations and in the drug, gem, and logging businesses as a means to become wealthy. Following a pattern familiar from other such dictatorships, they used their public positions for private gain.

SLORC and its successor, the SPDC, have stayed in power by oppressing any potential threats. Dissidents are jailed or executed, and peasants are forcibly relocated to areas where they will be less influential. Military officers have replaced civilians at all levels of the bureaucracy. The government allows no freedom of assembly; it has closed the universities and colleges, except for several engineering schools for children of the military; and it has banned political parties (the National Unity Party having long ceased to function). There has been no elected legislative body since the opposition's victory in 1990. The press is totally controlled by the generals, so that the daily paper *Myanmar Alin*, and its

English-language version, *The New Light of Myanmar*, are considered to be among the dullest in the world. The only news allowed is "good news" about the generals visiting various factories and pinning "hero" medals on workers. Burmese citizens are not allowed to subscribe to foreign publications, and all e-mail and Internet communication must go through the government. E-mail to foreign embassies as well as postal mail to ordinary citizens are intercepted by government authorities.

SLORC and the SPDC have also suppressed movements by the hill peoples for more autonomy, although ceasefire agreements have been signed with several of the hill people's armies. The military regime's policies against the hill peoples have forced some hundreds of thousands of persons into Thailand, where they have become refugees and languish in terrible living conditions. One Karen hill group, known as "God's Army," was led by 12-year-old twin brothers, Johnny and Luther Htoo, who were believed to have extraordinary powers and to see visions. In January 2000, members of God's Army attacked a hospital in Ratchaburi in Thailand, where they were then massacred by the Thai army, who had committed themselves to a policy of no tolerance against Burmese dissidents.

The major relationship between the generals and hill people leaders concerns the heroin trade: it seems likely (although again, conditions in Burma require wariness about using any figures) that around 70% of the world's supply of opium and heroin comes from there. The drug trade is undoubtedly responsible for generating huge amounts of money that are reinvested in banks, hotels, real estate, and business corporations, often directed by military officers. The money comes when opium enters Thailand, is refined into heroin, and then sold on to the world market. The refineries are operated by Burmese, often minority leaders such as Khun Sa, a Chinese-Shan drug lord who had close connections to the military. Indeed, when he retired, he was allowed to live in luxury in Rangoon as a guest of the military, despite his career in illegal activities.

Meanwhile, Suu Kyi remains under tight control in her home in Rangoon, even though SLORC technically removed the order of house arrest in July 1995. In reality, she has been kept from outsiders. Her words are not carried in the press or other media, and she is constantly vilified by the generals as a "lackey" of the western world. Only her worldwide prestige protects her from further imprisonment or death. During her house arrest she has managed to send her writings to the outside world, where they have been collected into books (notably the selection edited by her late husband and cited below).

Prospects for Democracy

The prospects for major change in Burma are bleak. The population has been socialized into a culture that teaches deference to authority, so that there is a high degree of fatalistic acquiescence to the regime in power. Many Burmese believe that those in authority deserve their high positions because of their good karma. More importantly, the overwhelming power of the generals keeps potential dissidents from organizing. The police state is so pervasive that the opposition cannot maneuver.

The Burmese army is huge. Because there are no external threats to the country, the army is mobilized to put down internal threats. It is believed that the army, having tripled in size during the 1990s (according to reports from the International Labor Organization), now numbers at least 400,000, and that about 40% of total government spending goes to the military. The leaders of the army have joined lucrative businesses, and Burma is now one of the world's most productive opium-growing areas.

The greatest chance for change comes from the international community communicating with the Burmese. Information seeps into Burma through tourists, the internet, clandestine radio programs, and the omnipresent gossip, to remind Burmese that their society is increasingly anachronistic in the modern world. To mitigate the rising frustration of the Burmese, SLORC began a program of "economic liberalization," which has eased the worst economic conditions. Despite this small opening toward a market economy, Burma remains isolated and impoverished. Moreover, the international community's disdain for the generals has kept outside investment and tourism, two traditional ways of enriching a society, to a minimum.

SPDC officials have claimed that the Asian economic crisis that began in 1997 did not adversely affect Burma because the nation had been kept out of the global capitalist system. They point out that Thailand, Malaysia, Indonesia, and the Philippines were affected more severely precisely because they no longer enjoyed true independence from nefarious international relations. This claim masks the obvious point that Burma's level of economic development is so low that there has been little to lose. Moreover, the economic crisis has kept Burma's neighbors from being able to invest in the economy. The crisis has undermined the investments from Burma's neighbors, and from Japan and South Korea, the major foreign investors in Burma. Burma's balance of trade deficit, unemployment rate, rate of inflation, and net foreign debt have all grown virtually every year. The value of the national currency is skewed so that, while the official rate is six kyat to one US dollar, the market rate is approximately 345 kyat to one dollar. The primary reason Burma can function economically is that the great majority of Burmese, probably about 70% of the adult population, are involved in agriculture and can grow enough food to exist on. However, even farmers have difficulty surviving. Rice is no longer plentiful, and malnutrition is a severe problem.

The governments of the United States, Canada, and most European nations have spoken harshly against the ruling regime, and have denied Burma foreign investment and development loans. The International Labor Organization has criticized the regime for the use of forced labor. ASEAN has pursued a different approach, accepting Burma into the organization in July 1997 and pursuing "private diplomacy" as a means to open Burma. ASEAN's decision to welcome Burma as a full member legitimated the regime as a functioning member of a respected international organization. ASEAN officials argued that membership would make it easier for neighboring nations to have a positive impact on Burma's government, including human rights policies. Western nations opposed ASEAN's decision, and the evidence thus far is that membership has not led to liberalization in Burma.

Burma's principal foreign ally has been China. With its huge size and 1.3 billion people, China is a giant neighbor. As Burma remains isolated from the world, China plays an increasingly important role. Crossborder trade has expanded and China has provided Burma with arms. Because China desires to establish a presence in the Indian Ocean for strategic purposes, China has also provided aid to Burma for road and rail construction. This aid has compensated for the contraction of investment from ASEAN as a result of the economic crisis.

Burma's isolation has kept meaningful political change from occurring. The nation's one newspaper continues to promote the military and deride those who favor democracy. The following topics are blacklisted: democracy, human rights, politics, Aung San Suu Kyi, criticism of the regime, prostitution, or anything positive about the West. The SPDC continues to justify its involvement in government as the only means to keep the union from disintegrating. The examples of anarchy in the former Soviet Union, Yugoslavia, and, more recently, in Indonesia are viewed as disasters that only the military can preclude.

The prospects for democracy rest to a great extent on the success of dialogue between Aung San Suu Kyi, as leader of the opposition, and the SPDC. She has continually spoken in favor of dialogue but has been rebuffed. Signs can be seen throughout the country stating that she is a tool of foreign powers. NLD members have been imprisoned and tortured, and some have been forced to speak against her leadership.

The prospects for democracy depend, then, on whether the ruling SPDC is willing to move toward new elections and a reduction in their power. The SPDC knows that Burma's economic crisis results from the nation's isolation and anachronistic economic policies. There is no chance that the SPDC will acknowledge the already elected government of Aung San Suu Kyi and the NLD. Elections would be for a new government under a new national constitution, which would doubtless provide a central role for the military. The plan is to have 25% of the seats in a new legislature, and control over key ministries, reserved for the army.

A second possibility is that the Burmese people will overthrow the SPDC, and install Aung San Suu Kyi and the NLD as the legitimate government. The fall of the Marcos and Suharto regimes may be instructive in this regard. However, almost no serious scholar of Burma believes that this can happen without a terrible civil war between the people and the military. The military know that if they allowed free elections, the opposition would once again triumph.

The military are particularly threatened by Burma's Buddhist monks, who constitute some 400,000 citizens, around the same number as the military. The monks are highly respected as teachers and religious leaders, and in the past their predecessors have toppled kings when the rulers were not meeting the needs of the people. Keeping the monks from rebellion is the main goal of the SPDC's policy of infiltrating the monkhood with their own agents, who report to the junta when dissidence arises.

Some "compromise" between continued dictatorship and democracy is the most feasible and likely eventuality. The SPDC will continue to have the "leading role" in ruling the country, at least until a new constitution is promulgated and elections are held. The military will continue to dominate the government and have the power of veto over civilian decisions. Aung San Suu Kyi and the NLD will play only a peripheral role in the new government, perhaps in some interest group that supports human rights. Civilian leaders will be frontmen for the military officials who will run the important ministries. The new government will call for an end to isolation, and the opening of the country to foreign investment and development. The minority hill peoples will be given autonomy to run their own affairs free from central control. ASEAN and the West will take over from China as the principal providers of aid and investment. This compromise is likely to commence within the next ten years.

In the meantime, Burma remains an anachronism, a beautiful country run by autocrats. Visitors to Rangoon speak about the magnificence of the Shwedagon Pagoda, perhaps the most ethereal place on Earth. The stunning temples at Pagan are among the most remarkable in the world. Most Burmese continue to wear the traditional *longyi*, a sarong-like skirt for men and women; the monks wear saffron-colored robes; the rivers and mountains are beautiful. Amidst all this grandeur is a government that abuses the people's rights, and is exploitative and corrupt. The educated people are in jail and the ignorant run the country.

Further Reading

Aung San Suu Kyi, *Freedom from Fear and Other Writings*, edited by Michael Aris, Harmondsworth and New York: Penguin, 1991, revised edition 1995; and *Letters from Burma*, London and New York: Penguin, 1998

Aung San Suu Kyi, the leader of Burma's democracy movement, has been under house arrest for many years. Her late husband, Michael Aris, brought together many of her important speeches and essays in the first volume cited here. The second supplements the previous book, providing more information on Suu Kyi's political and social principles, and the motives for her campaign against the military regime. Suu Kyi's writings also include *Aung San of Burma: A Biographical Portrait by His Daughter*, which is currently out of print but may be available from libraries.

Carey, Peter (editor), *Burma: The Challenge of Change in a Divided Society*, London: Macmillan, and New York: St Martin's Press, 1997

This is a varied collection of essays addressing the main issues facing the Burmese regime and people at the close of the 20th century.

Lintner, Bertil, *Outrage: Burma's Struggle for Democracy*, London: White Lotus, 1990; and *Burma in Revolt: Opium and Insurgency since 1948*, Boulder, Co: Westview Press, 1994

A distinguished journalist who has reported on Burmese politics for several years, Lintner has written extensively on the subject. The first book cited here is considered the authoritative account of the student movement against the military and the rise of the National League for Democracy. The second is an informative survey of the main developments among the ethnic minorities in the country since it achieved independence.

Marlay, Ross, and Bryan Ulmer, "Human Rights in Burma," paper delivered at the Association for Asian Studies, San Diego, California, March 2000

This is a useful and relatively up-to-date overview of the principal factors in the appalling human rights situation in the country, with a special focus on the relations between the Burman majority and the various ethnic minorities (including estimates of their numbers).

Maung Maung Gyi, *Burmese Political Values: The Sociopolitical Roots of Authoritarianism*, New York: Praeger, 1983

Maung Maung Gyi attempts to explain why Burma has had difficulty in achieving democratic government, suggesting that Burmese values are status-oriented and that the Burmese are fatalistic, acquiescing in their lot in life.

Neher, Clark D., *Southeast Asia in the New International Era*, third edition, Boulder, CO: Westview Press, 1998

This book covers the whole of Southeast Asia and includes a chapter on Burma. Each of the region's nations is discussed with emphasis on political and economic problems and prospects.

Rotberg, Robert I. (editor), *Burma: Prospects for a Democratic Future*, Washington, DC: Brookings Institution Press, 1998

This collection of essays by leading scholars and officials is probably the best single book on contemporary Burma. Written for the nonspecialist, the volume covers politics, values, foreign policy, the role of the military, the economy, health, and education.

Silverstein, Josef (editor), *Independent Burma at Forty Years: Six Assessments*, Ithaca, NY: Cornell University Press, 1989; and *The Political Legacy of Aung San*, Ithaca, NY: Cornell University Press, 1993

Silverstein, a strong supporter of the democratic opposition, has written extensively and critically on the Burmese military. The first of these books is an interesting collection of papers on Burma's prospects as they appeared at the end of the 1980s; the second, by Silverstein himself, focuses on the man who led Burma to independence.

Smith, Martin, *Burma: Insurgency and the Politics of Ethnicity*, London: Zed Books, 1991

Because Burma is ethnically diverse, this book is important for understanding the role of the various minority groups and the claims by the military regime that the army is the only organization capable of maintaining unity in the midst of this diversity.

Taylor, Robert, *The State in Burma*, London: Hurst, and Honolulu: University of Hawaii Press, 1987

This academic book, written by the leading scholar of Burmese history, has been controversial because of the pervasive implication that the present authoritarian government follows on naturally from a history of centralized, oppressive rule. Critics have argued that there is no relationship between the present regime's repression and past systems of rule.

Dr Clark D. Neher is Professor of Political Science, and a former Director of the Center for Southeast Asian Studies, at Northern Illinois University, DeKalb. He has written or cowritten ten books on the region, and has been a consultant for the US Department of State and the US Agency for International Development.

Facing the New Century

FACING THE NEW CENTURY

Chapter Fifteen
Nation-building, Ethnicity, and Politics

David Martin Jones and Kirsten E. Schulze

Southeast Asia encompasses a plurality of religions, ethnic groups, and languages, many of which can be traced back through precolonial, colonial and postcolonial migrations, invasions, and conversions. Despite the potential for volatility, the region long seemed to have escaped the worst effects of the turmoil that engulfed postcolonial regimes elsewhere. Indeed, with the integration of Vietnam, Cambodia, Laos, and Burma into ASEAN during the 1990s, the region as a whole seemed set fair to benefit from the export-oriented growth that had already brought wealth and somewhat uneven development to Singapore, Thailand, Malaysia, and, to a lesser extent, Indonesia and the Philippines, between 1975 and 1995. The relative absence of ethnic or religious conflict and the apparent success of nation-building in Southeast Asia were often attributed to a distinctive brand of cultural harmony, religious tolerance, syncretistic ideology, and notions such as "unity in diversity" (*Bhineka tunggal ika*, the official motto of Indonesia), euphemistically known as "Asian values."

In practice, however, the emerging Balkanization of states in Southeast Asia along ethnic and religious faultlines now strikingly resembles similar rifts that have destabilized southeastern Europe since 1991. The disparity between image and actuality can probably best be explained by the fact that, while there was extensive *state*-building in Indonesia, Malaysia, Thailand, Burma, the Philippines, and Singapore, *nation*-building was at best incomplete. While the new postcolonial states espoused unifying syncretistic ideologies, and elaborated common vocabularies and symbols to project national unity, they did not bridge the ethnic, religious, tribal, and cultural cleavages that already existed. The end of the Cold War, globalization, tentative political liberalization, and the economic crisis that started in 1997 thus allowed for the re-emergence of divisions that had only ever been superficially concealed by the corporatist enterprise, suppressed by authoritarian policies, and obscured by the conflict between the superpowers.

The Colonial Legacy and the Challenge for State-building

It has often been assumed, both by western modernization theorists and by members of postcolonial elites, that the extension of urbanization, schooling, and communication and transport facilities would assimilate ethnic attachments into the developing national enterprises in Southeast Asia. Yet the multiethnic heritage of the region showed little evidence of assimilation and integration. Indonesia, for instance, was characterized by pervasive regionalism from the outset. The nationalist movement was dominated by Javanese and Sumatrans, tensions damaged relations between Java and the outer islands, and separatist elements flourished in West Papua and Aceh, the latter resulting more than once in outright rebellion. Interethnic tensions persisted elsewhere, too, as Walker Connor observed in 1980 (Connor p. 9): in Vietnam, the ideological conflict between North and South obscured

> an active self-determination movement on the part of tribal hill peoples. In Thailand, the effectiveness of Bangkok's writ diminishes

rapidly when one leaves the. . . Chao-Phraya valley for the Lao-speaking Northeast, for the Karen-populated hills of the West, or for the Malay- and Chinese-populated regions of the Malay peninsula. In Malaysia, cohesion suffers from the antagonisms between Malays and a strong Chinese minority. . . In Burma, it has been estimated that Rangoon controls only half of the territory.

Religious and ethnic differences also underlie the separatist struggle in the Philippine provinces of southern Mindanao.

With the exception of Thailand, all the countries of Southeast Asia enjoyed the dubious benefits of European colonialism and have retained various institutional legacies from it. Significantly, the postcolonial states either remained within, or broke down along, the boundaries delineated by European administrators. The establishment of these boundaries and their propensity to induce ethnic alienation within the new states has often gone unremarked. Yet as Benedict Anderson has shown, the territorial definition of the postcolonial states had profound resonance for the potential politicization of ethnic attachments (see Anderson).

Building Indonesia

Across the 6,000 islands of the Indonesian archipelago, which are inhabited by around 300 ethnic and linguistic groups, the evolving but inchoate Indonesian nationalist movement sought to build "unity in diversity." Indeed, the interwar Partai Nasional Indonesia promoted the concept of a "greater Indonesia" that potentially embraced the whole of the Malay world. Consequently, the idea of postcolonial Indonesia represented what John R.W. Smail has called a "leap of the imagination," facilitated by the innovatory "transmutation of the Malay [language] of [some of] the islands into the national language" (Smail p. 309). Indonesia was then crystallized through a shared language and in resistance to Dutch colonialism, but within the boundaries determined, to a large extent, by Dutch rule.

From its inception, Indonesia manifested a propensity to elite patrimonialism. In the influential view of the anthropologist Clifford Geertz, Indonesia became divided along religious and cultural lines into mutually exclusive *aliran* or "streams" (see Geertz). Almost 90 per cent of Indonesians are Moslems, but they are divided between the devout, or *santri*, and those who are affiliated with a more syncretic *abangan* practice, which is particularly prevalent amongst the Javanese. There also existed an evolving cleavage, both economic and ethnoreligious, between the Javanese core and the peripheral minorities. The Javanese of Central and Eastern Java constituted 45% of the total population of the new country, and the Sundanese of West Java 15%: together, these groups dominated the economy. The remaining groups on the outer islands are much smaller. They include the Moslem Minangkabau, the Protestant Bataks, the Islamist Acehnese, the Moslem Madurese, the Catholic Florinese, the Catholic/animist Dayaks of Borneo (which Indonesians know as Kalimantan), the Hindus of Bali, the Protestant/animist tribes of West Papua, and the Moslem and Christian populations of Celebes and the Moluccas. A further major cleavage exists between the "low" culture of the *adat* tribes and the "high" cultures of the former Moslem and Hindu kingdoms, as well as between the "indigenous" communities and the Chinese "alien minority."

Indonesians therefore function within overlapping communities, reflecting various levels of *aliran*, linguistic, and locality affiliation. As David Brown has emphasized, each of these has constituted a potential "focus for politically salient ethnic consciousness, and the relationship of such ethnic communal consciousness to an overarching Indonesian identity has been . . . contingent" (Brown 1994 p. 309). As a consequence, Indonesian identity has remained fragile. Until Sukarno established his "Guided Democracy" (see Chapter 9), there were divisions between the Islamic party Masyumi, the Communists and Sukarno's brand of secular nationalism. The establishment of Suharto's New Order (1965–98) came to depend on the promotion of a "national ideology" of *pancasila* (see Appendix 2), the depoliticization of Islam, and the maintenance of stability by the armed forces, which had assumed a dual role (*dwifungsi*) in order to maintain the integrity of Indonesia.

Building Malaysia

Malaysia's unification was no easier than Indonesia's. In the aftermath of World War II, the British had undertaken various efforts to create a viable political arrangement that could unite its disparate Malay protectorates, a task made more urgent by the serious Communist insurgency on the peninsula. After a couple of false starts, these culminated in the Federation of Malaysia, founded in September 1963 and initially embracing the peninsula, Sarawak, and Sabah, as well as Singapore. While it solved the problem of disunity, the Federation turned the question of Malay identity and its relationship with an evolving national consciousness into a matter of political urgency. It was significant that the oil-rich sultanate of Brunei opted out of the Federation, while the Chinese-dominated city of Singapore departed from it in August 1965.

Initially, the federal Constitution facilitated political pluralism. In this context, parties tended to reflect ethnic and class interests. Initially, the United Malay National Organization (UMNO) ruled Malaysia, both at the federal level and, more uncertainly, in almost all the states, in an unequal alliance with the Malay Chinese Association (MCA) and the Malayan Indian Congress (MIC). The privileged but economically uncertain position of the majority Malay community, and growing interethnic conflict between Malays and Chinese, culminated in interethnic riots in Kuala Lumpur in May 1969. Subsequently, the UMNO elite altered the Constitution and, in the words of Norma Mahmood, succeeded in "removing issues considered sensitive from public discourse and gerrymandering electoral districts in favor of ethnic Malay constituencies" (Mahmood p. 34). By the mid-1970s, UMNO, together with its ethnic and regional affiliates in Peninsular and East Malaysia, which served as minor partners in a Barisan Nasional (National Front) coalition, dominated the state and the economy.

Elsewhere in the Region

Similar responses to the territorial units inherited from former colonial powers characterized the policy of the new national elites in the Philippines and Burma. The boundaries of the new states uncertainly resembled their colonial roots. Like Indonesia, the Philippines is an archipelago, with around 7,000 islands concentrated in three clusters and inhabited by an estimated 140 ethnic groups. As a consequence of 300 years of Spanish colonialism, it is the only state in Asia where a majority of the population are Christians. After staging the first nationalist revolution in Asia, in 1898, Filipino nationalists were "rewarded" with 50 years of US colonialism, and had to wait until July 1946 for independence.

The postcolonial elite was left groping for an identity. On the one hand, David Wurfel was only one of many observers who maintained that the Philippines enjoyed "a national cohesion greater than that found in most colonial states" (Wurfel p. 24). On the other hand, this incipient notion of nationhood seemed to embrace only the Christianized northern Philippines, and its overwhelmingly Catholic character served to alienate both the tribal peoples expelled from their traditional lands, such as the Bontec and the Karunga, and the Moslems of southern Mindanao. The Moslems, or "Moros" as the Spanish had termed them, had practised Islam well before the arrival of the Spanish in 1572 (see May). As Peter Gowing has observed, they were faced with the "dilemma of having to reconcile the demands of their rather traditionalist conception of Islam with the demand of citizenship in a developing non-Islamic state" (Gowing 1975 p. 28).

Analogously, Burma inherited the former British administrative territory and with it a precolonial understanding of patrimonialism, reinforced by Buddhist views of hierarchically ordered authority. Indeed, Buddhism became, in the words of David Brown, the "most important element of Burman identity and the cultural basis for an emergent Burman ethnonationalism" (Brown 1994 p. 43). The junta that came to rule this fissiparous state represented the interests of the lowland Burman and Mon peoples, at the expense of other ethnic groups, notably the various non-Buddhist upland peoples, the Shan, Kachin, Karen, and Chin, who had enjoyed a degree of autonomy under the pluralist colonial administration. Thus, when the new central state became identified with the majority Burmans, the Burmans began a practice of "ascriptive ethnicity" (Fee and Rajah p. 249). The ethnocratic Burmese state institutionalized ethnicity at a time when the concern with territorial integrity made it a matter of urgency to extend administrative control to the peripheries. Consequently, when an evolving state structure, dominated by ethnic Burman personnel and values, attempted to extend "Burmanization" beyond the core areas, a countervailing movement of Shan and Karen separatism took shape, leading to rebellions in 1948 and 1952.

Somewhat differently from Indonesia, Burma, the Philippines, and Malaysia, Indochina did not retain the boundaries bequeathed by the former imperial power. Nevertheless, the new states of Cambodia, Vietnam, and Laos did reflect the federal administration devised by the French, who had divided Indochina into five territories – the colony of Cochin China (southern Vietnam), and the protectorates of Annam (central Vietnam), Tonkin (northern Vietnam), Cambodia, and Laos – each possessing its own system of administration. In Tonkin and Annam, the old Vietnamese mandarinate remained in a vestigial form; the Cambodian monarchy also survived, as did the three main Laotian principalities. Only in 1928 did the French constitute a legislative assembly for the whole of Indochina, which came too late to play for Indochina the integrative role played in the Dutch East Indies by the Volksraad (People's Council). Unlike in Malaysia or Indonesia, France's *mission civilisatrice* in Indochina preserved the separate linguistic identities of Laos, Cambodians, and Vietnamese. Indochina therefore had no common language in which to imagine a single postcolonial nation-state.

Finally, Siam (Thailand), which escaped the experience of colonialism that gripped its neighbors, may also seem to have evaded such problems. Yet it was in response to the European colonial model that the modernizing Chakri dynasty developed a centralized state. Siam's independence survived the 19th century because the country constituted a useful buffer zone between the British and the French. By 1910, Siam had clearly defined borders on all sides. In other words, the protean Thai state attained its identity in reaction to the prospect of British or French colonialism. King Vajiravudh, who ruled from 1910 to 1925, was the first ruler to popularize the idea of a Thai nation, *chat*, as part of a trinitarian concept of nation, religion, and king (Fee and Rajah p. 252). After the first of many coups, in 1932, the territorial state became Thailand. This identification had important implications for the Chinese migrant population in Bangkok, ethnic minorities in the Isan region of the Northeast, and the southern Moslem minority in an officially Buddhist state.

Nation-building and the Management of Difference

Given the often arbitrary territorial boundaries inherited by the new states, the problem of building national identities always presented the possibility of generating countervailing ethnic or ethnoreligious separatisms. However, although separatist movements in Burma, Thailand, the Philippines, and Indonesia immediately made themselves apparent, the new states initially seemed capable of containing them. Their authoritarian character also made it easy for them to repress or depoliticize what the region's gerontocrats dismissed as "communalism."

Anti-Communism and Nationalism

This apparent success in securing national unity can also be explained, at least in part, by the shared response of the non-Communist states in Southeast Asia to their perceptions of the threat of Communism. The prospect of the fall of Vietnam, Laos, and Cambodia to Marxist nationalists provoked the formation in August 1967 of a loose anti-Communist grouping, ASEAN, based on the recognition of the unalterable territorial integrity of the founding member states, at a time when none of those states practised liberal democracy. Thailand oscillated between military coups and unstable multiparty coalitions, but showed a penchant for strongmen. In the Philippines, Ferdinand

Marcos had established his distinctive form of constitutional authoritarianism. Indonesia had created state-licensed parties that legitimized Suharto's evolving kleptocracy. In Malaysia and Singapore, dominant single parties effectively negated political opposition through manipulation of the respective Constitutions and a judicious use of internal security legislation. Meanwhile, outside ASEAN or any other grouping, a long-established military regime promoted the "Burmese road to socialism."

The uncertainty of the regional environment thus helped promote a "siege legitimacy" that sought to contain separatism and promote shared national values, as political dissent was equated with communal disunity and disloyalty (Brown 1985 p. 989). By linking external and internal elements to present a prospect of imminent disaster, the ruling elites could persuade their societies that they constituted garrisons under siege, and that they required active support for the sake of survival. The regional propensity to authoritarian political arrangements obviously possessed real advantages in combating separatist movements.

Nevertheless, each of the new states adopted quite different strategies for handling ethnic and religious minorities. This, in turn, had implications for the ways in which, on the one hand, states defined ethnicity, and, on the other, ethnic and religious minorities perceived themselves and their attachment to the state. In this context, official definitions of "the nation" had profound implications for perceptions of the overseas Chinese, whose "bamboo network" embraced the economies of all the states in the region, as well as for the self-definition of ethnic and religious minorities.

Malaysia

In the case of Malaysia, the very formation of the new state exacerbated regional tension. The state of emergency declared in June 1948 in the face of a Communist insurgency, and maintained until July 1960, was followed by "confrontation" (*Konfrontasi*) with Indonesia from January 1963 to August 1966. President Sukarno of Indonesia violently objected to the new entity, and thus fed a burgeoning siege mentality in the leaders of the new state.

Not only were the boundaries of Malaysia a source of anxiety, but the notion of what constituted a Malay, let alone a Malaysian, was equally unclear. To be Malay had traditionally meant to be *kerajaan*, or unconditionally loyal to a Sultan. However, from UMNO's formation in March 1946 its brand of populist nationalism had emphasized the shared and equal identity of the *bangsa Melayu* ("Malay nation"), while for many others to be Malay was to be Moslem. For such Malays, who eventually formed the backbone of the Parti Islam se-Malaysia (PAS), the fact that the Constitution of 1957 gave Islam a privileged status was an intimation of the possibility of an Islamic state some time in the future. Meanwhile, the pressing need for Malay unity tended to override these conflicting feudal, ethnic and religious definitions of allegiance.

There was also the problem of establishing the terms of interethnic engagement in the postcolonial state. Between 1955 and 1969, this took the form of an electoral alliance between UMNO, the MCA, and the MIC, a coalition of ethnic elites that presided uncertainly over a friable community. Their "consociationalism" initially permitted the political dominance of Malay aristocrats such as the first Prime Minister, Tengku Abdul Rahman, whose urbane style afforded considerable latitude for Chinese economic influence. The alliance was already falling apart, however, when the riots mentioned above erupted in Kuala Lumpur in May 1969. After a period of emergency rule, Malaysia's second Prime Minister, Tun Abdul Razak, announced the New Economic Policy and ever since the government has actively engaged in economic management to ensure that "Bumiputras" ("sons of the soil") directly participated in and benefited from economic growth.

Singapore

In Singapore, by contrast, where the majority Chinese existed in what the long-serving Prime Minister Lee Kuan Yew once called "a sea of Malay people," the ruling People's Action Party adopted a policy of ethnic neutrality. In order to achieve this, the state encouraged a process of depoliticization, and during the

1970s it promoted an official model of a society comprising "Malays," of various ethnic affiliations; "Chinese," of Fujianese, Teochew and Hakka derivation, together with those who had become assimilated with indigenous Malays to form the distinctive *peranakan* culture; "Indians," of Tamil, Sikh and Sinhalese provenance; and assorted varieties of "Eurasians." The government considered these four compartments to be internally homogenous and mutually distinctive. The policy, reinforced by education and the depoliticizing of religion under the Religious Harmony Act of 1992, has been aimed at reinforcing cultural attachments: in the words of one observer, Geoffrey Benjamin, it effectively puts "Chinese people under pressure to become more Chinese, Indians more Indian, and Malays more Malay in their behavior" (Benjamin p. 124).

The depoliticization of ethnicity and the "sanitization" of ethnic cultures did not proceed without problems, given the perceived need to develop national loyalty to an ethnically neutral and meritocratic managerial state, while at the same time promoting cultural attachments within an ethnically plural society (Brown 1994 p. 84). The government's response to this problem was to promote a garrison mentality by endlessly warning against the alleged dangers of communalism and inculcating psychological defense through state-sponsored bonding activities, such as the month-long celebration of independence each August.

In non-Communist Southeast Asia in the 1970s and 1980s, the evil that dared not speak its name was either Marxism or communalism. By the early 1990s, however, it had become the generically decadent West, against which a bland collocation of "Asian values" was to be the national and regional prophylactic. It was by no means accidental that Singapore and Malaysia, the states with the greatest uncertainty about their national identities, were at the forefront of the promotion of an "Asian renaissance" that, it was claimed, could syncretize Confucianism and Islam, Buddhism and Hinduism, in a formula of cohesion, community, and harmonious balance. In Singapore, for example, the government introduced a scheme for "Team MPs" (members of Parliament), later relabeled as members for "group representation constituencies": under this scheme, parties seeking to win such seats in Parliament have to put together multiethnic groups of candidates (Suryadinata p. 84) The scheme has had the further consequence of tightening the ruling party's grip on the state, as opposition parties have had considerable difficulty in recruiting candidates from the requisite range of ethnicities.

Thailand

Elsewhere in the region, states tended to promote official forms of nationalism that either assimilated minorities or rendered them problematic. This becomes evident when contrasting the status of the Chinese in Buddhist Thailand or the Catholic Philippines with their status in predominantly Moslem Malaysia and Indonesia. In both Thailand and the Philippines, national identity is defined in cultural rather than ethnic terms. Indeed, individuals of Chinese descent are often considered Thai or Filipino once they have become acculturated. This was by no means an inevitable process. Under the autocratic tutelage of Luang Plaek Phibunsongkhram ("Phibun"), Prime Minister from 1938 to 1944 and again from 1948 to 1957, the idea of a new Thai nation, *sang chat*, was defined as against the Chinese. Invoking the slogan of "Thailand for the Thais," Phibun's first government introduced a comprehensive code of measures intended to reduce the commercial influence of the Chinese (Wyatt p. 254), but after World War II the economic nationalism of the bureaucratic Thai polity engendered often corrupt clientelistic links between the Thai elites and the ethnic Chinese entrepreneurs of Bangkok. During the 1950s, ethnic Chinese, who constituted less than 10% of Thailand's population, included 70% of the owners of businesses in Bangkok (Laothamatas p. 197). Later governments, notably those led by Sarit Thanarat (1958–63) and Prem Tinsulanonda (1980–88), sought to develop the private sector rather than merely seeking rent from it. The gradual transformation of the ethnic Chinese into Thai nationals facilitated this process. As David K. Wyatt has put it, "The ladder to success in the Thailand of the 1960s and 1970s, at a time of rapid economic growth, consisted of Thai education, Thai surnames, Thai language and even intermarriage with Thai families" (Wyatt p. 292). Increasingly, Chinese business interests were assimilated into the Thai state. By 1973, 63% of the members of Chinese-dominated trade associations and 87% of their presidents held Thai citizenship (Laothamatas pp. 201–02). Intermarriage across ethnic lines further facilitated the integration of the elites, as relatives of the "indigenous" bureaucratic elite entered the ranks of the Chinese-dominated business elite in significant numbers.

While the Chinese, resident mainly in Bangkok and other cities, could assume a Thai identity with relative ease, the process of assimilation was less successful in the historically peripheral and economically underdeveloped North and Northeast of Thailand, and in the religiously distinctive South. In the North, the growing impoverishment of the lowland Khon Muang, who were wet rice cultivators, impelled many of them to migrate to upland areas populated by the Hmong, who practised dry rice cultivation (and also grew and traded opium). Intercommunal clashes followed, while efforts at acculturation into Thai society through education proved to have a disintegrative effect. From the mid-1960s, many of the Hmong, as well as members of other hill tribes, gave varying degrees of support to the Communist Party of Thailand. The fact that the Thai state regarded hill tribes as "aliens" requiring acculturation further exacerbated the problem. In the multiethnic Northeast, meanwhile, communal consciousness tended to remain focused at the level of villages, kin groups, and localities, and the development of an ethnoregional "Isan" consciousness owed much to the migration of the rural poor from the Northeast to the more prosperous Bangkok region, together with the government's attempts to educate Northeasterners into a Thai identity. Similarly, isolation from internal colonization by the Buddhist Bangkok elite generated regional separatism among the Moslem Malays of Pattani in the South.

The Philippines

The Philippines displays an analogous pattern, with the Chinese minority, who form less than 1.5% of the population (Suryadinata p. 35), being integrated into the nation-building enterprise, while alienation has spread among tribal and religious minorities. The members of the Philippine nationalist elite, the *ilustrados*, have been characterized by their mixed (*mestizo*) heritage from Indio, Spanish, American, and Chinese forebears ever since the late 19th century. The "People Power" movement that ended Ferdinand Marcos's martial law regime involved politically prominent *mestizo* Chinese, such as Corazon "Cory" Aquino and Jaime, Cardinal Sin. As elsewhere in Southeast Asia, such fully assimilated individuals of Chinese descent contrasted with migrants of more recent vintage who did not possess Philippine citizenship until Marcos introduced a mass naturalization scheme in 1975. Again as elsewhere in Southeast Asia, the ethnic Chinese played a prominent role in Philippine business. The most influential Chinese organization has been the Federation of Filipino-Chinese Chambers of Commerce (Shang Zong), established in 1954, which has maintained close links between the Catholic Filipino power elite and the Filipino-Chinese both under Marcos and since his fall from power (Suryadinata p. 37).

The integrationist policy pursued by successive governments reflected the norms of the Christian majority (Gowing 1979 p. 210), with an emphasis on the maintenance of order, the promotion of education, and the pursuit of economic development. A Commission of National Integration was established in 1957 to inculcate a sense of Filipino nationhood, but more specifically to integrate the Moslems of southern Mindanao into the body politic. However, the government's policy of mass population transfers, intended to develop and assimilate an underdeveloped land of "wild tribes," served only to radicalize Moro distinctiveness, especially when the government declared large tracts of the relatively underpopulated island of Mindanao public land and divided it among Christian settlers. State-sponsored resettlement displaced Moslems from their traditional lands and altered the demographic composition of Moroland, engineering non-Moro majorities in several provinces. Following the establishment of the Moro National Liberation Front (MNLF) in 1968, Islam became the badge of Moro selfhood and *bangsa Moro* ("Moro nation") emerged as an identifying name for the Moslems of Mindanao and Sulu, distinct from the *bangsa Filipino*. Christian communities in the region also organized armed militias to advance their interests, while the MNLF initially received sponsorship both from Libya and from Malaysia, at a time when the Malaysian government resented Ferdinand Marcos's revival of the Filipino claim on Sabah.

Indonesia and Malaysia

The perception among many Filipino Moslems that the government is seeking to impose Christianization is mirrored by the belief, widespread among Christians in Indonesia, that the government there is pursuing Islamicization. The similarity is particularly striking in view of Indonesia's transmigration (*transmigrasi*) program, which aims to relieve overpopulation in the inner islands, such as Java and Madura, by resettling thousands of families in the outer islands. In both Kalimantan (Borneo) and West Papua, where tribal land has been requisitioned by the government and handed over to settlers, this has resulted in conflict, not only between predominantly Christian/animist indigenous peoples and Moslem transmigrants, but also between the Javanese center and the outer islands. Islamicization is perceived to go hand in hand with Javanization. In West Papua, transmigration has strengthened the struggle for independence spearheaded by the Organisasi Papua Merdeka (Free Papua Organization, or OPM) and its predecessors since Indonesia took control of the territory in May 1963. In Kalimantan, it has led to open conflict between the indigenous Dayaks and migrant Madurese, and there has been a resurgence of Dayak headhunting traditions that were previously assumed to be extinct.

In both Indonesia and Malaysia, the Chinese minorities have to a large extent been excluded from the nation-building project. The fact that there are far more Chinese Christians than Moslems may have made assimilation easier in the Philippines than in either of these majority-Moslem states, yet the key barrier to integration has been not religion but ethnicity, and specifically ethnicity associated with perceptions of economic power. In Indonesia, where the Chinese represent less than 4% of the population, Chinese conglomerates accounted for two thirds of the output of Indonesia's private, urban economy by the 1990s, dominating the distribution network for food and other essentials, and controlling at least 80% of the 162 companies listed on the Jakarta Stock Exchange in mid-1993 (see Schwarz). In Malaysia, the numerically much larger Chinese population similarly occupied a disproportionately influential role in commercial life, and the largest conglomerates, such as Hong Leong or the Robert Kuok group, cultivated close ties with key figures in the ethnically Malay UMNO elite, ensuring that their business activities continued outside the realm of public scrutiny or comment (see Gomez).

Such opaque relationships between Chinese business and ethnically dissimilar political elites reflected the contingent historical legacies of the new states of Southeast Asia after 1945. During the colonial era, the Chinese performed ambivalent roles as compradors in the European empires to which they emigrated from southern China, and indigenous nationalist leaders in both Indonesia and Malaysia found it easy and expedient to accuse them of serving the interests of the colonial masters. In the postcolonial period, their status remained uncertain, even though they rapidly came to dominate business and finance in both countries, while the new governments deliberately cultivated distinctions between "indigenous" and "nonindigenous" citizens. The attitudes of the People's Republic of China and the Chinese Nationalist government of Taiwan further complicated the question of allegiance. The regime in Taiwan followed the practice of the Qing Dynasty and considered all those of Chinese descent living overseas as Chinese nationals; by contrast, it was not until September 1980 that mainland China clarified its position by adopting a citizenship law under which overseas Chinese who take up another nationality automatically lose Chinese nationality (Suryadinata p. 14).

The status of the Chinese communities of Malaysia and Indonesia was thus already disturbingly unclear when the ruling elites adopted policies that excluded them from political participation, after 1969 in the case of Malaysia (as described above) and after 1965 in Indonesia, where the "New Order" government of President Suharto curtailed the use of the Chinese language, required Chinese children to attend Indonesian-language schools, and prohibited the use of Chinese characters. Nevertheless, Chinese business conglomerates remained central to the rapid growth of both these economies after 1970. In Malaysia, the conglomerates acted as business proxies for the Malay elite, while in Indonesia conglomerates such as Lim Sioe Liong's Salim group, or the Apkindo group headed by the assimilated Chinese businessman Mohammed "Bob" Hassan, entered into *cukong* or clientelistic arrangements with Suharto or, in the course of the early 1990s, with the businesses run by Suharto's sons and daughters.

The paradox of being economically powerful but politically impotent seemed unimportant when most of the economies of the region were booming, but the crisis that started in 1997 was bound to leave wealthy Chinese unprotected from the politically and economically disaffected, with obvious repercussions for interethnic relations. In May 1998, Indonesia's Chinese community was targeted by organized groups of *primam* (vigilantes), who systematically murdered, raped and pillaged their way across the Chinese districts of Glodok in West Jakarta and Solo in East Java, triggering the flight of Chinese capital. B.J. Habibie, President from 1998 to 1999, did little to allay anxiety when he observed, in July 1998, that the place formerly occupied by the Chinese could be "taken over by others" (as quoted in *The Straits Times*, July 20, 1998). This remark reflected the commitment shared by both populist reformers, such as Amien Rais, and technocrats of Habibie's type to the establishment of Islamic values at the center of a reformed Indonesian polity.

Indeed, Habibie's brief presidency witnessed a proliferation of political parties that exacerbated political and racial unease. These new parties reflected the reemergence of the *aliran* that had characterized Indonesia's unstable parliamentary democracy in the years immediately following the achievement of independence in 1945. Thus, the Indonesian Communist Party (Partai Komunis Indonesia, or PKI) reappeared as the Prosperous Labor Union, under the leadership of Mochtar Pakpahan; Abdulrahman Wahid's Nadhlatul Ulama (NU), Indonesia's largest Moslem organization, abandoned its "nonpolitical" position to establish the National Awakening Party, which set out to represent the rural Islamic *pesantren* tradition; and Amien Rais's Mummadiyah inherited the reformist and modernizing Islamic nationalism represented in the 1950s and 1960s by Masyumi.

Political uncertainty and economic turmoil held little appeal for the Indonesian military, which remained ideologically committed to *dwifungsi* as the source of order and *pancasila* as the basis for an inclusive nationalism. It was always unlikely that the military would easily accept the uncertainties of multiparty democracy, the diminution of its political authority, or any form of autonomy for such troublesome territories as Aceh, West Papua, or East Timor. Consequently, elements in the army, some of them probably loyal to Suharto, have continued to operated in questionable and opaque ways, for instance by aiding the activities of the militias in East Timor and by escalating tensions between Moslems and Christians in the Moluccas.

The Resurgence of Ethnic Nationalism

Between 1970 and 1997, the states of Southeast Asia adopted nation-building strategies that necessarily problematized a variety of ethnoreligious attachments, and were thus incomplete at best. At the same time, the export-oriented growth strategies adopted by the region's more successful states ameliorated ethnic separatism, while authoritarian rule generally succeeded in stifling its more intransigent manifestations, both within each state and at the regional level, where ASEAN forswore involvement in the internal affairs of its member states.

The disintegration of Indonesian politics since 1997 is one example of the way in which ethnic tensions have been reawakened by economic crisis, undermining consensus within the states of the region and diminishing their capacity to resolve difficulties through export-oriented growth. Two cleavages have appeared. First, the more autocratic regimes, such as those of Indonesia, Singapore, and Malaysia, still consider the principle of noninterference in internal affairs to be sacrosanct, while more democratically accountable governments, such as the present government of Thailand, maintain that the crisis requires ASEAN to move toward what the Thai Foreign Minister Surin Pitsuwan has called "flexible engagement on issues that have a negative bearing on others in the region" (as quoted in *The Jakarta Post*, July 13, 1998). Across this emerging ideological faultline runs a less widely advertised cultural cleavage that increasingly affects the position of the overseas Chinese in Southeast Asia. The revival of Islamic identity has proved to have significant appeal for political reformers in Indonesia and Malaysia, and is contributing both to the increasingly hostile perception of Chinese minorities and to the difficulties that have recently emerged between Singapore and its neighbors.

Territorial conflicts have also remained intractable. They include the Philippines' outstanding claim to Sabah; competing Malaysian and Indonesian claims to the islands of Litigan and Sipidan; and the dispute between Malaysia and Singapore over the island of Pedra Branca. Simmering suspicion often assumes an ethnoreligious flavor. This is evidently the case with the often frayed relations between Malaysia and Singapore, since the governments of both countries seem incapable of refraining from comment on each other's internal affairs. In November 1986, for example, Prime Minister Mahathir Mohamad of Malaysia protested against a visit to Singapore by Chaim Herzog, then President of Israel, claiming that it offended Moslem sensibilities. He became even more indignant when Lee Hsien Loong, the Deputy Prime Minister of Singapore, publicly questioned the loyalty of ethnic Malays in the city state (Tan p. 9).

Elsewhere in the region, worrying signs of communal attachments, conflicting with the nation-building process, have appeared in the aftermath of the economic crisis of 1997–98. For example, the Malaysian government's relations with both Thailand and the Philippines have been damaged by its tacit support for Moslem separatist organizations in both countries, notably the Moro Islamic Liberation Front (MILF) in southern Mindanao, which was permitted to organize training camps in Sabah in 1980, and the Pattani United Liberation Organization (PULO), which has drawn resources from the increasingly Islamicized Malaysian states of Perak and Kelantan (Tan p. 40).

Such examples indicate that, contrary to expectations among many observers, the low-intensity conflicts that used to be obscured by the Cold War have not withered away. Instead, the growing influence of militancy, often adapted from the Middle East, on the formerly moderate and sometimes syncretic Islam of Southeast Asia has only been aggravated by the inability of the region's gerontocrats to relinquish power, and has fueled the rise of increasingly chiliastic millenarian Islamic sects, such as Al-Ma'unah (Brotherhood of Inner Power) in northern Malaysia or Abu Sayyaf (Father of the Sword) in southern Mindanao. Yet these groups are far from being unprecedented or wholly imported phenomena. Ever since 1945, the Moslem minorities in the Philippines and southern Thailand have found it increasingly difficult to reconcile what they see as the demands of Islam with the requirements of the new nation-building states, in which non-Moslems predominate (Gowing 1975 p. 28; see also Che Man).

Perhaps most disturbingly, Indonesia, the largest and probably the most significant member of ASEAN, has continued to be divided by conflicts over identity and religion, often provoked or exacerbated by dissident elements in the armed forces. The referendum on independence in East Timor was followed by the systematic destruction of the territory's infrastructure by militia groups linked to elements in the Indonesia military (see Chapter 10). In Aceh, the Free Aceh Movement (Gerakanan Aceh Merdeka, or GAM) has maintained a secessionist struggle since 1976. In West Papua, the OPM has derived significant encouragement for its independence struggle from the outcome of events in East Timor. In the Moluccas (the provinces of Maluku and North Maluku), Christian and Moslem communities that have traditionally followed a code of *pela gandong*, or nonviolence, plunged in the course of 1999 into a bloody confrontation that, by June 2000, had caused 3,000 deaths. Since May 2000, Moslem violence there has been orchestrated by the Lascar Jihad, a group that has links to elements in the armed forces and has been encouraged by Amien Rais, the leader of Mummadiyah (see above) who is now presiding officer in the legislature. President Abdurrahman Wahid's regime appears to be powerless to contain the dissolution of the Indonesian periphery, while conflicts within the political elite distract attention from the violence.

The broader regional situation seems no better. ASEAN has proved unable either to manage the economic crisis or to address the issue of intercommunal violence, displaying instead the ineffectiveness that is the corollary of its commitment to noninterference in the internal affairs of member states. Indonesia, Malaysia, and Singapore, the core members of the organization, retain their inflexible commitment to this policy, despite the fact that the violence transcends state boundaries, as, for example, in June 2000, when the leader of a GAM faction was killed in Kuala Lumpur, or in August 2000, when an attempt was made to assassinate Leonides Caday, the Philippine Ambassador in Jakarta, perhaps by Moro secessionists. Clearly, despite the rhetoric about shared "Asian values," nation-building and ethnicity remain sources of tension across much of Southeast Asia.

Further Reading

Anderson, Benedict, *Imagined Communities: Reflections on the Origin and Spread of Nationalism*, revised edition, London and New York: Verso, 1991

> Anderson contends that a nation exists not as a tangible or political fact but as an idea in the collective consciousness that has clearly defined boundaries in the minds of its adherents. He draws examples in support of his argument from many parts of the world, including Southeast Asia.

Benjamin, Geoffrey, "The Cultural Logic of Singapore's Multiracialism," in R. Hassan (editor), *Singapore: Society in Transition*, Kuala Lumpur, Oxford, and New York: Oxford University Press, 1996

Brown, David, "Crisis and Ethnicity: Legitimacy in Plural Societies," in *Third World Quarterly*, Volume 7, number 4, October 1985

Brown, David, *The State and Ethnic Politics in Southeast Asia*, London and New York: Routledge, 1994

> Brown provides a good introduction to, and overview of, the region's complex ethnic and religious politics. His book is the most comprehensive guide to the evolution of state strategies and their ethnic implications.

Che Man, W.K., *Muslim Separatism: The Moros of the Philippines and the Malays of Southern Thailand*, Singapore, Oxford, and New York: Oxford University Press, 1990

> This book offers a useful analysis of the historical, economic and social background of Moro and Malay identities, and their problematic incorporation into the respective larger societies and polities.

Connor, Walker, *Ethnonationalism: The Quest for Understanding*, Princeton, NJ: Princeton University Press, 1994

> This is an intelligent and thoughtful introduction to the confused area of identity politics, and an indispensable guide to thinking about the character of ethnonationalism.

Fee Lian Kwen and Ananda Rajah, "The Ethnic Mosaic," in Grant Evans (editor), *Asia's Cultural Mosaic*, Singapore, London, and New York: Prentice Hall, 1993

Geertz, Clifford, *The Religion of Java*, Glencoe, IL: Free Press, 1960

> In this influential text, as in his later works on the interpretation of cultures, Geertz holds that culture constitutes a kind of destiny and that ethnic attachments exist as primordial givens.

Gomez, E.T., *Political Business: Corporate Involvement of Malaysian Political Parties*, Townsville: James Cook University, 1994

Gomez affords an insight into the ethnic coalitionism of the National Front that rules Malaysia in the Malay interest.

Gowing, Peter, *Muslim Filipinos*, Manila: Solidaridad, 1975

Here and in the text cited below, Gowing explores and explains the dilemmas faced by Moslems in the Philippines.

Gowing, Peter, *Muslim Filipinos: Heritage and Horizon*, Quezon City: New Day, 1979

Laothamatas, Anek, "From Clientelism to Partnership: Business-Government Relations in Thailand," in Andrew Macintyre (editor), *Business and Government in Industrialising Asia*, St Leonards, NSW: Allen and Unwin, and Ithaca, NY: Cornell University Press, 1994

Mahmood, Norma, "Political Contestation in Malaysia," in Norma Mahmood and Zakaria Haji Ahmad (editors), *Political Contestation: Case Studies from Asia*, Singapore: Friedrich Naumann Foundation, 1990

May, R.J., "Muslim and Tribal Filipinos," in R.J. May and F. Nemenzo (editors), *The Philippines After Marcos*, London: Croom Helm, and New York: St Martin's Press, 1985

Schwarz, Adam, *A Nation in Waiting: Indonesia in the 1990s*, St Leonards, NSW: Allen and Unwin, and Boulder, CO: Westview Press, 1994

Schwarz provides a comprehensive account of the structure of the "New Order" regime and presciently explores the cronyism, corruption, and nepotism that were to contribute to its demise in 1998–99.

Smail, John R.W., "Indonesia," in D.J. Steinberg et al., *In Search of Southeast Asia: A Modern History*, Honolulu: University of Hawaii Press, 1987, and Sydney, London, and New York: Allen and Unwin, 1989

Suryadinata, Leo, *Chinese and Nation-Building in Southeast Asia*, Singapore: Singapore Society of Asian Studies, 1997

Leo Suryadinata has spent many years exploring the pattern and nature of Chinese settlement, business, and assimilation across Southeast Asia. Here he offers an interesting insight into the various state strategies that address the Chinese who settled in the region during the colonial era.

Tan, Andrew, *Intra-ASEAN Tensions*, London: Royal Institute of International Affairs, 2000

Tan suggests worrying ways in which ethnic and religious irredentism can constitute the basis for new terror in the economic meltdown and social disorder of Southeast Asia.

Wurfel, David, *Filipino Politics: Development and Decay*, Ithaca, NY: Cornell University Press, 1988

Wurfel's account of the political development of the Philippines, now somewhat dated, is squarely within the tradition of US political science, with an emphasis on the crises that the polity has encountered en route to modernization.

Wyatt, David K., *Thailand: A Short History*, New Haven, CT, and London: Yale University Press, 1984

A clear account of the evolution of Thai national identity

Dr David Martin Jones is Senior Lecturer in Government at the University of Tasmania. Dr Kirsten E. Schulze, a Lecturer in International History at the London School of Economics and Political Science, is researching and writing a book on ethnic conflict in Indonesia.

FACING THE NEW CENTURY

Chapter Sixteen

Environment, Resources, and Hazards

Greg Bankoff

The forest fires that blanketed a large part of Southeast Asia in thick haze during 1997 and 1998 are only the most visible sign of the environmental crisis that looms over the region, threatening to stifle the economic growth and rising living standards of the past few decades. A combination of drought conditions brought on by a particularly severe occurrence of the El Niño southern oscillation (ENSO), a periodic climatic phenomenon that affects the Pacific Ocean, with the use of fire in land preparation by rubber and oil palm plantations, set standing forests on the islands of Sumatra and Borneo ablaze. Fires raged uncontrollably between September and November 1997, and again in February and March 1998 after an abnormally short wet season. The pall of smoke extended over much of Indonesia, Malaysia, Singapore, southern Thailand, and the Philippines. At its height, on September 23, 1997, the Air Pollutant Index in the city of Kuching, in East Malaysia, recorded a staggering 839, when a reading of over 100 is regarded as unhealthy and one of over 300 as hazardous (Hiebert et al. p. 78). In all, it is estimated that fire-produced gases and particulates seriously affected the health of over 20 million people and caused damages in excess of US$4 billion (Levine et al. p. 2). The haze was even held partially responsible for the crash of a Garuda Indonesia passenger plane near Medan in Sumatra in September 1997, killing 234 people (Hiebert et al. p. 75) and for a ship collision in the Straits of Malacca that caused 29 fatalities (Levine et al. p. 12).

While Indonesia caught the attention of the media at the time, the devastation caused by similar uncontrolled fires in the Americas, Europe, Russia, and East Asia has made environmental sustainability an issue of increasing worldwide public concern. In Southeast Asia, the haze that choked much of the region's inhabitants reinforced the close association between biomass burning, greenhouse gases, habitat destruction, loss of biodiversity, air pollution, and public health. Moreover, it clearly linked the state of the environment, the rate of resource extraction, and the frequency of hazard with adverse effects on industry, agriculture, forestry, and tourism, making such matters into issues of national and, increasingly, international politics. The financial crisis that simultaneously enveloped the region's markets just as the haze enveloped its cities, precipitating significant falls in the value of currencies, was not entirely unrelated to environmental questions either (Savage et al. p. 9, Rigg p. 27). The fires certainly played a contributing role in the subsequent overthrow of President Suharto of Indonesia in 1998 (Cotton p. 347).

State of the Environment

The past 200 years have seen the transformation of Southeast Asia from a region of generally low population densities and ample resources to one where the present weight of humanity often places an inordinate strain on the carrying capacity of the environment. Until the 19th century, the majority of people lived on the rich alluvial valley soils of the principal river systems, in distinct cores that were also important cultural centers. These more intensively cultivated areas were bounded by extensive interior peripheries and highlands, covered in either tropical moist forests (mainly dipterocarp) or seasonally humid (deciduous) forests of generally lower population densities and more diverse cultural traditions. Above all, Southeast Asia was a region dominated by the sea: the degree of marine influence over its climate, environment, settlement, communications, and development of resources is considered unmatched in any other part of the world (Barrow p. 78). However, its increasing integration into a globalizing capitalist economy, initially under European colonialism and more thoroughly since, has made it into one of the principal sources of export-oriented commercial resource extraction in the world and introduced human-induced environmental changes on an unprecedented scale (Bryant and Parnwell pp. 4–8).

The increasing impact of human activity on the environment is most visible on the landscape. The natural vegetation cover of most of Southeast Asia is forest. Over half the land area was still classified as such in 1980, although mainland and maritime parts of the region exhibit distinctive profiles, reflecting the historical orientation of the former more toward intensive agriculture and the latter's greater engagement in trade-related activities. Deforestation increased significantly in the 1980s, but has slowed somewhat over the last decade due more to resource depletion than to conscious efforts at conservation. While the forests of Asia and the Pacific were reduced by a further 17 million hectares (42 million acres) between 1990 and 1995, deforestation was fastest in Southeast Asia, at 1.6% a year on the mainland and 1.3% a year in maritime areas (see GEO). Average annual forest loss ranged from 0.6% in Brunei to 3.5% in the Philippines (Rosenberg p. 128, Colchester p. 7). Already in 1995, Asia and the Pacific had a significantly lower forest cover, at 0.17 hectares (0.42 acres) per capita, than the world average of 0.61 hectares (1.51 acres) per capita (see GEO), and timber reserves were not estimated to last much beyond 2030 at present rates of resource depletion (Panayotou p. 2,270). The magnitude and pace of this deforestation have exposed vast tracts of naturally fragile land to erosion, exacerbating the tendency of many soils in the region to become easily waterlogged and flow down even the slightest of gradients (Barrow p. 86). Upper water catchment areas are particularly vulnerable in this respect and landslides have become commonplace.

Agricultural activities have often suffered from, and also caused, further environmental degradation. One of the most notable aspects of recent environmental change in Asia and the Pacific has been the enormous expansion in the area of croplands, which increased from 210 million hectares (519 million acres) in 1900 to 453 million hectares (1.12 billion acres) in 1995. In Southeast Asia, agricultural land use rose from 16.8% of total area in 1975 to 19.2% by 1992. Much of this expansion has been to provide land for commercial agriculture: in Malaysia, cultivated areas rose from 5,640 square kilometers (2,178 square miles) in 1900 to 48,060 square kilometers (18,556 square miles) in the 1990s, mainly to accommodate rubber and oil plantations (see GEO). The conversion of marginal land to intensive commercial cropping, especially in upland areas, has led to further deforestation and soil loss. Studies from Thailand suggest that erosion is two to three times greater when vegetation cover is reduced to 20% (Panayotou p. 2,270). Dwindling productivity has also prompted shifting cultivators to adopt shorter fallow periods, in an attempt to maintain their livelihoods, initiating a vicious cycle of falling production and declining fertility. Jonathan Rigg refers to farmers in northeastern Thailand as "mining" the uplands by cultivating nutrient-demanding crops, such as cassava, with virtually no inputs or land conservation measures (Rigg p. 255).

The fundamental shift from land abundance to land scarcity in many parts of the region has brought about a concomitant intensification of agriculture with a consequent expansion of irrigation systems, the extensive application of chemical pesticides and fertilizers, and, often, programs of dam construction. Vast expanses of formerly arable land are now either waterlogged or affected by salinization, while many watercourses have become choked with sedimentation, causing siltation and flooding downstream.

Water, too, like land, has become an increasingly scarce commodity: freshwater withdrawals have increased more in Asia during the past century than in any other part of the world (see GEO). Southeast Asia's current average annual renewable water resources of around 10,000 cubic meters (13,080 cubic yards) per inhabitant are substantially above the global average of 7,700 cubic meters (10,071 cubic yards) per person. However, such statistics mask the significant variation that exists within the region, from Singapore, at 172 cubic meters (225 cubic yards) of fresh water per capita, to Malaysia, at 21,000 cubic meters (27,468 cubic yards) per capita (see GEO, and Panayotou p. 2,271). The island republic is forced to meet its demands by importing water from the neighboring Malaysian state of Johore. Across the region, rising domestic and industrial requirements, the widespread development of hydroelectricity, increased irrigation, and the uncontrolled exploitation of aquifers have reduced groundwater reserves, lowered river levels, and depleted wetlands. Shallow lakes, such as Laguna de Bay in the Northern Philippines, are rapidly dwindling and declining as sources both of drinking water and of fresh fish (Severino p. 13).

Just as serious as the growing water shortage is the increasing contamination of most rivers. Pollution caused by industrial effluents, hazardous and toxic residues, runoff from land-based activities such as agriculture and mining, and untreated domestic sewerage and animal wastes threatens the quality of most water courses. In Malaysia, 42 rivers have already been officially declared "dead," while in Thailand 600,000 tons of hazardous wastes annually find their way into the nation's waterways (Panayotou p. 2,271).

Nor have Southeast Asia's coastal or marine environments fared any better. The region has a vibrant maritime tradition. Both Indonesia and the Philippines are archipelagic nations, while Malaysia, Burma, Thailand, and Vietnam all have extensive coastlines and important national interests in marine resources. Fishing is an important part of the regional economy and fish are the most important source of protein for most people. In recent decades, however, the number of fishing operators has increased rapidly. Indonesia now has over 1.5 million fisherfolk, a rise of 77% between 1979 and 1990, or a rate of growth twice that of the natural population increase (Bailey and Pomeroy pp. 192–93). As a consequence, marine fisheries production in Asia and the Pacific rose by an average of 2.9% annually between 1975 and 1995 (see GEO). Traditional marine stocks reached full exploitation in many areas by the early 1990s, and overfishing, aided by the introduction of new technologies such as the trawl and even the motorization of traditional craft, now threaten the diversity and quantity of fish stocks. Many of the region's most important fisheries, such as in the Gulf of Thailand, have already been seriously depleted.

The capacity of many marine stocks to sustain these pressures has been further undermined by the destruction and pollution of key coastal habitats. More than three million hectares (7.4 million acres) of mangroves, which serve as important nursery areas for demersal fisheries, have been cleared to make way for the expansion of aquaculture. Vietnam's mangrove forests shrank from 400,000 hectares (988,000 acres) to 252,000 hectares (623,000 acres) between 1950 and 1983, while the once extensive mangroves along Thailand's coasts dwindled from nearly 368,000 hectares (909,000 acres) in 1961 to only 160,000 hectares (395,000 acres) by 1996 (see GEO). Chemical pollution from urban, industrial and agricultural sources, and sedimentation due to erosion, have also reduced the productivity of fisheries, especially in estuaries and along coral reefs. Moreover, the rising level of nutrients pouring into coastal waters from land-based sources poses yet another serious marine issue for the region. The increasing eutrophication of inshore waters has made phytoplankton blooms or red tides, many of which produce toxins that are readily ingested by shellfish and present a serious health hazard to consumers, a seasonal event throughout much of the archipelago since their first appearance in the 1970s. In the Philippines, the spread of Pyrodinium bahamense var. compressa to encompass coastal waters from Zamboanga del Sur in the South to Zambales in the North has been accompanied by more than 50 recorded fatalities between 1983 and 1995 (Bankoff 1999b pp. 103–07 and 113). Even if not toxic, such blooms deplete oxygen from the water, causing the mass death of aquatic organisms, and leave marine "deserts" in their wake.

The dramatic decline in the extent of the region's fisheries and tropical forests has also been paralleled by a drastic loss in biodiversity, as both sylvan and marine habitats have come under intense pressure. Southeast Asia is noted for its particularly rich flora and fauna. The Indo-Malesian floristic region (which includes Papua New Guinea and the Solomon Islands) has more flowering plants than any other (Barrow pp. 93–94), while more than two thirds of all coral reefs are located in these waters (Panayotou p. 2,271). Indonesia and Malaysia are included among the world's 12 "megadiverse" countries, and their territories account for the bulk of the second largest rainforest system. Already it is estimated that two thirds of wildlife habitats and 70% of major vegetation types have been lost in the Indo-Malayan region, which includes South Asia (see GEO). Within Southeast Asia, original habitat loss has been in the order of 80% in the Philippines and Vietnam, and has also been acute in Thailand (Panayotou p. 2,271). Moreover, comparative research on the extensive forest fires connected with the last three major ENSO events, in 1972, 1982–83, and 1997–98, suggest that "biodiversity hotspots" – rainforests that, because of their relative ecological stability, support an especially rich number of species and endemics – are now at risk. Previously burned forests are more sensitized to future fires and so progressively extend the range of their impact. By spreading to previously unburned forest, the fires are now encroaching upon those areas that support the highest levels of biological diversity and uniqueness, on the Sunda Shelf (Taylor et al. p. 1,172). Wildlife is also under threat from traditional hunting practices that have become unsustainable in the region's remaining forests. In many areas, animal populations show consistent declines in density as species are locally extirpated or reduced to insignificant numbers, so creating the phenomenon described as "the empty forest" (see Redford). This condition is furthest advanced in the Philippines, but significant hunting still characterizes many of the forest areas in Indonesia and the East Malaysian states of Sarawak and Sabah (Robinson and Bodmer p. 6). Given the limited and incomplete understanding of the region's existing biodiversity, it is not yet possible to assess accurately how much of it is endangered, let alone the state of any particular species or ecological community.

No overview of the environment is complete without some consideration of the impact of urbanization on the region, despite the fact that even in 1995 only 34% of all Asians lived in urban areas. As human activity increasingly becomes a prime agent of climatic and geomorphologic change, the rapid growth of urban populations is having an increasingly detrimental effect on the forests, waters, and wildlife of Southeast Asia. Much of this growth has been in the nations' capitals, giving rise to the emergence of primate cities, three of which – Jakarta, Manila, and Bangkok – each now have a metropolitan population greater than 10 million. Most of these conurbations have expanded exceptionally quickly. Jakarta grew to 8 million residents within 15 years, one tenth of the time it took New York City to reach the same size (see GEO). Economic growth also occurs disproportionately in or near these centers. Bangkok, for example, which is home to 11% of Thailand's population, accounts for 37% of GDP (Bruestle p. 2,280). As urban incomes are generally two to three times those in the provinces, a high percentage of residents are recent migrants fleeing unproductive rural circumstances. Between 40 and 60% of the population of Jakarta and Manila are poor and live in marginal housing, without basic urban infrastructure. Increasingly the rapidity and form of this urbanization is having an impact on the physical environment: encroaching on agricultural land and forests; depleting the groundwater table, thus causing subsidence and salt water intrusion; releasing huge quantities of untreated waste into coastal waters; and filling the air with choking fumes and greenhouse gases.

Most of the region's large cities have serious atmospheric pollution problems. A concomitant feature of the economic growth, industrialization, and urbanization of recent decades has been the increase in energy consumption and the proliferation of car ownership. While global energy consumption fell by 1% a year between 1990 and 1993, Asia's grew by 6.2% a year (see GEO). It is estimated that total energy consumption in Southeast Asia rose by 75% in the years 1983–93 (Panayotou p. 2,272). Fossil fuels currently account for about 80% of energy generation and, although Asia's carbon dioxide emissions per capita are still only little more than half the world average (Fu et al. p. 312), industrial emissions have grown 60% faster than elsewhere (see GEO). Similar figures indicating rates of increased emissions higher than world averages have been collected in relation to sulfur dioxide, nitrogen oxide, methane, and chlorofluorocarbons (Hameed and Dignon pp. 159–63). Acid rain has already become a significant factor in Cambodia, southeastern Thailand, and southern Vietnam (see GEO).

Air quality has also been affected by the significant rise in car ownership in Asia. The number of registered vehicles rose from 52.3 million in 1980 to 127.3 million in 1996, an increase of more 240% or a doubling of the total every seven years (see GEO and Bruestle p. 2,282). The Philippines had more than one million vehicles in 1984, 471,000 of which were on the capital's roads; by 1996, the figure had risen to around 1.5 million vehicles operating in Metro Manila alone. Bangkok had a reported 600,000 vehicles in 1982, but the total rose by an average of 9% a year throughout the rest of the 1980s. Traffic has become so congested in the commercial centers of these two cities that it has slowed to an average speed of just 13 kilometers (eight miles) an hour in Manila and seven kilometers (four miles) an hour in Bangkok (Tongzon pp. 210–11). Around 70 to 80% of total emissions in Kuala Lumpur and Manila are now attributable to motor vehicles (Bruestle p. 2,282). The annual cost of all this is high, both financially and in consequences for health. Motor vehicles everywhere are major generators of carbon monoxide, nitrogen oxide, and lead, while a large proportion of cars in the region are old, use fuel inefficiently, are poorly maintained, and still use leaded gasoline and diesel. Air pollution costs on average around US$2 billion a year in Bangkok and around US$600 million in Jakarta (Tongzon p. 210). The level of suspended particulate matter in urban areas is also a major factor contributing to respiratory complaints, hypertension, heart attacks, strokes, and lung cancer. In Bangkok, around 1,400 deaths a year are directly attributed to poor air quality, and it has been estimated that, by the age of seven, the average child loses three to four IQ points (assuming that IQ means anything) through exposure to excessive lead levels (Bruestle p. 2,283).

Resource Use and Management

A certain level of resource depletion is an unavoidable cost of human activity. Any overview of the environment in Southeast Asia is likely to dwell on the consequences of such degradation, presenting at times what seems like an unmitigated litany of destructive practices. However, just as telling as any environmental appraisal is consideration of the attitudes, both historical and contemporary, that societies have manifested toward resource use and management. Historically, the state in Southeast Asia has been driven by what can be called a "resource frontier mentality." If the pressure of population grew too great in one area, then some of the population moved or was moved to another. If the soil was depleted in one area because of overuse, then new lands were cleared for cultivation. If the forest was logged out in one locality, then there were plenty of trees somewhere else to cut down. There was always a new resource frontier to provide the state with a continuous flow of revenue. So long as a resource frontier persisted – and it has persisted in most of Southeast Asia throughout historical times to the present – there has been no pressure on the state to seek alternative uses of the natural environment (see Bankoff 1995 pp. 17–37 and Bankoff 1997 pp. 81–100). Many of the languages of the region tend to differentiate between nature tamed and manipulated for human interests (Thai *thammachaat*, Malaysian and Indonesian *taman*, Burmese *thaba-wa*), and nature as a wild, rustic and untamed space, often associated with evil spirits and to be entered with care (Thai *pa thuan*, Malaysian and Indonesian *hutan*, Burmese *taw*) (Rigg pp. 46–48).

Evidence of this resource frontier mentality in operation can be seen in the history of the environmental damage inflicted on the region's forests. Much of the extensive forest loss can be attributed to export policies promoting unrestrained commercial logging. While only 12% of the world's tropical rainforests lie in Southeast Asia, the region accounts for around 80% of the global trade in tropical timber (Peluso et al. p. 196). Trade in precious woods, such as teak, mahogany, ebony or rosewood, has existed since colonial times (Boomgaard pp. 59–87), but the extensive exploitation of tropical timbers for commercial application in construction, framery, and plywood is a particular feature of recent decades. While stimulating periodic booms since World War II, legal and illegal logging has reduced and degraded much of Southeast Asia's old growth forests. In the East Malaysian states of Sarawak and Sabah, log production has so far exceeded sustainable limits that it threatens the very existence of the tropical forest as a sustainable ecological system. Between 1963 and 1985, loggers harvested around 30% of the total forest area of Sarawak, so that only 4 to 5 million hectares (10 to 12 million acres) of primary forest remained by 1990. Nor has Sabah fared any better, as the primary forest cover fell from 55% of total land area in 1973 to only 25% in 1983. By the early 1990s, loggers had cut more than 80% of the dipterocarp forests set aside for commercial harvesting and log production was averaging 9.60 million cubic meters (12.56 million cubic yards) a year between 1992 and 1994 (Dauvergne p. 1).

However, it is not just in the quantity of timber felled but equally in the manner and the rate of returns on the trade that the resource frontier mentality is evident. The region's postwar commerce in tropical wood has been dominated by the large Japanese corporate trading companies, the *sōgō shōsha*. Exports of logs to Japan rose from 116,654,000 cubic meters (152,583,000 cubic yards) between 1950 and 1969 to a staggering 216,627,000 cubic meters (283,348,000 cubic yards) in the years 1970 to 1979, before falling back some to 138,584,000 cubic meters (181,268,000 cubic yards) in the years 1980 to 1989; in all, Japan had imported more than 500,000 cubic meters (654,000 cubic yards) of tropical logs up to 1995 (Dauvergne pp. 186–87). What Peter Dauvergne calls Japan's "ecological shadow" fell first on the Philippines, but, as sources there became depleted and demand soared in the 1970s, the *sōgō shōsha* turned next to Indonesia and Sabah. As first one government and then the other implemented bans on log exports, beginning in the 1980s, these companies have increasingly relied on supplies from Sarawak, and more recently on Papua New Guinea and the Solomon Islands.

In return, unrealistic tax and royalty rates, forged export records and transportation documents, misrepresented harvest totals, and widespread circumvention of reforestation fees and duties ensure that Southeast Asian governments receive only a small fraction of the timber's real worth. Moreover, low fees and tax evasion, compounded by the granting of limited-term timber concessions, encourage inefficient and destructive logging practices, promote the export of logs and plywood at cut-price rates, and encourage disregard of reforestation programs. Governments in the region are both able and willing to participate in this resource scramble. Nation-states, as heirs to their colonial predecessors, have laid claim to the control of vast areas of the national estate under the guise and through the function of their respective forestry departments. These agencies currently administer 74% of the national territory in Indonesia, 55% in the Philippines, and 40% in Thailand (Colchester pp. 6–7). The money generated by the trade in tropical timber also provides governments with an important source of largesse from which to dispense official patronage through traditional mechanisms of patron-clientage, otherwise known as "crony-capitalism." Little consideration is extended to the people, numbering somewhere between 35 and 45 million, who have had their rights denied, their habitat destroyed, and their homes displaced in the course of this trade. Japan and the other more minor importers have consumed huge volumes of logs at prices and at rates that blatantly ignore the environmental and social costs of their actions.

Much the same attitude is apparent toward the marine environment. Fishing grounds are typically seen as open-access resources, with few if any restrictions on entry. The numbers of small-scale fisherfolk in Southeast Asia are rising despite declining catches, partly because they are being swollen by displaced agricultural laborers fleeing rural conditions even more desperate than those in coastal areas. The relatively small investment required to purchase a boat, access to loans in return for exclusive buying rights to catches, and the greater prospects of upward social mobility make the sea an attractive alternative to urban migration or marginal upland farming (Bailey and Pomeroy p. 194). All too often, the result has been overfishing of remaining stocks, employing techniques that devastate the marine environment and often preclude the possibility of natural regeneration. In particular, the use of poison and "blast fishing" are clearcutting the region's coral reefs in much the same fashion that loggers are decimating its forests (McManus pp. S121–S122).

Natural poisons found in leaves, berries, and roots such as derris, barringtonia, tephrosia, and wikstroemia can be considered standard equipment for the traditional fishing methods employed by coastal and riverine communities in Southeast Asia. However, as marine stocks decline and economic pressures rise, fisherfolk have increasingly turned to the use of industrial chemicals in their attempts to maintain the sizes of their catches. Commercial bleaches are poured into tidal pools and shallow waters, sodium cyanide solutions in squeeze bottles are used to stun fish taking refuge on coral reefs, and *chum*, composed of poisoned fish and shrimp bits, is broadcast on the surface of the sea (McManus p. S122). Moreover, there is increasing demand for brightly colored live reef fish to satisfy the tastes of aquarium enthusiasts, as well as the appetites of fashionable restaurant-goers in Hong Kong and Taiwan. Reef fish such as highfin grouper regularly command prices of US$180 a kilogram (or around US$82 a pound), while one special delicacy, humphead wrasse lips, can cost as much as US$1,000 a plate (Dayton p. 14). At the same time, the use of explosives made from fertilizers packed into bottles is also having an equally devastating effect on fish populations and coral reefs. Blasts usually destroy coral within a radius of one to two meters, while numbers of reef-associated pelagic fish such as casio spp., which have sensitive swim-bladders, decline rapidly. There is also substantial wastage, as most dead fish sink to the seabed, requiring retrieval by divers holding their breaths, who often find it difficult to locate the bodies among the coral rubble and in conditions of poor visibility (McManus p. S122). Nor is the prognosis optimistic. The recovery of coral reefs depends upon factors such as the rate of environmental disturbance, levels of siltation, and competition from other organisms such as seaweed, all of which are anticipated to increase (McManus p. S122). As with timber, an "ecological shadow" is falling upon Southeast Asia's marine environment. First, the rich waters off Indonesia were stripped bare of their coral fish during the late 1960s and now the Philippines is suffering the same fate (Dayton p. 14).

Hazard and the Environment

The state of any society's physical environment, and the attitudes it holds towards resource use and management, are closely related to the occurrence of hazard and disaster. Typhoons, floods, storm surges, droughts, earthquakes, landslides, and volcanic eruptions have historically posed more of a threat to some human societies than to others, causing great loss of life and extensive damage to property and infrastructure (Hewitt p. 59). In many cases, this is simply a question of geography, but in others it is also compounded by social and economic factors that have changed over time. Thus, Asia, and especially Southeast, Northeast and South Asia, have suffered disproportionately to other regions because of their location within a very active tectonic area, much of it subject to cyclonic disturbance and with many lowlying coasts. While Southeast Asia amounts to just 3% of the world's total landmass, it experienced more than 12% of all recorded hazards between 1900 and 1997. Moreover, the region appears to be increasingly vulnerable to the impact of hazards, despite the undoubted advances in scientific understanding and greater technological sophistication. In one respect, of course, this may simply be a matter of larger populations, denser urban areas, and more costly infrastructure than ever before. On the other hand, there are also indications that human activity affects the magnitude and frequency of such phenomena, that it adversely alters climate, and that it poses a serious threat to agricultural production in the region.

Mounting evidence indicates that human activity is altering the composition of the Earth's atmosphere, substantially increasing the percentage of heat-trapping greenhouse gases, especially carbon dioxide and methane. The atmospheric build-up of these gases is widely credited with having caused a rise in global temperature of around four degrees Celsius since the end of the last Ice Age (around 18,000 years ago), and of between one half a degree and two degrees in just the last 450 years. Conservative estimates suggest that temperature will rise by a further two and a half degrees by 2060, making the planet the warmest it has been in the past 2 million years (Mendelsohn p. 7). Global warming on such a scale will have considerable effects on a whole range of environmental mechanisms that bear importantly on human habitation. Significant alterations to sea levels, wind directions, ocean currents, the severity and frequency of storms, rainfall patterns, agricultural productivity, and the range of disease-bearing organisms can all be anticipated. Nor will the effects of these changes be necessarily the same throughout Southeast Asia. A projected doubling of carbon dioxide in the atmosphere may well cause a rise of three to four degrees in global temperature, a one-meter increase in sea level, and higher rainfall. In Indonesia, however, the temperature increase could be as low as one degree in Sorong and as high as four degrees in Bengkulu. Rainfall could vary from one extreme to the other even at the same location: Merauke might experience a 32% decrease as well as a 234% increase for different months. Only the rise in sea levels appears likely to be more uniform, with around 10,000 square kilometers (3,861 square miles) of land under threat of inundation across Indonesia, and a further 5,000 square kilometers (1,930 square miles) under threat in both Malaysia and Thailand (Handley p. 65).

More ominous still are the possible effects of global warming on the region's weather patterns, especially the ENSO (see above) and its influence on monsoon rains. Changes in temperature and ocean currents affect the frequency of ENSO events, and affect the future occurrence and magnitude of floods and droughts (Nicholls pp. 154–75). Also closely related to the ENSO phenomenon and likely to be affected by global warming are the number and severity of typhoons and storm surges (Pitcock et al. p. 160). The average wind speeds of typhoons have already doubled from 100 kilometers (62 miles) an hour in the 1930s (Hermoso p. 16), while the effects of storm surges will only be intensified by rising sea levels and the further destruction of mangroves, which have hitherto acted as coastal defenses. Lowlying farms and aquaculture ponds will probably be completely submerged or rendered unproductive through the frequent incursion of saltwater (Iglesias et al. p. 14). As rainfall patterns change, heavier falls may improve irrigation in some areas but may also greatly exacerbate erosion and leaching of soils, and increase the likelihood of landslides. Other areas may simply become too dry to support existing agriculture (Handley pp. 65–66). Southeast Asia is especially vulnerable in this respect, as it has the highest projected absolute increases in water demand of any region in the world (Arnell p. S33). Forest fires, too, will probably become more common as species distribution is affected by climate changes and ENSO-induced droughts increase the incidence of major conflagrations (Levine et al. p. 4). Moreover, the biological activity and geographical distribution of the malarial parasite and its vector are sensitive to changes in temperature and precipitation. A global mean rise of several degrees Celsius would increase the epidemic potential of the mosquito population twofold and have serious consequences for human health in the region (Martens et al. p. 463).

The direct and indirect effects of all these climatic changes on the region's agriculture are potentially very disquieting. In particular, rice yields are projected to decrease at lower latitudes but increase at higher ones, indicating a possible shift of rice-growing away from equatorial regions (Iglesias et al. p. 19). The incalculable social, economic and political consequences of this shift are suggested by a study of climate change in Indonesia, Malaysia, and Thailand, conducted on behalf of the UN Environment Program, which estimates that farmers may well lose income amounting to between US$10 and US$130 a year (see Parry et al.). Nor would these changes solely be felt on land areas. Higher water temperatures would cause weather patterns to change, which, in turn, would alter sea currents and marine resources. Many of the region's coral reefs – among the planet's most productive ecosystems, providing a substantial proportion of the total fish catch – are already being "bleached," a process involving the mass expulsion of the symbiotic algae known as zooxanthellae that are responsible for the coral's distinctive color, health, and growth rate (Palis pp. 1–2). Furthermore, aquacultural production would be threatened not only by encroaching saltwater inundation but also by increasing competition from developed countries, such as Canada, which, with the melting of the ice cap, would be able to use their extensive coastlines to farm tropical fish and seafood (see Castro). Warmer seas may also cause fish to move deeper or migrate entirely, severely affecting catches already under strain from overfishing.

Recent trends suggest that the sheer magnitude and frequency of all these factors are already having an increasing impact on the lives and property of the peoples of Southeast Asia. Moreover, the absolute cost of these and other hazards is not shared by society as a whole but falls disproportionately on those least able to bear them. The poor suffer most, and the poverty of their subsequent conditions often drives them to actions that further degrade their environment and further increase their vulnerability to such events. This is the cycle of poverty and hazard that afflicts many societies in the region. Just as there is a relationship between disasters and class inequalities at the national level, so there is also a relationship at the international level between meteorological and seismic conditions, on the one hand, and low per-capita income and "third world" status, on the other. The magnitude and frequency of disasters caused by natural hazards can affect both economic development and a nation's ability to attract foreign investment (Bankoff 1999a pp. 381–420).

Politics of the Environment

Issues relating to natural hazards, resource use, and environmental degradation are increasingly becoming matters of local, national and international politics, for ecosystems operate regardless of any political borders. Environmental questions have become prominent in Southeast Asia precisely because the rapid pace of development there in recent decades has exposed the contradictions between economic and environmental priorities more forcefully than in most other regions of the world (Hirsch and Warren p. 5, Bryant and Parnwell p. 2). Since the countries of the region attained independence, their governments, whatever their formal political persuasion, have all relentlessly pursued policies emphasizing economic growth over environmental protection, transforming largely agrarian societies into more urbanized and industrialized societies that are export-oriented and better integrated into global markets (Dixon pp. 149–216). For many years, little consideration was given to the environmental costs of these activities, despite demonstrable evidence of those costs, indicating how questions of access to, use of, and control over natural resources closely reflect the social relations and power structures within each country. That more prominence is now accorded to environmental issues suggests that broader transformations are taking place in these societies. New social interest groups have emerged in response to rapid change, while the ambivalent ecological implications of new development processes have become more manifest (Hirsch and Warren p. 11). The assumption that politics and the environment are everywhere inseparably linked gives rise to what Raymond Bryant calls a "politicized environment," so that the various dimensions of ecological change are considered in relation to everyday human practices, episodic natural hazards, and the byproducts of industrial activity (Bryant pp. 82–84). Such a categorization of the environment can be applied locally, nationally, and internationally.

At the local level, the interaction of politics and the environment largely involves disputes about the alienation of resources from one group by another, with the former incurring most of the environmental costs and the latter most of the resource benefits. In one respect, this may simply involve the outright encroachment upon a resource endowment through some legal mechanism, such as a license, a concession or even outright sale, that effectively deprives the former "owners" of its use. Disputes can equally be about the more indirect costs of resource extraction, such as increases in the frequency or magnitude of sedimentation, siltation, flooding, landslides, erosion, mine tailings, loss of fisheries, and the like (Hirsch and Warren pp. 11–12). Environmental advocacy in these instances is mainly through community-based organizations acting either independently or in concert with national and sometimes international non-governmental organizations. A notable feature of politics in Southeast Asia in recent decades has been the proliferation in the number and influence of such bodies. Some have achieved national prominence, including Sabat Alam in Malaysia, the Green Forum in the Philippines, and WALHI (the Indonesian Environmental Forum) in Indonesia (Rush pp. 55–96).

Often, however, the issue is more complex than resource use or abuse. Almost two thirds of the world's indigenous peoples live in Asia, and in Southeast Asia they comprise at least 30% of the population of Burma, 23% in Laos, and 16% in the Philippines, or altogether more than 32 million individuals (Colchester p. 6). Much of the resource extraction that has underpinned economic development in the region has taken place in the forests and uplands that these peoples regard as their ancestral domains. The physical environment not only offers shelter and livelihood, but is also an integral part of their heritage: loss of habitat is cultural as well as environmental destruction. Faced with this assault on their survival, these peoples have increasingly begun to assert their rights to territories and to self-determination as the indigenous or first peoples of the land. Where their claims have met with little more than government-directed violence, as in Burma, the Philippines, and West Papua (under Indonesian control), the level of conflict has risen to such a point that at times it has threatened the territorial integrity of the nation, as indigenous peoples have sought nothing less than independence and statehood. Where the policy has been less one of repression and more one of neglect, indigenous peoples have begun to affirm their influence on a national and even international stage. A Charter of the Indigenous Tribal Peoples of the Tropical Forest was formulated at the First Forest People's Conference held in Penang, Malaysia, in February 1992, demanding respect for traditional ownership, compensation for forest damage, and the right to self-determination. The environment in these instances is much more than just a matter of resource entitlement: it also affects issues of identity and human rights.

On the other hand, the interaction of politics and the environment at the national level tends to mean the mediation of conflicting interests over issues of resources and conservation. Many countries have developed substantial bodies of law and regulation dealing with the management of natural resources and the protection of the environment, largely as a result of the Conference on the Human Environment held in Stockholm in 1972. The individual country reports issued as a result of that meeting represented the first attempt at estimating the state of the environment at a global level – and they made generally dismal reading. In response to widespread public concern over resource depletion and pollution levels, and as a consequence of obligations under multilateral agreements and protocols, many governments in Southeast Asia established the necessary framework for implementing environmental legislation. The Malaysian Environmental Quality Act of 1974 (amended 1985), the Filipino Environmental Policy (Presidential Decree No. 1,151) of 1978, and the Indonesian Environmental Management Act of 1982, otherwise known as the Basic Law on the Living Environment, are examples. In Thailand, many environmental regulations have been made more binding by inclusion in the latest Constitution, promulgated in 1997. Some form of environmental impact assessment is now a standard requirement under such legislation and regulatory agencies have been established to ensure their implementation. Even in Cambodia, the Lao People's Democratic Republic, and Burma, such institutional frameworks are in their initial stages of development (see GEO). Enforcement, however, remains a major problem and monitoring of compliance is weak. In the Philippines, for example, circumvention of the regulations remains widespread: 80% of environmental impact certificates were still pending in 1990, but projects were permitted to proceed in their absence (IEDP p. 15). The challenge at the national level is to find ways to reconcile the promotion of liberal trade policies with the protection of the environment and natural resources (Intal pp. 94–97).

The most notable environmental initiatives in recent years have been at the international level, where governments have engaged sufficiently in diplomacy and dialogue to influence each other's domestic policies. This represents a major departure from previous norms of behavior in the region (Rosenberg p. 138). ASEAN has a history of environmental cooperation stretching back to 1977, but a more proactive approach was not implemented until 1989, when annual meetings of ASEAN Senior Officials on the Environment (ASOEN) began. The trend was confirmed with ASEAN's adoption of the Strategic Action Plan on the Environment in 1994. The Indonesian forest fires of 1991, 1994, and 1997–98 were particularly instrumental, encouraging member states to agree in June 1995 to an ASEAN Cooperative Plan on Transboundary Pollution, with separate programs for atmospheric, shipborne and hazardous waste, and, in December 1997, to a Regional Hazard Action Plan that provides for enhanced monitoring mechanisms and has improved firefighting capabilities (see ASEAN). Subsequently, in December 1998, the menace of forest fires persuaded ministers to adopt the Hanoi Plan of Action, in which they pledged to take all necessary steps to protect the environment (Cotton pp. 342–44). Despite the promising rhetoric in these and other documents, the practical extent of collective environmental responsibility is more accurately represented by the openness with which public criticism was directed at Indonesia's policies and President Suharto's unprecedented apologies to neighboring states for the smoke haze.

Conclusion

Southeast Asia faces major environmental challenges at the dawn of the 21st century. In this, of course, it is not alone: the region shares many of the problems caused by population growth, resource depletion, and global warming in other major regions of the world. Globalization, too, is increasingly emerging as an important factor: in particular, the environmental consequences of trade liberalization are uncertain at best. The World Trade Organization and regional trading accords such as the ASEAN Free Trade Agreement are supposed to promote more rational uses of resources, and greater economies of scale as between countries, but the policies that they embody may just as easily promote the further erosion of environmental standards and encourage the inappropriate shifting of resource pressures from one part of the world to another. Any approach that does not recognize that poverty is a root cause of environmental degradation and that fails to integrate resource considerations fully into development programs is likely only to exacerbate the present discouraging state of affairs. Without a fundamental change in prevailing attitudes to resource management and use, the frequency and magnitude of natural hazards are also likely to increase, so causing a further deterioration in the state of the environment.

Meanwhile, there is already ample evidence to suggest that the wealth of some nations is increasingly allowing these states to export many of their most pressing environmental problems offshore, to neighbors willing (or, effectively, compelled) to accept additional pollution as the price of additional employment. For the poorer countries of the region, the environmental future is much bleaker. Foreign debt – with the improbability that many governments will ever be in a position to reduce the amounts owed, or the proportion of national budgets required simply to service the interest on these borrowings – ties many states into a web of chronic dependency and underdevelopment. There is a symbiotic relationship between debt, environmental degradation, and disasters caused by natural hazards, a relationship that maintains these economies in a chronic condition of underdevelopment, encourages a resource frontier mentality, and increases their governments' overseas borrowing requirements as they seek to fund reconstruction and rehabilitation works. Moreover, as the burden of debt is bequeathed from generation to generation, it begets a vicious cycle whereby slow economic growth creates the conditions that further promote unsustainable development practices, which, in turn, increase the vulnerability of a society to disasters caused by natural hazards and so generate a deeper state of underdevelopment. The current precarious political stability of some states in contemporary Southeast Asia is also a reflection of their deteriorating physical environments.

Further Reading

Arnell, Nigel, "Climate Change and Global Water Resources," in *Global Environmental Change*, number 9, 1999

ASEAN: "Functional Cooperation, Environment, Plan of Action," on the ASEAN website at www.asean.or.id

Bailey, Conner, and Caroline Pomeroy, "Resource Dependency and Development Options in Coastal Southeast Asia," in *Society and Natural Resources*, number 9, 1996

Bankoff, Greg, 1995: "Coming to Terms with Nature: State and Environment in Maritime Southeast Asia," in *Environmental History Review*, Volume 19, number 3, 1995

Bankoff, Greg, 1997: "Europe's Expanding Resource Frontier: Colonialism and Environment in Southeast Asia," in Brook Barrington (editor), *Empires, Imperialism and Southeast Asia*, Clayton: Monash Asia Institute, 1997

Bankoff, Greg, 1999a: "A History of Poverty: the Politics of Natural Disasters in the Philippines, 1985–1995," in *Pacific Review*, Volume 12, number 3, 1999

Bankoff, Greg, 1999b: "Societies in Conflict: Algae and Humanity in the Philippines," in *Environment and History*, number 5, 1999

Barrow, Chris, "Environmental Resources," in Denis Dwyer (editor), *South East Asian Development: Geographical Perspectives*, Harlow: Longman, and New York: Wiley, 1990

Boomgaard, Peter, "Forests and Forestry in Colonial Java, 1677–1942," in John Dargavel, Kay Dixon, and Noel Semple (editors), *Changing Tropical Forests: Historical Perspectives on Today's Challenges in Asia, Australasia and Oceania*, Canberra, ACT: Centre for Resource and Environmental Studies, Australian National University, 1988

Bruestle, Arthur, "East Asia's Urban Environment," in *Environment, Science and Technology*, Volume 27, number 12, 1993

Bryant, Raymond, "Power, Knowledge and Political Ecology in the Third World: A Review," in *Progress in Physical Geography*, Volume 22, number 1, 1998

Bryant, Raymond, and Michael Parnwell, "Politics, Sustainable Development and Environmental Change in South-East Asia," in Michael Parnwell and Raymond Bryant (editors), *Environmental Change in South-East Asia: People, Politics and Sustainable Development*, London and New York: Routledge, 1996

Castro, Eddee, "Big Climate Changes Seen," in *Manila Bulletin*, July 3, 1989

Colchester, Marcus, "Introduction," in Minority Rights Group International (editors), *Forests and Indigenous Peoples of Asia*, London: MRG, 1999

Cotton, James, "The 'Haze' over Southeast Asia: Challenging the ASEAN Mode of Regional Engagement," in *Pacific Affairs*, Volume 72, number 3, 1999

Dauvergne, Peter, *Shadows in the Forest: Japan and the Politics of Timber in Southeast Asia*, Cambridge, MA: MIT Press, 1997

This thoughtprovoking study of the politics of timber extraction in Southeast Asia lays bare the chain of exploitation that stretches from the region's forests to the rapacious timber markets of Japan. Particularly strong with regard to Indonesia, Malaysia, and the Philippines, the book contains a comprehensive statistical appendix on tropical hardwood exports.

Dayton, Leigh, "The Killing Reefs," in *New Scientist*, November 11, 1995

Dixon, Chris, *South East Asia in the World Economy*, Cambridge and New York: Cambridge University Press, 1991

In this very thorough historical geography of the region since the advent of western commercial interests, the treatment of the colonial economy is particularly impressive, and is supported by ample statistical tables and diagrams.

Fu, C., J.-W. Kim, and Z. Zhao, "Preliminary Assessment of Impacts of Global Change on Asia," in James Galloway and Jerry Melillo (editors), *Asian Change in the Context of Global Climate Change*, Cambridge and New York: Cambridge University Press, 1998

Ganguli, Barin, *Breakthroughs in Forestry Development – Experience of the Asian Development Bank*, Manila: Asian Development Bank, 1995

The second half of this useful, if somewhat uncritical, account of the Asian Development Bank's recent shift towards sustainable forest management policies is devoted to case studies that include Indonesia, Malaysia, the Philippines, and Thailand.

GEO: "The State of the Environment," in *Global Environment Outlook 2000*, on the UN Environment Program website at grid2.cr.usgs.gov/geo2000/english/0020.htm

This is a comprehensive overview of the current state of the world's environment by major regions. Chapters two and three are of particular reference to Southeast Asia, further broken down into "Maritime" and "Mainland" subregions (the latter also including Yunnan province in China). The text is available either online at the address shown or through the UN Environment Program.

Hameed, S., and J. Dignon, "Global Emissions of Nitrogen and Sulphur Oxides in Fossil Fuel Combustion 1970–1986," in *Journal of Air Waste Management Association*, number 42, 1992

Handley, Paul, "Before the Flood: Climate Change May Seriously Affect Southeast Asia," in *Far Eastern Economic Review*, April 16, 1992

Hermoso, Christina, "Global Warming Affecting RP," in *Manila Bulletin*, August 3, 1993

Hewitt, Kenneth, *Regions of Risk: A Geographical Introduction to Disasters*, Harlow and New York: Longman, 1997

A culturally sensitive introduction to the whole question of hazard, disaster, and human vulnerability that, although not specific to Southeast Asia, still provides the best overview of the topic that there is for the region

Hiebert, Murray, S. Jayasankaran, and John McBeth, "Fire in the Sky," in *Far Eastern Economic Review*, October 9, 1997

Hirsch, Philip, and Carol Warren (editors), *The Politics of Environment in Southeast Asia: Resources and Resistance*, London and New York: Routledge, 1998

An informative series of recent case studies that chart the emergence of the environment as an issue for public debate in Southeast Asia (though there is somewhat limited coverage of Indochina), and the roles that various political actors play in conflicts over resource use and management

IEDP: *Private Investment and Trade Opportunities in Air and Water Pollution Control*, Economic Brief number 11, Honolulu: Institute for Economic Development Policy, East-West Center, 2000

Iglesias, A., Lin Erda, and C. Rosenzweig, "Climate Change in Asia: A Review of the Vulnerability and Adaptation of Crop Production," in *Water, Air, and Soil Pollution*, number 92, 1996

Intal, Ponciano, "Perspectives from the Philippines and ASEAN," in Simon Tay and Daniel Esty (editors), *Asian Dragons and the Green Trade*, Singapore: Times Academic Press, 1996

Levine, Joel, Tom Bobbe, Nicholas Ray, Ashbindu Singh, and Ronald Witt, *Wildland Fires and the Environment: A Global Synthesis*, Nairobi: Division of Environmental Information, Assessment, and Early Warning, UN Environment Program, 1999

This report on the significance of recent large-scale fires to the global environment, with special emphasis on the loss of biodiversity, property, and livelihoods in Indonesia during 1997–98, is available online at grid2.cr.usgs.gov/pubs/wildfire.pdf.

Martens, Willem, Louis Niessen, Jan Rotmans, Theo Jetten, and Anthony McMichael, "Potential Impact of Global Climate Change on Malarial Risk," in *Environmental Health Perspectives*, Volume 103, number 5, 1995

McManus, J.W., "Tropical Marine Fisheries and the Future of Coral Reefs: A Brief Review with Emphasis on Southeast Asia," in *Coral Reefs*, Supplement to number 16, 1997

Mendelsohn, Robert, "The Impact of Global Warming on Pacific Rim Countries," in Robert Mendelsohn and Daigee Shaw (editors), *The Economics of Pollution Control in the Asia Pacific*, Cheltenham and Brookfield, VT: Edward Elgar, 1996

Nicholls, Neville, "ENSO, Drought, and Flooding Rain in Southeast Asia," in Harold Brookfield and Yvonne Byron (editors), *Southeast Asia's Environmental Future: The Search for Sustainability*, Tokyo, New York, and Paris: United Nations University Press, 1993

Palis, Honorato, "Coral Discoloration an Indicator of Global Warming," in *Canopy*, Volume 19, number 1, 1993

Panayotou, Theodore, "The Environment in Southeast Asia: Problems and Policies," in *Environment, Science and Technology*, Volume 27, number 12, 1993

Parry, M., Blantran de Rosari, A. Chong, and S. Panich (editors), *Socioeconomic Impacts of Climate Change in Southeast Asia*, Nairobi: UN Environment Program, 1992

Peluso, Nancy, Peter Vandergeest, and Leslie Potter, "Social Aspects of Forestry in Southeast Asia: A Review of Postwar Trends in the Scholarly Literature," in *Journal of Southeast Asian Studies*, Volume 26, number 1, 1995

Pitcock, A., K. Walsh, and K. McInnes, "Tropical Cyclones and Coastal Inundation under Enhanced Greenhouse Conditions," in *Water, Air, and Soil Pollution*, number 92, 1996

Redford, K., "The Empty Forest," in *BioScience*, number 42, 1992

Rigg, Jonathan, *Southeast Asia: The Human Landscape of Modernization and Development*, London and New York: Routledge, 1997

A probing and thoughtprovoking assessment of economic progress in the region that gives prominence to indigenous notions of development, and the manner in which different people have responded to the challenges and opportunities of change

Robinson, John, and Richard Bodmer, "Towards Wildlife Management in Tropical Forests," in *Journal of Wildlife Management*, Volume 63, number 1, 1999

Rosenberg, David, "Environmental Pollution around the South China Sea: Developing a Regional Response," in *Contemporary Southeast Asia*, Volume 21, number 1, 1999

Rush, James, *The Last Tree: Reclaiming the Environment in Tropical Asia*, New York: The Asia Society, 1991

Although the coverage of this study is not confined to Southeast Asia, it remains one of the few attempts to present an overview of the environmental impact of historical developments in the region.

Savage, Victor, Lily Kong, and Warwick Neville, "The Naga Awakens: Political Imperatives and Economic Opportunities in Southeast Asia," in Victor Savage, Lily Kong, and Warwick Neville (editors), *The Naga Awakens: Growth and Change in Southeast Asia*, Singapore: Times Academic Press, 1998

Severino, Howie, "The Unabated Rape of Laguna de Bay," in *Manila Chronicle*, July 3, 1988

Taylor, D., P. Saksena, P. Sanderson, and K. Kucera, "Environmental Change and Rain Forests on the Sunda Shelf of Southeast Asia: Drought, Fire and the Biological Cooling of Biodiversity Hotspots," in *Biodiversity and Conservation*, number 8, 1999

Tongzon, Jose, *The Economies of Southeast Asia: The Growth and Development of ASEAN Economies*, Cheltenham and Northampton, MA: Edward Elgar, 1998

Dr Greg Bankoff is Senior Lecturer in Southeast Asian History in the Department of History at the University of Auckland.

FACING THE NEW CENTURY

Chapter Seventeen
Agriculture

John Minns

Despite the many profound changes that have taken place in the societies and economies of Southeast Asia since World War II, most people in the region are still either directly engaged in agriculture or are closely connected to it. To generalize about such a major activity, across a region of 450 million people, is, of course, extremely difficult, given the significant variations among its 11 economies. For example, the proportions of the labor force engaged in farming, fishing, and forestry range from as high as 80% in Cambodia and Laos, around 65% in Vietnam and Burma, and 54% in Thailand, to 45% in Indonesia, 40% in the Philippines, and as low as 16% in Malaysia (see CIA). Nevertheless, there are some important similarities between the agricultural sectors of each of the eight countries of Southeast Asia discussed in this chapter (excluding Brunei and Singapore, where there is little or no agricultural output, and East Timor, where agriculture has been as disrupted as the rest of the economy in 1999–2000).

First, the entire region can be characterized as having a rice-based culture. Rice is still the most important crop and the basis of the popular diet. Second, the long period of colonial rule in every country in the region, except Thailand, saw the development of significant cash crops, largely for export to the colonial powers. However, the colonialists were much less interested in, and tended to ignore, the food crops produced for local consumption on which most of the population depended for their diet and for employment. The main technological advances in agriculture during the colonial period therefore took place across a rather narrow range of cash crops, while the output of food crops languished until the "green revolution" began to have an impact on farming techniques from the 1960s (see below). A third similarity across the region is that over the past 30–35 years most agricultural production has been deeply affected by increasing involvement in local and global capitalist markets, entailing the transformation of social relations at the local level.

Southeast Asia's agricultural record over these last three decades has been impressive, at least statistically. Growth in output has been higher than in any other less developed region. In one extremely rapid spurt, during the ten years after 1972, agricultural production in Southeast Asia rose by nearly 60%, more than double the rate of increase in South Asia over the same period (Fryer 1990 p. 168). However, this growth has not brought good news for all the people of the region, for there have been wide variations in agricultural development, both between countries and within them. Although debates on this issue continue, it is highly probable that the absolute numbers of people living in poverty in rural areas have increased and that rural incomes have lagged far behind those of urban dwellers. Famine and outright starvation are now rare in Southeast Asia, but there are still many who eat less than is good for their health. In some places, such as parts of Java, the Philippines, Cambodia, Vietnam, and Laos, rural people often experience real hunger in the period just before harvest time, for which in Indonesia, for example, there is a specific term, *peceklik*. Both landlessness and rural indebtedness have increased in the past three decades.

Forms of Agriculture in the Region

Three major forms of agriculture are still practiced in Southeast Asia: "swidden" or shifting agriculture; *sawah* or settled agriculture; and cash crop production, both on large estates and by small cultivators.

Swidden (burned field) agriculture involves the preparation of fields by burning undergrowth or forest. Crops planted quickly after such burning are able to take advantage of the nutrients thus released into the soil. This fertilization creates the basis for good crops for short periods. When the yield declines, the swidden is abandoned except, perhaps, for some tree crops. Usually, short periods of growing crops in this way are then followed by much longer ones when the land is left to revive and the forest to take over, typically for eight to 12 years, but sometimes for considerably longer. Although swiddens are usually small plots (often between 0.5 and 1.5 acres), those who use them need access to large areas because of the need to shift cultivation and to leave most swiddens fallow at any one time. Swidden agriculture can therefore normally support only low population densities. If population pressure increases or swidden areas are restricted, overuse can damage the land (Fryer 1970 pp. 39–42). When swidden agriculture is free of these pressures, however, it can produce good yields with relatively small inputs of labor, while remaining ecologically sustainable.

In some parts of Southeast Asia, swidden agriculture is used to grow "dry" rice as well as a range of other vegetables and tree crops, often by people who also have access to permanent irrigated rice fields. However, swidden cultivators have been under enormous pressure from two sources. First, pressure to expand agricultural production has led to the destruction of many swiddens by plantation and other agriculturalists, who destroy the forest cover permanently, and by forestry companies, who exploit the swiddens as a one-off exportable resource. Second, governments of the newly independent states of the region have tended to see swidden cultivation as a problem in the process of state-building. Its practitioners are difficult to control, their product is difficult to tax, and it generates few if any export revenues (Fryer 1990 p. 171).

The main form of settled agriculture in Southeast Asia is known as *sawah*, from the Indonesian word for an embanked and flooded rice field. In the hot climates of Southeast Asia, soils are depleted by the unrelenting sun. Sawah farming covers the ground with water, fed either by rain or by elaborate systems of irrigation, protecting and enriching even quite poor soils and maintaining high productivity over long periods. If they are to remain covered by water, such fields must be as level as possible, requiring terracing on sloping ground. The requirements for labor in sawah agriculture are therefore much higher than in swiddens, with fields being constructed and maintained over generations. However, the yield is also very much greater. The key crop in sawah agriculture is wet (padi) rice, although household gardens producing fruit and vegetables supplement the diets of sawah farmers. So too did fish taken from the flooded fields, until the use of chemical fertilizers, herbicides, and pesticides became prevalent.

The third main form of agriculture in the region centers on the production of cash crops, including food and non-food crops, produced both on large estates and by small landholders. The most important cash crops in the region are rubber, palm oil, sugar, coconuts, coffee, copra, tea, tobacco, pepper, and other spices. Although plantation crops occupy only about one sixth of the land area under rice in Southeast Asia, they can be quite lucrative (Fryer 1990 p. 179). Rubber, for example, produces greater incomes than even the most high-yielding forms of rice, while palm oil produces more than either. Further, both these cash crops bear throughout the year, generating steady incomes and allowing the constant use of labor (Fryer 1990 pp. 180–81). Small-scale cash cropping could develop even in swidden agriculture, with crops such as coffee and rubber being planted near the end of the productive life of the swidden.

During the colonial period, cash cropping was conducted mainly to satisfy demand in Europe and North America. Some of the estates were owned by foreign planters, as in the British colonies in peninsular Malaya (now

West Malaysia) or in the Dutch East Indies (now Indonesia). Nevertheless, even in colonial times much of the development of cash crops took place under local ownership, in many cases by smallholders responding directly to the market or to various mechanisms by which the colonial authorities extracted wealth. In what is now Indonesia, for example, the Dutch imposed the *Cultuurstelsel* ("culture system") between 1830 and 1870. This system granted peasants remission of their land taxes in exchange for the cultivation of crops for the government on one fifth of their land or for 66 days of labor every year on government-owned estates (Geertz pp. 52–53).

The long-term effect of colonial cash cropping on agriculture in Southeast Asia has been a matter of intense debate. The American anthropologist Clifford Geertz argued in 1963 that the effect of the *Cultuurstelsel* on Java was to create a dual economy, with one section devoted to cash crops and the other to the growing of the staple food, rice. Sawah agriculture was starved of capital, and land for its expansion was restricted; it languished, yet was forced to absorb and support more and more people. As Geertz put it (Geertz p. 80):

> "The Javanese could not themselves become part of the estate economy, and they could not transform their general pattern of already intensive farming in an extensive direction, for they lacked capital, had no way to shuck off excess labor, and were administratively barred from the bulk of their own frontier, the so-called 'waste lands,' which were filling up with coffee trees. Slowly, steadily, relentlessly, they were forced into a more and more labor-stuffed sawah pattern . . . [with] tremendous populations absorbed on minuscule rice farms."

Thus, villages reliant on sawah fields became poorer, sharing greater poverty between their members.

Geertz's argument has since been attacked on many grounds, including his overestimation of the egalitarian aspects of village life and his underestimation of the likelihood that villagers would take up new production techniques. However, it is undoubtedly the case that dual economies were created throughout the region, with some cash crops attracting investment while staple food production on small, often sawah, plots remained undercapitalized and, mostly, extremely poor.

Main Trends of Government Policy

In the immediate postcolonial period, agriculture was rarely the major concern of most of the new states of Southeast Asia. During the 1950s, development economists and economic policy-makers tended to concentrate on the problem of industrialization. Agriculture was seen as, at best, a provider of low-cost labor that might aid the profitability of industry (see Lewis). One influential development theorist, Albert Hirschman, argued in 1958 that for development to take place, "linkages" between sectors of production were crucial (see Hirschman). Agriculture, it was claimed, has fewer "linkages" than manufacturing and therefore should not be emphasized in development efforts. When Raúl Prebisch suggested in 1959 that there was a long-term tendency for agricultural prices to fall relative to those of manufactured goods, and therefore for the terms of trade of largely agricultural countries to decline, the case for a major shift from agriculture to manufacturing seemed clear (see Prebisch). Thus, the relative neglect of agriculture by theorists was, at first, mirrored by policy neglect from governments. During the 1960s, this downgrading of agriculture began to change. At the level of theory, Theodore Schultz published his important text *Transforming Traditional Agriculture* in 1964, arguing for new technologies and capital to be made available to peasant farmers in poor countries.

In any case, despite the emphasis on planning for industrial development, the reality was that cash crops remained the major sources of export incomes in Southeast Asia. Ironically, importing the technology and capital goods that were required if manufacturing was to be developed necessitated some attention to agriculture, or at least to some sectors of it. Even more central to the reevaluation of agriculture was the realization, in Southeast Asia and elsewhere in what was then called the "Third

World," that industrial development was not happening nearly fast enough to absorb the massive and relatively young labor force being generated in rural areas but unable to find employment there. In the 1960s, the minds of policy-makers in Southeast Asia and of western strategists were even more concentrated by guerrilla wars supported by peasants in Vietnam, Laos, Cambodia, Thailand, the Philippines, and elsewhere. Declining rural incomes, rising unemployment, and increasing landlessness had fanned the flames of peasant discontent. Government resources and international aid funds began to return to agriculture, in attempts to undermine mass support for insurgents.

Another reason for increased government support for agriculture was the perceived need to consolidate the power of the newly independent and often fragile states of the region by transplanting people – "sowing peasants" – in areas where the state had limited control, where minority ethnicities seemed to pose problems for national integration, or where relocation removed a potential source of resistance (see De Koninck and Dery). Internal migration has been used for these purposes in Indonesia, Vietnam, Malaysia, and, most tragically, Cambodia under Pol Pot's regime (1975–79). In some cases, internal migration programs sponsored or directed by the state have been seen as providing a safety valve for rural discontent, providing land for the landless without tackling politically difficult questions of land reform. In Indonesia, for example, the practice of *transmigrasi* ("transmigration") has resulted in the settlement of, mostly, Javanese along the border with Malaysia in Borneo (Kalimantan), as well as in East Timor, West Papua, and elsewhere. In Vietnam, the government sent lowlanders to "New Economic Zones" created in upland areas after the reunification of the country in 1975–76. In what was then Malaya, the authorities first resettled indigenous people of the peninsula – collectively known as the *Orang Asli* – from areas where it was thought that they were supplying information and food to Communist insurgents during the "Emergency" of 1948–60 (see Nicholas p. 2). Then, after independence and the creation of Malaysia, vast land development schemes carried out by the Federal Land Development Authority (FELDA) and several other similar bodies were used to consolidate state control over internal regions of the peninsula. Between 1956 and 1990, FELDA resettled more than 119,300 families, totaling 715,000 people, in land development schemes promoting the production of rubber, palm oil, and cocoa (Sivalingam p. 101). The newly located peasantry were given land, but for tenure and their own protection they depended entirely on the regimes, and thus were bound to them. Resettlement was forced on millions of people for different reasons under the Khmer Rouge in Cambodia from 1975. The "peasantization" of almost the entire urban population was intended to remove possible opponents of the new masters of "Democratic Kampuchea."

Thus, from the 1950s and 1960s onward, government policies for agriculture reflected what were seen as the pressing needs of relatively poor, largely new, and often contested state regimes. In the minds of the policy-makers, these needs were, principally, for enhanced export income; for absorption of burgeoning rural populations within the countryside itself; for mitigation of rural discontents; and for political consolidation of territory and of ethnic minorities.

The Green Revolution

Just as the attention of governments and international agencies such as the World Bank turned back to agriculture in the 1960s, a solution to at least some of the problems of rural poverty seemed to have arrived at last. New agricultural technologies, which rapidly attracted the collective name of the "green revolution," were based at first on fast-growing or high-yielding varieties of rice and wheat, sometimes called simply "HYVs," but also known as "modern" varieties. These varieties were able to absorb more water and fertilizer without damage, and therefore to grow faster. It was possible, as a result, to achieve one or even more extra crops each year without expanding the area planted.

This solution seemed especially attractive since, by the 1960s and 1970s, most countries in the region were reaching the limits of expansion into areas that had previously been unfarmed. Farm incomes, it was thought, should rise. Further, since the production of basic staples, especially rice, would grow dramatically, food prices could fall, consumption could increase, and malnutrition could become a thing of the past. Finally, the new strains would employ more labor, soaking up the unemployed and underemployed rural population. More harvests and planting would increase the need for workers, and the new strains demanded more labor anyway, not only to apply fertilizers but also, and as a consequence, to weed the fields.

With official encouragement and, in some places, strong pressure, the new varieties took much of the region by storm. By 1975, high-yielding rice had been planted on 62% of the rice area of the Philippines, 41% of Indonesia's, and 36% of Malaysia's (Wong p. 95). There is no question that green revolution technology, especially in rice, did both increase yields and cut prices. Between 1967 and 1984, when the green revolution had its greatest effects, rice production in Asia as a whole increased at an average rate of 3.2% each year, and much faster in those countries where the new strains were introduced most widely. Most of the increase was due to higher yields per unit of land (acre or hectare), rather than expansion of the total area of land under cultivation (Dawe p. 948). What is more, rice prices did fall dramatically: in one short period, from 1981 to 1985, they declined by 62% and then stabilized (Dawe p. 948).

However, the early enthusiasm for the green revolution was quickly dampened by a wave of studies arguing that its effects were not entirely benign. A key charge made by the critics was that the technology was not "scale neutral": it was most likely to be used by richer farmers, and the green revolution could even be interpreted as a "landlord-based innovation" (Wong p. 104). The high-yielding varieties of rice needed higher amounts of fertilizers, pesticides, and water, all of which are relatively expensive additional inputs to production, more affordable to the relatively well-off than to the poor. It was also argued that those with relatively small amounts of land could not afford to take risks, whereas richer farmers could experiment with new varieties on parts of their holdings (see Rigg 1989). For these reasons, the take-up rates for the new varieties varied according to wealth (see Wong) and inequalities in the countryside became even wider. Farmers who were wealthier to start with had a greater competitive edge over their poor neighbors and eventually took over their land. Landowners seized opportunities to make profits by operating the land themselves with hired labor, rather than renting it to tenant farmers. At the least, the potential to make more money from farming led to an increase in rents. Some studies suggested that, for all these reasons, the green revolution had contributed to even greater landlessness and rural poverty, the opposite of the claims made by its advocates (see the discussion of these early studies in Eicher and Staatz p. 9). Finally, critics of the green revolution argued that, as the new techniques sacrificed traditional methods of crop rotation, greater pressure was being put on the soil, providing breeding grounds for pests and requiring even more chemical pesticides. The increased use of fertilizers added more chemicals to the water system, with serious effects on wildlife and human beings alike.

Just as there is no doubt that rice production grew and prices fell as a result of the green revolution, it is equally certain that many of the serious social and ecological consequences noticed by the critics followed as well. Shortly after the introduction of high-yielding varieties in the rice-growing areas of Central Luzon in the Philippines, there were waves of rent increases and evictions. Similar effects have been observed elsewhere in Southeast Asia and in other parts of the continent (see Ahmad, Kirkpatrick and Harris, and Patnaik). While the studies are not agreed on this question, it seems that the new technology tended to reduce rural employment (Kirkpatrick and Harris p. 327). The reason was not that the technology itself was less labor-intensive, but that the richer farmers, often using hired labor, who were most likely to take it up were also those most likely to introduce labor-replacing technology. Increased profits from the new strains gave them greater capacity to do so. Others have suggested that governments, in their enthusiasm for technological improvements in agriculture, including the green revolution, have reduced the real price of labor-saving equipment to farmers through subsidized credits, thus cutting the rural workforce (Jayasuriya and Shand p. 423; see also Herath and Jayasuriya).

Proponents of the green revolution have responded that, while some of these problems did exist with green revolution technology in the early stages of its introduction, they have been exaggerated and, perhaps, largely overcome more recently. Jonathan Rigg, for example, has argued that although smaller farmers did not at first take up the new varieties in the same proportions as the richer ones, they began to catch up in the 1980s (Rigg 1989 p. 146). In other words, the technology eventually did become "scale-neutral." Some have argued that the green revolution initially increased the need for rural labor, but that this effect has since disappeared (see Otsuka, Gascon, and Asano p. 103). A study of Central Luzon and Laguna in the Philippines by Robert Herdt directly attacked the major claims of the critics of the green revolution. He concludes that, over the whole period between 1965 and 1982, small farmers were not displaced, the agricultural workforce increased, and small-scale farmers and farm workers retained some of the benefits of the technology (Herdt p. 348). However, using either this study, or others concerned with other specific areas, to project a broadly positive picture of the impact of the green revolution in Southeast Asia seems far too optimistic. It appears that the weight of evidence is that the green revolution has had quite differential effects: there have been improvements for some, but it has created worsening situations for others. The best that can be said is, probably, that the green revolution has been a mixed blessing.

Indonesia

Rice dominates the agriculture of Java and Bali, which are still the most densely populated parts of Indonesia, and it is also very important in South Celebes and West Sumatra. Sumatra is the largest producer of cash crops, which are significant also in West Kalimantan, East Java, and Celebes.

By and large, Indonesian agriculture performed poorly in the 1950s and early 1960s. The pessimism of Clifford Geertz's position that Java's rural economy was trapped in a cycle of poverty seemed fully justified. One problem, recognized by Geertz and many others, was the tiny size of rice plots, especially in Java. The land shortage in Java was, at least, recognized by the government under the Basic Agrarian Law of 1960, which limited the amount of land that any single family could own and required those possessing more to have it distributed to the landless (Booth p. 136). However, landlordism and large estates were uncommon in Java, so in practice the amounts of land redistributed were strictly limited. Under the "New Order" regime, from 1965, land reform was largely ignored, as the efforts of the government shifted to increasing agricultural yields using green revolution technology (Booth pp. 137–38). Whatever the difficulties of that technology – and the undoubted brutalities of the Suharto regime – after 1970 and especially after 1979 rice production did in fact rise rapidly in Java, and even in relatively land-abundant areas such as the Kalimantan provinces on Borneo. As elsewhere, the major factor in this rise was the increase in yields per hectare (Booth, pp. 39–40). In the late 1970s Indonesia imported more rice than any other country in the world, but by 1985 it had achieved broad self-sufficiency in rice (Hill p. 125).

The situation has been rather different in the case of cash crops. During the first half of the 20th century, cash crops were grown either on large Dutch-owned estates or by smallholders, especially in such regions as Sumatra, Borneo, and Celebes. The smallholders' output of major crops such as rubber even surpassed that of the estates. The 1950s and most of the 1960s were years of poor growth of cash crops, and the nationalization of the Dutch-owned estates, in December 1957, caused exports of cash crops to slump even further. The production of cash crops was revived in the 1970s and 1980s, although less impressively than rice. Palm oil has been the outstanding performer, and Indonesia became the world's second largest producer of palm oil (after Malaysia) in 1992 (Hill p. 139). However, the development of cash crops has not always brought great benefits to many people in the producing regions, since world prices for most cash crops have been low.

Malaysia

Unlike in Indonesia, rice production in Malaysia has never been sufficient to meet the country's needs, although from independence in 1963 until the late 1980s self-sufficiency in rice was regarded as a national priority of the Malaysian state. The government provided substantial investment in irrigation, support for green revolution technology, and, at times, free fertilizers to padi farmers in order to boost rice production. The reasons for this emphasis are based in Malaysian nationalism and Malay ethnic sensitivities. The poverty of the rice growers, most of whom are Bumiputras (Malays or members of other "indigenous" groups), has been of major concern for the political alliances that have controlled Malaysia since independence and have been dominated by elite Malays. Only in the last decade has it been officially accepted that, since Malaysia is a relatively high-cost producer of rice, something less than total self-sufficiency in rice is acceptable.

It is plantation crops, especially palm oil and rubber, that constitute the major form of agriculture in Malaysia. At the time of independence, plantation estates were overwhelmingly foreign-owned, with perhaps only 25% in the hands of Malaysian citizens. During the 1960s, these began to diversify from rubber into other plantation crops, especially cocoa, tobacco and, above all, palm oil. Plantation agriculture has also been heavily supported by the state since the late 1960s. In part, the reason for this high level of state support was to maximize the export earnings that the sector could generate. In part, too, it was linked with the promotion of Bumiputra interests. From 1971, under the government's New Economic Policy, control of the large foreign-owned plantations began to pass either to Malaysian citizens of one ethnic group or another, or to the government itself, which was to hold these properties in trust until Bumiputras could take them over.

As mentioned above, the Federal Land Development Authority (FELDA) resettled people on land that was to be largely devoted to cash crops, especially rubber and palm oil. While these land development schemes have some aspects of smallholder agriculture, and the farmers consider themselves to be smallholders, they operate in many ways like estates, often with centralized processing, equipment, and accounting.

Since the mid-1980s, the policy of the Malaysian government has been to reduce state involvement in agriculture. Accordingly, there has been a degree of market deregulation, many government subsidies for farmers have been withdrawn, and there has been a retreat from the extensive land developments of the past.

Thailand

By international standards, Thai agriculture performed quite well from the 1950s, with output increasing by an average of nearly 2% each year in the 1950s, 5.4% between 1958 and 1973, and 3.9% between 1973 and 1984 (Siamwalla, Setboonsarng, and Patamasiriwat p. 81). Unlike in much of the rest of Southeast Asia, the reasons for this growth had little to do with investment in green revolution techniques.

Some important improvements did take place: the replacement of animal-drawn ploughs with mechanical ones; the use of other mechanical equipment; and large-scale investment in irrigation until the early 1980s in some areas, especially the established "rice bowl" of the Chao Phraya delta. However, the major reason for this good record was simply the expansion of the area under cultivation. Vast inland areas were opened by forest clearance and the expansion of road networks in the period of greatest agricultural expansion, between 1958 and 1973. Tractors were used to clear huge areas. The numbers of farming families grew in the new agricultural regions and the land available for each family was actually increased. These new areas were particularly suited to the growing of maize, kenaf, cassava, and sugar cane, most of which are exported (Siamwalla, Setboonsarng, and Patamasiriwat p. 82). Much of this new agricultural land is rain-fed, not irrigated. Since the capacity to produce crops, especially dry rice, throughout the year depends on water supplies, many families in the North and Northeast cannot survive on farm incomes alone but must look for paid employment elsewhere.

One major problem concerns security and land tenure. As many as one in five farming families have no clear land title, so that, often, they are technically squatters. Although the chances of eviction are low, their land cannot be used as collateral for loans; hence, much of Thai agriculture suffers from serious underinvestment and technological lag (Siamwalla, Setboonsarng, and Patamasiriwat p. 100).

The Philippines

Forms of landholding in the colonial Philippines were extremely varied, reflecting differences in the experience of colonization between regions and over time. At one end of the scale were large landed estates, a legacy of Spanish colonialism, with landowners collecting rents or a share of the crops from many tenants (see Cushner). In other areas, more modern plantations, largely owned by Filipinos, sprang up producing for the US market in the period before World War II. In yet others, peasant farmers struggled to maintain an existence on subsistence farms.

Against this background, the class struggle between large landowners and tenant farmers has periodically broken out into open armed conflict, with guerrilla war in the 1930s, Communist-led peasant rebellion after independence in 1946, and then the struggle of the New People's Army (NPA), the armed wing of the new, Maoist Communist Party of the Philippines, from the late 1960s. Weak attempts to undertake land reform by governments closely connected to large landed interests have done little to lessen the causes of discontent. In an attempt to overcome their labor problems, some landowners began to introduce labor-saving machinery in the 1950s, a move that increased the numbers of landless and worsened the condition of many rural dwellers (Hutchison p. 73).

The Philippines took up the green revolution in rice more rapidly than anywhere else.

By 1975, as we have seen, 62% of the rice area in the Philippines was planted with the new high-yielding varieties. Rice production rose accordingly, and so did the incomes of some farmers, especially those who were already better off, but many others found themselves deeper in debt to banks and local traders. Eventually losing their land, they joined the ranks of the low-paid contract workforce (Hutchison p. 74).

The production of sugar and coconuts, the two major export products of the Philippines since World War II, has faced serious problems. Coconut growing expanded in Mindanao, often on small farms, but its development was severely disrupted by the wars being waged against the central government by the NPA, in Eastern Mindanao, and by the Moros, in Western Mindanao and Sulu. Sugar production, which in Northwest Negros is largely carried out by extremely poor smallholder tenants and in the southern Visayas by wage-laborers on large farms, faced declining world prices, poor productivity, and strong competition from other sugar suppliers. By the 1980s, both major export crops were in serious difficulty. Many other agricultural exports have been developed more recently by large-scale, transnational agribusiness enterprises, including fruit, poultry, and other livestock.

Burma

Agriculture accounts for around two thirds of employment and more than half of GDP in Burma (Guyot p. 192). Teak, rubber, cassava, maize, pulses, and other crops are grown, but padi rice is by far the main crop. Indeed, Burma was once "the rice bowl of Asia," being the largest exporter of rice in the world before World War II and still a major exporter until the 1980s.

In 1962–63, 1.52 million tons of rice were exported, but by 1987–88 only 300,000 tons were exported, and the total for the following financial year was just 50,000 tons (Than p. 240). Forced government procurement of crops is the most likely main cause of the decline. For most of the period since 1962, the government paid below-market prices for rice in order to keep urban wages low (Than and Nishizawa p. 96). The result was that farmers reduced production, hoarded rice, and smuggled it illegally to India, Bangladesh, and China. Another reason is the starving of credit to farmers, as the government spent most of its limited resources on processing, manufacturing, and the military. By the late 1990s, the army absorbed around 40% of the government budget.

While green revolution varieties were introduced in the 1970s, the associated costs of improved irrigation, fertilizers, and hired labor could not be afforded by farmers who were still being forced to sell their rice at very low prices. Government credit to farmers did increase somewhat in second half of the 1970s and the early 1980s, and the official rice price was raised, contributing to increased rice production (Than and Nishizawa pp. 104–05). Growth in agricultural output averaged 6.6% a year between 1973 and 1983 (Than and Nishizawa p. 112).

After the present military regime, then under the name of the State Law and Order Restoration Council, seized power in 1988, it reintroduced a centralized procurement system, but toward the end of 1997 it was forced to liberalize the sale and production of rice once again. Burma's rice exports now face two additional problems. They are considered to be of relatively low quality; and there is growing competition from elsewhere in the region, including Thailand and, increasingly, Vietnam.

Vietnam

Around two thirds of the Vietnamese labor force are employed in agriculture. Collectivization of farming was established government policy in the Democratic Republic of Vietnam ("North Vietnam"). Although collectivization met some resistance, it was popular with many poorer peasants, giving them access to land, draft animals, and irrigation, and providing them with improved health and education services (Beresford p. 181). However, by the time of reunification, in 1975–76, peasants were spending more time on their household plots, the crops from which they were permitted to sell on the market. By the end of the 1970s, peasants were earning between 60% and 70% of their incomes from such plots (Beresford p. 184), while production on collective land stagnated. Greater divisions of wealth in the countryside began to develop and, despite various government attempts to crack down on it, a black market opened up.

Stagnant official production and growing production for the black market eventually forced changes to the official system, allowing greater market activities from the late 1970s and early 1980s. Farm land was increasingly decollectivized. Then, in 1986, the process known as *doi moi* ("new structure" or "renovation") was introduced. In 1988, a resolution of the Politburo, the main decision-making body in the ruling Communist Party, led to the effective reintroduction of small-scale capitalist farming. All output from land could be sold privately and land was to be distributed for 15 years (50 years in the case of perennial crops) to individual families (Beresford p.190).

There is little doubt that this move, as well as the more gradual recovery of the countryside from decades of war, helped to increase agricultural production dramatically. After two decades of being a rice importer, Vietnam became a rice exporter once again (Pingali and Xuan p. 697). Indeed, it has become one of the main sources of rice for Singapore, Malaysia, and Brunei, and, to a lesser extent, for Indonesia and the Philippines. As well as rice, the traditional export staple crop of Vietnam, significant increases have taken place in the production and export of rubber, coffee, and tea.

Laos

Agricultural cooperatives and state farms were also established in Laos, under the rule of the Lao People's Revolutionary Party, in 1978. This was perhaps one of the worst possible times to attempt such a major change in the countryside. Drought in 1977 and floods in 1978 had reduced an already impoverished agricultural sector, in one of the poorest countries in the world, to subsistence or below it. Except in the areas where war had so completely devastated their farms that they had no choice, peasants strongly resisted collectivization (Evans pps. 32–33). By the middle of

1979, production had been so severely disrupted that the government suspended its attempts to recruit the peasants to cooperatives. In 1986, it began a series of major liberalizing reforms, introducing market prices, allowing private production outside cooperatives, and ending preferential treatment for cooperatives and state farms. Since then, most farming has returned to private hands.

Rice is the main crop, but it is still mainly rain-fed rather than based on artificial irrigation. Thus, double-cropping is uncommon and limited irrigation means that the rice crop is very sensitive to weather patterns. Laos, despite being quite land-abundant, has rarely achieved self-sufficiency in rice. Meanwhile, the liberalization of trade, except in designated "strategic" products, has prompted an increase in the export of agricultural goods such as coffee (the most important cash crop), live cattle, sesame seeds, and the country's major export, logs and wood products.

Cambodia

At the end of the colonial period Cambodia agriculture was poor and undeveloped, the only major expansion of production being the introduction of large rubber plantations. By 1953, the year of independence from France, most peasants in Cambodia were heavily indebted (see Raymond). Rice and rubber provided the basis for some exports, but debt and the high level of "parcelization" (the division of each farmer's land into a number of small, widely dispersed plots) meant that very few peasants had the ability to invest in agricultural improvements. Production stagnated and, from 1964, it declined. After 1970, carpet bombing by the United States drove millions of farmers off their land.

The Khmer Rouge regime that seized power in 1975 saw the urban population as parasitic and a potential threat. It drove most of them to work in the countryside. Through execution, starvation and other means, at least 1.05 million Cambodians died, until Vietnam invaded on December 25, 1978. The new government installed with Vietnamese support at first attempted to continue with some degree of agricultural collectivization, but, in line with developments in Vietnam, this was gradually reversed in the 1980s.

Rice takes up 80% of the total land cultivated. Agricultural production slowly increased in the 1990s, but even today the country has not recovered from the shattering experiences of US bombing and Khmer Rouge government. Much of the irrigation infrastructure of the country was destroyed and Cambodia now has one of the lowest levels of irrigation in Southeast Asia. The war and its aftermath has left its mark on agriculture and on the people in another way: between five million and six million landmines were laid from the 1960s, many in some of the most fertile areas (East Asia Analytical Unit p. 200). Clearing them has been a slow, expensive and deadly process, which also limits the development of land for crops and pasture.

Change and Future Directions

The idea that, until recently, agriculture in Southeast Asia was based on rough equality, village solidarity, and some form of "moral economy" that redistributed wealth from rich to poor is, in part, a romantic fiction (for the debate on this question see Scott, Popkin, Rigg 1994, and De Koninck). Nevertheless, it is true that in the past few decades whatever forms of communal or patron-client relationships existed in the countryside in Southeast Asia have been largely extinguished by thoroughly capitalist market relationships. Subsistence farmers and swidden agriculturalists are disappearing, and being supplanted by capitalist farmers and wage or contract laborers. Those small-scale producers who survive often find themselves enmeshed in market networks dominated by large agricultural companies. The drawing of agriculture into the capitalist economy, a process that began long ago, is complete, or nearly so. Capitalism is now the force that determines what goods are produced and how they are produced (see Pincus).

As incomes rise for many in the region and exports play ever larger roles in the economies of the region, the share of rice in agricultural production has fallen from about 75% in the 1950s and 1960s to between 40% and 50%

today. In some areas, farmers find that they can get better returns from nontraditional products such as fruit or vegetables than from the traditional staples. It is also in these products that new export markets have opened up. Cash crops such as rubber and palm oil have remained important. In the case of palm oil and in the timber industry, deforestation and land degradation have been major side effects. Responding to new market opportunities, livestock industries are the fastest growing type of agriculture in Southeast Asia. They now make up around 10% of agricultural output. In a further response to the market, some basic staples are being supplied across national borders to a much greater degree than in the past. Vietnam has become a major rice supplier, while other countries shift from rice and attempt to move to higher-value agricultural production. Over the past two to three decades, agricultural production has diversified in most countries of the region (Taylor p. 267).

All of this complex process of change has created new opportunities and considerable riches for some in the countryside. However, we should not forget that, as capitalism has thoroughly permeated agriculture, there have been many losers as well.

Further Reading

Ahmad, Zubeida Manzoor, "The Social and Economic Implications of the Green Revolution in Asia," in *International Labour Review*, Volume 105, number 1, January 1972

An early survey generally supporting the green revolution

Beresford, Melanie, "Vietnam: The Transition from Central Planning," in Garry Rodan, Keith Hewison, and Richard Robison (editors), *The Political Economy of Southeast Asia: An Introduction*, Melbourne, Oxford, and New York: Oxford University Press, 1997

Beresford traces the introduction of liberalizing reforms in Vietnam from the late 1970s onward.

Booth, Anne, *Agricultural Development in Indonesia*, Sydney: Asian Studies Association of Australia in association with Allen and Unwin, 1988

This is a very thorough study of agriculture in Indonesia over the long term.

CIA: *The World Factbook 2000*, Washington, DC: US Central Intelligence Agency, 2000

This comprehensive reference work on states, other territories, and international organizations is available both in book form and online, at odci.gov/cia/publications/factbook.

Cushner, Nicholas, *Landed Estates in the Colonial Philippines*, New Haven, CT: Yale University Southeast Asia Studies, 1976

A historical study of the large estates of the colonial Philippines, emphasizing the Spanish influence on patterns of land tenure

Dawe, David, "Re-energizing the Green Revolution in Rice," in *American Journal of Agricultural Economics*, Volume 80, number 5, 1998

Dawe stresses the early positive effects of the green revolution in rice production and explores the reasons for the subsequent slowing down of increases in yields.

De Koninck, Rodolphe, *Malay Peasants Coping with the World: Breaking the Community Circle?*, Singapore: Institute of Southeast Asian Studies, 1992

In this detailed study of the Malay village, De Koninck argues that there has been a fairly recent process of "peasantization" or "sedentarization" as a tool of state-building and control.

De Koninck, Rodolphe, and Steve Dery, "Agricultural Expansion as a Tool of Population Redistribution in Southeast Asia," in *Journal of Southeast Asian Studies*, Volume 28, number 1, 1997

This paper deals with government-sponsored land colonization as a "safety valve" for agricultural problems, such as inequitable land distribution.

East Asia Analytical Unit of the Department of Foreign Affairs and Trade, *Subsistence to Supermarket: Food and Agricultural Transformation in Southeast Asia*, Canberra, ACT: Australian Government Publishing Service, 1994

A useful publication providing details of recent agricultural development in Southeast Asia, especially crop diversification and the effect of liberal reforms

Eicher, Carl K., and John M. Staatz, "Agricultural Development Ideas in Historical Perspective," in Carl K. Eicher and John M. Staatz (editors), *Agricultural Development in the Third World*, second edition, Baltimore, MD: Johns Hopkins University Press, 1990

The authors survey the development of theory concerning agriculture and its effect on policy.

Evans, Grant, *Agrarian Change in Communist Laos*, Singapore: Institute of Southeast Asian Studies, 1988

An outline of the background of agricultural reform during the 1980s (forming part of the basis of the longer study by Evans cited in Appendix 6 of this book)

Fischer, Paige, and Harlan Thompson, "Oil Palms and Sarawak's Forests," in *Earth Island Journal*, Volume 14, number 3, 1999

The authors detail the destruction of Sarawak's forests by the extension of the palm oil industry, and argue that the World Bank and the IMF are complicit in this environmental destruction.

Fryer, Donald W., *Emerging Southeast Asia: A Study in Growth and Stagnation*, New York, McGraw-Hill, and London: George Philip, 1970

A review of the early development of the region's economies after independence

Fryer, Donald W., "Agriculture and Fisheries," in Denis Dwyer (editor), *South East Asian Development: Geographical Perspectives*, Harlow: Longman, and New York: Wiley, 1990

A brief survey of agricultural development in Southeast Asia

Geertz, Clifford, *Agricultural Involution: The Process of Ecological Change in Indonesia*, Berkeley, CA: University of California Press, 1963

Geertz argues that the impact of the colonial system on Javanese agriculture was to entrench traditional and inefficient forms of production.

Guyot, James, "Burma in 1997: From Empire to ASEAN," in *Asian Survey*, Volume 38, number 2, February 1998

Guyot concentrates on political changes in Burma and recent liberal reforms to its agricultural market.

Herath, Gamini, and Sisira Jayasuriya, "Adoption of HYV Technology in Asian Countries: The Role of Concessionary Credit Revisited," in *Asian Survey*, Volume 36, number 12, 1996

In this review of the role of credit provided by governments to farmers for the purposes of using green revolution technology, the authors argue, against economic liberals, that this credit was useful in agricultural development.

Herdt, Robert, "A Retrospective View of Technological and Other Changes in Philippine Rice Farming, 1965–1982," in *Economic Development and Cultural Change*, Volume 35, number 2, 1987

Herdt expresses strong support for the green revolution in the Philippines, arguing that it has benefited farmers and farm workers there.

Hill, Hal, *The Indonesian Economy*, second edition, Cambridge, New York, and Melbourne: Cambridge University Press, 2000

This is a comprehensive overview of the Indonesian economy, with a major section on agricultural developments.

Hirschman, Albert O., *The Strategy of Economic Development*, New Haven, CT, and London: Yale University Press, 1958

Hirschman argues that, since industrial development has more "linkages" to other sectors of the economy, manufacturing rather than agriculture should be the focus of development planning in poorer countries.

Hutchison, Jane, "Pressure on Policy in the Philippines," in Garry Rodan, Keith Hewison, and Richard Robison (editors), cited under Beresford above

Hutchison assesses the political difficulties facing agricultural development in the Philippines and the effects of the green revolution there.

Jayasuriya, S.K., and R.T. Shand, "Technical Change and Labor Absorption in Asian Agriculture: Some Emerging Trends," in *World Development*, Volume 14, number 3, 1986

This paper suggests that government credit policies have made capital cheaper for farmers, causing displacement of rural labor.

Kirkpatrick, V. J., and Harris, G. T., "A Note on Trends in Poverty and Inequality in Rural Asia, 1950–1985," in *Journal of Contemporary Asia*, Volume 13, number 3, 1989

An assessment of the effect of the green revolution on landlessness

Lewis, W. Arthur, "Economic Development with Unlimited Supplies of Labor," in *Manchester School of Economic and Social Studies*, Volume 22, number 2, 1954

Lewis discusses the contribution to industrial development of low-paid labor originating in the agricultural sector.

Nicholas, Colin, *Pathway to Dependence: Commodity Relations and the Dissolution of Semai Society*, Clayton: Monash University Centre of Southeast Asian Studies, 1994

Nicholas deals with government attempts to control an indigenous group in Malaysia and the effect of the market economy on them.

Otsuka, Keijiro, Fe Gascon, and Seki Asano, "Green Revolution and Labour Demand in Rice Farming: The Case of Central Luzon, 1966–90," in *The Journal of Development Studies*, Volume 31, number 1, 1994

The authors suggest that the early use of green revolution strains of rice improved employment prospects in the countryside, but also argue that increased mechanization has since displaced rural labor.

Patnaik, Utsa, "Some Economic and Political Consequences of the Green Revolution in India," in Henry Bernstein, Ben Crowe, Maureen Mackintosh, and Charlotte Martin (editors), *The Food Question: Profits Versus People?*, London: Earthscan, and New York: Monthly Review Press, 1990

An overview of the effects of the green revolution in India, with some interesting implications for comparisons with Southeast Asia

Pincus, Jonathan, "Approaches to the Political Economy of Agrarian Change in Java," in *Journal of Contemporary Asia*, Volume 20, number 1, 1990

Pincus argues that capitalism has been the engine of change in rural Java.

Pingali, Prabu L., and Vo-Tong Xuan, "Vietnam: Decollectivization and Rice Productivity Growth," in *Economic Development and Cultural Change*, Volume 40, number 4, July 1992

Pingali and Xuan trace the effect of liberal economic reforms in Vietnam.

Popkin, Samuel L., *The Rational Peasant: The Political Economy of Rural Society in Vietnam*, Berkeley: University of California Press, 1979

This classic work argues against "the romanticism of the village."

Prebisch, Raúl, "Commercial Policy in the Underdeveloped Countries," in *American Economic Review*, Volume 49, 1959

Prebisch claims that the prices of agricultural products will tend to fall relative to those of manufactured goods.

Raymond, Chad, "Political Implications of Stagnant Agricultural Productivity in Cambodia," in *Journal of Contemporary Asia*, Volume 26, number 3, 1996

Raymond details the poor agricultural record and the war devastation that preceded the attempt by the Khmer Rouge to obliterate the urban population, and surveys changes since the overthrow of that regime.

Rigg, Jonathan, "The Green Revolution and Equity: Who Adopts the New Rice Varieties and Why?", in *Geography*, Volume 74, number 323, April 1989

Rigg suggests that, while smaller farmers did not at first take up green revolution techniques, they did so later.

Rigg, Jonathan, "Redefining the Village and Rural Life: Lessons from South East Asia," in *The Geographical Journal*, Volume 160, number 2, July 1994

Rigg surveys the debate on the nature and origins of the village in Southeast Asia.

Schultz, Theodore, *Transforming Traditional Agriculture*, New Haven, CT, and London: Yale University Press, 1964

An influential text arguing that governments in less developed countries should pay more attention to the development of agriculture

Scott, James C., *The Moral Economy of the Peasant: Rebellion and Subsistence in Southeast Asia*, New Haven, CT, and London: Yale University Press, 1976

Scott argues that the peasant village had a strong "moral economy" through which wealth was redistributed from richer to poorer members.

Siamwalla, Ammar, Suthad Setboonsarng, and Direk Patamasiriwat, "Agriculture," in Peter G. Warr (editor), *The Thai Economy in Transition*, Cambridge and New York: Cambridge University Press, 1993

This chapter in an interesting survey volume deals with changes in Thailand's agricultural sector over the past few decades, focusing on the expansion of of the area under cultivation as the main factor in its development and emphasizing the difficulties with land tenure.

Sivalingam, G., *Malaysia's Agricultural Transformation*, Selangor Darul Ehsan, Malaysia: Pelanduk Publications, 1993

An overview of the Malaysian government's agricultural policy and its effects

Taylor, Donald C., "Agricultural Diversification: An Overview and Challenges in ASEAN in the 1990s," in *ASEAN Economic Bulletin*, Volume 10, number 3, 1994

A summary of attempts in Southeast Asia to diversify agricultural production and exports

Than, Mya, "Agriculture in Myanmar: What Has Happened to Asia's Rice Bowl?", in *Southeast Asian Affairs*, 1990

The author addresses the massive decline in Burma's rice exports since the 1960s.

Than, Mya, and Nobuyoshi Nishizawa, "Agricultural Policy: Reforms and Agricultural Development in Myanmar," in Mya Than and Joseph L.H. Tan (editors), *Myanmar Dilemmas and Options: The Challenge of Economic Transition in the 1990s*, Singapore: Institute of Southeast Asian Studies, 1990

An overview of agricultural policy in Burma and its effects on rice production

Wong, John, *ASEAN Economies in Perspective: A Comparative Study of Indonesia, Malaysia, the Philippines, Singapore, and Thailand*, London: Macmillan, and Philadelphia: Institute for the Study of Human Issues, 1979

Many aspects of this survey have been overtaken by events, but it has a useful section on the green revolution in Southeast Asia, discussing take-up rates and interpreting the innovations as "landlord-based."

Dr John Minns is a Lecturer in History and Politics, and Coordinator of the Postgraduate Program in International Relations, at the University of Wollongong, New South Wales.

Chapter Eighteen
Financial Services and the Asian Crisis

Yoichiro Sato

The impressive record of growth and development sustained by the economies of East Asia throughout the 1980s and 1990s came to a sudden halt during the second half of 1997, when the financial crisis struck Thailand and then spread to the rest of Southeast Asia, to South Korea, and beyond. The crisis (or series of crises) did not remain simply a matter of financial difficulties or their economic effects, but had a significant impact on the political sphere, in particular by posing a threat to the stability of the political institutions of Indonesia and Malaysia. The various diagnoses of the crisis, and the remedies prescribed for it, both at the time and since, have revealed the diverse interests of the individuals and institutions that the crisis affected. These include, among others, US and Japanese financial institutions, to the extent that their responses did much to shape the differing policies of the US and Japanese governments respectively. In the end, however, the "contagion" effects of the crisis, which spread far beyond the region, alerted lenders and borrowers alike to the dangerous consequences of failing to cooperate.

From Speculative Bubbles to Financial Crisis

How the Bubbles were Formed

From the 1970s to the 1990s, the era of sustained growth in most of the economies of Southeast Asia was generally characterized by a pattern of development in which the key role was played by export-led growth in manufacturing industries, stimulated by steady infusions of foreign direct investment. However, economic growth also fostered the creation in most of these economies of expanding domestic sectors made up of various service industries. In particular, financial services, and the development of housing and office space, both expanded on a scale, and with a degree of sophistication, not seen in the region before. Such service industries were crucially dependent on the sustained success of the manufacturing industries producing goods for export. However, by the mid-1990s the maturing of the manufacturing sector and the decline in its profitability induced those who held and directed the accumulated domestic capital of the region's main economies to switch much of their interest and investment into these services, particularly in the form of real estate developments and equities.

Supported by the relatively high levels of domestic saving that have long distinguished these economies from most of the nations of the West, cheap money flowed into speculative trades, largely because it was easier to move in and out of equity investments, while real estate development required longer-term commitments and took significantly longer to achieve profitability. At the same time, however, many domestic banks in Southeast Asia took to financing such long-term development projects with short-term borrowing from foreign banks, on the optimistic assumption that rapid growth would continue and they would therefore be able to keep renewing these loans. The result of this piling into speculative development was an increasingly problematic "mismatch" between credits and debts. Investors in these areas of the region's economies also failed to hedge against the foreign exchange risk of their borrowings, once again displaying their optimism in assuming that national governments could indefinitely maintain their arrangements for pegging their currencies to the US dollar, or to a basket of currencies including the US dollar (Jomo pp. 3–4). In short, risky projects were pursued in the hope that, if things went right, investors would reap the rewards, but, if things went wrong, governments would bail them out.

The capital liberalization of the early 1990s was not accompanied by any serious endeavor to make governments' supervision of their financial services sectors more rigorous (Jomo p. 23). Even on the foreign lenders' side, as some Japanese banks have since revealed, the manipulation of accounts to conceal what had become bad credits was a common practice in the wake of the bursting of the bubbles (Dattel pp. 66–67 and 73). With domestic saving remaining high and short-term loans available on ever easier terms from foreign banks, the domestic banks of most countries in Southeast Asia poured money into real estate and the stock market, creating the mismatch of loan maturities that would eventually force many of these domestic banks into defaulting.

The Importance of Currency Pegs

For Thailand, Indonesia, and Malaysia in particular, the pegging of their currencies seemed amply justified because it provided stable trade and investment environments for their efforts to promote growth and development. This was particularly important because all three had been relatively quick to give up on earlier attempts at import-substituting industrialization (see Chapter 3) and to turn instead to pursuing export-led growth, using foreign direct investment to the fullest practicable extent. As foreign direct investment, largely led by Japanese corporations, built up a network of manufacturing parts suppliers across most of Southeast Asia, intrafirm trade came to account for a larger proportion of trade within the region. The locations of these new production facilities were selected primarily on the basis of relative cost advantages, promoting an intraregional division of labor. It was thus the stability of their pegged exchange rates that permitted these and other countries in Southeast Asia to maintain their expanding supply networks (Ayuz p.39; see also Katzenstein and Shiraishi, and Yamamura and Hatch).

Further, the stability of most of the region's exchange rates against the US dollar had also helped in the creation and maintenance of a stable export environment. Until 1995, the value of the Japanese yen against the US dollar had continuously appreciated. This had meant that the value of those currencies that were pegged to the US dollar, even if only partially, continuously depreciated against the yen, improving the export advantages of these countries over Japan and accelerating the process of production transfer by Japanese corporations into Southeast Asia (see Katzenstein and Shiraishi, and Yamamura and Hatch). However, as these countries liberalized their capital markets over the course of the 1990s, in hope of attracting more foreign capital but also in response to pressure from the financial service industries of the developed countries, maintaining the pegging arrangements for their exchange rates had become more difficult. Despite the declining rate of return in manufacturing by the mid-1990s, foreign capital had kept on flowing in, in the form of short-term loans and equity, as well as real estate investments. Thus, the presence of the currency pegs created a "moral hazard": borrowers resident in those countries that maintained pegging arrangements did not see that there was still a need to sustain adequate hedging of their foreign borrowings against exchange rate risks.

Variations from Country to Country

This general pattern should not, however, obscure the significant differences in policy and practice from country to country, which meant that, for example, Thailand, Indonesia, and Malaysia suffered relatively greater damage when the crisis came than Singapore or the Philippines did.

In Thailand, the booming stock market led to large increases in equity for many domestic companies, but these firms then used the equity to borrow more instead of reducing their dependence on debt (Punyaratabandhu pp. 162–63). Meanwhile, owners of real estate superficially inflated the values of their assets in order to borrow yet more money (Lauridsen p. 139). In Indonesia, as Judith Bird has observed, "outrageous prestige projects," such as "a bridge to Malaysia and the world's tallest tower in Jakarta," were proposed, worsening the problem of debts owed to the banks (Bird p. 172). In Malaysia, a series of massive infrastructure projects were in waiting, only to be postponed when the national currency, the ringgit, lost its value following the crash of the Thai baht (Chin p. 185).

By contrast, in Singapore the government introduced legislation in 1996 to curb speculative trades in the housing industry, and it was at least partly because of the successful implementation of the new law that the economic bubble was prevented from expanding as far in Singapore as in its neighbors (Rodan 1998 p. 178). In the Philippines, meanwhile, economic reforms introduced during the presidency of Fidel Ramos (1992–98), with the approval and support of the IMF, had helped to improve the position of the country's financial institutions as compared to those of most of the other economies in East Asia. As a result, the exposure of banks in the Philippines to the risks inherent in real estate development was effectively limited (Romero pp. 199–200).

Insolvencies in the Financial Sector

Nevertheless, things eventually did go wrong. In Thailand, stock market prices peaked in December 1996 and steadily declined throughout the first half of 1997 (Punyaratabandhu p. 161). The Thai central bank had previously raised interest rates, in an effort to cool the market down, but this turned out to be counterproductive as the higher rates attracted more short-term foreign funds into the economy (Lauridsen p. 138). As Thailand had liberalized its capital markets, the tools it retained for applying macroeconomic policy to highly mobile short-term capital were no longer effective. Thus, the stage was set for further misallocation of abundant capital and the eventual bursting of the bubble economy (Lauridsen pp. 138–39). Indonesia too had liberalized capital outflow from its economy by 1970, but the country enjoyed the benefits of a bias toward capital inflows, mainly because of its significant revenues from the exploitation of its oil and natural gas reserves. In 1982, when world oil prices fell, the government had removed controls on capital inflows, and relaxed its regulation of the domestic banks and stock markets in order to encourage domestic saving and foreign investment. This relaxation of regulations, combined with the legacy of corrupt lending practices, contributed to the creation of asset bubbles and the financial insolvencies that followed (Montes and Abdusalamov pp. 163–71).

Such market-distorting inflated the asset bubbles in both countries and inevitably required larger corrections than would have been needed in more transparent market conditions (Jomo p. 14). Several domestic financial institutions went bankrupt, and foreign investors, anticipating the collapse of the market, rapidly withdrew their short-term capital. The outflow of money then put downward pressure on the pegged domestic currencies, the baht and the rupiah, which in turn became an open invitation to hedge fund speculators to move toward the final stage of the crisis. The Thai central bank initially attempted a defense of the currency peg, but, having rapidly exhausted its foreign currency reserves, it was compelled to abandon the peg and shift to a floating rate system. The crisis that developed from the collapse of the baht soon spread to trading in the Indonesian rupiah. Indonesia had gradually depreciated the value of the rupiah in an controlled manner, but when the central bank came to face the downward pressure in July 1997, it first decided to broaden the trading band and then, in mid-August, followed the example of its Thai counterpart and allowed the currency to float (Bird p.173).

It was too late. Just as in Japan in the late 1980s and the early 1990s, the speculative bubbles in both the stock markets and the real estate markets had burst, causing a chain reaction of market crash, financial insolvencies, desertion of short-term foreign capital, and downward pressure on currencies. The activities of the speculative currency traders, who took devalued positions on the region's currencies in the future markets, only worsened the currency crisis.

Devaluations and Consequences

The immediate consequence of the devaluation of these two currencies was an increase in the debt (denominated mostly in US dollars) of Thailand and Indonesia. In theory, these and other countries hit by the crisis should have been able to restore their programs of debt repayment in the long term, on the assumption that the balance of trade in each case could be improved by a revival of exports and an increase in the prices of imports, which would tend to reduce demand for them. In theory, too, devaluation should also have made domestic assets cheaper for foreign investors, restoring the inflows of investment while discouraging investment outflows, provided that the exchange rate could be stabilized once again at its newly devalued level.

However, reality is more complex than theory. Investors' confidence in any given economy is not based exclusively on the macroeconomic fundamentals, as theory suggests, but is also affected by psychological factors, as well as by perceptions of the political environment in the country in question. In the cases of Thailand and South Korea, the national currencies were soon reestablished at parities with the US dollar, at devalued levels, resulting in a relatively rapid return of foreign investment to both economies. In the case of Indonesia, by contrast, no such parity could be established between the rupiah and the US dollar after the initial crisis of 1997, and the currency went through a second crisis, with its attendant instability, from the summer to the fall of 1998. This second crisis reflected investors' anxieties about the delay in the implementation of economic reforms, the evident failure of the government and the IMF to agree on the appropriate response to the situation, and the sharp rise in political instability within the country. The country also suffered from massive capital flight, which was even more rapid and dramatic than it would have been if Indonesia's capital markets had not been so extensively liberalized.

Like Thailand and Indonesia before it, Malaysia too initially attempted to defend its currency peg through interventions by the central bank in the markets, but it too was compelled to abandon the attempt after losing at least US$8 billion (Jomo p.184). The harsh criticisms leveled by Prime Minister Mahathir Mohamad and other Malaysian politicians against what they saw as damaging activity by international currency speculators served to worsen the situation. This in turn led to calls for the Prime Minister to resign, even from business leaders closely associated with his own party, the United Malay National Organization (Chin p. 186). In response to signs of a renewed weakening of the ringgit throughout the first half of 1998, Malaysia adopted a set of moderate capital controls, exercising its sovereign right to take measures to maintain exchange rates in a way that the IMF had to accept but would not necessarily have recommended.

Another factor in the crisis that strict economic theory tends to underestimate is that political leaders always have to deal with the short-term consequences of economic problems. Both Thailand and Indonesia experienced changes of government after the crisis, but the process took very different forms in each case. In Thailand, parliamentary democracy, restored as recently as 1992, had since been consolidated to the extent that it was capable of sustaining a smooth transition from one civilian leadership to another. In Indonesia, however, the transition from authoritarian government to a more democratic system has been disrupted by interethnic violence, rebellions on the periphery, and the unruliness of the armed forces. In Malaysia, Mahathir Mohamad has remained in power, and has since won yet another general election, but the prolonged crisis has provided additional momentum for his opponents to challenge his leadership. In all three countries, as elsewhere in the region, there have thus been distinctive political responses to the crisis, confirming the salience of noneconomic factors that may sometimes be overlooked by economic theory but always have some influence in the decisions of investors.

Damage to the Real Economy

Following the collapse of their financial sectors, the capacity of the countries hit by the crisis to repay their debts came to depend upon the performance of the real (nonfinancial) economy. The manufacturing industries that produce the region's exports have therefore continued to play the key role in economic recovery. Although, as discussed above, it is likely that the devaluations will have a positive impact on the balance of trade of each country in the long term, their short-term impact on the real economy was disruptive.

For commodity exporters, devaluation should, again in theory, be immediately translated into improved export performance. However, the leading economies of Southeast Asia have relied on regional division of production, with the result that a large proportion of trade within the region is in manufacturing components. The disruption in supplies of such components that resulted from the instability of exchange rates hindered recovery and reduced the ability of these countries to repay their debts. Consequently, volumes of trade, notably of Thailand and Indonesia, suffered for an extended period of time.

Contagion and External Responses

When the crisis started in Thailand, policy-makers in the United States and, to a lesser extent, their counterparts in western Europe, were reluctant to commit either their own resources or those of the IMF to rescuing Thailand. It was widely considered that the crisis was primarily a problem for the Japanese banks: it had started in East Asia and should be solved in East Asia. It was possible for the US government and US financial institutions to take this attitude because, despite the presence of US banks in Southeast Asia, their exposures to the risks associated with local borrowers were limited. They tended to focus on earning commissions on foreign exchange transactions while undertaking considerably less lending to local borrowers than either Japanese or European banks did (Jomo p. 20). Even when the crisis spread to Indonesia, the United States focused its attention largely on corruption, human rights violations, and the problem of East Timor, and limited its concern for the economy to prescribing standard *laissez-faire* remedies.

Meanwhile, Japanese banks faced a serious threat from possible defaults on their loans by domestic banks in the countries affected by the crisis. Japanese banks had already spent years coping with another crisis, ever since the stock market and real estate bubbles had burst in Japan in the late 1980s and the early 1990s respectively. The government had been pumping public money into schemes intended to rescue those institutions that were at risk, and to promote mergers between the stronger and weaker institutions in order to save the latter from insolvency. In order to meet the capital adequacy requirements laid down by the Bank for International Settlements, these banks tightened the conditions for their loans across the board, failing to discriminate between safe and risky loans, and choking off credit not only to borrowers overseas but also to domestic borrowers (Nakao pp. 74–77).

In these circumstances, the Japanese government faced a dilemma in considering how to respond to the financial crisis in Southeast Asia. On the one hand, more cases of insolvency of Japanese banks were anticipated, in the absence of public finance for the countries in crisis. On the other hand, domestic public opinion was becoming ever more critical of the government's attempts to bail out domestic financial institutions, especially after the exposure of several cases of collusive relations between those institutions and the bureaucrats at the Ministry of Finance. The Japanese government hoped that the IMF would provide large rescue packages, on the expectation that the other six members of the Group of Seven would agree to increase their contributions to the IMF in order to enlarge its rescue capacity. The United States was reluctant to support this proposal, for several reasons. First, as we have seen, its financial services sector was less exposed to the immediate crisis in Asia. Second, the prevailing economic ideology among US policy-makers and leaders of the financial sector was less interventionist than tended to be the case among their Japanese and European counterparts (see CFR). Third, even if US President Bill Clinton had been more sympathetic to the idea of allocating larger funds to the IMF, the Republican majorities in both houses of the US Congress would have blocked an increase in the US contribution anyway.

In the face of US reluctance to intervene, the Japanese government attempted to take the lead in extending the public financing of Thailand in particular. Japanese manufacturers had undertaken a great deal of foreign direct investment there, establishing a network of "transplant" and local suppliers whose well-being depended on continuous access to credit. For these suppliers, the solvency of the Thai banks, and of the Japanese banks linked with them, was crucial. The Japanese government therefore proposed a separate Asian Monetary Fund (Jomo p. 20). Although Japan temporarily backed down from the idea when the US government expressed strong opposition to it, Japan revived it after the IMF rescue packages had been agreed, and it remains a subject of discussion and debate even now. In addition, Miyazawa Ki'ichi, a veteran Japanese politician who was Minister of Finance at the time of the crisis, launched a proposal for "Miyazawa bonds," yen-denominated bonds to be issued by other Asian governments but guaranteed by the Japanese government (see MFA). Both proposals were indicative of Japan's frustration at the initial lack of commitment on the part of the United States, its own ambitions to establish a more independent financial policy in relation to the rest of Asia, and its desire to maintain stable foreign exchange regimes and balance of payments environments in the region.

The crisis eventually spread to Indonesia, South Korea, Russia, and Brazil, where US and European banks had significantly greater exposure to bad debts (Sakakibara pp. 188–191). Indeed, despite the commonly held perception that western Europe has been shielded from the crisis in Asia, European banks were exposed directly to the region and, indirectly, to possible contagions in Latin America, as well as in central and eastern Europe (Ayuz p.41). The attitude of the US authorities also changed, and the Group of Seven began to converge toward a common response. With Russia's economic reforms and democratization apparently facing serious risks, policy-makers both in the United States and in western Europe actively solicited cooperation from their respective domestic financial institutions, with a view to extending loans to the countries hit by the crisis.

As the successive rescue packages for Thailand, Indonesia, and South Korea were formulated, it became increasingly clear that the IMF's capacity to respond to multiple crises and contagion was distinctly limited. This led to calls from developing countries, and from the banking industries in the developed countries, for an upscaling of the IMF's lending capacity. Despite the warnings of neoclassical economists against the associated moral hazard – that banks will keep lending recklessly to risky countries and projects, knowing that the IMF will bail them out – the IMF eventually agreed to increase its capital base in the aftermath of the crisis.

The Role of the IMF

In the end, then, both Thailand and Indonesia switched to floating exchange rates, and accepted loans from the IMF in order to overcome their balance of payments problems. Acceptance of the loans in turn compelled the two countries' governments to follow the IMF's recommendation for further liberalization of their capital markets in the hope that this would help them to win back the confidence of foreign investors. Meanwhile, Malaysia erected controls against outflows of short-term capital, minimizing the downward pressure against its currency and deterring the speculators.

Somewhat mistakenly, the IMF also prescribed its standard remedy of fiscal restructuring to Thailand and Indonesia in the initial stages of their balance of payments crises. By the late 1990s, the IMF's own institutional culture was still deeply influenced by the experience of imposing fiscal tightening on indebted populist governments in Latin America throughout the 1980s, in an attempt to combat galloping inflation. This predisposition, combined with poor initial assessments of the crisis, appears to have led the IMF to prescribe the same deflationary fiscal tightening to the countries in Southeast Asia where government deficits had not presented serious problems before the crisis began.

The measures of fiscal restructuring also appeared to be justified by the stereotypical view that the region as a whole was characterized by "crony capitalism," which misallocated resources into inefficient industries owned by political clients or allies of the ruling regimes. Although such views certainly had some basis in the facts, as was clear in such cases as that of President Suharto and his cronies in Indonesia, or some of the Bumiputra firms in Malaysia (see Chapter 5), it was not at all clear that the measures prescribed by the IMF would necessarily be more effective than the main alternative approach advocated by many other economists and some politicians. They recommended jumpstarting these economies first, through a more expansionary fiscal policy related more accurately to each country's unique circumstances, including the size of its debt, the degree of corruption actually prevailing, and the overall efficiency of the economy.

Further, the rapid tightening of government expenditure contributed, as many observers had warned that it would, to an increase in political instability in Thailand, Malaysia, and Indonesia alike. In Indonesia and Malaysia, the sudden removal of various subsidies and monopoly protections for the industries owned by the political cronies of their respective leaderships threatened the foundations of these longstanding regimes. In the case of Indonesia, the weakness of the rupiah pushed import bills and consumer prices to much higher levels, demoting the lower middle class into absolute poverty. The IMF's conditions for extending its rescue loan shook the foundations of the Suharto regime, by demanding increased transparency on the part of domestic financial institutions (see below), as well as rationalization of the manufacturing and mining industries, and threatening the supply of various subsidies to domestic industries. There followed an intensification of the conflicts over the control of natural resources between the central government and the provinces, which in turn fanned the flames of interethnic tensions (Murphy p.47).

The reform packages prescribed by the IMF were also intended to address the problem of the easy and excessive "bad" loans that had led to the formation of the speculative bubbles in the first place. While the growth in these loans was partly (and again, somewhat glibly) blamed on "crony capitalism," equal or greater blame was attached to the lack of transparency in the financial institutions of the region. The disclosure of just how far these institutions had become exposed to risky investments therefore became a principal objective of the reform.

The expectation among neoclassical economists, both inside the IMF and beyond, is that strengthened supervision of financial institutions will prevent any future speculative bubble from inflating too quickly or to excessive sizes, thereby minimizing the damage to be expected when the bubble eventually bursts. In short, the market will correct itself, and in a more orderly and less drastic manner. However, critics argue that bubbles can and will appear, and expand, even under conditions of transparency, on the grounds that financial markets in general are "increasingly divorced from the real economy" (Jomo p. 10) and that the US stock market in particular is characterized by a series of just such bubbles (Nakao p.49).

Such critics of the "Washington consensus" (the orthodox economic views prevalent at the IMF, the World Bank, and the US Treasury, all based in Washington, DC) do not, of course, oppose reforms that succeed in reinforcing the supervision of financial institutions. On the other hand, they still reject what is often the next stage in the argument of neoclassical economists, who tend to conclude that the blame for the financial crisis should be placed wholly or mainly upon the inadequacy of the domestic financial institutions of Southeast Asia. On this view, a "perfect" market would have prevented the bubbles from expanding to too large a size in the first place, thereby correcting itself. This view ignores the fact that an economic bubble can be formed even in the most liberalized financial markets, such as those of the United States. Nevertheless, to the extent that market imperfections worsened the crisis, the neoclassical argument does have some validity.

Extensive collusion between at least some of the domestic banks and the government officials who were charged with supervising their activities was also a source of ineffective bank supervision or even government-guided capital flows, which were often politically motivated and economically irrational, most notably in Indonesia and Malaysia. Increased competition under transparent government supervision was therefore high on the lists of policy priorities agreed between the IMF and the national governments concerned. Increased competition was also expected to bring about increased efficiency in banking, contributing to a revival of confidence not only among domestic depositors but also among foreign investors.

Conclusion

The Asian economic crisis provided definite answers to some questions about the management of national and global economies, but left others as subject as they have ever been to debate, controversy, and even conflict.

First, it has become sufficiently clear that the pegging of national currencies and the liberalization of capital markets do not go well together. Thailand and Indonesia (as well as South Korea) liberalized their capital markets and were then forced to give up their pegging arrangements. Malaysia, by contrast, defended its peg by resorting to capital controls. As the process of recovery depends on renewed infusions of foreign capital into these economies, Malaysia's option seems to be at odds with the majority trend. China and Hong Kong, for example, have also maintained their pegs, while China's program of capital liberalization will be more cautious as it absorbs the lessons of the Asian crisis.

Second, a crisis in one region can spread to others, even in the absence of direct or deep economic ties between them. The contagion from East Asia to Russia and even Brazil suggests that global financial markets are not as "rational" as many, perhaps most, economists have supposed. This has implications for any attempts in the future to reform the workings of the IMF. The United States continues to be noninterventionist, preferring to leave global finance to the care of private institutions.

Meanwhile, there are signs that regional arrangements for emergency financing facilities will become more common and more effective. However, the debates between the advocates of continuing *laissez-faire* and those who favor varying degrees of interventionism, like the debates between supporters of globalization and those who might be called "regionalists," have tended to focus on the scope and content of the remedies applied after the crisis happened. The more fundamental question of whether blind liberalization of the capital markets should be pursued remains unaddressed.

Third, the "flying geese" pattern of manufacturing development that has been pursued by countries in Southeast Asia seeking to expand their real economies (see Chapter 3) has been shown to be highly vulnerable to the shocks caused in the financial sector. This lesson has implications, in particular, for Japan's policy toward the region.

Fourth, an economic crisis may be transformed, with ease and rapidity, into a political crisis that makes purely economic remedies ineffective. Events in Indonesia have amply demonstrated how this kind of downward spiral can develop, as political instability, lack of confidence among investors, and economic stagnation interact to make a bad situation worse. Although countries facing financial crisis may need drastic austerity measures, the blind application of this formula on the basis of crude generalizations, as previously prescribed by the IMF, can have devastating consequences. The lesson here is that debtor governments and, perhaps even more importantly, the people they claim to represent need to be listened to more closely.

Fifth, the crisis has left unanswered the important question of whether injections of domestic and foreign capital instigated primarily by governments have been effective in promoting development in East Asia. One possible answer is that the artificial injection of capital into the domestic economies of the countries in question, accompanied by capital controls and pegged exchange rates, created an illusion of growth that was not based on any real improvements in productivity. An alternative answer, however, is that, while these economies were truly competitive, they still suffered from the impact of economic bubbles that were exported into them through short-term investments by foreign investors based in developed countries. The damage done to the more open economies of the region, such as Singapore and the Philippines, seems to support the latter view, whereas the prolonged crisis in Malaysia and Indonesia, previously closed economies that still display some resistance to liberalization, seems to attest to the former.

Finally, the devaluations of the national currencies of Thailand, Indonesia, Malaysia, and the Philippines have increased the burden of debt repayment facing these countries, while their neighbor and partner Singapore (the remaining member of the group of five countries that constituted ASEAN from 1967 to 1984) is still a creditor country. Whether the benefits of the reforms will outweigh the effects of this increased debt burden is yet another question that awaits a definitive answer.

Further Reading

Ayuz, Yilmaz, "The East Asian Financial Crisis: Back to the Future," in Jomo, K.S. (editor), cited below

Bird, Judith, "Indonesia in 1997: The Tinderbox Year," in *Asian Survey*, Volume 38, number 2, February 1998

CFR: Council on Foreign Relations Task Force on the Future of the International Financial Architecture, "Report: The Future of the International Financial Architecture," in *Foreign Affairs*, Volume 78, number 6, November/December 1999

Chin, James, "Malaysia in 1997: Mahathir's *Annus Horribilis*," in *Asian Survey*, Volume 38, number 2, February 1998

Dattel, Eugene R., "Reflections of a Market Participant: Japanese and Asian Financial Institutions," in Karl D. Jackson (editor), cited below

Jackson, Karl D. (editor), *Asian Contagion: The Causes and Consequences of a Financial Crisis*, Boulder, CO: Westview Press, 1999

This collection of essays convincingly demonstrates the linkages and similarities between Japan's financial crisis, beginning in the early 1990s, and the broader Asian financial crisis that began later in the decade. In addition to the paper by Eugene R. Dattel cited above, the book includes papers by David Asher and Andrew Smithers on Japan, by Richard F. Doner and Ansil Ramsay on Thailand, by Nicholas R. Lardy on China, by Ross H. McLeod on Indonesia, and by Manuel F. Montes on the Philippines.

Jomo, K.S. (editor), *Tigers in Trouble: Financial Governance, Liberalisation and Crises in East Asia*, London: Zed Books, and Hong Kong: Hong Kong University Press, 1998

This edited volume provides critical assessments of the mainstream, neoclassical interpretation of the Asian economic crisis (or crises). Its country studies, including three papers cited elsewhere in this list, show the diverse causes and consequences of the crisis, and the book also contains the editor's own introductory survey of financial governance, liberalization, and crises in East Asia.

Katzenstein, Peter J., and Shiraishi Takashi (editors), *Network Power: Japan and Asia*, Ithaca, NY: Cornell University Press, 1997

The chapters in this book illustrate the increasing economic interdependence between Japan and other countries in Asia, and explore the "Asia shift" in Japan's foreign policy.

Lauridsen, Laurids S., "Thailand: Causes, Conduct, Consequences," in K.S. Jomo (editor), cited above

MFA: Ministry of Foreign Affairs of Japan, "Asian Economic Crisis and Japan's Contribution," at www.mofa.go.jp/policy/economy/asia/crisis0010.html, October 2000

Miyashita, Akitoshi, and Yoichiro Sato (editors), *Japanese Foreign Policy in Asia and the Pacific: Domestic Interests, American Pressure, and Regional Integration*, New York: St Martin's Press, forthcoming

This edited volume on Japan's policies toward its neighbors includes chapters on trade, monetary relations, and financial policies.

Montes, Mauel F., and Muhammad Ali Abdusalamov, "Indonesia: Reaping the Market," in K.S. Jomo (editor), cited above

Murphy, Dan, "Jakarta's Cash Crunch," in *Far Eastern Economic Review*, December 9, 1999

Nakao Shigeo, *Doru Shihai wa Tsuzukuka?* [Will the Dominance of the Dollar Continue?], Tokyo: Chikuma Shobō, 1998

This book provides what amounts to a conspiracy theory of the financial crisis in Asia.

Punyaratabandhu, Suchitra, "Thailand in 1997: Financial Crisis and Constitutional Reform," in *Asian Survey*, Volume 38, number 2, February 1998

Rodan, Garry, "Singapore in 1997: Living with the Neighbors," in *Asian Survey*, Volume 38, number 2, February 1998

Rodan, Garry, Kevin Hewison, and Richard Robison (editors), *The Political Economy of South-East Asia: An Introduction*, Melbourne, Oxford, and New York: Oxford University Press, first edition 1997, revised edition 2000

This collection introduces readers to the strengths and weaknesses of the contemporary economies of Southeast Asia. It includes, among others of special relevance to the financial crisis, papers by Kevin Hewison on Thailand, by Rajah Rasiah on Malaysia, and by Richard Robison on Indonesia.

Romero, Segundo E., "The Philippines in 1997: Weathering Political and Economic Turmoil," in *Asian Survey*, Volume 38, number 2, February 1998

Sakakibara Eisuke, *Kokusai Kinyū no Gemba* [The Frontline of International Finance], Tokyo: PHP Kenkyusho, 1997

Sakakibara, formerly a high-ranking civil servant in Japan, presents a practitioner's counterthesis to the abstractions of the neoclassical economists. Some useful discussions of the Asian financial crisis are offered.

Yamamura, Kozo, and Walter Hatch, *Asia in Japan's Embrace : Building a Regional Production Alliance*, Cambridge and New York: Cambridge University Press, 1996

This book explores and illustrates the increasing role of Japanese investment in Asian countries and the integration of the region's economies through the development of complex procurement networks.

Dr Yoichiro Sato is a Lecturer in the Department of Political Studies at the University of Auckland.

Relations with the Wider World

RELATIONS WITH THE WIDER WORLD

Chapter Nineteen
Relations with Northeast Asia

Stephen Hoadley

Since the early 1970s, the countries of Southeast Asia have substantially broadened and deepened their links with Northeast Asia. As a reciprocal process is also taking place, the two regions are increasingly interdependent.

Indeed, Northeast Asia is fundamental to Southeast Asia and always has been. It is no more possible to understand Southeast Asia without reference to Northeast Asia than to understand the Mediterranean countries without reference to Western Europe, or Latin America wholly separately from North America. In precolonial times, Southeast Asia was a borrower from Sinic cultures, as well as from Hinduism and Islam, and a point of convergence of trade routes from the North and the West. The colonial era brought migrations from China, as well as from India, and, later, military occupation by Japan. In the first few postcolonial decades, when the entities of Southeast Asia had little in common before the establishment of ASEAN in August 1967, it was convenient to define Southeast Asia simply as the geographic region South of Northeast Asia and East of South Asia. Thus, Southeast Asia exists not apart from, but in indissoluble juxtaposition to, Northeast Asia.

Nevertheless, the relationship must be understood as a reciprocal one. Just as traders from Southeast Asia have found opportunities in Northeast Asia, and the governments of the region have had to react to threats from the North, so entrepreneurs and governments in Northeast Asia have looked South, and have become deeply engaged in Southeast Asia. In this regard, Southeast Asia has now resumed much of the importance to Northeast Asia that it enjoyed before the colonial period.

Both regions are characterized by vibrant diversity in cultures, economic bases, and political institutions, making a region-to-region analysis vulnerable to challenge. On the other hand, since the states of Southeast Asia have begun to interact in regional multilateral institutions, a regional approach is becoming tenable. The most visible manifestation of regionalism is ASEAN. Increasingly, the region's leaders, not outsiders, are defining the convergent aspirations and undertaking the common programs of their region, not least with regard to Northeast Asia. ASEAN's offshoots, including the ASEAN Regional Forum (ARF), the Asia-Europe Meetings (ASEM), and the East Asia Cooperation (ASEAN+3), have been designed deliberately to embrace participation by states in Northeast Asia. As these institutions develop, it is becoming possible to speak more confidently of a relationship between Southeast Asia and Northeast Asia that transcends the myriad bilateral links.

In contrast to Southeast Asia, however, Northeast Asia remains one of the least institutionalized regions on the globe. Its constituents span the division established during the Cold War, from the Communist regimes of North Korea and China, through the fledgling democracies of Mongolia, Hong Kong, Taiwan, and South Korea, to the mature and western-leaning democracy of Japan. Until the end of the 1980s, these seven entities had not one regional institution in common. APEC, ARF, ASEM, and ASEAN+3 have now emerged, and subregional arrangements have also sprung up, such as the Four-party and Trilateral talks on the future of Korea. Thus, while Northeast Asia remains diverse and is still best suited to analysis country by country, it is possible to glimpse the beginnings of a regional self-consciousness, and nascent intergovernmental institutions, perhaps analogous to those of Southeast Asia in the early 1960s.

Given the emerging coherence (however tenuous) of these two neighboring regions, the relationship between them may be analyzed in terms of two principal aspects, one strategic, the other economic. In these cases, the principal linkages are between Southeast Asia and China and Japan, respectively: for much of the past 50 years, China has posed a strategic challenge to Southeast Asia, while Japan has offered economic opportunities. These vectors are becoming blurred now that the Cold War has ended, but they are still discernible.

Strategic Perceptions and Interregional Relations

The dominant strategic fact for Southeast Asia is the presence of China at its northern borders. Imperial China colonized Vietnam for more than 1,000 years and threatened its independence for another 1,000 years after that. The People's Republic of China made incursions into Burma and Laos soon after its foundation in 1949. China's attack on Vietnam in February 1979, its dispatch of weapons to the Khmer Rouge in the 1980s, and the clashes on the common border and over islands in the South China Sea are all recent manifestations of China's readiness to play an assertive role in Southeast Asia, using military means when they appear to advance its interests.

Other states in Northeast Asia have also played military roles in Southeast Asia. Japan's invasion and occupation (1941–45) are not forgotten and such minor Japanese initiatives as the transit of minesweepers through the Malacca Straits, or the dispatch of peacekeepers to Cambodia, are monitored with apprehension by older leaders. South Korea fought in South Vietnam under US guidance and Taiwan offered troops (although the offer was declined).

However, over the whole period since World War II no state in Southeast Asia has ever fought in, based its armed forces in, or directly menaced the territory of any state in Northeast Asia, with the partial exception of the clashes between Vietnam and China. There has been no alliance aimed at any state in Northeast Asia since the Southeast Asia Treaty Organization (SEATO) was disbanded in June 1977; even before that, the SEATO Treaty was widely considered inoperative, and only two of the states in the region, Thailand and the Philippines, ever actively participated in it. The Five-Power Defense Arrangements that link Singapore and Malaysia with Britain, Australia, and New Zealand are relatively loose, and entail only occasional joint exercises. The United States has treaties with the Philippines and Thailand, and bilateral logistics, training and exercise arrangements with Singapore, Malaysia, and Indonesia, but there are few US forces in the region and there are no longer any US bases there. Armed conflict or lawlessness in the Malacca Straits or other regional seaways would pose a threat to oil supplies from the Persian Gulf to Northeast Asia, as well as to other forms of shipping, but such events are highly unlikely to be the result of deliberate decisions by any government in Southeast Asia.

One widely held view of the interregional strategic relationship focuses on potential threats by China and defensive responses by the states of Southeast Asia (Morrison pp. 59, 110; see also Buszynski). China's maintenance of the world's largest standing army (with 2.5 million troops), its rising defense budgets, its weapons modernization program, and its history of using armed force against South Korea, the United States, India, the Soviet Union, Vietnam, and Taiwan have all stimulated strategic speculation in Southeast Asia, particularly in Vietnam, the Philippines, and Indonesia. The acquisition of new warships and fighter aircraft by some of the states of Southeast Asia during the early 1990s was one response (see Huxley and Willett, and *Economist*).

The stark pictures that result from such speculation need to be qualified, however. Chinese leaders ceased their overt support of Communist insurgencies in Southeast Asia in the 1960s and abandoned their championing of overseas Chinese communities in the following decade. They have professed the "five principles of peaceful coexistence" and accepted international law as guide to their foreign policy. They have acknowledged the validity of ASEAN's Declaration of the Zone of Peace, Freedom, and Neutrality, its Treaty of Amity and Concord, and its Nuclear Weapons Free Zone. China has established diplomatic relations, as well as trade, investment, and aid links, with all the states in Southeast Asia. Exchange visits and border demarcation with Vietnam have proceeded smoothly, and relations between these two historical adversaries are thawing, although they are not yet warm (Morrison p. 159; see also Amer, and Chang). A case may be made that China's vigorous defense upgrading is proportionate to its size, is necessary to guard its extensive borders, and is directed toward raising its status as a major power.

However, the governments of Southeast Asia are inclined to err on the side of caution. This was especially so after China staked its claim to large portions of the South China Sea, in direct challenge to claims made by Vietnam, Malaysia, Brunei, the Philippines, and Indonesia. China has backed up its claims with action. In January 1974, China wrested the Paracel Islands from South Vietnam. After indecisive deployments during 1982, in March 1988 Chinese warships sank three Vietnamese gunboats to prevent Vietnam from setting up a garrison on the Fiery Cross Reef, and China set up a garrison of its own. In May 1992, China awarded an oil concession to a US firm in an area claimed by Vietnam, and in March 1997 it carried out exploratory drilling nearby; on both occasions Chinese warships deployed to the disputed areas. In 1995–98 China occupied and erected structures on the Mischief Reef, which is well within the area of the Philippine claim, and fended off Philippine approaches with warship maneuvers (see Storey). Strategic analysts began forecasting armed conflict between China and one or more of the states of Southeast Asia, with intervention by Japan and the United States if vital oil and trade routes were interdicted.

A more sanguine view holds that China has behaved with restraint. Since 1991, China has sent representatives to informal workshops on managing potential conflict in the South China Sea, an initiative by Indonesia now conducted by scholars at the University of British Columbia (see Townsend-Gault). In July 1992, China joined its rival claimants in signing the Manila Declaration, a pledge to seek a resolution of the conflicting claims by peaceful means, in accordance with the UN Convention on the Law of the Sea of 1982, which China subsequently ratified. In August 1995, China, the Philippines, and Vietnam agreed to negotiate a code of conduct regarding movements and new construction in the disputed areas, in order to avoid armed conflict. Finally, China acquiesced to the inclusion of the issues around the South China Sea on the agendas of the ARF and the Council for Security and Cooperation in the Asia Pacific (CSCAP), an unofficial forum. Dialogue seems to have prevailed over confrontation, at least for the time being (see Lee, and Foot). The proposed code of conduct has not been agreed, but in November 1999, when Vietnam, the Philippines, and Malaysia became more assertive in making their claims, China predictably rebuked them, but still avoided armed clashes.

Thus, depending on one's point of view, the South China Sea dispute is either a war waiting to happen or one that has been avoided and will continue to be avoided. Ultimately, the likelihood of war depends more on China's assessment of the importance of the area to her security than on preventive diplomacy or measures to resolve conflicts (see She). In any case, the dispute has brought Southeast Asia face to face with the power of China, and China's rise to military predominance in the region seems inevitable. The questions that tax strategists in Southeast Asia are whether this will occur sooner or later (see Karmel), and whether China will be a participant or an adversary. The stability of Southeast Asia will rest on entente between the three major powers active in Northeast Asia: China, Japan, and the United States. Accordingly, the South China Sea will remain one of the world's potential flashpoints, along with the Korean Peninsula and the Formosa Strait.

Investment

Economic growth obviously requires capital (see Tan Chapter 4 for a demonstration of the correlation between growth and investment), but World War II and the Japanese conquest cut off capital flows from the European colonial powers and eroded the infrastructure they had laid down. Further, the anticolonial rhetoric and state-interventionist policies of the governments in the newly independent states proved inhospitable to new capital. Singapore and the Philippines were the exceptions, because they adopted liberal investment regimes and actively sought foreign investment, and US investors in particular responded favorably. Indonesia's oil reserves and, later, its tropical timber were attractive in spite of the government's nationalistic policies, and European, US and Japanese investors entered into deals with state monopolies to exploit these resources.

By the late 1960s, the severity of domestic insurgencies was being moderated and the region's governments could begin to concentrate on economic issues rather than security. Coincidentally, there was a broad shift in emphasis among development economists. Economic nationalism and import-substituting industrialization, previously thought by dependency theorists to be the remedy for dependence on western capital, gave way to the idea that export-promoting industrialization was more likely to lead to economic growth. The rapid rise to prosperity of Japan and the "newly industrialized countries" (NICs) – South Korea, Taiwan, Hong Kong, and Singapore – seemed to provide convincing evidence for the effectiveness of the new model (see Tan Ch. 9). Consequently, governments one by one adopted policies more attractive to foreign direct investment (FDI) and began to court foreign investors. A surge of FDI followed in the ensuing decade. The leading investing nations were no longer Britain and other European powers, but the United States and, not far behind, Japan and the NICs. Singapore provided not only a destination for capital and a center for financial services, but also a conduit for capital from the rest of Asia, North America, and Europe into neighboring countries, particularly Malaysia and Indonesia.

By the end of the 1980s, the total stock of new FDI exceeded the capital destroyed or foregone over the years since 1941. The value of FDI in 1989 was estimated at US$77 billion, of which more than one quarter, and rising, was from Japan (Lindblad p. 26; Lim p. 165). However, the sources, destinations, and uses of FDI were by no means uniform across the region. Indonesia attracted the lion's share (46%), of which three fourths was applied to extraction, mainly of oil and natural gas. Singapore attracted the second largest share (34%), but 57% of this went to service industries, including financial services, and 42% to manufacturing, including oil refining. Malaysia, Thailand, the Philippines, and Vietnam received the rest (in that order), and used it mostly for manufacturing and services.

Cambodia, rent by civil war, Laos, still under Communist rule, and Burma, under an isolationist military dictatorship, participated hardly at all in the new capital inflow from Northeast Asia until the later 1990s. All three eventually adopted capital investment laws, but their economies have proved unattractive to most investors, who have ample opportunities to lodge their funds in less politically risky countries. Only China's state firms have shown much interest in them, more for historical and strategic reasons than with any expectation of profit. Small Chinese enterprises operate in the manufacturing and marketing sectors of these countries, as throughout Southeast Asia, but do not contribute substantially to economic development.

FDI from western Europe has never again reached the level it attained before World War II. By 1989 it accounted for less than one sixth of new capital, except in Singapore and Malaysia, where Britain retained a strong presence. The United States became the leading investor in the Philippines, Singapore, and Indonesia, while Japan moved up to second place in each of these countries and in Vietnam, to third place in Malaysia, and to first place in Thailand (Watanabe p. 138). By the early 1990s, Japan had risen to become equal to the United States in supplying investment in Malaysia (Rao p. 134). In 1994, Japan invested a total of US$5.1 billion in Southeast Asia (see MFA 1995).

The NICs of Northeast Asia, transformed from capital-importing to capital-exporting economies, now emerged as substantial investors in the region. Taiwan and Hong Kong became the first and second largest investors in Vietnam (Nguyen p. 105; and see Ku), while Taiwan became Japan's equal as the largest investor in Malaysia (Rao p. 134). Taiwanese investors, urged by their government to reduce dependence on mainland China, had invested more than US$20 billion in Southeast Asia by 1994, principally in Malaysia and Indonesia (Chan p. 210). The *chaebols*, the conglomerates of South Korea, began investing in subcomponent and other labor-intensive manufacturing, and throughout the 1990s South Korea invested more than US$2 billion a year in Southeast Asia, around one fifth of all its investment around the world (Ro p. 212). Overall, South Korea, China, Taiwan, and Hong Kong provided nearly one fifth of Southeast Asia's new capital in the 1990s.

China, traditionally a capital-importing country, has played the least significant role in investment in Southeast Asia, with the exception of a small number of state trading companies and banks established more for diplomatic than for economic reasons. During the 1990s, however, many small-scale enterprises were set up by overseas Chinese with financial backing from businessmen transferring their profits from the burgeoning commercial centers of China.

The variety of motives for and modes of investment in Southeast Asia should be acknowledged. Japanese firms initially invested in resource exploitation, financial services, and sales outlets, reflecting Japan's combination of a resource deficit with a surplus of capital and manufactured goods. The exploitation of Indonesian and Malaysian tropical hardwood forests by Japanese firms, in collusion with the authorities and landowners in those countries, is the most notorious example (see Dauvergne). However, by the 1980s, following the product cycle and responding to rising labor costs at home and the appreciation of the yen abroad, Japanese firms had begun investing heavily in manufacturing plants, particularly for electronics and motor vehicles. This pattern made use of low labor costs and also avoided the region's formidable trade barriers, because wholly owned subsidiaries qualified as domestic firms (Watanabe pp. 130–35). The pattern evolved further, from manufacture for Southeast Asia to export for Japanese and international markets, as Japanese firms in Southeast Asia began to produce a substantial portion of the region's exports of components and semifinished items as well as of finished products. Firms from South Korea, Taiwan, and Hong Kong followed much the same pattern in the 1990s (see Singh and Siregar).

Some investment flows have begun from Southeast Asia into Northeast Asia, but they remain small-scale by comparison to the flows the other way (Chia pp. 209–11; and see Ramasamy and Viana). They are channeled mainly by multinational firms based in Singapore to joint ventures in Hong Kong, Shanghai, or the special economic zones of southern China. Smaller-scale investments are also made by overseas Chinese, many of them living in Thailand, and encouraged by the government of mainland China. However, to the extent that, apart from China, Northeast Asia tends to have a capital surplus, while Southeast Asia remains a capital-consuming region, the predominant flows will doubtless remain from North to South.

Trade

As with capital investment, Southeast Asia's trade relationship with Northeast Asia links the region principally to Japan and secondarily to the NICs. Only recently has China become a significant element in interregional trade.

As a resource-poor country, Japan has depended on the import of resources ever since the beginning of its modernization and industrialization in 1868. The bulk of these resources have come from elsewhere in Asia, and, increasingly, from Southeast Asia. One of Japan's motives for proclaiming the "Greater East Asia Co-Prosperity Sphere" and conquering Southeast Asia in 1941–42 was to secure sources of raw materials. This historical relationship persisted after World War II, and was manifested in a pattern of provision of commodities such as food, fuel, rubber, and fibers, from Southeast Asia in exchange for

Japanese provision of manufactured goods. Because the latter generally command higher prices, the terms of trade have tended to shift in favor of Japan and the rest of Northeast Asia. In 1968, Japan achieved an overall positive balance of trade with Southeast Asia (Lim p. 156). By the 1980s only Indonesia enjoyed a trade surplus with Japan, and all the other countries in Southeast Asia were in deficit, an imbalance that persisted throughout the 1990s. In 1987, nearly 44% (by value) of Indonesia's exports, mainly of oil and natural gas, went to Japan (Lim p. 149). In the mid-1990s, Brunei and Vietnam similarly achieved trade surpluses by exporting oil and natural gas to Japan.

Overall, by the mid-1990s more than one fifth of all ASEAN exports went to Japan and another tenth or more to the three NICs of Northeast Asia, South Korea, Taiwan, and Hong Kong (Rao p. 123). In 1995, the seven members of ASEAN (then excluding Cambodia, Laos, and Burma) sent 31% of their exports to Northeast Asia (Garnaut p. 221). In a few cases, exports to the three NICs approximate those to Japan. For example, in 1998 Thailand sent 11% of its exports to Taiwan, South Korea, Hong Kong, and China, approaching the 14% that it sent to Japan (see IMF). Nevertheless, all exports from Southeast Asia together constitute less than one sixth of Japan's imports, for Japan's dependence on Southeast Asia for raw materials has given way to a dependence by Southeast Asia on trade with Japan. A similar pattern has appeared in trade with South Korea, which imports labor-intensive electrical components and oil, and exports parts, finished manufactures, and chemicals. South Korea thus earns a surplus from all its partners in Southeast Asia except Indonesia and Brunei (Ro pp. 208–10). Taiwan and Hong Kong trade along similar lines, and the asymmetry is compounded by dependence on capital and aid from Northeast Asia.

At various points during the 1980s, regional leaders, notably Mahathir Mohamad, Prime Minister of Malaysia, Ferdinand Marcos, former dictator of the Philippines, and Tony Tan, Minister of Finance in Singapore, made public complaints about this asymmetry (Lim p. 119), but the trade relationship was already being altered, in quality if not balance. US pressure was forcing Japan to open its markets in selected commodities, increasing opportunities for exporters in Southeast Asia. The rising value of the yen was making the region's products more attractive to Japanese consumers on price grounds. Japanese manufacturing firms were increasingly relocating to Southeast Asia to escape high wage costs, with the result that manufactured exports grew as a proportion of Southeast Asia's total exports. By 1985, Japanese firms obtained more intermediate manufacturers via intrafirm trade with Southeast Asia than from supplier industries in Japan itself (see Shirai). Jobs, technology, and amenities for the host economies ensued, and overseas market links were formed. Consequently, the economic interdependence of Southeast Asia and Japan increased, to the benefit of both (Basu and Miroshnik pp. 57–62). More recently, the complaints have come, not from the leaders of Southeast Asia, but from Japanese critics of "globalization," who fear that the Japanese economy will be "hollowed out," with a loss of jobs and tax revenues to support the unemployed and aged (Watanabe pp. 146–47).

Nevertheless, in 1996 Japan still enjoyed a trade surplus with the region of around US$23 billion, in spite of obtaining 14.9% of its imports from there (*Asahi Shimbun* pp. 114 and 115). A similar pattern obtains with respect to China and South Korea: the value of imports into Southeast Asia from both these economies is roughly double the value of exports, with Indonesia being the only country to achieve a surplus (Singh and Siregar pp. 100 and 101). The economies of Southeast Asia have been obliged to make up the deficit by international borrowing, much of it from Japan and the NICs of Northeast Asia, and by continuing to solicit FDI. Economic aid has also helped to relieve the trade deficit to a minor degree (see below).

Aside from small-scale border trade, China's trade relations with Southeast Asia were governed by political and strategic goals until relatively recently. Favored partners were granted "friendship trade" status or offered other incentives such as subsidies, special prices, or accelerated licenses. At various times,

China has devoted special attention to Burma, Malaysia, and Thailand, but in the 1970s and 1980s Singapore emerged as China's strongest trade partner in Southeast Asia, not least because it is an entrepot for produce from neighboring countries (Chia p. 206).

The Aid Relationship

Southeast Asia, a region of 11 developing countries, has long been a recipient of aid from richer countries. Initially, aid came from the colonial and former colonial countries, and was motivated by the need for amenities such as infrastructure, public health, and education that colonial enterprises and their staffs could enjoy, but also by well-meaning paternalism and humanitarianism. The next wave of aid was strategic, being given by the United States and Britain as part of their efforts to reduce the appeal of Communism and to bolster governments in the region in their fight against Communist insurgents or, as in Indochina, to guard against real or perceived threats from North Vietnam and China. A third wave of aid came in the form of the reparations given by Japan as it became financially stronger, which were intended to atone for the destruction that it caused during World War II and to cement good diplomatic relations with the new governments. Since then, aid has been focused on supporting economic development, providing for the basic needs of the poor, and administering technical assistance for good governance and public services.

After the end of the Vietnam War and the waning of US interest in Southeast Asia, Japan emerged as the principal aid donor. In 1980, Japan provided US$861 million to Southeast Asia, representing 44% of Japan's official development assistance (ODA) to all recipients around the world. By 1995, Japan's ODA to Southeast Asia had climbed to nearly US$2.23 billion: the main recipients were Indonesia (US$892 million), Thailand (US$667 million), the Philippines (US$416 million), and Vietnam (US$170 million) (see MFA 1995 and 2000). In the 1980s, grants to Singapore and Brunei, and, in the 1990s, grants to Malaysia were phased out in favor of technical assistance as these recipients became more prosperous.

Japan also gives aid to Burma, Laos, and, particularly, Cambodia, following the dispatch of Japanese peacekeeping forces in the mid-1990s, as well as to regional projects via multilateral organizations.

Japan's ODA has been criticized, however, especially as the bulk of aid takes the form of loans requiring repayment. In 1995, loans made up 58.7% of Japan's ODA, while grants comprised only 13.4% (MFA 1995). The appreciation of the yen has made loan aid harder to repay (Lim p. 229). Because technical assistance aid is tied to Japanese services and products, and Japanese contractors execute most large projects, critics allege that Japanese aid is a form of export promotion. They believe that aid in the forms given by Japan displaces self-help efforts, entrenches elites, discourages reforms, disadvantages small entrepreneurs, and increases debt dependency (Tan pp. 78–81). Nevertheless, Japanese aid remains an indispensable source of financial and technical resources for the poorer countries of Southeast Asia, and their governments continue to welcome it while negotiating to improve its quality and reduce their indebtedness.

Taiwan and South Korea have also offered concessionary loans and technical assistance to governments in Southeast Asia. South Korea has followed the model of Japan by offering infrastructure projects to be executed by its own firms (Ro p. 214). Taiwan's aid has been more modest, and China, not to be outdone by Taiwan, has joined the ranks of aid donors; both incline toward less expensive, low-key technical assistance projects, often in agriculture. Hong Kong has not become a significant donor except in education and training exchanges.

Cultural Exchanges

Japan in particular has allocated large amounts of money to cultural promotion programs, which are organized by the Ministry of Education, the Japan Foundation, and other government and private agencies. In 1984, Japan's Ministry of Foreign Affairs initiated the "Friendship Program for the 21st Century": under this scheme, and the subsequent Youth Invitation Program, it has invited 800 young people from ASEAN countries to visit Japan every year. In 1990, Takeshita Noboru, then Prime Minister, set up the ASEAN Cultural Center to coordinate and focus exchanges with the region. Three of Japan's six overseas centers for teaching Japanese are located in Southeast Asia. The other states of Northeast Asia also have cultural exchanges with Southeast Asia. China has used them sporadically to promote "friendship diplomacy," Taiwan has used them to reinforce its drive for wider recognition, and South Korea has joined in since it became more affluent and self-confident in the 1980s. The states of Southeast Asia and the ASEAN Secretariat fund counterpart exchanges and programs.

Tourism is both an economic and a cultural relationship. Individuals from Southeast Asia have long visited Northeast Asia, and vice versa, and this private travel has accelerated as affluence has increased in both regions. The Japanese are the principal visitors to Southeast Asia and spend the most money there, followed by tourists from Taiwan, Hong Kong, South Korea, and China. Tourists from Southeast Asia, who tend to be fewer in number because of the lower level of affluence, come mainly from the Philippines, Thailand, Malaysia, and Singapore. It should be noted that the difficulty in distinguishing between a tourist and a businessperson taking time off for sightseeing makes statistics misleading.

Migration is another interregional relationship that has both an economic and a cultural dimension, and sometimes a strategic dimension as well. Migration from South to North is restricted by stringent immigration controls, especially in the case of Japan, but a substantial number of Filipinos find their way into the North to work in construction or as domestic servants, particularly in Taiwan. Temporary migration for small-scale trade or farming takes place from Burma, Laos, and Vietnam into the southern provinces of China, and refugees, mainly from Vietnam, have become a political problem for Hong Kong.

Migration from North to South was encouraged by the British and the other colonial powers, and resulted in the creation of large overseas Chinese communities in every country in Southeast Asia. In the postcolonial period, their presence facilitated trade and investment, but also illegal immigration, organized crime, and, during periods of Chinese advocacy of wars of national liberation, gun-smuggling and subversive activities in support of Communist insurgents. In September 1980, however, China outlawed dual citizenship, virtually surrendering the claim that Chinese governments had long maintained to protect overseas Chinese. Since then, China has actively encouraged overseas Chinese to return to China or to send remittances or make investments (see Suryadinata). China has not come to the aid of Chinese communities suffering severe discrimination or even mob attack in the last two decades, for example in Cambodia, Vietnam, Malaysia, and Indonesia.

The overseas Chinese remain a conduit for economic relations and migration between Southeast Asia and China, including Hong Kong and Taiwan, and are active in local politics and social affairs, but they have not significantly affected interstate diplomatic or strategic relations since the 1970s. Small numbers of people have migrated from Japan and South Korea to Southeast Asia, usually in connection with trade or investment, or accompanying their spouses, but their influence is insignificant compared with that of the overseas Chinese.

Diplomatic Relations and Interregional Institutions

A complex web of bilateral diplomatic relations links Southeast Asia with Northeast Asia, similar to those linking any two regions of the modern world. Following the establishment of relations between Malaysia and China in May 1974, China rapidly became part of the diplomatic web, but, as China entered it, Taiwan was excluded from it. A few other anomalies persisted. For example, Singapore delayed setting up direct links with China until October 1990, out of respect for the anxieties of its neighbors, particularly Indonesia. Cambodia and Laos have been diplomatically peripheral because of internal preoccupations, while North Korea and Burma remained relatively isolated until recently.

Asian states have also become full participants in multilateral intergovernmental organizations, most prominently the UN, the IMF, the World Bank, the Asian Development Bank, and the General Agreement on Tariffs and Trade (GATT), now subsumed by the World Trade Organization (WTO). Delegations from Southeast and Northeast Asia now work together in the Asia Group within the UN General Assembly and other UN organs and affiliates. Adhering to the "one China policy," they recognize only the People's Republic, and have presented a common front against proposals to admit Taiwan to the UN.

Compared to western Europe or the Americas, Asia has a dearth of intergovernmental institutions, reflecting its greater cultural diversity, the unique colonial and postcolonial experiences of each state, the suspicions engendered by the Cold War, the unevenness of economic development, and, in many countries, the fragility of governing institutions. Asian leaders have generally been more concerned, first, to consolidate their own state's fragile institutions; second, to secure their state's place in the region and the world; and, third, to develop their own economies. The cautionary principles of state sovereignty and noninterference prevail over any tendency to take advantage of longer-term opportunities for cooperation, with all its attendant risks.

Such institutions as appeared in the 1950s were motivated by fear of the Soviet Union and the Asian Communist states. SEATO, set up in September 1954, and the Asia Pacific Council (ASPAC), set up in June 1966, both faded away as soon as the Communist threat passed. It was only after the failure of two alternative proposals – for an Association of Southeast Asia (ASA) and for "Maphilindo" (a federation of Malaysia, the Philippines, and Indonesia) – that five of the states in the region felt able to establish ASEAN, in August 1967, and then only by eschewing discussion of political and security issues. Nevertheless, ASEAN's importance increased as it sponsored economic cooperation and, later, coordinated its members' policies in relation to the Cambodian civil war and the Vietnamese invasion. ASEAN also generally attracted foreign approbation as the most promising indigenous regional organization in Asia.

In July 1991, ASEAN stepped up to a new plane when it initiated its "post-ministerial conferences" (PMCs) with major external partners, including Japan, China, and South Korea. Because Northeast Asia has never been able to set up a counterpart institution, the ASEAN PMCs emerged by default as the first institution linking the two regions. In June 1994, ASEAN took another initiative to link the two regions by organizing the ARF. Its Asian members, other than the ASEAN states themselves, include Japan, China, South Korea, Mongolia, and, from 2000, North Korea, but Taiwan and Hong Kong have been excluded.

Three other institutions have emerged in the economic sphere. The first was the East Asian Economic Caucus (EAEC), proposed by Mahathir Mohamad of Malaysia in December 1990. The idea was opposed by the United States, Canada, Australia, and New Zealand, which were excluded, by Japan, because of its special relationship with the United States, and by Singapore, because of its protectionist overtones, and the EAEC was never institutionalized. However, its exclusively Asian membership was reflected in the list of states invited to consult with the EU at the first ASEM in March 1996: this list comprised the seven states that then belonged to ASEAN, China, Japan, and South Korea. While the decisions reached at the first ASEM were not substantial, the event has been taken to symbolize the emergence of an interregional self-awareness that had seldom been manifested previously, and ASEM has become a regular event. Thus, the EAEC survives as a caucus within the larger ASEM, APEC and WTO groupings, assuring that Asian interests will be voiced and taken into account by the western economic powers.

The most recent expression of interregionalism is the East Asia Cooperation or "ASEAN+3" summit, the first taking place in Manila in November 1999 partly in reaction to the intrusive policy prescriptions proffered by the IMF during the financial crisis of 1997–98. The discussions at the summit covered such issues as intra-Asian financial cooperation, currency swaps, and, more tentatively, an Asian Monetary Fund to counterbalance the influence of the IMF.

None of these interregional institutions – whether the ASEAN PMCs, ASEAN+3, the

ARF, or ASEM – is authoritative or comprehensive (see Wanandi). Each is voluntary and consultative in nature, partial in membership, and selective in focus. Delegates express diverse opinions, avoid criticism of their fellows, and reach consensus only on generalities. Policy convergence will therefore continue to be very gradual and uneven. Nevertheless, the idea of an "Asian" approach to security and economic issues is beginning to become entrenched, and that serves to motivate exchanges of views and information. The potential for convergence between Southeast Asia and Northeast Asia is growing apace.

Prospects

The paradox that globalization is accompanied by rising local self-consciousness is showing itself in Asia as it is elsewhere. Much of the economic interaction between Southeast Asia and Northeast Asia is a manifestation of the worldwide liberalization of markets, while global and regional interconnections have been highlighted by the rapid spread of the Asian financial crisis. Yet at the same time, and in part as a reaction to globalization, the two regions are becoming more self-aware and their relations more institutionalized.

In speculation about the future, three general patterns are plausible. The first is economic globalization, presupposing the continued growth of interconnections within, and between, Southeast and Northeast Asia, in investment, trade, aid, tourism, migration, and other areas, and concomitant interconnections with the rest of the world. It presupposes also continued liberalization, international and regional peace, and political stability. Its inadvertent consequence, however, might be a widening development gap between the two regions as Northeast Asia interacts more successfully with the rest of the world. This gap is already visible in the wake of the Asian financial crisis (see Holland). Its worsening could generate resentment and protectionism in Southeast Asia, retarding interregional integration, or generate internal disorder, further slowing growth and widening the gap. Political and strategic rivalry could follow.

Because political aspirations and strategic fears can trump economic rationality, a second set of outcomes is plausible: there could be a growth of strategic tensions between Southeast Asia and Northeast Asia, particularly China. Anxious that the United States may continue drawing down its military forces from Asia as China modernizes its armed forces, the states of Southeast Asia may strengthen their own defense capabilities and search for new allies. One such potential ally, already working militarily with Burma and Vietnam, is India, which has been a rival of China ever since both regained independence in the late 1940s. Other potential allies might be Russia, Taiwan, South Korea, or Japan. The outcome would be the pitting of much of the rest of Asia against China, with consequent potential for destabilizing shifts in the balance of power and the outbreak of armed conflict as states maneuver for advantage.

The third plausible set of outcomes represents a continuation of current trends, depending on multifaceted explorations by states, firms and individuals of relative advantages, within frameworks of political and international stability. The implicit assumptions are that the United States will remain actively engaged in Northeast Asia; that India and Russia will not meddle in Southeast Asia; that China will maintain its engagement with all other Asian states; and that the states of Southeast Asia will persist in pragmatic liberalization and regional cooperation. This prospect is neither dramatic nor elegant, but it is more optimistic than the two alternatives, and thus commends itself to regional leaders until a better one emerges.

Further Reading

Amer, Ramses, "The Territorial Disputes between China and Vietnam and Regional Stability," in *Contemporary Southeast Asia*, Volume 19, number 1, June 1997

Asahi Shimbun: The Asahi Shimbun *Japan Almanac 1998*, Tokyo: The *Asahi Shimbun* Publishing Company, 1997

Basu, Dipak R., and Victoria Miroshnik, *Japanese Foreign Investments, 1970–1998*, Armonk, NY: M.E. Sharpe, 2000

This is an analysis by specialists, mainly aimed at specialists, and backed up with a wealth of statistics.

Buszynski, Leszek, *ASEAN: Security Issues of the 1990s*, Canberra, ACT: Australian National University Security and Defence Studies Centre, 1988

This competent analysis of security matters as they stood in 1988 retains its relevance, as the main trends have continued since then.

Chan, Gerald, "Taiwan's Economic Growth and its Southward Policy," in Ray F. Watters and Terry G. McGee (editors), *Asia-Pacific: New Geographies of the Pacific Rim*, Wellington: Victoria University Press, and London: Hurst, 1997

Chang Pao-Min, "Vietnam and China: New Opportunities and New Challenges," in *Contemporary Southeast Asia*, Volume 19, number 2, September 1997

Chia Siow-yue, "China's Economic Relations with ASEAN Countries," in Joyce K. Kallgren et al. (editors), *ASEAN and China: An Evolving Relationship*, Berkeley: University of California Institute of East Asian Studies, 1988

Dauvergne, Peter, *Shadows in the Forest: Japan and the Politics of Timber in Southeast Asia*, Cambridge, MA: MIT Press, 1997

This is a sobering critical exposé of collusion between Japanese firms and Indonesian officials.

Economist: "Asian Security: East Asia's Wobbles," in *The Economist*, December 23, 1996

Foot, Rosemary, "China in the ASEAN Regional Forum," in *Asian Survey*, Volume 38, number 5, May 1998

Holland, Tom, "Asia's New Fissure," in *Far Eastern Economic Review*, June 29, 2000

Huxley, Tim, and Susan Willett, *Arming East Asia*, London: International Institute of Strategic Studies, 1999

This thoughtful and informative analysis recommends caution but concludes that there is no arms race in East Asia.

IMF *Country Reports* at www.imf.org

Karmel, Solomon M., *China and the People's Liberation Army: Great Power or Struggling Developing State*, London: Macmillan, and New York: St Martin's Press, 2000

Karmel documents the modernization of the Chinese army and concludes that internal problems will preclude a turn to expansionism by China for at least another ten years.

Ku, Samuel C.Y., "The Political Economy of Taiwan's Relations with Vietnam," in *Contemporary Southeast Asia*, Volume 21, number 3, December 1999

Lee Lai To, *China and the South China Sea Dialogues*, Westport, CT, and London: Praeger, 1999

Lim Hua Sing, *Japan's Role in Asia*, second edition, Singapore: Times Academic Press, 1999

This is a comprehensive survey, including analyses of selected forms of linkage, although it occasionally lacks detail and recent statistics.

Lindblad, J. Thomas, *Foreign Investment in Southeast Asia in the Twentieth Century*, London: Macmillan, and New York: St Martin's Press, 1998

Lindblad covers the entire century in considerable detail, providing statistics and pointing out some interesting continuities.

MFA 1995: Ministry of Foreign Affairs Economic Cooperation Bureau, *Japan's ODA: Official Development Assistance Annual Report 1994*, Tokyo: Ministry of Foreign Affairs, March 1995

MFA 2000: Ministry of Foreign Affairs of Japan website at www.infojapan.org, 2000

Morrison, Charles E. (editor) *Asia Pacific Security Outlook 1998*, Tokyo: Japan Center for International Exchange, 1998

This collection combines accounts of recent events in specific countries with a regional overview.

Nguyen Quang Thai, "Priority Areas to Attract Foreign Investment Capital to Vietnam," in Tran Van Hoa (editor), *Economic Development and Prospects in the ASEAN: Foreign Investment and Growth in Vietnam, Thailand, Indonesia, and Malaysia*, London: Macmillan, and New York: St Martin's Press, 1997

Ramasamy, Bala, and Venus T. Viana, *ASEAN's Foreign Direct Investment into the People's Republic of China*, Palmerston North: Massey University School of Applied and International Economics, 1995

A brief and sketchy study with useful statistics

Rao, Narhari, "Intra-Asian Trade," in Ki'ichiro Fukasaku (editor), *Regional Cooperation and Integration in Asia*, Paris: Organization for Economic Cooperation and Development, 1995

Ro Jaebong, "Economic Cooperation between ASEAN and Korea," in *The Indonesian Quarterly*, Volume 28, number 2, 2000

She Poon Kim, "The South China Sea in China's Strategic Thinking," in *Contemporary Southeast Asia*, Volume 19, number 4, March 1998

Singh, Daljit, and Reza Y. Siregar (editors), *ASEAN and Korea: Trends in Economic and Labour Relations*, Singapore: Institute of Southeast Asian Studies, 1997

A collection of specialized studies without much political analysis

Shirai Sayuri, "The Pattern of International Trade Between Japan and the Pacific Basin Countries: A Comparison between 1975 and 1985," in IMF *Country Reports* at www.imf.org

Storey, Ian James, "Creeping Assertiveness: China, the Philippines and the South China Sea Dispute," in *Contemporary Southeast Asia*, Volume 21, number 1, April 1999

Suryadinata, Leo, *China and the ASEAN States: The Ethnic Chinese Dimension*, Singapore: Singapore University Press, 1985

Although somewhat dated, this remains an authoritative study by a respected academic analyst.

Tan, Gerald, *The Economic Transformation of Asia*, Singapore: Times Academic Press, 1997

An ambitious and largely successful attempt to comprehend and describe a vast and rapidly changing subject

Townsend-Gault, Ian, "Preventive Diplomacy and Proactivity in the South China Sea," in *Contemporary Southeast Asia*, Volume 20, number 2, August 1998

Wanandi, Jusuf, "Regionalism in the Asia Pacific Region," in *The Indonesian Quarterly*, Volume 28, number 2, 2000

Watanabe Susumu, "Regional Integration of East and Southeast Asian Economies: The Role of Japan," in Duncan Campbell et al. (editors), *Regionalization and Labour Market Interdependence in East and Southeast Asia*, London: Macmillan, and New York: St Martin's Press, 1997

Dr Stephen Hoadley is Associate Professor of Political Studies at the University of Auckland.

RELATIONS WITH THE WIDER WORLD

Chapter Twenty

Worlds in Collision: Southeast Asia and the West

Mark Beeson

The development of contemporary Southeast Asia has been profoundly influenced by its contact with the complex forces of modernity conveniently subsumed under the rubric of "the West." Indeed, the very idea of "Asia" in general, let alone Southeast Asia in particular, is very much the product of a comparatively recent interaction between Europe and Asia, in which people in different parts of the world have come to think of themselves as distinct and different from some alien "other" (see Korhonen, and Saïd). As capitalist forms of economic organization and associated patterns of social relationships spread inexorably from their European heartland into the rest of the world, Asia found itself drawn into an emerging world order dominated by external powers and imbued with very different views about the ways in which commerce, society, and politics could, and perhaps should, be organized. This collision of disparate worlds has been one of the central dynamics that has shaped the development of Southeast Asia, and is a force that continues to animate its interactions with "the West". Nevertheless, while the countries in the region, established as sovereign states comparatively recently, remain dependent on and vulnerable to external forces, they retain some capacity for shaping their own destinies and exerting a modest influence on the world around them.

Historical Background

The contemporary nation-states of Southeast Asia have been shaped by fairly recent historical events. The very fact that we see discrete sovereign entities exercising claims to political authority over clearly demarcated geographical areas is itself a manifestation and product of developments that have occurred within the last century or so, and owe much to the region's historical interaction with Europe in particular. Although powerful political and social systems existed in certain areas, such as Java, Thailand, Cambodia, and Vietnam, before Europeans began to penetrate the region, the coming of the array of commercial practices and social relationships associated with western capitalism inaugurated a period of profound change for the region that we now think of as Southeast Asia. As economic relations between Asia and Europe were consolidated and intensified, no part of the region, with the exception of Thailand, escaped the experience of colonization by and subordination to a western power (see Chapters 1 and 2 of this book). The Dutch, the British, the French, the Spanish, the Portuguese, and even the Americans, in the Philippines, established direct economic and political ties with Southeast Asia that have given them continuing influence and concern with the region's affairs.

A couple of aspects of this historical interaction merit particular emphasis. First, as imperial economic relationships became formalized in colonial control, the shape of these relationships reflected the interests of the colonizers, not the colonized. Western domination established patterns of economic dependence that not only shaped the structure of nascent domestic economies throughout the region, but also had a profound impact on the subsequent evolution of political structures and social relations. For example, Malaysia's current ethnic mixture of Indians, Chinese, and Malays owes its distinctive form to the imperatives of Britain's imperial economic needs, and the labor-intensive nature of resource extraction in the colonies. Likewise, the sprawling, heterogeneous composition of modern Indonesia owes its shape to the exigencies of Dutch colonial rule, rather than any indigenous political order preceding it. The second point to make is that the nations of Southeast Asia are also a product of the withdrawal of the colonial powers in the wake of World War II, on the one hand, and of the western system of international states, on the other (see Jackson). The very existence of such sovereign states as Indonesia and Malaysia owes much to the influence of the West in general and the colonial experience in particular (see Tarling).

A final point to be emphasized at the outset is that, since the colonial period, Southeast Asia has been ever more deeply integrated into an increasingly global political economy that remains dominated by the former colonial powers of Europe and, latterly, the United States and Japan. The fact that one of these countries is Asian does not alter the underlying basic reality: on a world scale, the economies of Southeast Asia are small, their political systems are relatively fragile, and their position in the international system is consequently vulnerable, constrained, and marked by continuing uncertainty. The recent crisis in East Asia, which hit Southeast Asia especially badly, was dramatic confirmation of just how exposed the region remains to external forces over which it has little control. This amalgam of historically determined vulnerabilities, contradictions, and uncertainties provides the backdrop against which the region's contemporary relations are played out.

The Impact of US Hegemony

The establishment of the nations of Southeast Asia as sovereign, independent states in the wake of decolonization coincided with the rise of the United States to hegemonic status in the international system. US ascendancy had an immediate impact on the region. Indeed, one of the reasons why the former colonists allowed the peoples of Southeast Asia to achieve independence was precisely that they were encouraged to do so by the United States. The goal of establishing independent capitalist states in the region not only resonated with the dominant values of US politics and culture, but was seen as a crucial part of the intensifying ideological and strategic struggle with the Soviet Union. It is important to remember that US policy toward Asia in the aftermath of World War II was shaped by the desire to contain the perceived danger of Communist expansion, which was then seen as a highly credible threat and alternative to capitalism. Japan was one of the first beneficiaries of the US policies that ultimately played a large part in underpinning a more generalized "East Asian miracle" (see Cronin).

To understand contemporary relations between the United States and Southeast Asia, it is necessary to say something more about the sort of postwar international order that the United States tried to establish, and the impact that this had on Southeast Asia in particular. It is the transformation of this order and its underpinning strategic rationale that, more than anything else, has redefined US policy toward the region and presented difficult challenges for Southeast Asia's ruling elites.

To describe the United States as "hegemonic" is to say something about the quality and extent of US power, not to make a judgement about the purposes to which such power is put. Throughout modern history the most powerful states of any era have played a role shaping the international system of which they are members. The rules, institutions, and organizations that govern international economic and political activity are the products of complex processes of international interaction, in which the most powerful nations attempt to create an international order that furthers their interests and reflects their values (see Agnew and Corbridge, as well as Gilpin). In the postwar era, the United States has found itself in an increasingly powerful position, able to exert a "hegemonic" influence across military, economic, political, and even ideational or cultural spheres. In the wake of the demise of the Soviet Union, its only credible strategic rival, the United States finds itself in an unparalleled position of unipolar ascendancy (see Matsanundo). The United States, therefore, has the ability and, increasingly the desire, to influence the course of political and economic development in Southeast Asia.

During the Cold War, US policy was overwhelmingly shaped by security considerations. Within the overarching, transnational strategic framework that emerged as a consequence of confrontation with the Soviet Union, the liberal order that the United States helped to create accommodated a range of illiberal political entities (see Latham). The principal attraction of such regimes, of course, was that they were anti-Communist. This strategy of turning a blind eye to, or actively encouraging, the consolidation of authoritarian governments throughout the region was justified by the imperatives of superpower contestation and the perceived need to shore up fragile political structures (see Huntington). It is important to recognize that the way in which the United States approached its strategic relations with East Asia generally gave a particular cast, not only to its relationships with individual countries, but also to the sorts of relationships Asian countries had with each other. One of the most distinctive qualities of US foreign policy during the Cold War was the establishment of a series of "hub and spoke" relationships throughout the region: *bi*lateral rather than multilateral strategic relationships were the preferred vehicle for its hegemonic ambitions. As a consequence, the nations of East Asia were locked into a vertically ordered security regime in which they were often forced to conduct their relationships with each other via the United States (Cumings pp. 154–55).

Seen in this context, it becomes easier to understand the logic that underpinned the emergence of ASEAN, which was very much a product of the Cold War environment and the maneuverings of the major powers in the region, especially the United States and China. ASEAN embodied the desire of some of the nations in the region to create a political organization that was designed to ameliorate regional tensions – including those between ASEAN states – by institutionalizing relations among the region's anti-Communist states and, simultaneously, providing a structure within which they might address their anxieties about the possible actions of the major powers (Frost pp. 2–4). Indeed, it is worth emphasizing that ASEAN's consolidation and increased effectiveness as an organization were driven by the reunification of Vietnam and the partial strategic withdrawal of the United States from the region. Significantly, even a diminution of US power within the region had a profound effect on the potentially vulnerable ASEAN states.

Despite the fact that the United States now has a relatively unconstrained ability and, as we shall see, willingness to use its power in the region, Southeast Asia has enjoyed some major advantages from its relationship with the United States, especially during the Cold War. Because the United States was preoccupied with the military aspects of its confrontation with Communism, it was prepared to overlook behavior that might have been expected to provoke unfavorable reactions, and indeed have provoked such reactions in less ideologically charged circumstances. As far as the capitalist states of East Asia were concerned, US hegemony created the perfect environment in which to accelerate economic expansion. The "open" or liberal world economic order that the United States helped to establish, notably with the creation of the Bretton Woods institutions (the World Bank and the IMF) in 1944, presented uniquely favorable circumstances for a number of nations. Not only did countries such as Japan receive a massive boost from US military spending during the Korean War, but they enjoyed, at least until relatively recently, unfettered access to the massive domestic market of the United States.

The rapidly industrializing nations of Southeast Asia have attempted to follow a similar strategy, yet the ASEAN nations have enjoyed a good deal less political and economic leverage with the United States than Japan has. The ASEAN states have been further constrained by the fact that the economies of Northeast Asia (with the obvious exception of North Korea) have occupied some of the more lucrative niches in the global economy, effectively locking the nations of Southeast Asia into a more subordinate role in the regional division of labor (see also the previous chapter in this volume). Nevertheless, the ASEAN nations have also benefited from, and have come to rely on, access to the lucrative markets of the United States and, increasingly of late, on US investment capital (see Gangopadhyay, and Akyuz et al.). Access to US markets and capital have helped the nations of Southeast Asia to pursue the same export-oriented industrialization strategies that proved so successful in Northeast Asia. What has changed, however, is that there is a much less benign international economic environment in which to pursue such strategies. Not only have interstate and intercorporate rivalries continued to intensify, in an increasingly global external economy, but the international political order has become much less permissive toward authoritarian and mercantilist regimes in the wake of the ending of the Cold War.

One of the most important political and strategic developments within East Asia generally has occurred as a direct result of the ending of the Cold War, and has been largely driven by the ASEAN nations: the increasing importance of the ASEAN Regional Forum (ARF). The creation of the ARF, in July 1994, marked an attempt by the ASEAN nations to come to terms with their transfigured security environment. ASEAN leaders wanted to develop a regional security architecture that addressed their strategic concerns in a manner that was an extension of the "ASEAN way," favoring consensus and emphasizing noninterference in domestic affairs (see Acharya). The ARF was an attempt to keep the United States militarily engaged in the region while simultaneously maintaining Japan's low military posture and attempting to socialize China into nonbelligerent behavior (see Leifer). Significantly, it was an initiative that was strongly supported, and effectively facilitated, by the United States and Australia. The complex configuration of power between the major regional actors, especially when dealing with China, and the need to contain possible tensions both meant that it suited the interests of all parties to allow ASEAN to make the running in establishing a new multilateral security system.

It is not clear whether this model will be able to contain possible tensions within the region, especially between the United States and China. It is also far from certain whether the ARF will be able to influence the policies of such major powers unless it is in the latter's interest to allow it to do so. What is clear, however, is that, now that the United States is relatively unconstrained by wider strategic considerations, it appears willing to use its power to pursue longer-term objectives in other areas.

The economic crisis that has so badly affected Southeast Asia since 1997 has demonstrated that the United States still enjoys massive economic and political influence in the region, and that it is prepared to use it in pursuit of a wider reformist agenda. Although this new agenda is nominally about economic reform, and apparently driven by nonpartisan intergovernmental organizations such as the IMF, in reality it reflects US interests and values, and is a direct threat to existing economic and political practices within the region. Not only does the United States enjoy a preponderant influence over the IMF and its policies, but the substance of the initiatives that the IMF is pursuing is directly aimed at breaking up the close relationships between government and business that are such a distinctive aspect of regional political economies, and that have come in for heavy criticism in the wake of the recent crisis (see Beeson 1999).

Although the causes of the crisis are complex and not restricted to the shortcomings of "crony capitalism" (see Beeson and Robison), the view that the structural relationships embedded in the region's forms of capitalism are the cause of the region's problems, and should be the main targets of reform, has rapidly become conventional wisdom among policymakers in the West. Whatever the merits of this contention, its effect has been to place even greater pressure on the governments of Southeast Asia in particular to institute "appropriate" reforms in line with the injunctions of the United States and its institutional ally, the IMF. In an environment where the United States is no longer constrained by security considerations, and the nations of Southeast Asia have suffered a concomitant diminution in their strategic importance, the United States can exert direct pressure on them to reform themselves in a way that resonates with US values, and that may be expected to further its interests. In short, the crisis has afforded the United States a possibly unique opportunity to consolidate political and economic liberalism in a part of the world with which it has experienced diplomatic friction and massive trade deficits.

What, then, does a resurgent and activist United States mean for Southeast Asia? In the short term, it is probable that pressure designed to encourage market reforms and decrease the influence of governments in determining economic outcomes will continue. Such a development would also have major political implications for the states in the region. Ensuring compliance with IMF reforms, for example, requires much more intrusion into the sovereign affairs of individual states. The possibility of insulating or retaining control over national political and economic space in such circumstances – the essence of the "ASEAN way" – is clearly diminished for the small economies and comparatively fragile political structures of Southeast Asia. The continuing vulnerability of the ASEAN states in the face of external pressures from international actors, including not just the United States and the IMF, but also the increasingly powerful international financial markets, is clearly a major factor behind the tentative expansion of the ASEAN grouping to include the larger economies of Japan, China, and South Korea (see Beeson, forthcoming). Whether this will prove a viable way of reflecting and promoting common "Asian values," or whether such an expansion would provide the basis for institutionalizing an effective counterweight to the United States and the EU, is a moot point at this stage. What is clear is that, as we shall see in the next section, the larger "ASEAN + 3" grouping (that is, ASEAN minus Burma, Cambodia, and Laos, but plus Japan, China, and South Korea) has already achieved *de facto* status in the region's dealings with the EU.

Southeast Asia and the EU

Although the United States currently exerts the most powerful overall influence on Southeast Asia, the influence of "Europe" – meaning, in practice, various nations in western Europe – has been of much longer duration. The experience of European colonization and decolonization has profoundly influenced the economic structures and political development of East Asia in general, and Southeast Asia in particular. The recent upsurge of interest and interaction between the EU and the states of East Asia, which culminated in the first Asia-Europe Meeting (ASEM) in 1996, may have been primarily a product of a new international order in which East Asia had become a significantly more important economic player, but the "rediscovery of Asia" was overlaid by enduring historical perceptions. The possibility that the potential advantages of western Europe's longstanding relationship with the region might be lost, especially relative to the United States, provided a powerful incentive for revitalizing interest in East Asia. It is testimony to the increasing economic and political importance of the links between Asia and "Europe" that potentially problematic lingering animosities and resentments were subsumed by the perceived need to institutionalize interregional ties.

The increasing economic importance of the links between East Asia and western Europe was the most obvious force encouraging closer dialogue. The scale of the rapid growth in trade between the two regions over the past couple of decades was highlighted by the fact that trade between western Europe and the Asia-Pacific region surpassed trans-Atlantic trade for the first time in 1991 (Hilpert p. 57). For both sides, this is a rapidly developing and increasingly important economic relationship that merits greater political attention. However, it is also important to recognize that, for both parties, closer mutual links also offer one way of exerting pressure on, and acting as a counterweight to, the United States. A more substantial and formalized relationship between "Asia" and "Europe" has the additional potential benefit of keeping the United States engaged in a range of multilateral economic and security relationships (see Segal). Not only do closer ties between the two regions represent a potentially important curb on US unilateralism and the temptations that accrue to relatively unconstrained power, but they actually provide an incentive for the United States to remain constructively engaged, with Southeast Asia in particular, in the face of diplomatic activism on the part of the EU, its member states, and other nations in Europe.

From the perspective of western Europe, then, ASEM offers a potential method of countering US dominance in East Asia. Yet Europe as a whole shares some common interests and values with the United States, and these have led them to develop a number of similar policies toward East Asia, especially in the wake of the recent financial crisis. Given the difference in political structures and normative values between East Asia and "the West," this is, perhaps, unsurprising. From the inception of the ASEM process, issues of rights and values, and possible differences over questions of human rights and labor standards in particular, have been impossible to avoid. While these are issues over which the United States has also expressed concern, they are especially contentious in the context of relations with Europe (and its various components), as they have been brought directly into the ASEM process. What is most significant, however, is that at the first ASEM, in 1996, when the economies of East Asia were still performing strongly – especially as compared to the allegedly "sclerotic" economies of the EU – questions about potentially fundamental clashes of values were set aside and incorporated into a continuing dialogue process. Indeed, until the crisis began, a number of prominent EU politicians, such as Tony Blair, the Prime Minister of the United Kingdom, claimed to believe that there was much to be learned from the way in which business activities and social relations are organized in East Asia (see *The Guardian*, January 6, 1996). By contrast, at the ASEM in 1998, when East Asia was gripped by crisis and serious questions were being asked about both the sustainability of its forms of capitalism, and the propriety of the relationships between its business and political elites, a very different approach was taken.

Critics claim that the crisis has provided grounds for the EU and the United States to make common cause against East Asia (see Cammack and Richards). In the wake of the crisis, EU leaders also see an opportunity to encourage the adoption of neoliberal reforms of precisely the same sort that have been advocated by the IMF and the United States. In this manner, it is claimed, the EU is seeking to use the ASEM process to promulgate and then encourage the institutionalization of reforms that will effectively entrench modes of governance that are much closer to those of the Anglo-American political economies. Clearly, the atmosphere surrounding the first ASEM summit was very different from that surrounding the second, for the countries of Asia, and especially the smaller economies of Southeast Asia, had suffered a significant diminution in their status and influence. Despite this comparative decline, however, there are some important implications of this new development for interregional relations.

Perhaps the most important long-term consequence of the formalization of relations between East Asia and western Europe is its impact on East Asia's sense of itself as a region. In the ASEM process, "Asia" essentially means the ASEAN + 3 grouping (referred to above), and is thus a *de facto* expression of the longstanding proposal by Mahathir Mohamad, Prime Minister of Malaysia, for an East Asian Economic Caucus (EAEC). Although the notion of the EAEC has been actively discouraged by the United States and other prominent countries outside East Asia, such as Australia and New Zealand, the ASEM process is the most important expression of an "Asians only" political grouping yet realized. Certainly, its initial impulse and rationale may have owed much to changes in the economic structure of the region, but the formalization of relations between East Asia and western Europe is essentially a political process that has the potential to consolidate and inculcate a sense of a shared regional identity in the future.

Somewhat ironically, therefore, western Europe continues to exert a powerful influence on the evolution and identity of Southeast Asia. What is of greater significance now, however, is that political actors from Southeast Asia itself are more directly involved in shaping this process. Rather than simply being the victims of external forces over which they have no control, the nations of Southeast Asia have the potential, especially when part of a wider grouping that involves the nations of Northeast Asia as well, to exert some influence over the course of events. This increased political and strategic power, and the additional economic leverage and resources that such an expanded grouping potentially offers, present the countries of Southeast Asia with a way of increasing their influence and independence that they could not hope to achieve on their own. If this initiative proves successful, it augurs badly for those countries that find themselves at the literal and figurative edge of regional affairs: Australia and New Zealand.

Life on the Edge: Australia and New Zealand

For both Australia and New Zealand, Southeast Asia has historically represented a complex mixture of threats and, increasingly in recent years, opportunities. For much of the history of these two products of British colonialism, Southeast Asia was viewed with a good deal of concern. The transplanted, alien Anglo-Saxon cultures of these two countries were perceived to be at odds with those of the vast and impoverished nations to the North, which in turn were perceived as threatening. The specter of the "yellow peril" loomed large, especially in Australia, and not just in popular opinion. Discriminatory immigration policies based on "race" were a central component of public policy until the 1960s.

Despite this unpromising historical legacy, over the last couple of decades there has been a profound change in attitudes to Asia in Australia and New Zealand, particularly among members of the political and economic elites. At one level, this reflects the emergence of more independent foreign policies and (arguably) more progressive domestic social policies. In 1972, the recognition of the People's Republic of China by the Australian government headed by Gough Whitlam was taken to symbolize the beginning of this transformation in elite opinion regarding the region generally. Yet there has also been a more mundane and ultimately powerful incentive to rethink relations with East Asia. The rapid industrialization of the region to Australia and New Zealand's North has meant that the direction of both countries' trade has shifted inexorably toward their immediate region, and away from earlier colonial links with Britain and other European countries (see Palat). Not only have the countries of East Asia become more important to Australia and New Zealand as direct economic partners, but their superior economic growth has galvanized policy-makers into rethinking their approaches to foreign and domestic policy.

In this regard, Australia and New Zealand have been at the forefront of an international trend, especially among the Anglo-American economies, toward embracing competitive, market-based reforms domestically and less protectionist policies internationally (see Beeson and Firth, as well as Kelsey). What is especially significant about this change as far as relations with East Asia are concerned is that Australia in particular has attempted to link its domestic reforms and its newfound enthusiasm for East Asia into an ambitious policy of regional engagement. The centerpiece of the engagement strategy, for both Australia and New Zealand, has been the establishment of the Asia-Pacific Economic Cooperation forum (APEC).

Originally proposed by an Australian Prime Minister, Bob Hawke, with the active support of Japan, APEC was primarily driven by Hawke's successor, Paul Keating. For Keating, APEC presented a number of crucial opportunities. First, APEC offered a way to continue developing closer political links with a region that had become an increasingly important component of Australia's long-term economic welfare. Simply put, Australia – and, indeed, New Zealand – could not afford to risk being marginalized in a world that threatened to fracture into regional economic groupings. The growing economic weight of the EU and the inauguration of the North American Free Trade Agreement (NAFTA) seemed to demonstrate how real such a possibility was. Second, APEC provided a potential mechanism with which to export Australia and New Zealand's preferred model of economic integration and development. In other words, if APEC succeeded it would not only place Australia in particular at the center of an important regional body, but it would encourage the hitherto neomercantilist states of East Asia to open up their economies in a way that would benefit both Australia and New Zealand (see Keating).

In reality, however, APEC has proved something of a disappointment, even to its most ardent supporters. Not only was APEC virtually invisible during a crisis in which it was uniquely well-placed to play a crucial management and mediating role, but, as we have seen, the United States has discovered more effective and unilateral means through which to pursue its foreign policy goals in the region (see Beeson). APEC has been hamstrung by internal opposition, particularly from Malaysia, and the need to rely on persuasion and consensually determined policy outcomes when trying to implement and develop its ambitious reformist agenda. In the Australian context, APEC has been further diminished by the election of a government that is much less committed to and enthusiastic about engagement with East Asia than Keating's was.

Since John Howard's Liberal-National Party coalition government was elected in 1996, older ties with the United States and Britain have been reinvigorated. Whereas the active diplomacy of the former Labor government was at the center of initiatives such as APEC, and the potentially even more important ARF, Howard's government has displayed much greater ambivalence toward the region. Revealingly, Australia's most conspicuous interaction with the region under Howard has been Australia's military intervention in East Timor during 1999. While this occurred under UN auspices and was generally considered a success, the suggestion that Australia might perform the role of "deputy sheriff" to the United States in maintaining regional security helped to sour relations with Indonesia and undermined Australia's chances of becoming an integrated and accepted part of the region (Brenchley pp. 22–24).

The policies pursued by Australia, by far the larger economic and political entity, necessarily have major implications for New Zealand. The recently elected Labor government led by Helen Clark has been heavily criticized by Australia for reducing defense spending, which in Australia itself, revealingly, is the only government activity that has received an increased budgetary allocation under Howard's government. Whatever its intentions, New Zealand may find it difficult to separate itself from the increasingly negative perceptions of Australia that have developed over the last few years in parts of East Asia. The Australian Foreign Minister, Alexander Downer, has recently suggested that the only form of regionalism Australia can take part in is a "practical" sort, one that is not based on a "culture" that Australia does not share (see Downer). In some ways, Australian policy appears to have come full circle, reviving stereotypes about "Asia" that show little appreciation of the complexity of the region or the very different regimes that constitute it. At the very least, such statements seem certain to exclude Australia – and, by association, New Zealand – from potentially significant long-term projects such as the ASEAN + 3 process. The one area in which "practical" regionalism might produce important benefits for Australia is in the proposed linking of the ASEAN Free Trade Area (AFTA) with the Australia-New Zealand Closer Economic Relations (CER) arrangement (see Lloyd, as well as Edlin). If such an initiative comes to fruition, Australia and New Zealand's futures may yet be more comprehensively integrated with those of their neighbors.

Conclusion

The development of Southeast Asia has been significantly affected by its relations with western powers. The United States and western Europe in particular have exerted a direct influence on the region through the colonial experience and, latterly, through Southeast Asia's incorporation into the global strategic confrontation between capitalism and "socialism." Their influence has not disappeared in the wake of the ending of the Cold War. On the contrary, in many ways the apparent secular diminution of the importance of security issues in the calculations of the major powers has meant that they have the opportunity and, often, the desire to pursue their perceived national interests (or, in the case of the EU, regional interests) unencumbered by strategic constraints. In short, the nations of Southeast Asia remain extremely vulnerable to external pressures that they have only limited capacities to control.

This vulnerability helps to explain the attractiveness of an expansion of ASEAN and the possible institutionalization of ASEAN + 3, which appears to be gaining momentum. While this process is clearly driven by the nations of the region themselves, especially Malaysia, the Philippines, and China, it is important to recognize that external pressure has also been a major incentive to pursue greater regional cooperation and collective self-reliance. The extensive and intrusive nature of US and, to a lesser extent, European reformist initiatives has engendered resentment and concern among a significant number of members of regional political elites (see Higgott). One way of providing a possible counterweight to such pressures is by attempting to establish a broader grouping excluding countries that are not in East Asia. Whether such an initiative proves to be successful remains to be seen.

For Australia and New Zealand, such a development provides a particularly difficult foreign policy challenge. Unless Australia and New Zealand are able to reinvigorate and consolidate their ties with Asia generally, and particularly with the crisis-hit countries of Southeast Asia, they risk being permanently marginalized from what may prove to be the region's most important political organization.

As far as Southeast Asia itself is concerned, perhaps the greatest challenges it faces are those subsumed under the all-encompassing rubric of "globalization." On the one hand, the region must to try to come to terms with the implications of being ever more deeply integrated into global economic structures that not only dwarf the individual economies of Southeast Asia, but expose their economic strategies and even their political systems to the judgment of external market forces. As the experience of Malaysia suggests, while governments in Southeast Asia may not like this situation, their capacity to resist it is limited. At the same time, however, the current generation of political leaders may find their position undermined by a subtler and more insidious threat. The spread of "western" ideas of political and economic liberalism, when combined with the pervasive influence of US values and lifestyles, may mean that Southeast Asia will either gradually converge on a more "western" endpoint, or be forced to adopt a more insular and exclusive form of regionalism in order to maintain its identity. Either way, the region will continue to be profoundly influenced by its interactions with the West – for good or ill.

Further Reading

Acharya, Amitav, "Ideas, Identity, and Institution-building: From the 'ASEAN Way' to the 'Asia-Pacific Way'?," in *Pacific Review*, Volume 10, number 3, 1997

Agnew, J., and S. Corbridge, *Mastering Space: Hegemony, Territory and International Political Economy*, London and New York: Routledge, 1995

This book adopts a critical political economy perspective, linking a theoretical framework derived from Antonio Gramsci to contemporary geopolitics.

Akyuz, Y., H.-J. Chang, and R. Kozul-Wright, "New Perspectives on East Asian Development," in *Journal of Development Studies*, Volume 36, number 4, 1998

Beeson, Mark, 1999: "Reshaping Regional Institutions: APEC and the IMF in East Asia," in *Pacific Review*, Volume 12, number 1, 1999

Beeson, Mark, forthcoming: "ASEAN: The Complexities of Organizational Reinvention," in Mark Beeson (editor), *Reconfiguring East Asia: Regional Institutions and Organizations After the Crisis*, London: Curzon Press, forthcoming

Beeson, Mark, and Ann Firth, "Neoliberalism as a Political Rationality: Australian Public Policy since the 1980s," in *Journal of Sociology*, Volume 34, number 3, 1998

Beeson, Mark, and Richard Robison, "Introduction: Interpreting the Crisis," in Richard Robison et al. (editors), *Politics and Markets in the Wake of the Asian Crisis*, London and New York: Routledge, 2000

Brenchley, Fred, "The Howard Defense Doctrine," in *The Bulletin*, September 28, 1999

Cronin, James E., *The World the Cold War Made: Order, Chaos, and the Return of History*, London and New York: Routledge, 1996

A very useful overview of the evolution and lasting impact of the Cold War

Cumings, Bruce, "Japan and Northeast Asia into the 21st Century," in Peter J. Katzenstein and Takashi Shiraishi (editors), *Network Power: Japan and Asia*, Ithaca, NY: Cornell University Press, 1997

Downer, Alexander, Opening Speech to the Asia Leaders Forum, Beijing, April 23, 2000

Edlin, Bob "CER – Unfinished Business," in *Independent Business Weekly*, March 15, 2000

Frost, Frank, "Introduction: ASEAN since 1967 – Origins, Evolution and Recent Developments," in A. Broinowski (editor), *ASEAN into the 1990s*, London: Macmillan, and New York: St Martin's Press, 1990

Gangopadhyay, P., "Patterns of Trade, Investment, and Migration in the Asia-Pacific Region," in G. Thompson (editor), *Economic Dynamism in the Asia-Pacific*, London and New York: Routledge, 1998

Gilpin, Robert, *The Political Economy of International Relations*, Princeton, NJ: Princeton University Press, 1987

This is a seminal work, in the "realist" tradition of studies of the international political economy, that explains and analyzes the emergence of the postwar international order.

Guardian, "Mr Blair Feeds the Tigers," January 6, 1996

Higgott, Richard, "The Asian Economic Crisis: A Study in the Politics of Resentment," in *New Political Economy*, Volume 3, number 3, 1998

Hilpert, Hanns G., "The Economic Setting," in H. Maull, G. Segal and J. Wanandi (editors), *Europe and the Asia-Pacific*, London and New York: Routledge, 1998

Huntington, Samuel P., *Political Order in Changing Societies*, New Haven, CT, and London: Yale University Press, 1968

An important, if dated, text from a highly influential author, which considers political development in the Third World

Jackson, Robert, *Quasi-states: Sovereignty, International Relations, and the Third World*, Cambridge and New York: Cambridge University Press, 1990

A highly original reading of the integration of the developing world into the international system of states

Keating, Paul, *Engagement: Australia Faces the Asia-Pacific*, Sydney: Macmillan, 2000

A former Prime Minister of Australia gives his own account of his attempts to "engage Asia."

Kelsey, Jane, *The New Zealand Experiment: A World Model for Structural Adjustment*, Auckland: Auckland University Press, 1997

An influential and important critique of New Zealand's experiment with neoliberalism

Korhonen, Pekka, "Monopolising Asia: The Politics of Metaphor," in *Pacific Review*, Volume 10, number 3, 1997

Latham, Robert, *The Liberal Moment: Modernity, Security, and the Making of Postwar International Order*, New York: Columbia University Press, 1997

An original and sophisticated analysis of US foreign policy, and of the influence of liberalism in international affairs

Leifer, Michael, "The ASEAN Regional Forum," Adelphi Paper 302, London: International Institute for Strategic Studies, 1996

Lloyd, P.J., "Should AFTA and CER Link?," in *The Australian Economic Review*, 3rd Quarter, 1995

Matsanundo, Michael, "Preserving the Unipolar Moment: Realist Theories and US Grand Strategy after the Cold War," in *International Security*, Volume 21, number 4, 1997

Palat, Ravi, "Up the Down Staircase: Australasia in the 'Pacific Century'," in *Thesis Eleven*, number 55, 1998

Saïd, Edward W., *Orientalism*, New York: Pantheon, and London: Routledge, 1978

A seminal, culturally grounded analysis of the discursive construction of "the Orient" by Europeans

Segal, Gerald, "Thinking Strategically about ASEM: The Subsidiarity Question," in *Pacific Review*, Volume 10, number 1, 1997

Tarling, Nicholas, *Nations and States in Southeast Asia*, Cambridge and New York: Cambridge University Press, 1998

A short and accessible introduction to the countries of the region, by a noted specialist

---

Dr Mark Beeson is a Lecturer in International Political Economy at the School of Asian and International Studies, Griffith University, Brisbane, Queensland.

Appendices

Appendices

APPENDICES

Appendix 1
Chronology

This listing is focused on political, economic and military events that continue to affect the countries of Southeast Asia as they pass from the 20th century into the 21st. It should be noted that many topics mentioned elsewhere in the appendices are not covered here.

1940	**May**	Nazi Germany conquers the Netherlands, but officials and troops responsible to the Dutch government in exile in London retain control of the Dutch East Indies (now *Indonesia*).
	June	Following the French surrender to Nazi Germany, the French possessions in Indochina – *Vietnam* (divided into Tonkin, Annam, and Cochin China), *Cambodia*, and *Laos* – are assigned to the control of the collaborationist Vichy government.
	September	Southeast Asia is drawn directly into World War II as Japan completes its takeover of the French possessions in Indochina. As an ally of Nazi Germany and therefore of Vichy France, Japan leaves French officials and troops in charge in all three countries, but takes over transportation systems, supports *Thailand*'s claims to parts of *Laos* and *Cambodia*, and makes Saigon, capital of Cochin China (southern Vietnam), its military headquarters for further incursions into Southeast Asia.
	November	The Nambo Uprising, a series of riots and rebellions in southern *Vietnam*, is suppressed by the French colonial authorities and disowned by the Indochinese Communist Party (founded 1930), which then starts to build a united front with other anti-French organizations.
		Phibun Songkhram, military dictator of *Thailand* since 1938, launches an invasion of *Laos* and *Cambodia* in pursuit of historic Thai claims to a number of border provinces.
1941	**January**	The navy of Vichy France defeats the navy of *Thailand* at Ko Chunag island. Japan then brokers a ceasefire between the two countries (see also May 1941).
	February	After 29 years in exile, Ho Chi Minh returns to *Vietnam* to begin preparations for a nationalist rising to be led by his Indochinese Communist Party.
	April	Norodom Sihanouk becomes King of *Cambodia* for the first time, at the age of 18; he is crowned in September.
	May	Under the terms of the Tokyo Convention, Vichy France cedes two provinces in *Cambodia* (Battambang and Siamrap) and two in *Laos* (Champasak and Xaignabouri) to *Thailand*, which once ruled them as well as other parts of these countries.

In *Vietnam*, Ho Chi Minh launches the Viet Minh as a united front against French colonial rule. The Viet Minh's links with the Indochinese Communist Party, also founded and led by Ho, are not publicized. Most of its early members are non-Communists, and it receives funds and arms from the Chinese Nationalists, as well as their western allies, throughout the remainder of World War II.

December The other colonies of the European powers in Southeast Asia are also drawn into the war, as Japan launches attacks on the Dutch East Indies (now *Indonesia*); the British colonies and protectorates in *Malaya*, North Borneo (now Sabah), Sarawak (all three now comprising *Malaysia*), and *Brunei*; the British colony of *Singapore*; and the US territory of the *Philippines* – as well as Pearl Harbor in Hawaii.

The government of *Thailand* decides to make peace with Japan after Japanese troops arrive in the capital, Bangkok.

Dutch and Australian troops land in the Portuguese colony of *East Timor*, disregarding Portugal's policy of neutrality (see February 1942 and January 1943).

Aung San, a prominent nationalist, establishes the *Burma* Independence Army, which is intended to fight against both the British and the Japanese.

1942 January Japanese forces capture Manila in the *Philippines*, Kuala Lumpur in *Malaya*, most of the island of Borneo, the Dutch-held island of Celebes, and much of the Moluccas, a Dutch colony defended by Australian troops.

Thailand declares war on the United Kingdom and the United States. The UK government reciprocates; the US government, interpreting the declaration as a result of Japanese pressure, does not. Up to August 1945, around 150,000 Japanese troops are stationed in Thailand, while Pridi Phanomyong, the Regent for the young King Ananda Mahidol (who is at school in Switzerland), secretly creates a force of around 50,000 "Free Thai" troops, with US support, to prepare to fight against the Japanese.

February Japanese forces capture *Singapore* from the British, for whom it had been their main center of operations in Southeast Asia. The new rulers change the city's name to Shōnan and use its facilities – notably its naval base, completed only four years earlier – to launch their attempt to create a "Greater East Asia Coprosperity Sphere" by force.

Japanese forces also enter the Portuguese colony of *East Timor*, and aid risings against the Dutch in Aceh and North Sumatra (but see November).

Ethnic Chinese Communists and other opponents of the Japanese occupation of *Malaya* and *Singapore* form the Malayan People's Anti-Japanese Army, which harasses the occupiers up to the end of World War II.

March Japanese forces capture Batavia (now Jakarta), the capital of the Dutch East Indies (now *Indonesia*), and complete the conquest of most of the other Dutch-held territories.

April In the *Philippines*, following the capture of Bataan by the Japanese, Communist and other resistance groups in central Luzon form the People's Anti-Japanese Army (Hukbo ng Bayan Laban sa Hapon, nicknamed Hukbalahap or Huk), led by Luis Taruc. The Hukbalahap target both the Japanese and their Filipino collaborators; by the end of the war, in August 1945, they control most of central Luzon.

	May	Japanese forces enter the British colony of *Burma*, partly in order to interdict supply lines from the Allies to China along the Burma Road. Their advance is resisted, unsuccessfully, both by forces loyal to the British and by the Burma Independence Army under Aung San, which is reorganized and renamed the Burma Defense Army.
	July	Japanese forces suppress the Angganita movement, one of a series of millenarian uprisings that have erupted in West Papua over the generations, becoming vehicles for protest against Dutch, Japanese, and, more recently, Indonesian control.
	November	Japanese forces suppress provincial uprisings in Aceh and North Sumatra, having aided them when they were aimed against the Dutch.
1943	**January**	After 11 months of fighting in *East Timor*, the Japanese expel the last remaining Australian and Dutch troops, and carry out mass executions of Timorese said to have collaborated with western forces.
	March	The Japanese military administration now in control of most of the Dutch East Indies (now *Indonesia*), as well as *East Timor*, establishes a collaborationist organization known as Putera, under the chairmanship of an Indonesian nationalist, Sukarno.
	July	Japan transfers to *Thailand* the Malay states of Perlis, Kedah, Kalantan, and Terengganu, which have been controlled by the British since 1909, as well as two of the 34 Shan states that have long been in association with *Burma*.
	August	Japan grants nominal independence to *Burma* under a government headed by U Ba Maw as Prime Minister, and including Aung San, the former head of the Burma Independence Army, as Minister of War, as well as U Nu as Foreign Minister. Its independence is recognized by Nazi Germany and its allies.
	October	The Japanese administration in *Indonesia*, working with Sukarno and other nationalist politicians, creates an auxiliary armed force known as Peta and a Moslem organization known as Masyumi, but also imposes requisitions of rice and of forced labor.
		Aung San, Minister of War in the government of *Burma* created by the Japanese, initiates secret contacts with Lord Mountbatten, commander of the Allied forces in Southeast Asia, with a view to cooperating against the Japanese.
1944	**June**	The US Navy defeats the Japanese Imperial Navy in the Battle of the Philippine Sea, which marks a turning point in the conflict over Southeast Asia.
		Phibun Songkhram, military dictator of *Thailand*, accepts responsibility for the country's damaging alliance with Japan and is removed from public life (but see November 1947 and April 1948).
	October	US forces under General Douglas MacArthur recapture Leyte, in the *Philippines*, from the Japanese, beginning the reconquest of the country.
	December	The Viet Minh launches the first of its "armed propaganda teams" inside *Vietnam*, under the leadership of Vo Nguyen Giap.
1945	**January**	US forces reopen the *Burma* Road, and restart the supply of materiel and funds to the Chinese Nationalists.

March	US forces retake Manila, the capital of the *Philippines*, from the Japanese. (Because the city has been largely destroyed in the fighting, the suburban community of Quezon City becomes the temporary capital of the Philippines, informally from 1945 and formally from 1948, up to 1976.)

In *Burma*, the Antifascist People's Freedom League and the Burma National Army, both led by Aung San, launch an uprising against the Japanese with Allied support.

After four and a half years of collaboration with French colonial officials, the Japanese take direct control of *Cambodia*, *Laos*, and *Vietnam*. They then permit Bao Dai, Emperor of the French protectorate of Annam since 1925, to proclaim Annam reunited with Tonkin and Cochin China as a single Empire of Vietnam under Japanese supervision. They grant a similar "independent" status to the kingdoms of Cambodia and Laos.

In *Cambodia*, a number of political groups opposed to both the Japanese and the French form the Khmer Issarak (Free Khmer) movement, an unstable coalition that engages in sporadic insurgency against Sihanouk's government up to 1954.

In *Indonesia*, the BPUKI, an assembly of nationalists created by the Japanese, announces plans for a postwar federation embracing not only the former Dutch East Indies but also *East Timor* and the British dependencies of *Malaya*, Sabah, and Sarawak.

May	US and UK forces jointly retake Rangoon, the capital of the British colony of *Burma*, from the Japanese.
June	In *Indonesia*, the leading nationalist politician, Sukarno, proclaims the doctrine of Pancasila – the "five principles" of nationalism, internationalism, democracy, social justice, and belief in one God – which becomes the official ideology of the country following independence (and see December 1984, January 1985).

Dutch forces return to *Indonesia*, starting with a landing in North Sumatra.

Australian forces take *Brunei* from the Japanese, restoring its prewar status as a sultanate protected by the British.

British colonial forces, Burman nationalists, and military units from the ethnic minorities are temporarily united in a victory parade through Rangoon, the capital of newly liberated *Burma*.

August	At the Potsdam conference, the United States, the Soviet Union, and the United Kingdom agree that *Vietnam* will be provisionally divided at the 17th parallel between a southern section policed by British forces and a northern section policed by Chinese Nationalist forces.

Shortly before the Japanese surrender and the end of World War II, the Japanese military authorities in Indochina compel Emperor Bao Dai of *Vietnam* to abdicate his throne, become an ordinary citizen (as Nguyen Vinh Thuy), and recognize the Democratic Republic of Vietnam created by the Viet Minh under Ho Chi Minh. Meanwhile, in Saigon and other southern cities, a mass movement led by Tran Van Thach, Ta Thu Thau, and other Trotskyists organizes "people's committees" opposed both to foreign intervention and to Ho's regime.

In *Laos*, the Japanese secure the appointment of Prince Phetsarath Ratanavongsa, leader of the nationalist Lao Issara (Free Laos) movement, as Prime Minister of a nominally independent government.

Shortly after Japan's surrender and the end of World War II, Sukarno declares the Dutch East Indies independent, as *Indonesia*, and proclaims a republican Constitution. The Netherlands continues to claim sovereignty, and to make sporadic attempts to enforce it, until December 1949.

The government of *Thailand* denounces its alliance with Japan, announces its intention to return to *Laos* and *Cambodia* the provinces taken in May 1941, and retrospectively nullifies the declarations of war on the United States and the United Kingdom of January 1942.

The British colonial authorities restored to power in *Burma* briefly imprison the nationalist leader Aung San, but release him in response both to protests inside Burma and to pressure from the Colonial Office in London.

September In a ceremony in *Singapore*, the British commander Lord Mountbatten takes the surrender of all Japanese forces in Southeast Asia. The British then establish a temporary military administration for Singapore and the Malay states (to April 1946).

Ho Chi Minh, leader of the Indochinese Communist Party and of the Viet Minh, declares the independence of the Democratic Republic of *Vietnam*, 83 years after the country first ceded territory to France and 57 years after its incorporation into French Indochina. After four years of aiding the Viet Minh against the Japanese and the Vichy French, the British and Nationalist Chinese governments change policy, deploying forces, including some released Japanese soldiers, to restore the French colonial administration. During the ensuing conflicts, Communist agents dismantle the "people's committees" in the South and kill numerous Trotskyist opponents of Ho Chi Minh.

Prince Phetsarath, Prime Minister of *Laos* since August, declares the country unified and independent, 52 years after it was incorporated into French Indochina.

British forces take charge of Java and much of the rest of *Indonesia*, but reject Dutch demands to arrest Sukarno and other leaders of the Republic.

British forces complete the reoccupation of *Malaya*.

October The government of the Republic of *Indonesia* creates an army to fight against the Dutch, British, Indian and Japanese troops occupying the country under British command.

French forces restore the prewar Protectorate of *Cambodia* in the face of opposition from King Sihanouk.

King Sisavang Vong of *Laos*, seeking to improve relations with the returning French colonial authorities, dismisses Prince Phetsarath from the prime ministership. In response, Phetsarath, his half-brothers Souphanouvong and Souvanna Phouma, and other leaders of the Lao Issara (Free Lao) movement form a new government, and announce that the King has been deposed. The Lao Issara government then enters into conflict with royalist, French, British, and Chinese Nationalist forces, as well as an army led by the pro-French Prince of Champasak, Boun Oum (see March 1946).

	November	The army of the Republic of *Indonesia* seizes the city of Surabaya from the British with the aid of defecting Indian troops, and fights Japanese units under British command elsewhere in Java.
	December	French forces drive the Vietminh out of Hanoi and temporarily secure control of most of *Vietnam*.
1946	**March**	The United Malay National Organization (UMNO), which is to be the dominant political force in *Malaya* and then in *Malaysia*, is formed in Kuala Lumpur by Onn bin Jaafar, Prime Minister of the state of Johor, and other Malay politicians.
		Thailand signs a peace treaty with the United Kingdom under which the Malay and Shan states seized by Thailand in July 1943 are returned to British control.
		France recognizes *Vietnam*, under the Viet Minh government, as a free state within the French Union. French troops arrive in northern Vietnam as Chinese Nationalist troops withdraw.
		French troops force Prince Phetsarath Ratanavongsa to resign as Prime Minister of *Laos*; he and other leading members of the Lao Issara movement flee the country. The French government then recognizes Laos as a free state within the French Union, under King Sisavang Vong.
	April	The United Kingdom abolishes its colony of the Straits Settlements, makes *Singapore* a separate crown colony, and unites the other two Settlements, Malacca and Penang, with the nine other Malay states in the Malayan Union. The Union is opposed by the leading political movement, UMNO, and by some of the states' traditional rulers (see February 1948). It excludes three other British-controlled and mainly Malay states, all on the island of Borneo: British North Borneo (now Sabah), *Brunei*, and Sarawak. The United Kingdom also takes direct control of Sarawak from the Brookes, a British family that has ruled the territory since it was ceded to them by Brunei in 1841.
	May	*Thailand* reverts to the name "Siam," dropped in 1939, for use in international forums (but see May 1949).
		Admiral Georges Thierry d'Argenlieu, French High Commissioner in *Vietnam*, proclaims a Republic of Cochin China (southern Vietnam), apparently without first consulting the French government, and appoints General Nguyen Van Xuan as its Prime Minister (see June 1948).
	June	The young King Ananda Mahidol of Siam (*Thailand*) is found shot dead in his palace, in circumstances that remain mysterious, six months after his return from schooling in Switzerland. His brother King Bhumibon Adulyadej succeeds to the throne at the age of 18. (Bhumibon has since become the longest-reigning living monarch in the world, apart, arguably, from Sihanouk of *Cambodia*, whose reign has been interrupted by himself and others.)
	July	Under the terms of the Tydings-MacDuffie Act, passed by the US Congress in 1934, the *Philippines* achieves independence from the United States under President Manuel Roxas and a legislature dominated by the Liberal Party, which prevents supporters of the Hukbalahap movement (see April 1942) from taking their seats. US forces continue to be stationed in the country.

		The British formally surrender control of *Indonesia* to the Dutch, but the Republic of Indonesia retains control of Java and Sumatra. The last British troops in Indonesia depart in November.
	August	In the *Philippines*, the murder of Juan Feleo, leader of the National Peasant Union (Pambansang Kaisahan ng mga Magbubukid or PKM), sparks off the Hukbalahap or Huk rebellion against President Roxas's government. Starting in central Luzon, the rebellion spreads rapidly to other provinces, drawing in other former members of the wartime People's Anti-Japanese Army (see April 1942), which is renamed the People's Liberation Army (Hukbong Mapagpalaya ng Bayan).
	September	In the first legislative elections in *Cambodia*, the Democratic Party, which seeks immediate independence from France and a reduction in the power of the monarchy, wins a majority of seats. Political conflict between the Democrats and King Sihanouk's supporters goes on until 1955, while the Khmer Issarak insurgency continues.
	November	In *Vietnam*, following the collapse of negotiations between Ho Chi Minh and the French government (in Paris, May to September), the French navy bombards Haiphong, initiating what is variously known as the First Indochina War, the French Indochina War, or the First Vietnam War (to May 1954). Most of the funding for the French war effort is provided by the US government.
		The Netherlands and the Republic of *Indonesia* sign the Linggajati Agreement, brokered by the British, which formally creates the Republic of the United States of Indonesia (RUSI), an entity under the Dutch crown comprising the Republic (Java, Sumatra, and Madura) and autonomous states in the rest of the former Dutch East Indies.
		Siam (*Thailand*) returns to *Laos* the provinces on the west bank of the Mekong River that it seized in May 1941.
1947	**April**	Elections to a Constituent Assembly in *Burma* are won by a coalition of nationalist parties led by Aung San, but are boycotted by the Communist Party, with its base among the Shan minority, and by groups representing the Karen minority.
	July	In their first "police action" in *Indonesia*, Dutch troops, disregarding the Linggajati Agreement of November 1946, occupy large parts of Java and other islands.
		Aung San, Prime Minister of *Burma* and leader of the Antifascist People's Freedom League, is assassinated, along with six of his colleagues, apparently on the orders of U Saw Maung, leader of the National Opposition. Aung San is succeeded by U Nu.
	November	The first National Assembly elected in *Laos* endorses the appointment as Prime Minister of Prince Souvannarath, a half-brother of the exiled Lao Issara leaders Phetsarath, Souphanouvong, and Souvanna Phouma.
		In Siam (*Thailand*), the military seize power once again, with the backing of the former dictator Phibun Songkhram.
1948	**January**	The Union of *Burma* attains independence under the leadership of Prime Minister U Nu, 11 years after being detached from the Indian Raj and made into a separate British colony. The Red Flag faction of the Communist Party immediately launches an uprising against the new state.

		The Netherlands and the Republic of *Indonesia* sign a ceasefire agreement brokered by the UN.
	February	The Malayan Union created in April 1946 is replaced by the Federation of *Malaya*, in which the traditional rulers of most of its states are given greater autonomy and the leading political movement, UMNO, agrees to cooperate with the British colonial authorities.
	March	President Roxas of the *Philippines* declares the insurgent Huk movement (the Hukbong Mapagpalaya ng Bayan, or People's Liberation Army) an illegal organization (see May 1954).
		The White Flag faction of the Communist Party of *Burma*, which has significant influence among the Shan minority, launches an uprising against the Burmese government. Rebellions by military units representing other ethnic minorities follow throughout the spring and summer. These conflicts continue sporadically for decades, but many are ended in a series of ceasefire agreements signed between April 1989 and January 1996.
	April	Phibun Songkhram, military dictator of Siam (*Thailand*) from 1938 to 1944, returns to power once more after another military coup; he remains in power until September 1957.
	June	The government of the Federation of *Malaya* declares a state of emergency as troops from the United Kingdom and other Commonwealth countries begin to combat an armed uprising by the Malayan People's Liberation Army. This group, largely composed of ethnic Chinese, led by Chin Peng, and dominated by the Communist Party of Malaya, includes former members of the wartime Malayan People's Anti-Japanese Army (see February 1942). Around 11,000 people are killed during the Emergency, which continues until July 1960.
		In *Vietnam*, General Nguyen Van Xuan, Prime Minister of the French-controlled Republic of Cochin China, joins Emperor Bao Dai in creating a new government of (South) Vietnam, which is declared an Associate State of the French Union, while Ho Chi Minh's government, based in the North, continues its war against French forces.
	September	In *Indonesia*, forces loyal to the republican government suppress a shortlived uprising led by pro-Communist soldiers in and around the city of Madiun. Despite the Communist Party's denials, then and later, that it sanctioned the uprising, several party officials are executed and around 35,000 supporters are arrested.
	December	Having completed the creation of 16 federated states in the parts of *Indonesia* not controlled by the Republic, the Dutch launch their second "police action" in the name of the United States of Indonesia, attacking parts of Java without warning, occupying its capital, Batavia (Jakarta), and arresting the Republic's government. The whole of the former Dutch East Indies returns to Dutch control, with the exception of Aceh. The UN Security Council imposes a ceasefire and the United States suspends aid to the Netherlands under the Marshall Plan.
1949	**January**	The Karen National Union, based among the Karen minority in southeastern *Burma*, launches an uprising against the central government, and many Karens serving in the army mutiny or desert in sympathy. The uprising has continued sporadically ever since.

	February	In *Indonesia*, republican forces execute Tan Malaka, a former Communist whose leftist guerilla movement had rivaled President Sukarno's forces in numbers and effectiveness.
	May	*Thailand*, known abroad as "Siam" since May 1946, reverts to using the name "Thailand" once again.
	July	*Laos* becomes an Associate State of the French Union, taking full responsibility for its internal affairs but leaving security and foreign policy in the hands of the French government (but see August 1950 and October 1953).
		President Sukarno and other members of the government of the Republic of *Indonesia* return to the temporary capital, Jogjakarta, after being released from Dutch custody. Most of the Indonesian states created by the Dutch agree to join the Republic.
	October	After a brief occupation of parts of *Burma* by Chinese Nationalist troops fleeing from the newly proclaimed People's Republic of China, the government of Burma reiterates its policy of neutrality, denouncing the United States for its support of the Chinese Nationalists and rejecting all foreign aid.
		Most members of the Lao Issara government deposed in March 1946 return to *Laos* under an amnesty. The returnees include Prince Souvanna Phouma, but exclude his half-brothers Phetsarath, who remains in Thailand, and Souphanouvong, who goes to Vietnam to seek the support of the Viet Minh.
	November	Under an agreement signed by King Sihanouk of *Cambodia* and the French government, but never ratified by the National Assembly of Cambodia (then dominated by Sihanouk's opponents), the country becomes an associate state of the French Union, taking full responsibility for its internal affairs but leaving security and foreign policy in the hands of the French government (but see November 1953).
	December	The independence of the United States of *Indonesia*, the former Dutch East Indies, is recognized by the Netherlands, 330 years after the first Dutch incursions into the archipelago. Indonesia adopts a federal Constitution giving considerable autonomy to its states, which are renamed provinces (but see August 1950), and also formally revives the historic name of the capital city, Jakarta (known as Batavia under the Dutch).
1950	**January**	Ho Chi Minh's Democratic Republic of *Vietnam*, which controls the North of the country, is recognized by the Soviet bloc, China, and Yugoslavia.
	February	Bao Dai's Empire of *Vietnam*, which controls the South of the country, is recognized by the United States and other western countries.
	March	Following the establishment of the Communist regime in China in October 1949, the US government adopts the "domino theory" (the view that Communism is likely to spread through Southeast Asia country by country), and begins to provide military aid to *Thailand, Cambodia, Laos*, and Bao Dai's government in *Vietnam*.
	April	In *Indonesia*, a mainly Christian rebel movement proclaims the creation of the Republic of the South Moluccas; it is not completely suppressed by the Indonesian army until October 1952.

		In *Cambodia*, the Khmer Issarak movement holds its first national congress, which launches the United Issarak Front led by Son Ngoc Minh, a clandestine member of the Indochinese Communist Party.
	June	Sir Omar Ali Saifuddin III succeeds his brother Ahmad Tajuddin as Sultan of *Brunei*, which remains a British protectorate.
	August	*Indonesia* replaces the federal Constitution of December 1949 with a new, centralized Constitution, and becomes the Republic of Indonesia. No nationwide elections are held until 1955: meanwhile, Sukarno retains the presidency and the "provisional" Assembly remains in being.
		At a secret meeting in *Vietnam*, Prince Souphanouvong, exiled from *Laos* in March 1946, establishes the Pathet Lao (Land of the Lao) movement to fight, in alliance with the Viet Minh, for Laos's complete independence from France.
1951	**February**	The Indochinese Communist Party, founded by Ho Chi Minh in 1930, is formally dissolved, and replaced by the *Vietnam* Workers (later Communist) Party, the *Cambodia* People's Revolutionary Party, and, in *Laos*, the Lao Itsala (see March 1955).
	November	In *Laos*, Prince Souvanna Phouma, a half-brother of two former Prime Ministers (Phetsarath and Souvannarath), and of Souphanouvong, leader of the Pathet Lao insurrection, becomes Prime Minister, forming the first of the several governments that he will lead up to 1975.
1952	**June**	In *Cambodia*, King Sihanouk, claiming that it is impossible for him to work with the Democratic Party majority in the National Assembly, suspends the Constitution and appoints himself Prime Minister.
	October	In *Indonesia*, following a display of strength by soldiers in armored vehicles around the presidential palace, the government abandons plans to demobilize sections of the armed forces. The position of the military is further strengthened by their success in crushing the rebellion in the South Moluccas and capturing the rebel leadership, which had declared itself the provisional government of the area.
	November	The military regime in *Thailand* launches a sustained campaign against the Chinese minority – detaining prominent Chinese, and closing many Chinese schools and associations – and justifies these actions on the grounds of opposition to the Communist regime in China.
1953	**January**	Sihanouk, King and Prime Minister of *Cambodia*, completes the imposition of direct rule begun in June 1952 by introducing martial law and appointing his father, Suramarit, as Regent.
	April	The Pathet Lao movement, formed by exiles in northern *Vietnam* in August 1950, invades *Laos* with support from the Viet Minh.
	September	A nationalist rising against the government of *Indonesia* begins in the province of Aceh.
	October	*Laos*, under King Sisavang Vong, becomes an independent state, but defense matters remain in the hands of the French. The Pathet Lao movement, allied to the Viet Minh, continues its insurrection.

Chronology

	November	*Cambodia*, under its King and Prime Minister Sihanouk, becomes an independent state, 86 years after it first became a French protectorate. Sihanouk continues to rule most of the country by decree, while the insurgent Khmer Issarak movement, now dominated by the People's Revolutionary Party (Communists), takes control of large areas.
1954	**May**	The Viet Minh's defeat of French forces after two months of fighting at Dienbienphu brings the First *Vietnam* War to an end. An international conference on Korea and Indochina opens the next day in Geneva.

In the *Philippines*, the surrender of Luis Taruc, the leader of the People's Liberation Army (Hukbong Mapagpalaya ng Bayan), brings the Huk rebellion to an end, after eight years of conflict. However, isolated Huk groups launch sporadic rebellions against the government up to the 1970s.

June — Bao Dai, head of the government of *Vietnam* recognized by the West but opposed by the Viet Minh, appoints Ngo Dinh Diem as Prime Minister, and then departs for France (where he remains until his death, aged 84, in 1997).

July — The international conference in Geneva ends with the signing of Accords on Indochina by France, the United Kingdom, China, the Soviet Union, and (reluctantly, under Soviet and Chinese pressure) the Viet Minh government – but not by the United States or Ngo Dinh Diem's government in South *Vietnam*. Laos and *Cambodia* are each to be united, independent and neutral, although the Pathet Lao movement retains autonomy in part of northeastern Laos (see May 1961). In Vietnam, the 17th parallel (see also August 1945) is to be the demarcation line between Viet Minh territory and the territory ruled by Ngo Dinh Diem's government until nationwide elections have been held.

In this and ensuing months, the US Navy and the French army fulfill one of the conditions of the Geneva Accords by assisting around 900,000 people, two thirds of them Catholics, to move from North *Vietnam* to the South.

Son Ngoc Minh and most other leaders of the Communist/Khmer Issarak insurgency in *Cambodia* go into exile in North Vietnam. Others remain in Cambodia to take part in elective politics, as the Pracheachon Party. Still others, opposed to collaboration with North Vietnam, form the nucleus of what will become the "Khmer Rouge" (see September 1960).

September — The United States, the United Kingdom, France, Australia, New Zealand, and Pakistan join the *Philippines* and *Thailand* in signing the Manila Pact, creating the Southeast Asia Treaty Organization (SEATO), which proves ineffective and is dissolved in June 1977.

1955 January — The US government begins to provide aid and military training to Ngo Dinh Diem's government in *Vietnam*, while the Viet Minh government begins preparing for guerilla warfare. The Second Indochina War, also known as the (Second) Vietnam War, has begun (though some date it as from 1959), between entities that come to be widely known as "South Vietnam" and "North Vietnam," although each of their governments claims to represent the whole of the country.

March — King Sihanouk of *Cambodia* abdicates the throne in favor of his father, Suramarit, and, as Prince Sihanouk, establishes a new political party, the Sangkum Reastr Niyum (Popular Socialist Community).

		Prince Souphanouvong and other leading figures in the Pathet Lao insurrection against the government of *Laos* establish the Phak Pasason Lao (Lao People's Party), replacing the Lao Itsala (see February 1951) but still allied to the Communist Party of Vietnam (see also January 1956).
	April	The Bandung Conference in *Indonesia*, attended by delegations from many of the new nation-states of Africa and Asia, launches the "nonaligned" movement (formally established in 1961), and, among other resolutions, declares support for Indonesia's claim to the Dutch colony of West New Guinea (West Papua).
		In South *Vietnam*, President Diem orders the suppression of the Binh Xuyen, an organized crime syndicate that dominates much of the economy in and around Saigon. His government also eventually suppresses the private armies of the Coa Dai and Hoa Hao sects.
	July	The US government endorses the decision of the government of South *Vietnam* that it will not participate in nationwide elections – and such elections are never held.
	August	In the *Philippines*, Congress approves President Ramon Magsaysay's plan for land reform, based on redistributing large estates to tenant farmers.
	September	The first nationwide elections ever held in *Indonesia* result in a legislature dominated by Moslem parties (the Nahdatul Ulama and the Masyumi) and nationalists (the PNI and others); the Communist Party (the PKI) comes fourth. The Moslem and nationalist parties form a government. Sukarno, who has not submitted himself to direct election and never will, remains President; he derides the new coalition government as a "three-legged horse," and presses for the inclusion of the PKI.
		In *Cambodia*, Prince Sihanouk's party, the Sangkum Reastr Niyum (Popular Socialist Community), wins all the seats in legislative elections, a feat repeated in all three subsequent elections (1958, 1962, and 1966) up to Sihanouk's overthrow in March 1970. He serves as Prime Minister until June 1960.
	October	In South *Vietnam*, following victory in a referendum in which he apparently won 98% of the votes cast, Ngo Dinh Diem declares Emperor Bao Dai deposed and becomes President of the Republic of Vietnam.
1956	**January**	The Pathet Lao movement, at war with the royal government of *Laos*, establishes the Lao Patriotic Front, a political organization dominated by the Lao People's Party.
	November	In North *Vietnam*, farmers in Nghe An province and elsewhere rise up in protest against the land reform imposed from 1955. The uprising is suppressed, but Truong Chinh, General Secretary of the ruling Workers (Communist) Party, is blamed for the "errors" in the land reform and dismissed. Ho Chi Minh temporarily takes over the secretaryship, and orders the release of around 12,000 detained farmers, as well as the posthumous rehabilitation of around 50,000 people killed during the land reform.
1957	**February**	In *Indonesia*, following coup attempts by sections of the armed forces in October and November 1956, President Sukarno denounces the divisiveness of party politics and calls for "Guided Democracy."
	March	Faced with continuing unrest, particularly in the outer islands, President Sukarno of *Indonesia* imposes martial law. The rebellion that began in Aceh in September 1953 ends in a ceasefire agreed by the rebels and the Indonesian armed forces.

	June	In the *Philippines*, Congress bans the Communist Party and imposes the death penalty as the maximum punishment for membership.
	August	The Federation of *Malaya*, comprising 11 former British dependencies, becomes an independent state within the Commonwealth, under a constitution that grants equal citizenship rights to all ethnic groups but also encodes special rights for Malays and other groups seen as indigenous (see July 1963).
	September	Marshal Sarit Thanarat takes power in a bloodless coup in *Thailand* and imposes martial law.
	October	North *Vietnam* secretly begins organizing armed units for action in the South.
	November	Rival political factions in *Laos* form a coalition government, led by Prince Souvanna Phouma, in an attempt to settle the longrunning conflict between the state and the leftwing Pathet Lao movement (but see August 1958).
	December	President Sukarno of *Indonesia* nationalizes most Dutch-owned estates and enterprises, and expels around 46,000 Dutch citizens. Many of the nationalized enterprises pass under the control of the armed forces.
1958	**January**	Nasution, head of the armed forces of *Indonesia*, makes a speech introducing the influential concept of *dwifungsi* ("double function"): the armed forces are to act both as guardians of national unity and security, and as leading players in economic and social development (see May 1982).
	February	Yet another rebellion breaks out in *Indonesia*, led this time by a "revolutionary government" of dissident soldiers and Islamic activists in Sumatra. The rising is suppressed after a few months, but provides the rationale for attempts by President Sukarno, the armed forces, and the Communist Party (the PKI) to secure greater power for themselves, through an unstable coalition against both the elected legislature and movements for regional autonomy or independence.
	May	Following the crash landing of a US Air Force B-26 in the South Moluccas, President Sukarno of *Indonesia* accuses the United States of supplying arms to the rebels in that region and steps up his campaign against western intervention in Southeast Asia.
	August	The coalition government formed in *Laos* in November 1957 is replaced by a rightwing government opposed to compromise with the leftwing Pathet Lao movement. The Pathet Lao relaunches its insurrection.
1959	**May**	North *Vietnam* secretly launches an armed campaign for the reunification of the country, a change of policy that arguably contravenes the Geneva Accords of 1954 and is kept secret until after reunification is achieved in 1975 (see also December 1960).
	June	The United Kingdom grants internal self-government to its colony of *Singapore*, where the People's Action Party, founded in November 1954 and led by Lee Kuan Yew, has formed its first government. It has remained in power ever since.
	July	President Sukarno of *Indonesia* abandons the "provisional" Constitution of 1950, reimposes the independence Constitution of 1945, and begins replacing elected officials with his own appointees.

As the government of *Laos* accepts increasing amounts of US aid and increasing numbers of US military advisers, the government of North *Vietnam* assists the leftwing Pathet Lao movement in consolidating its control of provinces in the Northeast of the country.

August The armed forces of *Indonesia* begin moving ethnic Chinese from the countryside into the cities, while around 100,000 members of this minority leave the country.

In *Laos*, yet another phase in the civil war begins, and members of the legislature representing the leftwing Lao Patriotic Front, linked to the Pathet Lao insurrection, are arrested and imprisoned.

September *Brunei* adopts a Constitution for the first time, curtailing the powers of the British High Commissioner (who doubles as Governor of Sarawak).

November President Sukarno of *Indonesia* launches another campaign against the Chinese minority, blaming its most prominent members for the economic crisis and ordering the deportation of around 40,000 Indonesian Chinese to China.

1960 February In *Burma*, the faction of the Antifascist People's Freedom League headed by former Prime Minister U Nu wins a decisive majority in legislative elections.

March President Sukarno of *Indonesia* dissolves the legislature after it has rejected his budget proposals, and appoints a new legislature, excluding the Islamic party Masyumi but including nationalists, Communists, members of the moderate Moslem party the Nahdatul Ulama (NU), state officials, and army officers.

June Following the death of King Suramarit of *Cambodia*, in April, his son Prince Sihanouk, who has been Prime Minister since September 1955, is elected Head of State by Parliament.

July The government of *Malaya* declares that the Emergency, which began in June 1948, has ended, and relaxes the special legal provisions introduced to combat the Communist insurrection (see also December 1989).

August *Indonesia* breaks off diplomatic relations with the Netherlands, which has rejected its claim to take over the Dutch colony of West New Guinea (West Papua).

In *Laos*, a military coup led by Captain Kong Le leads to the reinstatement of Prince Souvanna Phouma as Prime Minister (but see December 1960).

September Twenty-one young radicals meet secretly in *Cambodia* to establish the Workers Party, a Communist organization aligned with China and opposed to cooperation with North *Vietnam* (see February 1963).

December The guerilla units sent by the government of North *Vietnam* to attack the forces of South Vietnam are organized into the National Liberation Front for South Vietnam, which Ngo Dinh Diem, President in the South, nicknames "Viet Cong" (a shortened form of "Viet Nam Cong San," "Vietnamese Communist").

The government of *Laos*, led by Prince Souvanna Phouma and defended by Captain Kong Le's forces, flees the capital, Vientiane, in Soviet aircraft as troops loyal to General Phoumi Nosavan and other rightwing politicians seize control, in what has become known as the Battle of Vientiane.

1961	**January**	The civil war in *Laos* intensifies as Communist countries recognize and aid the coalition government led by Prince Souvanna Phouma, while western countries recognize and aid a rival rightwing government led by Prince Boun Oum.
	May	The royal government of *Laos*, under Savang Vatthana (King since 1959) and Prime Minister Souvanna Phouma, begins negotiations with the insurgent Pathet Lao movement on the future of the country, under the auspices of another international conference in Geneva. Agreement is reached in July on neutrality, the withdrawal of all foreign troops, and the formation of a second broadly based coalition government.
		In the course of a speech given in *Singapore*, Tunku Abdul Rahman, Prime Minister of the Federation of *Malaya*, launches his proposal for a new country, *Malaysia*, comprising Malaya, Singapore, Sabah, Sarawak, and (it was then hoped) *Brunei* (see January and September 1963).
	December	Community leaders and politicians in West New Guinea (West Papua), a Dutch colony claimed by *Indonesia*, issue a unilateral declaration of independence. The declaration is disregarded by the Netherlands, Indonesia, the United States, and the UN, which are preparing for negotiations on the future of the territory, but it has been reissued and commemorated on several occasions.
1962	**March**	Legislative elections take place in *Brunei* for the first and last time to date. Sheikh Ahmad Azahari and his Parti Rakyat Brunei (Brunei People's Party) win a majority of seats on a program of democratization (but see December 1962 and January 1963).
		The government of *Burma*, headed by U Nu, is overthrown in a military coup led by his former ally (and also former deputy to Aung San), General Ne Win. The Constitution is suspended, and U Nu is detained (until 1966).
		The Pathet Lao movement extends its control from parts of northeastern *Laos* into parts of the Northwest of the country.
	May	Citing the threat of incursions by the Pathet Lao movement from its bases in *Laos*, the military regime in *Thailand* invites the US government to station troops on Thai territory (see July 1976).
	July	In *Laos*, following years of fighting among the major political factions, a "National Union" government is formed by pro-western politicians loyal to King Savang Vatthana, the leftwing Pathet Lao movement, and a group of "neutrals" led by Prince Souvanna Phouma. This settlement endures for only two years (see February 1973).
		The new military regime in *Burma* responds to student protests by ordering the shooting of hundreds of protesters and having the Students Union building at Rangoon University blown up.
	October	The UN dispatches a Security Force (the UNSF) and a "temporary executive authority" (UNTEA) to West New Guinea (West Papua), a Dutch colony claimed by *Indonesia*, which has already stationed forces in the territory under the command of Major General (later President) Suharto. The UN personnel are due to remain in the territory for six years, but withdraw in April 1963.
	December	The government of *Brunei*, aided by British and Gurkha troops sent from *Singapore*, suppresses a rebellion led by Sheikh Ahmad Azahari and his Parti Rakyat Brunei (Brunei People's Party), and aimed at uniting the country with the British colonies

of Sabah and Sarawak in a new state, Kalimantan Utara (North Borneo). It appears that the rebels received secret support from the government of *Indonesia*. The Brunei government declares a state of emergency, suspends several provisions of its 1959 Constitution, and bans the Brunei People's Party (PRB).

1963 January   It is announced that the Federation of *Malaya* is to be united with the British dependencies of Sabah, Sarawak, and *Singapore* as the Federation of *Malaysia* (initially proposed in May 1961; and see September 1963). *Brunei* declines an invitation to join the new federation, while President Sukarno of *Indonesia*, pursuing his country's claims to Sabah and Sarawak, denounces the plan, and launches a campaign of *Konfrontasi* ("Confrontation") against it, which extends to nationalization of many British-owned enterprises. Indonesian guerilla units enter Sabah and Sarawak, and clash with Malaysian, British, Australian and New Zealand troops there up to August 1966.

February   In *Cambodia*, Saloth Sar, later to be known as Pol Pot, who probably arranged the mysterious disappearance of Tou Samouth in July 1962, takes Tou's place as General Secretary of the Workers Party, which is renamed the Communist Party in September 1966, and is nicknamed the "Khmer Rouge" by Prince Sihanouk.

May   Following the departure of the UN Security Force and officials in April, *Indonesia* assumes control of West New Guinea, which amounts to 20% of Indonesia's total area, and which is officially renamed Irian Barat and then, in 1973, Irian Jaya (but see December 1999). The new administration replaces the elected territorial council with an appointed assembly and bans all local political parties (see also July 1965, July 1969, and May 1977).

June   In South *Vietnam*, Thich Quang Duc becomes the first of five Buddhist monks to burn himself to death in protest against the regime's alleged persecution of Buddhism and favoring of Catholicism.

July   The governments of *Indonesia* and the *Philippines* begin negotiations with the provisional government of *Malaysia* on a proposal by President Diosdado Macapagal of the Philippines to form a confederation, to be known as "Maphilindo." The negotiations end in failure in March 1964.

September   The Federation of *Malaya* joins with Sabah, Sarawak, and *Singapore* to form *Malaysia* (but see August 1965), with the encouragement and support of the United Kingdom, which still has forces stationed in the new federation. The government of *Indonesia* responds by arranging for a crowd to gather and burn down the British Embassy in Jakarta.

November   Ngo Dinh Diem, self-appointed President of (South) *Vietnam*, and his brother Ngo Dinh Nhu, head of the security police, are overthrown and killed by a military faction led by General Duong Van Minh, with the support of Henry Cabot Lodge, US Ambassador to Diem's regime. The military then form a provisional government.

Cambodia decides to reject all aid from the US government in protest against its policies on Indochina.

1964 January   General Nguyen Khanh seizes power in South *Vietnam*.

March   The *Burma* Socialist Program Party, created by the military regime after it seized power in March 1962, is declared the only legal political organization.

	May	Following significant gains by the Pathet Lao movement, aided by North *Vietnam*, in its civil war with the royal government of *Laos*, the United States launches secret bombing raids over Pathet Lao-controlled regions.
	August	Following reports of two attacks on a US destroyer in the Gulf of Tonkin by patrol boats from North *Vietnam*, in circumstances that are still disputed, the US Senate passes a resolution giving President Lyndon B. Johnson the power to retaliate without declaring war. It is later revealed that the resolution had been drafted months before the attacks took place – if they ever did. The resolution is repealed in January 1971, but the war continues.
1965	**January**	In a protest over the admission of *Malaysia* to the UN, *Indonesia* steps up its incursions into Malaysian territory and becomes the first country ever to leave the UN (though it rejoins within a year).
	February	The US Air Force begins the first of its massive bombing campaigns over North *Vietnam*, known from March onward as "Operation Rolling Thunder." The People's Republic of China, accusing North Vietnam of siding with the Soviet Union in the Sino-Soviet dispute, rejects requests to intervene.
		Faced with a US naval blockade of its coasts, the government of North *Vietnam* steps up its use of the "Ho Chi Minh Trail" (the name, apparently invented by US journalists, for what was in fact a changing number of routes) in order to supply arms and aid through *Cambodia* and *Laos* to the National Liberation Front (the "Viet Cong") in the South.
	May	The government of *Cambodia*, which has refused US aid since November 1963, breaks off diplomatic relations with the United States as a protest against US Air Force bombing raids on Cambodian villages near the border with South *Vietnam*. Relations are resumed in June 1969.
	June	Nguyen Cao Ky forms yet another military regime in South *Vietnam*.
	July	Inhabitants of West Papua opposed to control of the territory by *Indonesia* form the Organisasi Papua Merdeka (Free Papua Organization, or OPM) and launch the first of several campaigns of violent protest.
	August	Following disputes over the pace of economic development, the distribution of government spending, and the allocation of parliamentary seats, *Singapore* leaves *Malaysia* to become an independent state under Queen Elizabeth II of the United Kingdom (but see December 1965). Singapore continues to be allied to Malaysia under the terms of the Anglo-Malayan Defense Agreement, and British forces remain in the country (but see April 1971).
	October	In *Indonesia*, the corpses of seven generals are found in a well, apparently the victims of mutinous officers probably linked to elements in the pro-Chinese Communist Party (the PKI). President Sukarno denounces what appears to be a coup attempt and gives Major-General Suharto emergency powers to suppress it. The army then launches a campaign of murder and intimidation against the PKI, using intelligence from both the US CIA and the Soviet KGB to trace its members, and giving arms – some supplied secretly by the CIA, as "medicines" – to local Moslem and nationalist groups. Over the course of the following year, at least 250,000 people, not all of them Communists, are massacred throughout Indonesia. Thousands more are detained without trial.
	December	Four months after attaining independence, *Singapore* becomes a republic.

1966	**March**	President Sukarno of *Indonesia* transfers emergency executive powers to General Suharto, as head of a government dominated by the military's political organization, Golkar (formed two years earlier). It bans the Communist Party (the PKI) and removes voting rights from around 1.1 million individuals (restored in April 1996).
	August	*Malaysia* and *Indonesia* sign a peace agreement, ending the period of Konfrontasi ("confrontation") that began in January 1963.
	October	Under pressure from General Suharto, President Sukarno of *Indonesia* introduces a program of liberal economic reforms.
		At a conference in Manila in the *Philippines*, six countries – the Philippines itself, *Thailand*, South Korea, the United States, Australia, and New Zealand – agree to continue their armed intervention in *Vietnam*.
	December	The government of *Indonesia* reasserts its policy of nationalism and centralization by executing Soumokil, the leader of the independence movement in the South Moluccas, and ordering the closing of all Chinese-language schools.
1967	**March**	The appointed legislature of *Indonesia*, dominated by representatives linked to the armed forces, appoints General Suharto as Acting President.
	April	A peasant uprising in the Samlaut region of *Cambodia*, swiftly suppressed by government forces, sees the first open activity by the dissident Communist group later to be known as the Khmer Rouge.
	May	The legislature of *Indonesia* enacts the Basic Forest Law, declaring forests to be "national assets," and thus permitting them to be appropriated and exploited regardless of the wishes of local populations.
	August	*Indonesia, Malaysia*, the *Philippines, Thailand*, and *Singapore* establish ASEAN at a summit meeting in Bangkok.
	September	Nguyen Van Thieu becomes the first (and last) elected President of South *Vietnam*.
	October	The Sultan of *Brunei*, Sir Omar Ali Saifuddin III, abdicates in favor of his oldest son, Hassanal Bolkiah. The state of emergency proclaimed in December 1962 is renewed.
1968	**January**	The government of North *Vietnam* and the National Liberation Front (the "Viet Cong") jointly launch a large-scale offensive throughout South Vietnam, known as the Tet offensive with reference to the Vietnamese word for the lunar New Year.
	March	The People's Consultative Assembly of *Indonesia* elects Acting President Suharto as President. He remains in the post, winning six subsequent elections in which no other candidate runs, until his resignation in May 1998.
	April	In *Singapore*, the People's Action Party, led by Prime Minister Lee Kuan Yew, wins all the seats in Parliament for the first time. It retains this monopoly in subsequent elections up to 1981.
	May	The US government suspends the bombing of North *Vietnam* as peace talks open; the suspension continues to December 1971, the talks to December 1972.
	December	The deployment of US troops in *Vietnam* reaches its peak, at 536,000 soldiers.

1969	**March**	US President Richard M. Nixon and his National Security Adviser, Henry Kissinger, launch the secret and illegal bombing of *Cambodia*, intending to disrupt the supply lines through that country to North *Vietnam*, but succeeding mainly in fatally disrupting Cambodian society.
	May	Kuala Lumpur, the capital of *Malaysia*, is disrupted by outbreaks of violence between Malays and Chinese following parliamentary elections. The government declares a state of emergency, which is maintained until 1971, and begins work on the New Economic Policy, which is primarily aimed at improving the economic position of the Malays and is implemented from 1971.
	June	In southern *Vietnam*, the "Viet Cong" announces the formation of a Provisional Revolutionary Government of South Vietnam, which continues in existence until shortly after unification in April 1975.
	August	The government of *Indonesia*, having promised the UN that it would organize a referendum in West New Guinea (West Papua) on whether that territory would remain under Indonesian control, instead conducts what is described as an "act of free choice": 1,025 leaders of indigenous groups, known beforehand to support Indonesian control, are asked to indicate their support. The territory is then declared to be a province of Indonesia.
1970	**March**	In *Cambodia*, following demonstrations organized by nationalist groups against the influence of North *Vietnam* over Prince Sihanouk's government, Sihanouk is overthrown (while on a visit to Moscow) in a military coup led by the Prime Minister, General Lon Nol, and supported by Sihanouk's cousin Prince Sirik Matak. Sihanouk, now visiting Beijing, forms a coalition with the pro-Chinese Communist faction that he had formerly sought to suppress and had nicknamed the "Khmer Rouge." This government is recognized by China, North Vietnam, and North Korea.
	April	Soldiers from the United States and South *Vietnam* enter *Cambodia* openly for the first time, in order to attack North Vietnamese bases established there under Prince Sihanouk. The new Cambodian leader, Lon Nol, denies all knowledge of this incursion.
	October	Lon Nol's military regime in *Cambodia* renames the country the Khmer Republic and continues the civil war between the regime's forces and those of the Khmer Rouge, which are receiving aid from China.
		Brunei Town, the capital of *Brunei*, is renamed Bandar Seria Begawan, honoring the former Sultan, Sir Omar Ali Saifuddin III, who took the title "Paduka Seri Bagawan Sultan" following his abdication in October 1967.
1971	**February**	South *Vietnam*, aided and supported by US President Richard M. Nixon, launches an invasion of *Laos*, which responds by seeking aid and troops from North Vietnam.
	April	Following the withdrawal of British forces from *Singapore*, the Anglo-Malayan Defense Agreement, covering Singapore and *Malaysia*, is replaced by a Five-Power Defense Arrangement, allying Singapore and Malaysia with the United Kingdom, Australia, and New Zealand.
	July	The first nationwide elections in *Indonesia* since September 1955 result in a legislative majority for the government-sponsored party Golkar, which it retains in subsequent elections (May 1977, May 1982, April 1987, June 1992, and May 1997).

	October	Lon Nol, the military ruler of *Cambodia*, imposes martial law, and denounces "the game of democracy and freedom."
	November	The United Kingdom grants full internal self-government to *Brunei*, the final stage before independence (see January 1984).
		Thanom Kittikachorn, head of the military regime in *Thailand*, annuls the Constitution, dissolves the National Assembly, and imposes martial law.
	December	The US Air Force resumes the bombing of North *Vietnam* suspended in May 1968.
1972	**April**	General Ne Win and other military officers in the Revolutionary Council that rules *Burma* retire from the army: henceforth, Ne Win is known as U Ne Win (with the addition of a civilian honorific).
	June	Lon Nol wins the only presidential election ever held in *Cambodia*; independent observers agree that the election process was extensively manipulated in his favor.
	September	Ferdinand Marcos, President of the *Philippines* since November 1965, declares martial law, claiming that emergency measures are needed to deal with armed uprisings by a Moslem separatist movement, the Moro National Liberation Front, and a Maoist organization, the New People's Army, active since 1968 and 1969 respectively.
1973	**January**	A ceasefire agreement negotiated by Henry Kissinger, US Secretary of State, and Le Duc Tho, Foreign Minister of North *Vietnam*, is signed in Paris (see October 1973). The US government ceases the bombing of North Vietnam and begins withdrawing forces from South Vietnam. Over the course of the conflict, 58,000 US personnel have been killed, while at least 2 million Vietnamese have been killed, and more than 100,000 tons of toxic chemicals have been deposited over the country.
		The government of *Indonesia* reorganizes the country's political parties to form three legally registered groups: the existing government party Golkar, which becomes the only party permitted to function outside the cities, and which government employees are required to support; the PPP, the result of a forced merger by four Islamic groups; and the PDI, which absorbs five nationalist and Christian groups.
	February	In a letter to Prime Minister Pham Van Dong of North *Vietnam*, US President Richard M. Nixon promises to provide US$3.25 billion for reconstruction. No such aid has ever been provided.
		After a further decade of fighting in *Laos*, both among the country's major political factions and in response to incursions by the forces of South Vietnam, North Vietnam, and the United States, another ceasefire agreement is made and a third broad coalition government is established under Prince Souvanna Phouma (but see December 1975).
	July	President Marcos of the *Philippines* secures victory in a referendum permitting him to remain in office beyond the expiry of his second term (in November) and to continue martial law. He goes on to secure victories in similar referendums in February 1976, October 1976, and December 1977.
	August	Following the revelation of the secret bombing of *Cambodia* by the United States, Congress compels President Richard M. Nixon to end the campaign, which has left large areas of Cambodia devastated.

	October	The military regime in *Thailand*, led by Thanom Kittikachorn, orders the army to use force against student protests and resigns after the orders are disobeyed. With the support of King Bhumibon, an interim government of officials, led by an academic, Dr Sanya Dharmasakti, holds office until after elections are held in January 1975.
		The Nobel Peace Prize is awarded jointly to Henry Kissinger of the United States and Le Duc Tho of North *Vietnam* (see January 1973). Le rejects the Prize, pointing out that peace has not yet been achieved in his country.
1974	**January**	Following a referendum held in December 1973, which was manipulated in the government's favor, *Burma* adopts a new Constitution for the Socialist Republic of the Union of Burma (suspended September 1988).
		Chinese forces occupy the Paracel Islands (Hoang Sa in Vietnamese, Hsisha in Chinese), previously controlled by South *Vietnam*.
		President Suharto of *Indonesia* responds to riots and other unrest over rising petroleum prices and economic problems by abandoning liberalization in favor of state-led development policies.
	May	China opens diplomatic relations with *Malaysia* (and then with *Thailand* and the *Philippines* during 1975), signaling the end of its support for Communist insurgencies in the ASEAN region as then constituted.
		Following the Portuguese revolution and the overthrow of Marcello Caetano's dictatorship in April, the new provisional government announces that it will grant independence to all the country's colonies, including *East Timor*, where two rival independence movements, the UDT and Fretilin, are engaged in hostilities with Apodeti, which seeks integration into *Indonesia*.
1975	**April**	During celebrations of the traditional New Year festival, President Lon Nol of *Cambodia* flees into exile in Hawaii as the Khmer Rouge conquer the capital, Phnom Penh, establish a new "government of Democratic Kampuchea," and begin to force city dwellers into the countryside.
		President Thieu of South *Vietnam* goes into exile as the combined forces of North Vietnam and the National Liberation Front (the "Viet Cong") capture Saigon, reunifying the country and ending the second Vietnam War after 20 years of devastating combat. Over the following months, the unification process creates disaffection and resentment in the South, both among non-Communists, of whom around 300,000 are detained for "reeducation," and among some Communists, particularly over the abolition of the Provisional Revolutionary Government of South Vietnam (established in June 1969), which had enjoyed a degree of autonomy.
	May	The government of newly reunified *Vietnam* merges Saigon, the largest city in the South, with its suburbs and nearby cities to form the new metropolis of Ho Chi Minh City.
		The armed forces of the Khmer Rouge regime in *Cambodia* clash with those of the Communist government of newly reunified *Vietnam* along their common border and around disputed offshore islands, disproving the tenacious myth that they are allies. The Khmer Rouge announces that it intends to assert Cambodia's claim to what it calls "Kampuchea Krom" ("downriver Cambodia," meaning the Mekong Delta and the region around Ho Chi Minh City).

	August	As Portuguese officials and troops depart from *East Timor*, the UDT, one of two rival independence movements, launches a coup against the other, Fretilin, and seizes control of the territory.
	September	Fretilin, a movement favoring independence for *East Timor*, defeats its rival, the UDT, and seizes control of the territory.
		Prince Sihanouk, still formally Head of State of *Cambodia*, returns to his country, which is now controlled by his supposed coalition partners, the Khmer Rouge.
		China reasserts its claim to control the Spratly Islands (Nansha in Chinese, Truong Sa in Vietnamese), which are mostly under the control of *Vietnam* but are also claimed, in whole or in part, by *Brunei*, *Malaysia*, the *Philippines*, *Indonesia*, and Taiwan.
	October	*Indonesia*, which has claimed *East Timor* since 1945, sends commando units into the territory.
	November	Fretilin, the leading pro-independence group in *East Timor*, unilaterally declares the independence of the Democratic Republic of East Timor; Portugal, the former colonial power, then recognizes it.
		The Communist-led Lao Patriotic Front wins a majority in legislative elections in *Laos*. King Savang Vatthana abdicates.
		Following repeated clashes between border troops, the government of *Laos* closes its border with *Thailand*.
	December	With the prior knowledge and acquiescence of the US and Australian governments, among others, the armed forces of *Indonesia* mount an invasion of *East Timor*. Between this date and 1999, around 200,000 East Timorese are killed by Indonesian forces and around 150,000 Indonesians are resettled in the territory.
		After months of controversy, following revelations in May that Pertamina, the State Oil and Natural Gas Corporation of *Indonesia*, has debts of more than US$10 billion, President Suharto dismisses its director, Ibnu Sutowo, and orders that its activities be limited to the hydrocarbons industry.
		In *Laos*, following many months of political dispute within the coalition government formed in February 1973, and the departure of many non-Communist politicians into exile, the Pathet Lao movement takes power in a bloodless coup and proclaims the Lao People's Democratic Republic. This new regime, allied to the government of *Vietnam*, is controlled by the Lao Patriotic Front (renamed the Lao Front for National Construction in February 1979), which in turn is dominated by the Lao People's Revolutionary Party (Communists).
1976	**January**	The government of *Vietnam* orders all ethnic Chinese resident in the South to register their citizenship. Those who declare themselves citizens of China are dismissed from their jobs and have their rations reduced. By the end of the year, most Chinese-language newspapers and schools have been closed, and around 70,000 people have emigrated to China.
	April	The Khmer Rouge regime in *Cambodia* compels Prince Sihanouk to resign as Head of State and places him under house arrest.
	July	*Indonesia*, defying international law, and ignoring protests from Portugal and other countries, declares *East Timor* to be its 27th province, under the name Timor Timur.

		The government of *Vietnam* announces the completion of reunification, the abolition of the provisional government bodies in the South, and the establishment of the Socialist Republic of Vietnam.

The US Military Assistance Command, stationed in *Thailand* since May 1962, is ordered to leave the country. Thailand and *Vietnam* reopen diplomatic relations.

October Amid mounting social unrest, the government of *Thailand*, headed by Prime Minister Seni Pramoj, fails to prevent the army from massacring protesting students at Thammasat University. The government is then overthrown in a military coup, and is replaced by the National Administrative Reform Council, composed of generals, and a cabinet headed by a judge, Thanin Kraivixien.

December Groups seeking independence from *Indonesia* for the province of Aceh, which has seen previous uprisings in 1942 and in 1953–57, announce the formation of the Free Aceh Movement (Gerakanan Aceh Merdeka, or GAM).

1977 May The Organisasi Papua Merdeka (Free Papua Organization, or OPM), launches a rebellion against *Indonesia* in pursuit of the unification of "Irian Jaya" (West Papua) with neighboring Papua New Guinea. The rebellion is suppressed by December 1979, but there are occasional outbreaks of anti-Indonesian violence in subsequent years (see also March 1991 and September 1998).

July *Vietnam* and *Laos* sign a treaty of friendship that, among other things, formalizes the stationing of Vietnamese troops in Laos after many years in which they have been present in the country, largely along the "Ho Chi Minh Trail" (see February 1965).

The first of several military clashes between *Thailand* and the Khmer Rouge regime in *Cambodia* occurs on their common border.

October In *Thailand*, another military coup overthrows the military regime installed 13 months earlier and establishes the relatively liberal regime of General Kriangsak Chamanand.

November In the *Philippines*, a military tribunal finds former Senator Benigno Aquino (detained since 1972) guilty of murder, subversion, and illegal possession of firearms, and sentences him to death. President Marcos orders a stay of execution on Aquino, a former leader of the opposition to Marcos's regime.

December *Vietnam*, accusing the Khmer Rouge regime in *Cambodia* of repeated violations of their common border, breaks off diplomatic relations.

1978 March The government of *Vietnam* nationalizes the rice trade in the South of the country, effectively depriving thousands of ethnic Chinese of their livelihoods. The emigration of some of these Chinese traders with their families forms the first wave of "boat people," migrants fleeing poverty and/or discrimination in Vietnam who mostly end up in refugee camps elsewhere in East Asia over the following decade.

May A section of the Khmer Rouge in eastern *Cambodia*, close to the border with Vietnam, leads a regional uprising against the Khmer Rouge regime, which responds by massacring around 100,000 people, including both Khmers and members of the Vietnamese minority.

In *Vietnam*, the introduction of a new single currency, the dong, has the effect of destroying the value of many forms of private saving, prompting thousands more people, especially from the South, to leave the country.

In the *Philippines*, former Senator Benigno Aquino (see November 1977) is released from prison and permitted to go to the United States for medical treatment.

June — *Vietnam* joins the Council for Mutual Economic Assistance (CMEA or Comecon), the economic organization of the Soviet bloc. China interprets this move in favor of the Soviet Union as an act of hostility to itself.

October — The government of *Indonesia* bans the use of Chinese characters in all public displays and printed matter.

November — *Vietnam*, isolated from international agencies by the United States and its allies, and spurned by China, signs a treaty of friendship with the Soviet Union.

December — *Vietnam*, citing continued incursions across its border by forces of the Khmer Rouge regime in *Cambodia*, invades its neighbor from bases both in Vietnam itself and in *Laos*.

1979 January — The Khmer Rouge regime in *Cambodia*, which has been responsible for the deaths of around 2 million people through massacres, starvation, and disease, is overthrown by forces invading from Vietnam. A provisional government of the People's Republic of Kampuchea, composed partly of former Khmer Rouge members, takes power with Vietnamese support. It is recognized by the Soviet Union and its allies, as well as by India, but the European Community, Japan, and the IMF suspend aid and loans to both Vietnam and Cambodia.

The governments of *Thailand* and *Malaysia* announce that they will no longer accept the entry of "boat people" (migrants from Vietnam).

The governments of *Thailand* and *Laos* agree to restore relations. Laos ceases to give aid and refuge to the Communist Party of Thailand, while Thailand ceases to give aid and refuge to Lao royalists.

February — China launches a small-scale and largely ineffectual invasion of *Vietnam*, in the first of numerous border clashes between the two countries.

June — At an ASEAN summit meeting in Bali, the governments of the five member states (see August 1967) condemn the invasion of *Cambodia* by *Vietnam*, agree to defend each other in case of attack, and commit their countries to significant increases in military spending.

December — Pol Pot, leader of the Khmer Rouge regime overthrown in *Cambodia* in January, is formally replaced by Khieu Samphan, while apparently retaining dominance, notably through control of troops and refugees in northwestern Cambodia, and across the border in *Thailand*.

1980 February — *Thailand* officially closes its border with *Cambodia*, but Khmer Rouge forces continue to operate across it with aid from elements of the Thai military, as well as from the United States and China.

September — The People's Republic of China abandons its longstanding policy of permitting dual citizenship as between itself and other countries, effectively compelling overseas Chinese, in Southeast Asia and elsewhere, to choose between Chinese citizenship and citizenship of their country of residence.

1981 January — President Marcos of the *Philippines* lifts most of the martial law provisions imposed in September 1972, but retains special powers to rule by decree.

	June	In an election characterized by extensive government manipulation, Ferdinand Marcos is re-elected President of the *Philippines*.
	July	At a special UN conference on *Cambodia* in New York, the five member states of ASEAN (see August 1967) call for the disarming of the Khmer Rouge, but are outvoted by China, the United States, and their allies.
	November	U Ne Win, paramount leader of *Burma* since March 1962, retires from the presidency and the chairmanship of the Council of State, but retains decisive power as Chairman of the ruling Burma Socialist Program Party (see July 1988).
1982	**May**	The government of *Indonesia* makes the policy of *dwifungsi* ("double function"), first announced in January 1958, legally binding: the armed forces are to continue acting both as guardians of national unity and security, and as leading players in economic and social development, with their own representatives in the legislature as well as a range of business enterprises.
	June	The Khmer Rouge, the Khmer People's National Liberation Front led by Son Sann, and Sihanouk's royalist movement, which are all opposed to the Vietnamese-backed government in *Cambodia*, form the Coalition Government of Democratic Kampuchea in exile, initially based in Kuala Lumpur, the capital of *Malaysia*. The coalition is recognized by the western powers, China, and the five member states of ASEAN (see August 1967), and is given Cambodia's seat at the UN.
	October	The *Philippines* formally abandons all claims to Sabah, a former British dependency in northern Borneo that is now a state in the Federation of *Malaysia*.
1983	**February**	President Suharto of *Indonesia* formally reopens Borobudur, the ninth-century temple complex that is the world's largest Buddhist monument, after the completion of a 10-year restoration program sponsored by Unesco.
	August	The Sultan of *Brunei* creates the Brunei Investment Agency to manage the country's enormous portfolio of shareholdings and other investments.
		Former Senator Benigno Aquino, returning to the *Philippines* from the United States, is shot dead at Manila Airport, and Rolando Galman, allegedly his killer, is also shot dead by soldiers (see September 1990).
1984	**January**	The Sultanate of *Brunei* achieves independence after 96 years as a British protectorate, becoming a sovereign state with membership of the UN, the Commonwealth, the Organization of the Islamic Conference, and ASEAN (as its sixth member state). The state of emergency declared in December 1962 is maintained; the Sultan appoints himself Prime Minister, Finance Minister, and Home Affairs Minister; and the new government, which also includes two of the Sultan's brothers as well as their father, declares an exclusive fishing zone, including part of the disputed Spratly Islands (see September 1975).
	December	All three legal political parties in *Indonesia* are required to conform to the state ideology of Pancasila (see June 1945 and January 1973).
1985	**January**	All social organizations in *Indonesia* are required to conform to Pancasila, and all labor unions are merged into the government-controlled Serikat Pekerja Seluruh Indonesia (All Indonesia Workers Union, or SPSI).
	May	A political party is registered in *Brunei* for the first time since 1962: the Parti Kebangsaan Demokratik Brunei (Brunei National Democratic Party) is largely

		composed of moderate Moslem businesspeople who proclaim their loyalty to the Sultan and exclude ethnic Chinese from membership.
	August	Australia recognizes the incorporation of *East Timor* into *Indonesia*, but the UN and most other countries continue to reject it as a breach of international law.
1986	**February**	Following the exposure of widespread fraud in the presidential election in the *Philippines*, President Ferdinand Marcos resigns and departs for exile in Hawaii; the rival candidate, Corazon Aquino, widow of Benigno Aquino, takes his place. She then orders the release of all political prisoners.
		The Sultan of *Brunei* approves the formation of a second political party, the Parti Perpaduan Kebangsaan Brunei (Brunei National Solidarity Party), which is closer to the government than its rival (see May 1985) and open to all citizens regardless of ethnicity.
	April	Organizations seeking independence for *East Timor* form the Conselho Nacional de Resistencia Maubere (Council of National Maubere Resistance CNRM), which becomes the leading body in the struggle against the occupation of the territory by *Indonesia*.
	November	In *Laos*, the ruling Lao People's Revolutionary Party launches a policy of economic liberalization known as *Jintanagan Mai* or, in English, the New Economic Mechanism.
	December	In *Vietnam*, the ruling Communist Party selects a new General Secretary, Nguyen Van Linh, and launches a policy of economic liberalization known as *doi moi* ("new structure").
1987	**February**	The *Philippines* promulgates a new Constitution, completing the restoration of liberal democracy.
	August	In pursuit of an agreement with the IMF, *Vietnam* begins to introduce economic reforms, reducing subsidies and other forms of government spending, changing the currency exchange and pricing systems, and imposing fees for health and education services.
	September	Riots and demonstrations break out in *Burma* as students and others protest against rice shortages; further riots follow in March and June 1988; all are brutally crushed.
	October	The government of *Malaysia* carries out "Operation Lalang," closing three newspapers, banning political assemblies, and ordering the detention without trial of 106 politicians, lawyers, journalists, and others alleged to have promoted discord between the Malay majority and the Chinese minority. All are released by April 1989.
	November	The government of *Malaysia* imposes further restrictions on the press and broadcasters, in a campaign against "false news" and in favor of "Malaysian values."
		Thailand and *Laos* start a brief war (ending with a ceasefire in February 1988) over a district straddling their common border that was left undelineated in a Franco-Thai treaty of 1907 and has since become a center for opium distribution.
1988	**February**	The Parti Kebangsaan Demokratik *Brunei* (see May 1985) calls for democratic elections and the withdrawal of the Sultan from government; it is immediately dissolved and its leaders, Abdul Latif Hamid and Abdul Latif Chuchu, are detained until March 1990.

		In *Malaysia*, at the end of a case brought by dissident members of UMNO, the leading party in the ruling National Front (Barisan Nasional), the High Court declares the legislative elections of April 1987 to be null and void since there were irregularities in the registration of participants in UMNO's internal selection procedures. The Prime Minister, Mahathir Mohamad, announces the formation of UMNO Baru (New UMNO) in order to regularize his position; the new party's assets are frozen until its legal status is resolved in September 1994.
	June	President Aquino of the *Philippines* promulgates a land reform program that is welcomed by many farmers' groups but criticized by others, who note that, like most other large estates, the President's own landholdings are not affected by the reform.
	July	Following repeated protests and riots throughout *Burma*, in September 1987, March 1988, and June 1988, the paramount leader Ne Win resigns from the chairmanship of the ruling Burma Socialist Program Party. His successor, Sein Lwin, imposes martial law and thousands of opponents of the regime are massacred.
	August	In *Burma*, Sein Lwin, the paramount leader of the regime, is faced with a general strike and demonstrations by up to 1 million protesters, and resigns after 17 days in office. He is replaced by Maung Maung, who rescinds martial law and allows the establishment of the All-Burma Students Union.
	September	In *Burma*, the ruling Burma Socialist Program Party decides to organize free elections and to ban state employees, including members of the armed forces, from joining political parties. In response, the armed forces, led by Saw Maung but apparently influenced by the former dictator Ne Win, seize power, suspend the 1974 Constitution, and establish the State Law and Order Restitution Council (SLORC). The former Prime Minister U Nu proclaims an "alternative government."
	October	*Indonesia* launches a program of extensive economic liberalization, initiating a period of rapid growth hailed by the World Bank and other observers as a major part of the "East Asian miracle" (but see October 1997 onward).
1989	**February**	In *Indonesia*, in this and subsequent months, there are numerous clashes between villagers and soldiers, as well as student demonstrations in several cities, arising from the government's perceived failure to provide adequate compensation to those who are compelled to relocate to other provinces and whose land is expropriated.
		Warships from China and the *Philippines* exchange fire in the waters around the disputed Spratly Islands (see September 1975).
	March	For the first time since the revolution in December 1975, the government of *Laos* conducts nationwide elections, within the framework of the one-party state.
	April	The Communist Party of *Burma*, which controls parts of the country inhabited mainly by the Wa people, is disrupted by internal feuding. Its leaders flee to China, while most of its armed followers surrender to the government and join a militia fighting against the Mong Tai Army (see January 1996). Meanwhile, the Chinese government agrees to withdraw its longstanding support for insurgents inside Burma and begins supplying arms to the military regime there.
		Cambodia reinstates Buddhism as the national religion.
	July	Aung San Suu Kyi (daughter of Aung San) and Tin Oo, the leaders of the National League for Democracy in *Burma*, are placed under house arrest by the military regime (see March and July 1995).

	September	*Vietnam* completes the withdrawal of almost all of its troops stationed in *Cambodia*.
		The military regime in *Burma* makes a ceasefire agreement with the Shan State Progressive Party, one of the armed groups representing ethnic minorities that have been fighting the regime and its predecessors.
	October	A visit to *East Timor* by Pope John Paul II arouses a favorable response from its largely Catholic population and helps to draw the attention of the outside world to the illegal occupation of the territory by *Indonesia*.
		The Khmer Rouge steps up its military campaign against the Vietnamese-backed government of *Cambodia*, seizing control of Pailin, a mining center on the border with Thailand.
	November	A majority of voters in the provinces of the *Philippines* where the Moro National Liberation Front has been active since 1968 approve a government plan to grant a special autonomous status to their majority-Moslem region (see September 1996).
	December	Tin Oo, a leading figure in the National League for Democracy in *Burma*, is sentenced to three years imprisonment by the military regime.
		The last remaining fighters of the Communist Party of Malaya, most of them resident in *Thailand*, call an end to their sporadic armed struggle against the governments of *Malaya* (until 1963) and then *Malaysia*, which began in June 1948 (see also July 1960).
		In defiance of UN policy, the government of Australia effectively recognizes the seizure of *East Timor* by *Indonesia* through its signing of a treaty with Indonesia on the exploitation of oil reserves in the Timor Gap.
		The government of the *Philippines* suppresses an attempted coup by a section of the military – the sixth since 1986 – with help from the US Air Force.
1990	**January**	The military regime in *Burma* places U Nu, a former Prime Minister and now head of an "alternative government," under house arrest, and bars Aung San Suu Kyi, leader of the National League for Democracy, from participating in the legislative elections due in May, on the grounds of her marriage to a British citizen, Michael Aris.
	May	In *Burma*, following the overwhelming victory of the National League for Democracy in elections for the People's Assembly (in which it wins 396 seats out of 485), the military regime refuses to allow the Assembly to meet, and forces the newly formed National Coalition Government, led by Aung San Suu Kyi's cousin Sein Win, to flee the capital.
	September	The Vietnamese-backed government in *Cambodia* and the opposition Coalition Government of Democratic Kampuchea (comprising the Khmer Rouge and two non-Communist groups) agree to form a Supreme National Council, which will take up Cambodia's UN seat, and to work toward ending the eight-year civil war between them (see June 1991). The Soviet Union agrees to stop arms supplies to the government in Phnom Penh and China agrees to stop supplying the Khmer Rouge.
		In the *Philippines*, 16 soldiers are found guilty of murdering Benigno Aquino and Rolando Galman in August 1983.

	November	Prime Minister Lee Kuan Yew of *Singapore*, the longest-serving head of government in Southeast Asia at this point, retires after 31 years in power.
1991	**January**	The Sultan of *Brunei* bans the importation of alcoholic beverages.
	February	The military take power in a bloodless coup in *Thailand*, establishing a National Peacekeeping Council under General Sunthorn Kongsompong.
		The United States establishes an office in Hanoi, the capital of *Vietnam*, to conduct inquiries into cases of US troops missing in action in that country, but continues to deny recognition to the Vietnamese government.
	March	The military regime in *Burma* makes a ceasefire agreement with the Pa-O National Organization, one of the armed groups representing ethnic minorities that have been fighting the regime and its predecessors.
		After being extradited from Papua New Guinea, Melkianus Salossa, leader of the OPM (see May 1977), is tried and sentenced to life imprisonment in *Indonesia*.
	May	The military regime in *Burma* makes a ceasefire agreement with the Palaung State Liberation Organization, one of the armed groups representing ethnic minorities that have been fighting the regime and its predecessors.
	June	A permanent ceasefire is declared between the rival factions in the civil war in *Cambodia*.
	August	*Laos* promulgates a new Constitution that permits private ownership, and liberalizes parts of the economy, but preserves the monopoly of power held by the Lao Front for National Construction, dominated by the Lao People's Revolutionary Party.
	October	A final peace settlement is agreed in Paris by the rival factions in *Cambodia*, under the auspices of the UN, which sends an Advance Mission (Unamic) to prepare for international administration of the country during and after elections.
		The Kampuchean People's Revolutionary Party, in control of most of *Cambodia* since 1979, renames itself the Cambodian People's Party, removes Heng Samrin from the chairmanship of the party, and introduces measures aimed at economic liberalization.
		ASEAN announces plans for an ASEAN Free Trade Area, to be implemented in full by 2006 (a target later brought forward to 2001).
		Aung San Suu Kyi, leader of the National League for Democracy in *Burma*, is awarded the Nobel Peace Prize.
	November	Troops from *Indonesia* attack the congregation at a funeral service in the Santa Cruz cemetery in Dili, the capital of *East Timor*, and kill around 270 Timorese. Subsequently, the Indonesian government imprisons many survivors of the incident, and disciplines 19 soldiers and policemen for their "excessive" response to a "riot."
		Prince Sihanouk returns to *Cambodia* and is made provisional Head of State. Khieu Samphan, who was President of the Khmer Rouge regime from 1975 to 1978, also returns, with three other Khmer Rouge officials; they rapidly depart for Thailand after being beaten up by relatives of some of the Khmer Rouge's victims.
		The government of the *Philippines* takes control of Clark air base, formerly operated by the US Air Force.

	December	The military regime that took power in *Thailand* in February replaces the 1978 Constitution with a new Constitution giving the army the right to appoint all the members of the Senate.

Thailand and *Laos* sign an agreement on border cooperation.

In the face of demonstrations in support of Aung San Suu Kyi, whose family are receiving the Nobel Peace Prize on her behalf, the military regime in *Burma* closes all universities and colleges, and dismisses thousands of teachers.

The Sultan of *Brunei* bans public celebrations of Christmas.

1992 March The UN Transitional Authority for *Cambodia* (Untac) assumes full control over the country, as the first such institution in history, alongside a peacekeeping force of 22,000. It begins implementing a refugee repatriation program. It also attempts to enforce the ceasefires among the country's rival parties, which are frequently violated, notably through Khmer Rouge massacres of members of the Vietnamese minority, and attacks by supporters of the People's Party (led by Hun Sen) on supporters of FUNCINPEC (led by Sihanouk).

Parties linked to the military regime in *Thailand* win a majority in legislative elections, which are followed by mass demonstrations against the regime (and see May 1992).

April The governments of *Burma* and Bangladesh reach agreement on a program to repatriate from Bangladesh more than 200,000 members of the Moslem Rohingya minority who have taken refuge there, in 1976–78, and again in 1991, in response to the Burmese government's policy of destroying their villages and resettling their homeland with members of the majority Burman people.

May Following mass demonstrations against the military regime in *Thailand*, King Bhumibon intervenes to compel the army and civilian politicians to agree on forming an interim government and restoring the 1978 Constitution.

The granting of a loan by the IMF to the government of *Laos* marks the beginning of the end of that country's isolation from contact with most countries other than *Vietnam*.

June *Vietnam* introduces a new Constitution incorporating economic reforms but retaining the Communist Party's monopoly of political power.

July ASEAN grants observer status to *Laos*.

August The UN General Assembly passes a resolution condemning violations of human rights in *East Timor* by *Indonesia*.

September Legislative elections in *Thailand* are won by a coalition of civilian parties, which form a government, led by Chuan Likphai, which is committed to reducing the role of the military in public life.

The government of the *Philippines* takes control of the facilities at Subic Bay formerly used by the US Navy.

November José Xanana Gusmão, leader of the pro-independence forces fighting against *Indonesia*'s illegal occupation of *East Timor*, is captured by Indonesian troops and jailed in Indonesia.

	December	Control of the naval base at Subic Bay in the *Philippines*, the last US base remaining in Southeast Asia, passes to the government.
1993	**January**	A constitutional convention appointed and supervised by the military regime in *Burma* begins work on drafting a new constitution (still unfinished to date).
	May	Legislative elections supervised by the UN Transitional Authority for *Cambodia* (Untac), but boycotted by the Khmer Rouge, are followed by the formation of a coalition government bringing together the royalist party FUNCINPEC and the Cambodian People's Party, while Prince Sihanouk is declared by the new Constituent Assembly to have been *de jure* Head of State ever since 1960.
	July	The IMF, the World Bank, and the Asian Development Bank provide US$721 million in loans and credits to *Vietnam*, ending its isolation from the international financial system.
	September	*Cambodia* promulgates a new Constitution, largely based on the Constitution of 1953, with reduced powers for the (elected) King, Sihanouk, and enhanced powers for the National Assembly. The UN Transitional Authority for Cambodia (Untac) winds up its activities. The Khmer Rouge announces that it does not accept the authority of the new government.
	October	The military regime in *Burma* announces that it has made a ceasefire agreement with the Kachin Independence Organization, one of the armed groups representing ethnic minorities that have been fighting the regime and its predecessors.
		The government of *Cambodia* announces a program of structural reform and economic liberalization; the IMF and other international agencies resume lending to the government.
	December	The government of the *Philippines* signs a ceasefire agreement with the Communist-led New People's Army.
		Against the wishes of the government of *Indonesia*, the PDI, one of the three legal parties in the country, chooses Megawati Sukarnoputri, daughter of former President Sukarno, as its leader (see June 1996).
1994	**February**	The US government lifts the ban on trade with, and investment in, *Vietnam* imposed in 1975.
	April	*Laos* and *Thailand* celebrate the opening of the Friendship Bridge, which has been completed with Australian aid and is the first bridge over the section of the Mekong River dividing the two countries.
	May	The military regime in *Burma* makes a ceasefire agreement with the Karen National People's Liberation Front, one of the armed groups representing ethnic minorities that have been fighting the regime and its predecessors.
	June	The Abu Sayyaf movement, a Moslem rebel group opposed both to the government of the *Philippines* and to the Moro National Liberation Front, carries out the first of its many kidnappings in Mindanao. Similar incidents continue into the year 2000.
		The government of *Cambodia* outlaws the Khmer Rouge but offers an amnesty to any of its fighters who surrender (see January 1995).

	July	The military regime in *Burma* makes a ceasefire agreement with the Kayan New Land Party, one of the armed groups representing ethnic minorities that have been fighting the regime and its predecessors.
	October	The military regime in *Burma* makes a ceasefire agreement with the Shan State Nationalities Liberation Organization, one of the armed groups representing ethnic minorities that have been fighting the regime and its predecessors.
1995	**January**	As the amnesty for Khmer Rouge fighters expires in *Cambodia*, the government announces that around 7,000 have surrendered while around the same number remain engaged in armed opposition.
	February	The sole remaining legal party in *Brunei*, the Parti Perpaduan Kebangsaan Brunei (see February 1986), holds its first convention and elects Abdul Latif Chuchu (see February 1988) as its President. The goverment compels Chuchu to resign the post.
	March	The military regime in *Burma* releases Tin Oo and other leading members of the National League for Democracy from house arrest.
	May	The US government lifts the ban on trade with, and investment in, *Laos*, imposed in 1975.
	June	The National Assembly of *Cambodia* expels one of its members, Sam Rainsy, a former Finance Minister who has made allegations of extensive corruption among government ministers and officials, even though expulsion is not mentioned in the Constitution. The Assembly also passes a law imposing penalties for press reports that infringe the country's national security or political stability.
	July	The military regime in *Burma* signs a ceasefire agreement with representatives of the New Mon State Party, and formally releases Aung San Suu Kyi, leader of the National League for Democracy, from house arrest.
		ASEAN admits *Vietnam* as its seventh member state, and grants observer status to *Cambodia*.
	August	The government of *Indonesia* marks the 50th anniversary of independence by releasing three prisoners detained since 1966, ordering the removal of the official code for "former political prisoner" from identity papers, which had exposed around 1.3 million individuals to various restrictions, and relaxing the law on permits for public gatherings.
		The governments of China, the *Philippines*, and *Vietnam* agree to negotiate a code of naval conduct aimed at reducing the possibility of confrontation over the disputed Spratly Islands (see September 1975).
	November	The UN investigator into human rights in *Burma* confirms numerous media reports that the military regime in that country is using forced labor on a large scale, notably for infrastructure, conservation and tourism projects.
1996	**January**	The military regime in *Burma* accepts the surrender of the Mong Tai Army, a force composed mainly of fighters from the Shan minority and led by Khun Sa, who is placed under house arrest in Rangoon. The regime announces that the surrender constitutes a major victory in its campaign to eliminate the production and trading of opium.

	May	The military regime in *Burma* orders the arrest of more than 260 members of the National League for Democracy and ensures that only 18 people are able to attend its first congress. It also launches an offensive against separatist forces in the Shan states, causing thousands of refugees to flee to Thailand.
	June	In *Indonesia*, the government arranges for the PDI, one of the three legal parties, to remove Megawati Sukarnoputri from its leadership; she and her supporters form a rival PDI. Outbreaks of violence between riot police and protesters follow the announcement of Sukarnoputri's removal, notably in the capital, Jakarta.
		The governments of *Indonesia* and *Malaysia* agree on new measures to reduce the numbers of Indonesians migrating illegally into Malaysia in search of work.
	July	ASEAN grants observer status to the military regime in *Burma*.
	August	The government of *Cambodia* accepts the surrender of Ieng Sary, a leading member of the Khmer Rouge, and two divisions of troops under his command. All are given amnesties by King Sihanouk, despite evidence that they are still in control of the districts of Pailin and Malai, and are resisting the army's efforts to remove them.
	September	The government of the *Philippines* and the Moro National Liberation Front bring 24 years of armed conflict to an end by agreeing on a special autonomous government, with limited revenue-raising powers, for a region comprising the provinces of Lanao del Sur, Maguindanao, Sulu, and Tawi Tawi.
	October	The Nobel Peace Prize is awarded to two citizens of *East Timor*, José Ramos Horta and Bishop Carlos Belo, who have both been prominent in opposing the illegal occupation of their homeland by *Indonesia*.
	December	Hundreds of settlers transferred by the government of *Indonesia* from Madura to West Kalimantan are massacred by indigenous Dayaks; fighting between the two groups continues until March 1997.
1997	**February**	The Sultan of *Brunei* dismisses his brother Prince Jefri from his post as Minister of Finance, but leaves him in charge of the Brunei Investment Agency (see July 1998).
	March	In Phnom Penh, the capital of *Cambodia*, demonstrators supporting the dissident politician Sam Rainsy (see June 1995) are attacked by unknown assailants who throw hand grenades, killing 19 of them. It is alleged that the assailants were supporters or agents of the People's Party, led by Co-Prime Minister Hun Sen.
		The government of *Thailand* imposes emergency measures aimed at restoring confidence in the country's financial institutions, as Finance One and other leading institutions incur heavy losses.
	May	It is officially declared that Golkar, the political party created and favored by the government and the military in *Indonesia*, has won nearly 75% of the votes cast in elections for the House of Representatives, which have been marked by numerous outbreaks of violence, leaving at least 275 people dead.
		In *Cambodia*, Pol Pot, still the leader of the Khmer Rouge, orders the executions of his rival Son Sen and his entire family. Another rival, Ta Mok, arranges for the pursuit and capture of Pol Pot.

	July	A financial crisis in *Thailand* results in the sharp devaluation of the currency, the baht, and the abandonment of fixed exchange rates. Under an agreement with the IMF, the government imposes higher interest rates, energy prices, and consumption taxes, as well as cuts in public spending. The crisis spreads rapidly to other countries in the region.

Burma and *Laos* become the eighth and ninth member states of ASEAN.

Violence breaks out in Phnom Penh, the capital of *Cambodia*, between supporters of the rival Co-Prime Ministers, Hun Sen, leader of the People's Party, and Prince Ranariddh, son of King Sihanouk and leader of FUNCINPEC. Hun Sen arranges for Ranariddh, who has left the country, to be replaced in his post by another FUNCINPEC politician, Ung Huot (see March 1998).

Also in *Cambodia*, Ta Mok's faction of the Khmer Rouge stages the trial of Pol Pot, their former leader, who takes the opportunity to tell the US journalist Nathaniel Thayer that his regime committed no atrocities.

August As the financial crisis spreads from *Thailand* to the rest of the region, Thailand and the IMF agree on a rescue plan, and speculators bring down the value of the rupiah, the currency of *Indonesia*, and of the ringgit, the currency of *Malaysia*. The government of Indonesia abandons fixed exchange rates.

September Seven hundred people gather at the home of Aung San Suu Kyi, leader of the National League for Democracy in *Burma*, to attend the League's second congress.

In reaction to the financial crisis spreading across the region, the government of *Malaysia* introduces cuts in public spending and other economic reforms.

In this month and in October, smog caused by land clearance fires in Sumatra and the three Kalimantan provinces in *Indonesia* spreads into *Malaysia*, the *Philippines*, *Singapore*, and *Thailand*. The government of Malaysia declares a state of emergency in Sarawak, where the resulting pollution has reached levels extremely hazardous to health.

October King Bhumibon of *Thailand* formally promulgates a new Constitution.

As the rupiah, the currency of *Indonesia*, continues to be sharply devalued, the government seeks loans from the IMF in return for further economic reforms (but see January 1998).

The government of *Vietnam* devalues its currency, the dong.

November SLORC, the military regime in *Burma*, renames itself the State Peace and Development Council.

Prime Minister Chaovalit Yongchaiyut of *Thailand* resigns in the face of demonstrations over the continuing economic crisis, and is replaced by Chuan Leekpai, who begins talks with the IMF on a reform program.

1998 January In *Malaysia*, Prime Minister Mahathir Muhamad comes into open conflict with Deputy Prime Minister and Finance Minister Anwar Ibrahim, who refuses to implement Mahathir's proposal to cut interest rates in the hope of stimulating the economy.

President Suharto of *Indonesia* defies the IMF by announcing a budget proposal containing provision for high levels of state spending. Riots and protests begin in

Java, partly in response to rising food prices, partly in protest against IMF-inspired cuts in government spending, and partly in favor of democratization. They spread throughout Indonesia in the following months.

March The People's Consultative Assembly of *Indonesia* elects President Suharto, unopposed, for what is expected to be his seventh five-year term in office, and accepts his nomination of Bacharuddin Jusuf Habibie as Vice President. Riots and protests continue.

Prince Ranariddh, former Co-Prime Minister of *Cambodia*, returns to the capital, Phnom Penh, and is convicted of illegally importing weapons and conspiring with the Khmer Rouge. Under the terms of a peace agreement among the main political parties, brokered by the Japanese government in February, he is immediately granted a pardon by his father, King Sihanouk.

April Representatives of pro-independence groups in *East Timor* meet in Portugal to establish the National Council of Timorese Resistance (CNRT).

Pol Pot, the former leader of the Khmer Rouge, is found dead at the organization's base in Anlong Veng in northwestern *Cambodia*.

May The government of *Indonesia* imposes a 70% increase in fuel prices. Rioting spreads throughout the country and at least 2,000 people are killed, including many from the ethnic Chinese minority. President Suharto resigns his office, after 32 years in power, and is succeeded by his Vice President, Bacharuddin Jusuf Habibie.

President Habibie of *Indonesia* announces that he will submit himself for re-election in 1999, rather than serve the remaining five years of Suharto's term (see October 1999); he also announces that new political parties may be formed, provided that they uphold Pancasila (see June 1945 and January 1973).

Lieutenant General Dao Trong Lich, the head of the armed forces of *Vietnam*, is killed, along with other military officers, in an airplane crash in northwestern *Laos*, amid unconfirmed reports that Vietnam is helping Laos to suppress a renewed uprising by the Hmong minority.

July The Amedeo Development Corporation, controlled by Prince Jefri, brother of the Sultan of *Brunei*, collapses following losses eventually assessed at US$16 billion. The Sultan then removes Prince Jefri from the chairmanship of the Brunei Investment Agency. Prince Jefri's successor as Chairman, Abdul Aziz, announces that a significant sum in government funds has been misappropriated (see March 2000).

August The International Labor Organization confirms numerous reports that the use of forced labor by the military regime in *Burma* is "pervasive" (see also November 2000).

September Prime Minister Mahathir Mohamad of *Malaysia* dismisses the Deputy Prime Minister and Minister of Finance, Anwar Ibrahim. Soon afterwards, Anwar is arrested, charged with corruption and sodomy, and (according to an official inquiry completed in March 1999) assaulted by police officers while in custody.

The government of *Malaysia* introduces selective exchange controls aimed at reducing speculative movements of capital.

The government of *Indonesia* signs a ceasefire agreement with the OPM, which has been fighting for autonomy for the province of Irian Jaya (West Papua) since May 1977.

October	In *East Timor*, the former Portuguese colony illegally occupied by *Indonesia* since 1975, paramilitary groups funded, trained, and armed by the Indonesian Army begin a campaign of intimidation and massacre against individuals and groups who support the territory's claim to independence.
	Following legislative elections in July, Hun Sen is reappointed Prime Minister of *Cambodia*. Amid outbursts of violence between the rival political parties, it is alleged that the government has not only manipulated the elections – even though they have been certified free and fair by international observers – but has also tortured and executed opponents. Hun Sen completes the formation of a new coalition government in December.
December	The government of *Vietnam*, hosting an ASEAN summit, admits a delegation from *Cambodia* as representing a member state. Cambodia's membership is formally confirmed in April 1999, bringing the total number of member states to ten.

1999

	January	A street brawl between Christians and Moslems in Ambon, the capital city of the province of Maluku (South Moluccas) in *Indonesia*, inaugurates the intercommunal violence that had caused at least 3,000 deaths by July 2000.
	February	In *Indonesia*, José Xanana Gusmão, leader of the *East Timor* independence movement, is transferred from the jail cell that he has occupied since 1992 to house arrest in Jakarta. The Indonesian government continues talks with Portugal and the UN on the future of East Timor.
	March	Following agreement between Portugal and *Indonesia*, the UN Secretary General, Kofi Annan, announces that a "direct ballot" will be held in *East Timor* on the autonomous status proposed for the teritory by Indonesia. Rejection of the proposal will count as approval of independence.
	April	Anwar Ibrahim, former Deputy Prime Minister of *Malaysia*, is found guilty on four charges of corruption and sentenced to six years in prison. The announcement of the verdict is followed by violent demonstrations in Anwar's favor.
	July	In *Indonesia*, at least 62 people are said to have been massacred by soldiers in the province of Aceh, where the army is engaged in combating a movement for independence; investigations into the incident continue into 2000.
	September	In *East Timor*, amid violent clashes between supporters of independence and militias linked to the armed forces of *Indonesia*, 78.5% of those voting in a ballot administered by the UN reject autonomy within Indonesia and thus choose independence. The militias then rampage through the territory, slaughtering their opponents, driving thousands to take refuge in West Timor and elsewhere, and forcing the UN to evacuate almost all its personnel.
		In *Indonesia*, President Habibie overrules his military advisers and decides to release José Xanana Gusmão, leader of the *East Timor* independence movement, from house arrest in Jakarta. While East Timor technically remains a province of Indonesia until the ballot result has been ratified by Indonesia's People's Consultative Assembly, President Habibie concedes control of security in the territory to the International Force in East Timor (Interfet), comprising troops from Australia, *Malaysia*, the *Philippines*, *Singapore*, *Thailand*, *Cambodia*, Canada, the United States, the United Kingdom (including Nepalese Gurkha troops transferred from *Brunei*), and several other countries (see February 2000).

	October	President Habibie of *Indonesia* resigns; the People's Consultative Assembly ratifies the election of Abdurrahman Wahid as his successor, and of Megawati Sukarnoputri, whom Wahid defeated in the election, as Vice President.
		The Assembly also ratifies the result of the ballot in *East Timor*, which consequently passes under the control of the UN Temporary Administration in East Timor (UNTAET).
	November	A summit meeting of ASEAN leaders declares support for the stability and unity of *Indonesia* (minus *East Timor*), and calls for the development of closer ties with Northeast Asia (China, Japan, Taiwan, and the two Koreas).
		The government of *Thailand* decides that up to 600,000 foreign workers must return to their homes in *Burma*. Thai police start transporting Burmese to the border, although it remains closed.
		In Banda Aceh, the capital of the province of Aceh in *Indonesia*, around 1 million people (around 25% of the province's population) take part in a rally to demand a referendum on independence. President Abdurrahman Wahid promises that a referendum will be held, but then announces that the idea will first have to be approved by the legislature and the military.
		In general elections in *Malaysia*, the ruling National Front retains a large majority in the federal House of Representatives, but loses control of two states, Terengganu and Kelantan, to the PAS, an Islamist party.
	December	The army takes over control of security matters from the police in the Moluccas, the two provinces in the extreme East of *Indonesia* where violent confrontations between Christians and Moslems have left at least 1,000 people dead and thousands more homeless over the course of the year.
		In West Papua, ruled by *Indonesia* as Irian Jaya, Theys Eluay, a former member of the Indonesian ruling party Golkar, announces that he is now "President of Papua" and raises the independence flag associated with the banned Free Papua Movement (OPM). There is evidence that Eluay is receiving funding and support from the Indonesian armed forces and from Golkar, and that members of his militia, Satgas Papua, have attacked and intimidated ethnic Chinese and other non-Papuans in the province. The Indonesian government responds to Eluay's announcement by legalizing the flying of the independence flag, but only below the Indonesian flag, and by accepting a change in the province's name from "Irian Jaya" to "Papua." (See also June and November 2000.)
		After decades of disputes, *Vietnam* and China sign an agreement on the delineation of their common land border, which is around 1,200 kilometers (750 miles) long.
2000	**January**	Violent clashes between adherents of different religions continue and spread in at least eight of the 27 provinces of *Indonesia*, with Moslems fighting Christians in the Moluccas, Celebes, and elsewhere, and Christians, Hindus, Chinese, and Moslems fighting each other in Lombok. President Abdurrahman Wahid publicly blames the violence on "provocation" by religious fanatics and military officers hostile to his government. Elsewhere in Indonesia, the army launches a new offensive against the Free Aceh Movement (Gerakanan Aceh Merdeka, or GAM), which seeks independence for the province of Aceh.

General Maung Aye, a leading member of the military regime in *Burma*, makes an official visit to India, initiating cooperation on defense issues and economic development that many observers see as indicating a shift away from Burma's long-standing dependence on China.

Officials of the UN administration in *East Timor* join with Timorese and Australian representatives to revise and legitimize the Timor Gap treaty (see December 1989), allowing the former signatory, *Indonesia*, to be replaced by East Timor as Australia's partner in developing offshore oilfields.

February Amid rumors of military unrest, President Wahid of *Indonesia* suspends General Wiranto from the post of Senior Security Minister for as long as Wiranto is under investigation in connection with atrocities committed in *East Timor* during 1999, when he was head of the armed forces.

Responsibility for security in *East Timor* is transferred from Interfet (see September 1999) to UN peacekeepers.

March The government of *Brunei* sues Prince Jefri, the youngest brother of Sultan Hassanal Bolkiah, and others over improper use of government funds (see July 1998); the case is settled out of court in May.

The IMF delays the transfer of a loan of US$400 million to *Indonesia* on the grounds that agreed economic reforms have not been fully implemented.

In the *Philippines*, Communist rebels attack oil depots in the province of Negros Oriental and the headquarters of the energy ministry in the capital, Manila.

The member states of ASEAN agree with China to work toward a code of conduct on handling rival claims to the Spratly Islands (see September 1975).

Following the first direct elections to the Senate in *Thailand*, 78 individuals – more than one third of those elected – are disbarred from taking their seats on grounds of vote-rigging.

April An explosion in a restaurant in Vientiane, the capital of *Laos*, injures 13 people; the government attributes it to a bomb planted by unspecified "terrorists." As more explosions follow at other sites during the rest of the year, opinions differ as to whether they are linked to unrest among the Hmong minority, rivalries within the ruling party or, perhaps, the activities of neighboring countries.

In Jakarta, the capital of *Indonesia*, a militant Moslem group, the Laskar Jihad, mounts protests against what it sees as the favorable treatment of Christians in the Moluccas, while at Bogor in West Java other members of the group organize training for a "holy war" against the Moluccan Christians.

In the *Philippines*, units of the armed forces launch assaults on bases of the Moro Islamic Liberation Front, and on those of a rival rebel group, Abu Sayyaf, which is holding 27 hostages. Members of Abu Sayyaf then abduct 21 more hostages from a resort in Sabah in *Malaysia* and take them to the southern Philippines.

May At a conference in Chiang Mai in *Thailand*, the 10 ASEAN countries agree with China, Japan, and South Korea on proposals to enhance cooperation among their central banks in pursuit of regional financial stability.

Violent riots erupt in Jakarta, the capital of *Indonesia*, on the second anniversary of the downfall of President Suharto; they are aimed primarily at businesses owned by ethnic Chinese.

Representatives of the government of *Indonesia* and of the Free Aceh Movement (GAM) meet in Geneva, and agree on a "humanitarian pause" in the 25-year struggle over the province of Aceh, to be implemented from June. Meanwhile, an Indonesian court convicts 24 low-ranking soldiers and one civilian of the murders of 57 unarmed civilians in the province of Aceh in 1999; the highest sentence imposed on any of the 25 is 10 years imprisonment.

The High Court of *Singapore* declares the bankruptcy of J.B. Jeyaretnam, leader of the Workers Party, opening the way to his expulsion from Parliament.

General Wiranto, suspended as Senior Security Minister of *Indonesia* in February, resigns the post after being criticized by officials investigating atrocities in *East Timor* during 1999.

Also in *Indonesia*, thousands of Moslems travel from Surabaya to Ambon in the Moluccas to launch a "holy war" against the Christian residents of the region, while fighting between members of the two faiths also erupts in central Celebes.

June

Violent clashes between Moslems and Christians continue in the Moluccas and Celebes in *Indonesia*: buildings are burned on the island of Halmahera and elsewhere, and at least 160 people, most of them Christians, are killed. With more than 3,000 dead since January 1999, President Abdurrahman Wahid declares a state of emergency in the province.

Sjahril Sabirin, the head of the central bank in *Indonesia*, is charged with corruption in connection with reports that he authorized the illegal transfer of funds to the former ruling party, Golkar, during 1999.

At a conference of Papuan comunity leaders in Jayapura, capital of West Papua – the province of *Indonesia* formerly known as Irian Jaya, also known simply as Papua – Theys Eluay, self-appointed "President of Papua" since December 1999, wins general support for opening negotiations with the Indonesian government on increasing autonomy and moving toward independence. (See November 2000.)

July

Violent conflicts continue in various provinces in *Indonesia*: between Acehnese and the Indonesian military in Aceh, between Papuans and non-Papuans in West Papua, between Dayaks and Madurese in Kalimantan, and between Christians and Moslems in the Moluccas.

One bomb explodes and two others are defused at the Jakarta offices of the Attorney General of *Indonesia*, amid rumors that supporters of former President Suharto are seeking to disrupt investigations of his personal finances.

In *Burma*, Brigadier General Zaw Tun, Deputy Minister for National Planning and Economic Development, states publicly that official economic statistics have been falsified; he is dismissed and placed under house arrest.

Vietnam and the United States sign a wideranging trade agreement, under which (subject to ratification) Vietnamese producers will gain greater access to the US market, while US and other western enterprises will be able to trade and invest more freely in Vietnam.

An uncertain number of Karens from *Burma* take refuge in *Thailand* as fighting resumes between the forces of the Burmese military regime and the Karen National Union (see January 1949 and November 2000).

Tensions rise between *Laos* and *Thailand* as the Thai authorities refuse to repatriate to Laos a number of Hmong rebels who have fled across their common border.

August A bomb explodes at the Jakarta residence of Leonides Caday, Ambassador of the *Philippines* to *Indonesia*, killing at least three people; it is unclear who placed the bomb, or why.

General Suharto, former President of *Indonesia*, is charged with corruption in respect of his seven charitable foundations, but his lawyers announce that he is suffering from brain damage and unable to respond to the charges (see September 2000).

At the annual meeting of the People's Consultative Assembly in *Indonesia*, President Abdurrahman Wahid agrees to transfer additional powers to the Vice President, Megawati Sukarnoputri; delegates vote to allow appointed representatives of the armed forces to remain in the Assembly until 2009, instead of 2004 as previously decided; and the Assembly calls on the government to accelerate economic liberalization and suppress the violent conflicts in several provinces. The Assembly also amends the Constitution to prevent new legislation from having a retroactive effect, thus shielding the military and officials from prosecution for human rights abuses.

Following repeated attacks on UN officials aiding around 120,000 refugees from *East Timor* in camps in West Timor, a province of *Indonesia*, the Office of the UN High Commissioner for Refugees (UNHCR) temporarily suspends operations on the island, and criticizes the Indonesian military for its failure to prevent the attacks. International aid agencies also suspend their operations.

September In West Timor, the province of *Indonesia* bordering on *East Timor*, militias continue attacks on UN personnel, killing three and forcing others to withdraw. The UN Security Council passes a resolution calling on Indonesia to disband and disarm the militias, restore law and order, and promote repatriation of refugees to East Timor.

After 25 years, the government of *Malaysia* unilaterally ends an agreement with one of the Malaysian states, Terengganu, under which Terengganu received part of the royalties from offshore oilfields, on the grounds that the Islamic party PAS, in power in Terengganu since November 1999, cannot be trusted to spend the revenue wisely.

A bomb explodes at the Jakarta Stock Exchange in *Indonesia*, killing at least 15 people. President Abdurrahman Wahid orders the arrest of Hutomo Mandala Putra, better known as Tommy Suharto, a son of the former President, but the police announce that they have no evidence that he was involved. Later the same month, Tommy is convicted by the Supreme Court of misappropriating government-owned land and sentenced to 18 months in prison, but remains at large. Meanwhile, the police arrest 27 people, most of them Acehnese, in connection with the bombing.

Mahathir Mohamad, Prime Minister of *Malaysia* since 1981, announces that he is transferring some of his powers to the Deputy Prime Minister, Abdullah Ahmad Badawi, and that he will retire from the Prime Ministership in 2004.

In *Vietnam*, *Cambodia*, *Laos*, and *Thailand*, the annual flooding of the Mekong River and its tributaries becomes more extensive and more damaging than it has been for 70 years. At least 150 people are killed, and tens of thousands are left without homes or livelihoods.

In *Burma*, Aung San Suu Kyi, leader of the National League for Democracy, risks imprisonment (under the terms of decrees issued by the military regime) with an announcement that the League is to draft a new constitution for the country. The regime responds by banning the League's annual commemoration of its establishment, closing its offices, and restricting Suu Kyi and eight other members of its Central Executive Committee to their homes.

The government of *Malaysia* announces that the new federal capital, Putrajaya, started in the state of Selangor in 1996 and due to be completed in 2010, will be administered directly by federal authorities and not by the state government.

A court in *Indonesia* announces that the corruption case pending against former President Suharto is to be closed, and his house arrest revoked, on the basis of medical evidence that he is mentally unfit to plead. Prosecutors announce that they will appeal against the decision, while violent demonstrations by supporters and opponents of Suharto errupt in Jakarta and elsewhere (and see November).

October In the town of Wamena in West Papua, violent clashes, apparently sparked off by attempts to raise the independence flag associated with the banned Free Papua Movement (OPM), end in the deaths of at least 30 migrants from other provinces of *Indonesia* and several members of indigenous groups. The government then rescinds its decision to permit the flying of the flag (granted in December 1999) and bans the militant group known as the Papua Taskforce.

In the *Philippines*, following claims by Luis Singson, Governor of Ilocos Sur, that he gave President Joseph Estrada P414 million (US$8.5 million) raised from gambling and from tobacco taxes, the Vice-President, Gloria Macapagal Arroyo, resigns as Secretary for Social Welfare; Jaime, Cardinal Sin, Archbishop of Manila and head of the Catholic Church in the Philippines, calls on the President to leave office; numerous members of Estrada's party in Congress defect to other parties; and both the stock market and the exchange rate of the peso move sharply downward.

As revealed in January 2001, Aung Sang Suu Kyi, leader of the National League for Democracy, begins direct negotiations on the future of *Burma* with members of the military regime.

November President Jiang Zemin becomes the first leader of China ever to visit *Laos*. He also visits *Cambodia*, where police disperse student groups and others seeking to respond to his visit by protesting against China's sponsorhip, arming, and funding of the Khmer Rouge.

Khamsay Souphanouvong, a minister in the government of *Laos* and a son of the former President Prince Souphanouvong, is granted political asylum in New Zealand, amid conflicting reports of divisions within the Lao leadership.

The High Court in Jakarta, *Indonesia*, orders the reopening of proceedings against former President Suharto, stayed in September, on the grounds that his presence in court is not required.

At least 39 supporters of the independence movement in the province of Aceh in *Indonesia* are killed in clashes with the security forces as they attempt to enter the provincial capital, Banda Aceh, in order to mark the anniversary of the rally held in November 1999 to demand a referendum on independence. The Free Aceh Movement (Gerakanan Aceh Merdeka, or GAM) announces that it will boycott further negotiations with the Indonesian government, scheduled to be held in Geneva later in the month.

In the *Philippines*, opponents of President Joseph Estrada, unable to obtain the support of more than two thirds of the House of Representatives for impeachment proceedings against him – despite numerous defections from his party – initiate the proceedings on the basis of a petition signed by more than one third of the members of the House. Estrada thus becomes the first Asian head of state or government ever to be impeached, in a trial starting in the Senate in December.

The Karen National Union, which has been in rebellion against successive governments of *Burma* since January 1949, defeats government troops in a battle near the border with *Thailand*, while other government troops are deployed against Rakhine (Arakanese) rebels near the border with Bagladesh.

The International Labor Organization calls for sanctions to be applied against the military regime in *Burma* because of its extensive use of forced labor. This is the first time that the organization has called for sanctions against any state since it was founded in 1919.

The government of *Singapore* signals its impatience with ASEAN's slow process of multilateral talks on free trade by signing a free trade agreement with New Zealand and beginning talks on similar agreements with other countries outside the region.

December In Phnom Penh, the capital of *Cambodia*, around 50 members of the Cambodian Freedom Fighters, a group that has denounced the government as a tool of the Communist regime in *Vietnam*, attack the Ministry of Defense and an army barracks, but are swiftly defeated, with seven of their number killed. Richard Kiri Kim, born in Cambodia but now a US citizen, is arrested and charged with leading the rebel group.

In West Papua, the province incorporated into *Indonesia* in August 1969, police detain Theys Eluay, self-appointed "President of Papua" since December 1999, Thaha Hamid, Secretary General of the Presidium Council led by Eluay, and around 60 other leaders of the independence movement, charging some of them with "subversion." Rallies commemorating an earlier declaration of independence (see December 1961) are held throughout the territory and 10 people are killed by security forces, while in Jakarta there are violent clashes between the police and West Papuan demonstrators.

President Abdurrahman Wahid of *Indonesia* announces the introduction of Sharia law in the province of Aceh, for the first time anywhere in Southeast Asia, and with the apparent approval of the Free Aceh Movement, which seeks independence for the province.

UN prosecutors in *East Timor* charge 10 members of a militia group and one Indonesian army officer with crimes against humanity in respect of forced deportations and killings committed during 1999, and announce plans to issue several more such indictments.

APPENDICES

Appendix 2
Glossary

Place Name Equivalents

Place names used in this book are on the left; the alternatives preferred by the governments of the countries in question are on the right.

Burma	Myanmar
Celebes (island in Indonesia)	Sulawesi
Malacca (city in Indonesia)	Melaka
Moluccas (island group in Indonesia)	Maluku (single province) until 1999; Maluku and North Maluku (two provinces) since then
Rangoon (capital city of Burma)	Yangon
Sumatra (island in Indonesia)	Sumatera

Special Terms

abangan: this Indonesian term refers to Javanese, usually engaged in agricultural work, who practice a "mixed" approach to Islam, combining with it many elements from animism and Hinduism.

aliran: an Indonesian term, literally meaning "stream," that refers to long entrenched sociocultural groups. It was introduced into English-language studies of Indonesia by the distinguished US anthropologist Clifford Geertz.

beraja: this is the Malay term for "royalty," "kingship" or "monarchy."

bumiputra/bumiputera: see entry in Appendix 5

Cultuurstelsel: a system for taxing peasants imposed by the Dutch colonialists in Java between 1830 and 1870

dependistas: a term first used in Latin America (hence its Spanish form) to refer to dependency theorists – and not always with wholehearted approval of their ideas and recommendations

doi moi: "new structure" or "renovation," the official name for the series of liberalizing economic reforms launched in Vietnam since 1986

dwifungsi: this term, combining Dutch and Sanskrit vocabulary, refers to the doctrine intended to legitimize the "dual function" of Indonesia's armed forces in governance and security. Elements in the military continue to defend the doctrine tenaciously, even as Indonesia attempts to democratize its politics.

gazetting: the practice in Singapore, Brunei, and Malaysia of printing in the government's official daily publication the title of each foreign publication that is to be banned or permitted only restricted circulation. All three countries have inherited the practice from the days of British colonialism.

Jintanagan Mai: this Lao term, literally meaning "new imagination" or "new thinking," is conventionally translated as "New Economic Mechanism". It refers to the liberalizing economic reforms introduced in 1986.

kekeluaragaan: the ideologues of Suharto's "New Order" in Indonesia claimed that this principle, which may be translated roughly as "familism," was deeply rooted in the country's sociocultural traditions, thus helping to justify the government's corporatist strategies of rule.

kongsi: Chinese "clan" networks

mestizo: people of "mixed" (usually Eurasian) descent

Pancasila: these "five principles" were established by President Sukarno of Indonesia in 1945, and have been upheld as the guiding ideology of his country, at first informally and then, since 1984–85, by legal requirements. The five principles are nationalism, internationalism (or humanitarianism), democracy, social justice, and belief in one God.

peceklik: an Indonesian term referring to the period of shortages and hunger before harvest time

pribumi: see under "Bumiputra/Bumiputera" in Appendix 5

priyayi: Javanese aristocrats, heirs to a combination of pre-Islamic courtly arts and Islamic tradition, who became civil servants in the parallel "native civil service" established by the Dutch. Those from humbler origins who rose up through the ranks became known as the "new *priyayi*."

rakyat: this is the Malay term for "subjects," "the people" or "commoners."

santri: this term is used in Indonesia to refer to devout Moslems who practice their religion in traditionalist ways in rural areas or in modernist ways in urban settings.

sawah: an Indonesian word for an embanked, flooded ricefield

swidden: a field that has been prepared for agriculture by burning. After some crops have been grown, the field is left to regenerate by returning to native growth. The term has also come to be used, loosely, as the name of an agricultural system in which swiddens are used.

Tengku: in Malaysia, this title indicates that its holder is descended from or related to one of the country's royal families; it is therefore equivalent to "Prince" in western countries.

transmigrasi: "transmigration," the Indonesian government's program of state-sponsored internal migrations, intended to relieve pressure on resources in overpopulated areas and to gather labor for exploiting underpopulated areas. The program, taken over from the Dutch colonial system, was officially abandoned in 2000, but the officials who oversaw it remain in place, partly to attend to the needs of refugees from the various interethnic conflicts that the program may have helped to inflame.

Tun: in Malaysia, this nonhereditary title is conferred by the paramount ruler, the Yang Di-Pertuan Agong, on individuals who have distinguished themselves on the national stage.

APPENDICES

Appendix 3
Personalities

Heads of Governments since the Attainment of Self-rule

The individuals listed here have been the paramount leaders in *de facto* governments. It should perhaps be stressed that none has enjoyed complete supremacy. The leaders of one-party states and military regimes have generally been members of collective leaderships; in other cases there have always been various formal and informal constraints to be observed (or evaded); and even the Sultans of Brunei have had to rely on ministers, officials, and experts (see also Appendix 4). Further, no comment or judgment is implied as to any claims or counterclaims about *de jure* governments, notably in the cases of Burma, Cambodia, Laos, and Vietnam. As for Thailand, while it is unique in the region for having continuously had forms of self-government since long before the modern era, we have started the list of its Prime Ministers from the point when they began exercising real powers of their own, following the abolition of the absolute monarchy. Finally, this appendix excludes East Timor, which has been under UN administration since the Indonesian occupation ended in 1999.

Sultans of Brunei

1950 to 1967	Sir Omar Ali Saifuddin III, 28th Sultan and Yang Di-Pertuan
1967 to date	Haji Sir Hassanal Bolkiah, 29th Sultan and Yang Di-Pertuan

Rulers of Burma

1947–58	U Nu	Prime Minister
1958–60	General Ne Win	Prime Minister
1960–62	U Nu	Prime Minister
1962–74	General Ne Win, known as U Ne Win from 1972	Chairman of the Revolutionary Council
1974–81	U Ne Win	President
1981–88	General San Yu	President
1988	Sein Lwin	President
1988	Dr Maung Maung	President
1988–92	General Saw Maung	Chairman of the State Law and Order Restoration Council
1992–	General Than Shwe	Chairman of the State Law and Order Restoration Council (1992–97), then of the State Peace and Development Council

Rulers of Cambodia

1941–70	Norodom Sihanouk	King (1941–55); Prime Minister (1955–60); Head of State (1960–70)
1970–75	General Lon Nol	head of military regime (1970–72); President of the Khmer Republic (1972–75)
1975–79	Pol Pot	Prime Minister of Democratic Kampuchea
1979–85	Heng Samrin	President of the People's Republic of Kampuchea
1985–93	Hun Sen	Prime Minister of the People's Republic of Kampuchea (to 1990), then of the State of Cambodia

1993–97	Prince Norodom Ranariddh and	First Co-Prime Minister
	Hun Sen	Second Co-Prime Minister
1997–98	Hun Sen and	Co-Prime Minister
	Ung Huot	Co-Prime Minister
1998–	Hun Sen	Prime Minister

Presidents of Indonesia

1945–68	Sukarno [deprived of his powers in 1967]
1968–98	Suharto [Acting President from 1967]
1998–99	Bacharuddin J. Habibie
1999–	Abdurrahman Wahid

Prime Ministers of Laos

1946–47	Prince Kindavong
1947–48	Prince Suvannarat
1948–50	Prince Boun Oum
1950–51	Phoui Sananikone
1951–54	Prince Souvanna Phouma
1954–56	Katay Don Sasorith
1956–58	Prince Souvanna Phouma
1958–59	Phoui Sananikone
1960	Kou Abhay
1960	Prince Sovanith
1960	Prince Souvanna Phouma
1960–62	Prince Boun Oum
1962–75	Prince Souvanna Phouma
1975–91	Kaysone Phomvihane
1991–98	Khamtai Siphandon
1998–	Sisavat Keobounphan

Prime Ministers of Malaya and (from 1963) Malaysia

1957–70	Tengku Abdul Rahman
1970–76	Tun Abdul Razak
1976–81	Tun Hussein Onn
1981–	Dr Mahathir Mohamad

Presidents of the Philippines

1946–48	Manuel Roxas
1948–53	Elpidio Quirino
1953–57	Ramon Magsaysay
1957–61	Carlos Garcia
1961–65	Diosdado Macapagal
1965–86	Ferdinand Marcos
1986–92	Corazon Aquino
1992–98	Fidel Ramos
1998–	Joseph Estrada

Prime Ministers of Singapore

1959–90	Lee Kuan Yew
1990–	Goh Chok Tong

Prime Ministers of Thailand

1938–44	General Phibun Songkhram
1944–45	Khuang Aphaiwong
1945–46	Seni Pramoj
1946	Khuang Aphaiwong
1946	Pridi Phanomyong
1946–47	Luang Thamrongnawasawat
1947–48	Khuang Aphaiwong
1948–57	Marshal Phibun Songkhram
1957	Pote Sarasin
1958–59	General Thanom Kittikachorn
1959–63	Marshal Sarit Thanarat
1963–73	General (later Marshal) Thanom Kittikachorn
1973–75	Dr Sanya Dharmasakti
1975	Seni Pramoj
1975–76	Kukrit Pramoj
1976	Seni Pramoj
1976–77	Thanin Kraivixien
1977–80	General Kriangsak Chamanand
1980–88	General Prem Tinsulanonda
1988–91	General Chatichai Choonhaven
1991–92	Anand Panyarachun
1992	General Suchinda Kraprayoon
1992	Anand Panyarachun
1992–95	Chuan Likphai
1995–96	Banharn Silapa–Archa
1996–97	Chavalit Yongchaiyudh
1997–	Chuan Likphai

Prime Ministers (to 1981) or Chairmen of the Council of Ministers of Vietnam

1945–55	Ho Chi Minh
1955–87	Pham Van Dong
1987–88	Pham Hung
1988	Vo Van Kiet
1988–91	Do Muoi
1991–97	Vo Van Kiet
1997–	Phan Van Khai

Rulers of South Vietnam

1946–48	Nguyen Van Xuan	Prime Minister of the Autonomous Republic of Cochin China
1948–50	Nguyen Van Xuan	Prime Minister of the Empire of Vietnam
1950–54	Tran Van Huu	Prime Minister of the Empire of Vietnam
1954–55	Ngo Dinh Diem	Prime Minister of the Empire of Vietnam
1955–63	Ngo Dinh Diem	President of the Republic of Vietnam
1963–64	General Duong Van Minh	Chairman of the Revolutionary Council
1964	General Nguyen Khanh	Commmander in Chief
1964–65	Phan Khac Suu	President
1965–67	Air Marshal Nguyen Cao Ky	Prime Minister
1967–75	General Nguyen Van Thieu	President
1975	Huynh Tan Phat	President
1975	General Duong Van Minh	President

Anand Panyarachun (1932–): Prime Minister of Thailand from 1991 to 1992, and later in 1992. Anand was educated at Bangkok Christian College, Dulwich College in London, and Trinity College, Cambridge, where he studied law. He entered the Ministry of Foreign Affairs in 1955, and rose to become Ambassador to the United States in 1972, Deputy Foreign Minister in 1976, and Ambassador to West Germany in 1977. After retiring from the diplomatic service, he entered the world of business; he has been Chairman of an influential pressure group, the Thai Industrial Federation, since 1990. He owed his two brief terms as Prime Minister to the military and their appointees in the legislature, and his governments' measures included the banning of labor unions from state-owned enterprises. However, since leaving office Anand has gained a reputation as one of the strongest advocates of liberal democracy in Asia, mainly because he played a central role in designing Thailand's present Constitution.

Aquino, Corazon "Cory" (1933–): President of the Philippines from 1986 to 1992. Mrs Aquino was born Corazon Cojuangco, a member of a prominent landowning family of Chinese descent. Her father was a member of Congress, her mother's father was once a candidate for the vice-presidency, and two uncles were politicians too. In 1946, she went with her family to the United States, where she was educated in convent schools and Catholic universities, although she returned to the Philippines to continue her education. She has remained a devout follower of the most popular and powerful religious faith in the Philippines ever since. She married Benigno "Ninoy" Aquino, Jr, a rising young politician, in 1954, and then divided her time between managing the Aquino family's home and serving as treasurer of the Cojuangco family's sugar corporation. She first entered politics in 1972, when her husband was imprisoned under President Marcos's martial law regulations (on a warrant signed by General Fidel Ramos, her eventual successor as President). Mrs Aquino maintained contacts between Ninoy and those of his colleagues in the opposition who were still at liberty, and then accompanied him into exile in the United States in 1980, to begin what she has since called the happiest years of her life. After Ninoy's assassination in Manila in 1983, she initially resisted pleas from his supporters that she take his place, but in 1985 she joined the United Nationalist Democratic Organization and announced that she would run against Marcos in the presidential election the following year. Her victory in that election was eventually recognized after the "People Power" demonstrations induced Marcos to resign. As President, she succeeded in arranging ceasefires with the Communist New People's Army and the Moslem separatist Moro National Liberation Front, weathered at least six major coup attempts by disaffected sections of the military, oversaw the introduction of a new Constitution, promulgated a fairly extensive program of land reform, and reached agreement with the United States on the future of its bases in the Philippines. She also attracted criticism, from opponents and supporters alike, for her occasional resort to rule by decree, her inability, or reluctance, to challenge the privileges of the landowning elite from which she came, and her tendencies to indecisiveness. Nevertheless, her six years in power showed that it was possible for a woman with little political experience to become a relatively effective leader of a multiethnic and deeply divided society, and that democracy could be restored after two decades of dictatorship, albeit with many economic and social problems still unaddressed. Since she left office, Mrs Aquino, one of the few former heads of state in the region to remain politically active, out of jail, and widely respected, has participated in projects to promote democracy elsewhere in Asia. She has also offered her support to Aung San Suu Kyi of Burma (see below); to Wan Azizah Ismail, the wife of Anwar Ibrahim, the former Deputy Prime Minister of Malaysia now in jail on questionable grounds; and, in 2000, to the movement to impeach President Estrada.

Aung San (1915–47): Burmese nationalist leader, Prime Minister from 1945 to 1947. Aung San first came to prominence in 1936 as the leader of a group of 30 young Burmese nationalists who adopted the honorific *Thakin* ("master") in defiance of the prevailing custom of limiting its use to British colonial officials. Having organized a student strike called by the All-Burma Students Union, he then took over the leadership of Dohbama Asiayone ("We Burmese"), a largely peasant-based nationalist movement that had been founded in 1930. In 1939, Aung San joined Ba Maw, who had been dismissed from the prime ministership by the colonial authorities, in forming the Freedom Bloc, within which Aung San represented the Burmese Communist Party. In December 1941, Aung established the Burma Independence Army, which was intended to fight against both the British and the Japanese, if and when they invaded. Nevertheless, Aung San and other nationalists received military training in Tokyo in the interim. In May 1942, their organization became the Burma Defense Army. The incoming

Japanese easily defeated it and then, like other nationalist leaders elsewhere in the region, Aung and his colleagues decided to collaborate with the Japanese. Aung established a third military body, the Burma National Army, and served as War Minister in the government created by the Japanese in August 1943. Within a few months, however, he had made contact with the British, and begun to create clandestine units of his Antifascist Organization, an alliance of nationalists, Communists, and others against the Japanese. In March 1945, this alliance, expanded and renamed the Antifascist People's Freedom League, launched an uprising with Allied support and in liaison with the Burma National Army. After victory was achieved, the returning British authorities briefly imprisoned Aung and some of his colleagues, but it became clear even to the British that Burma was ungovernable in their absence. In September 1946, Aung San joined the colonial government of Burma and it was announced that the country would prepare for independence. In October, he expelled his former allies in the Communist Party from the League. In January 1947, he signed an agreement on the future of Burma with the British Prime Minister, Clement Attlee, in London, and in April the League won a majority in elections to the Constituent Assembly. Three months later, Aung San and six of his colleagues were assassinated, apparently on the orders of Saw Maung, a leader of the National Opposition, who was later tried and executed. Aung San's daughter Aung San Suu Kyi (see below) has become leader of the National League for Democracy.

Aung San Suu Kyi (1945–): Burmese political activist, leader of the National League for Democracy since 1988, and winner of the Nobel Peace Prize in 1991. Suu Kyi (whose full name is often prefixed by the honorific *Daw*) was just two years old when her father Aung San, then Prime Minister, was assassinated. She was educated in India and Britain, and then worked for the UN, but she returned to Burma in 1988 to take part in founding and leading the League in its campaign against the military regime that had taken power earlier that year. In July 1989, she and her colleague Tin Oo were placed under house arrest. In January 1990, she was banned from taking part in legislative elections because of her marriage to a foreigner, the British academic Michael Aris. The League won an overwhelming electoral victory in April 1990, but has been prevented from taking power ever since. Suu Kyi was awarded the Nobel Peace Prize in 1991 but chose not to receive it in person, as the regime threatened not to allow her to return if she left Burma to do so. She was released from house arrest in July 1995, but the regime has continued to restrict her activities. It was in protest against these restrictions that in July and August 1998, and again in August and September 2000, Suu Kyi spent several days living in her car at the side of roads she had been forbidden to travel on. The regime has attempted to coerce Suu Kyi into leaving Burma, notably at the time of her husband's death in Britain in March 1999, but she insists on remaining there. Some Burmese oppositionists, as well as some of their supporters abroad, have criticized Suu Kyi's commitment to nonviolent methods, which is informed by her Buddhist faith as well as by her sense of what is politically possible. However, there seems little doubt that she and the League continue to enjoy the support of most Burmese.

Goh Chok Tong (1941–): Prime Minister of Singapore since 1990. Goh studied economics at the University of Singapore before joining the administrative service of the Singapore government in 1964. He was first elected to Parliament in 1976 and was appointed Senior Minister of State for Finance in 1977. In 1985, he became First Deputy Prime Minister. Goh has broadly maintained the economic and social policies of his predecessor and patron, Lee Kuan Yew (see below). However, he is widely credited with generating a new interest in developing more formal ties between western Europe and Asia. As the leader of a state still closely associated with the controversial idea of "Asian values," Goh is well-placed to influence the course of regional development.

Gusmão, José Alexandre, alias Kay Rala Xanana (1946–): Commander in Chief of the National Liberation Armed Forces of East Timor (Falintil) from 1981 to 2000, and President of the National Council of Timorese Resistance (CNRT) since 1998. Gusmão, the son of a schoolteacher, was born in Manatuto and educated in a Jesuit seminary. However, instead of becoming a priest, he worked as a chartered surveyor and then as a schoolteacher in Dili, the capital of East Timor, then still a Portuguese colony. He became a member of the Revolutionary Front for an Independent East Timor (Fretilin) in 1974. Following the Indonesian invasion of December 1975, he joined other Fretilin fighters in mounting resistance to the invaders, becoming leader of Fretilin's armed wing, the Falintil, in 1981, and head of the National Council of Maubere Resistance (CNRM) in 1986. As early as 1983, Gusmão, against the advice of many of his comrades, had entered fruitless negotiations with the Indonesian occupiers. His critics believed that their fears had been justified when, in 1992, Gusmão was captured by Indonesian troops and jailed in Jakarta. He remained prominent in the liberation movement even so, becoming President of the CNRT, which replaced the CNRM, and occasionally entering talks with the Indonesian government. Since his release from detention, in 1999, he and his colleagues on the CNRT have cooperated with the UN in administering East Timor. His decision in August 2000 to retire from the leadership of Falintil and focus on civilian political activity was widely interpreted as preparing the way for an attempt to become East Timor's first President, in elections scheduled for early in 2001.

Ho Chi Minh (1890–1969): founding General Secretary of the Indochinese Communist Party from 1930 to 1951, President of (North) Vietnam from 1945 to his death; Prime Minister from 1945 to 1955, and General Secretary of the Workers Party from 1956 to 1960. Ho was born Nguyen Sinh Cung, son of a traveling scholar in Nghe Thinh province in Annam (central Vietnam), then under French control. After leaving school in Hue, the capital of Annam, he too became a teacher for some years, but in 1912, already committed to national liberation and using the first of a series of aliases to conceal his activities, he left Vietnam for the West. He worked as a crewman on various ships, possibly visiting the United States, up to 1914. He then spent three years, under the name Ba, as a pastry cook in the Carlton Hotel in London. As Nguyen Tat Thanh, and later as Nguyen Ai Quoc, he spent the years 1917–24 in Paris, working as a retoucher of photographs and a maker of fake Chinese antiquities, becoming a founder member of the French Communist Party (1930), and helping to create the Intercolonial Union, a Communist front organization active in most of France's numerous colonies. After a year being trained as a revolutionary in Moscow, he moved around between Canton, where he founded the Association of Vietnamese Revolutionary Youth, Bangkok, and Hong Kong, where he founded the Indochinese Communist Party in 1930. By 1932, he was in Shanghai, probably liaising with the Chinese Communists, who then favored an independent Indochinese federation comprising Cambodia, Laos, and Vietnam. He then spent at least a few years in Moscow once again. In 1941, or perhaps 1940, he returned to Vietnam, from China, to found the Viet Nam Doc Lap Dong Minh (the Vietnamese Independence League, or Viet Minh), a nationalist front. In China once again, seeking aid from the Chinese Communists, he was jailed by the Nationalist government from 1942 to 1943. By then, the Vietnamese nationalist known as Nguyen Ai Quoc had become the Communist leader Ho Chi Minh (the last of at least 17 "party names"), but the fact that they were one and the same person was concealed until 1958. This gave rise to persistent rumors that Ho was someone else entirely, perhaps even a foreign agent – of the United States, China, Japan or the Soviet Union, according to taste. By 1945, Ho was back in Vietnam once more. (It was there that, in July, Paul Hoagland, a physician with the OSS, the forerunner of the US Central Intelligence Agency, treated Ho for dysentery and malaria, probably saving his life.) In Hanoi in August 1945, Ho declared Vietnam independent. In March 1946, he apparently came close to failure in trying to persuade his party, and the broader Viet Minh, to accept the creation of an Indochinese Federation within the French Union. In any case, this arrangement collapsed, as Ho realized that even the French Communists wanted to use it to restore colonial rule, and by December France and Vietnam were at war. Both during this conflict (1946–54), and during the war against South Vietnam and its ally, the United States (1955 onward), Ho appears to have left strategy largely in the hands of Vo Nguyen Giap and other military experts. As both President and Prime Minister up to 1955, and then as President and General Secretary of the Communist Party, he concentrated on initiating industrialization in the North, imposing land reform, and maintaining courteous but largely distant relations with the two leading Communist powers, the Soviet Union and China. It is generally believed that Ho would have preferred not to have to choose between them when they became rivals. Nevertheless, by the time Ho died, Vietnam was still not unified, while the state that he had created and led was on increasingly close terms with the Soviet Union. Like his exemplar Lenin, "Uncle Ho" did not want to be commemorated after his death, but, like their predecessors in the Soviet Union, Ho's comrades disregarded their late leader's wishes and built a grandiose mausoleum for him.

Hun Sen (1950–): Prime Minister of Cambodia from 1985 to 1993 and since 1998, Co-Prime Minister in the interim. Hun Sen, who came from a poor peasant background in Kompong Cham, joined the resistance to Lon Nol's military regime in 1970. He lost his left eye in 1975 while fighting as a Khmer Rouge field commander. He appears to have risen rapidly within the ranks of the Khmer Rouge up to 1977, when he defected to the opposition and escaped to Vietnam. There, he helped to group and train Cambodian fighters who accompanied the Vietnamese invasion of Cambodia in 1979. He then served as Foreign Minister until 1985, when he became the world's youngest Prime Minister. Between 1991 and 1993, he cooperated with the UN Transitional Authority in Cambodia (UNTAC), abandoning previously held positions to promote liberalization of the economy and compromise with other political groups, although most observers have concluded that he did so for reasons of political expediency rather than conviction. His Cambodian People's Party, as the Communist Party was now known, came second in the UNTAC-supervised elections in 1993, and Hun Sen became Second Co-Premier in a coalition with the royalist party FUNCINPEC. By 1997, however, he had successfully maneuvered to obtain overall control of the government, partly through intimidation of rivals and manipulation of the media.

Kaysone Phomvihane (1920–92): Secretary General of the Lao People's Party, later renamed the Lao People's Revolutionary Party, from 1955 to 1992, Prime Minister of Laos from 1975 to 1991, and President from 1991 to 1992. Kaysone was born in a village in the southern province of Savannakhet, the son of a Vietnamese father and a Lao mother. He studied law in Hanoi, but dropped out to engage in revolutionary activities. He later became active in the Lao Issara independence movement in the southern region of Laos, where he gained valuable military experience in coordinating guerilla raids against the French as they returned to their colony after World War II. In 1950, Kaysone was appointed Minister of Defense in the Pathet Lao resistance government and in 1955 he became secretary general of the movement. He was to retain this post for 37 years. Following the declaration of the Lao People's Democratic Republic in 1975, Kaysone became Prime Minister and the highest-ranking member of the party leadership. During the 1980s, Kaysone revealed his pragmatism by supporting major economic reforms that enabled Laos to become a "transitional" economy, with a mixed system of market capitalism and state socialism. A museum has recently been built in Vientiane, the Lao capital, to honor Kaysone.

Lee Kuan Yew (1923–): Prime Minister of Singapore from 1959 to 1990, Senior Minister since then. Lee, generally known as Harry Lee until 1950, was born the grandson of Hakka immigrants from China. As a subject of the British empire, he studied, in English, at Raffles College, but also learned Malay and Cantonese; as a subject of the Japanese empire during the occupation (1942–45), he took a minor official post, but appears to have used it to gather information for the British. He then studied law at the University of Cambridge. He returned to Singapore in 1950 as a convinced socialist; there are different accounts of how and when he became committed to market capitalism and "Asian values." In 1954, he became one of the founders of the People's Action Party (PAP), which pressed for social reforms as well as independence from the United Kingdom. After the PAP secured a large majority in the colonial legislature, in 1959, Lee became the first Prime Minister of Singapore to enjoy full powers of internal self-government, although the British retained security and foreign policy powers. In 1963, Lee led Singapore into the new Federation of Malaysia; two years later, with apparent reluctance, he led his country to full independence. For the remainder of his years in office, during which he became the world's longest-serving Prime Minister, he dominated Singaporean politics, pursuing his vision of a nation-state blessed with harmony among its various ethnic groups, to be secured by state intervention, and prosperity for all, to be derived from promoting a free-market economy. Thus, for example, most Singaporeans now live in homes rented from the state, which has deliberately created multiethnic neighborhoods to replace monoethnic communities; most are educated by the state according to principles laid down by Lee; and most acquiesce in the application of detailed rules on public behavior, family planning, the use of languages, and many other matters. Lee has stamped his personality on his people: not surprisingly, opinions differ as to the benefits and drawbacks of his achievement.

Mahathir bin Mohamad (1925–): Prime Minister of Malaysia since 1981. Mahathir, born in Kedah, was educated in Malaya, Singapore, and the United States, and is a fully qualified physician. He was among the earliest members of the United Malays National Organization (UMNO) from its inception in 1946. He served as a government medical officer until 1964, when he was elected to the House of Representatives for the first time. A vocal advocate of special rights for Malays and other Bumiputras, he was briefly expelled from UMNO, and his book, *The Malay Dilemma*, was banned. In 1972, however, Prime Minister Tun Abdul Razak brought Mahathir back into UMNO, and he joined the Cabinet two years later, as Minister of Trade and Industry. In 1976, Prime Minister Tun Hussein

Onn named him Deputy Prime Minister. He succeeded Onn in 1981. Since then, Mahathir has presided over accelerated industrialization and economic growth, interrupted only by the crisis of 1997–98, and has enforced his own idiosyncratic versions of "Malay values" and "Asian values," both within UMNO and through an authoritarian style of government. He has also became influential beyond Malaysia as the major proponent of an "Asians only" economic grouping for the region. He has been much criticized by opponents, including his former Deputy Prime Minister Anwar Ibrahim, who is only the most prominent of many UMNO politicians who have been promoted and then demoted by Mahathir. Nevertheless, his government has retained its large majorities in Parliament throughout his many years in office.

Marcos, Ferdinand Edralin (1917–89): President of the Philippines from 1965 to 1986. Marcos, whose activities during World War II remain unclear and controversial, was first elected to Congress in 1949. By 1963, he was the majority leader in the Senate. From this post, he passed to the presidency two years later, being hailed at the time as a proactive modernizer. In the event, his administration proved unable to satisfy the demands either of the established elite or of the Maoist and Moslem insurgents who opposed them. In 1972, as the extent of the corruption within his circle of relatives and cronies became increasingly apparent, Marcos resorted to a declaration of martial law, partly to acquire powers to deal with the insurgencies but, more importantly, to avoid having to leave office at the end of his second term (in 1973). Supported by the western powers and much of the Philippine elite, who saw his regime as a bulwark against Communism, Marcos remained in office for 14 more years, as the economy collapsed and the Cold War moved toward its close. It is possible that if one of his opponents, ex-Senator Benigno Aquino, had not been murdered, probably on Marcos's orders but, fatefully, in front of television cameras, his rule might have lasted even longer. It is also possible that he came close to defeating Aquino's widow Corazon in the presidential election that he was compelled to call in 1986: accounts of the election process and its results are still contested. In any case, faced with an outpouring of popular opposition in Manila and elsewhere, Marcos resigned and found refuge with his family in Hawaii, where he died three years later. His corpse was returned to the Philippines in 1993 and his flamboyant widow Imelda was elected to the House of Representatives in 1995. Both these events have been interpreted as indicating that the "Marcos myth" still has some appeal to sections of the population.

Ne Win (1911–): Prime Minister of Burma from 1958 to 1960, Chairman of the Revolutionary Council from 1962 to 1974, President from 1974 to 1981. "Ne Win" was originally a *nom de guerre*; the future dictator was born Shu Maung. As a young man, he was a supporter of Aung San (see above) and became prominent in the Burmese Socialist Party, sharing its hostility to the Communists and its distrust of Aung San's occasional collaboration with them. In 1943, the Japanese occupation authorities appointed Ne Win as chief of staff of the Burma National Army, apparently against Aung San's wishes. Retaining this position, Ne Win maneuvered to make the Socialist Party the dominant component of the nationalist front, the Antifascist People's Freedom League, once the Communists had been expelled from it in 1946. He was appointed commander in chief of the Burmese armed forces by Prime Minister U Nu in 1949, and led the futile campaign to suppress the numerous insurgencies of Communists and ethnic minorities against the Burmese government. Having tasted supreme power between 1958 and 1960, when U Nu agreed to give him emergency powers to deal with the country's economic and political crises, Ne Win finally turned against his patron in 1962, establishing the military regime that has governed Burma, in various forms, ever since. Under Ne Win's leadership – first as head of a military junta, then as head of a partly civilian regime, and finally as a shadowy figure behind the scenes from 1981 – the regime has isolated Burma from the outside world, ruthlessly suppressed the democratic opposition while failing to control or satisfy the ethnic minorities, presided over an immense increase in the trade in narcotics, imposed forced labor on thousands of its subjects, and forcibly resettled tens of thousands of others. Ne Win, whose faith in military repression and numerological mysticism has long outweighed any commitment he ever had to "socialism" or democracy, is probably no longer actively involved in decision-making, but the present condition of Burma can be largely attributed to him.

Ngo Dinh Diem (1901–63): Prime Minister of (South) Vietnam from 1954 to 1955, President from 1955 to 1963. Diem's father was an adviser to Emperor Than Thai when Diem, his third son, was born, but turned to farming after the French deposed the Emperor in 1907. Diem grew up a devout Catholic (he was christened Jean-Baptiste), and was a student at Hanoi University at the same time as his future opponent Vo Nguyen Giap (see below). In 1926, Diem was appointed a provincial governor under the French. In 1933, he became Interior Minister in the government formed by the French in the name of the young Emperor, Bao Dai. By then, however, Diem believed that the French should have started ceding more powers to Vietnamese. He left the government after only three months and withdrew to the family home in Hue. In September 1945, he was detained by the Vietminh, whose agents assassinated one of his brothers, Ngo Dinh Khoi, a provincial governor, shortly afterward. Diem then rejected Ho Chi Minh's request that they collaborate against the French. In 1954, after four years spent largely in seminaries in the United States, Belgium, and France, he accepted Bao Dai's offer of the prime ministership. His first major initiative was the suppression of a powerful criminal syndicate, the Binh Xuyen; his second was the deposition of Bao Dai and the creation of a republic. During his eight years as self-appointed President, Diem collaborated with the United States in a secret program of military strikes against the North. He also created allegedly secure "strategic hamlets," to which villagers were forced to move; openly favored the Catholic minority, led by his brother Ngo Dinh Thuc, Archbishop of Hue; and fostered an authoritarian system policed by the security services and the Revolutionary Labor Party, both led by yet another brother, Ngo Dinh Nhu. Amid protests by Buddhist organizations, the military assassinated both Diem and Nhu, and seized power in Saigon with the blessing of at least some sections of the US administration.

Norodom Ranariddh, Prince (1943–   ): First Co-Prime Minister of Cambodia from 1993 to 1997. Prince Ranariddh, a son of Sihanouk and thus a member of the Norodom branch of the Cambodian royal family, spent 16 years studying and lecturing in France before joining his father's resistance movement in Bangkok in 1983. He became commander of the royalist army, ANKI, on the Thai-Cambodian border, but after the signing of the Paris Peace Agreement in 1991 he returned to Phnom Penh. He led the royalist party FUNCINPEC to victory in the elections of 1993, but was forced to share equal power with Hun Sen in a coalition government. He fled to Bangkok after factional fighting broke out in July 1997 but returned, under UN protection, to lead the largest and most successful faction of FUNCINPEC in the elections of 1998. He has since served as presiding officer of the National Assembly.

Norodom Sihanouk: see Sihanouk

Nu, U (1907–95): Prime Minister of Burma from 1947 to 1958, and from 1960 to 1962; later a prominent opponent of the present military regime. Nu, the only democratically elected head of government in Burma's history, came from a relatively prosperous background, as the son of a rice merchant ("U" is an honorific epithet for mature men, not a name). While he was studying at the University of Rangoon, he joined Aung San (see above) and other nationalists in leading a students' strike. From 1936 to 1940, he divided his energies between political campaigning, notably in the Dobama Asiayun movement, and literary activities, as a writer and head of a publishing company. Jailed by the British colonial authorities in 1940, he was released in 1942 to take office in the government created by the Japanese occupation forces under the leadership of Ba Maw. In 1946, Nu became a Vice President of the principal nationalist organization, the Antifascist People's Freedom League; in 1947, he became Speaker of the Constituent Assembly; and then, after the assassination of Aung San and six of their colleagues, he reluctantly accepted the prime ministership. Over the next 11 years, Nu attempted to govern an increasingly fractured nation by balancing policies of social reform, influenced by his devout Buddhism, with campaigns of military repression against ethnic minorities and Communist rebels. He agreed to vacate the prime ministership for two years so that the commander in chief of the armed forces, General Ne Win (see above), could use emergency powers to secure civil peace. When that move failed, Nu returned to office, after winning an overwhelming majority in elections, only to be overthrown by Ne Win after two more years of ineffectual rule. In 1967, Nu was allowed to leave Burma, where he had been placed under house arrest. He spent 13 years in exile, often on retreat in monasteries. From 1980, when he returned to Burma, he avoided political involvement, concentrating instead on literary studies, but in 1988 he helped to found the National League for Democracy and accepted the post of "alternative" Prime Minister in opposition to the military regime. He was promptly detained, and was not released until 1992. Since his death, three years later, the regime has made every effort to eliminate his name and deeds from the consciousness of the Burmese. Nu was always a reluctant leader, and proved incapable of controlling the military or conciliating his country's ethnic minorities. Nevertheless, for those Burmese who have been able to learn about him, he remains a potent symbol of resistance to the present regime.

Pham Van Dong (1906–2000): Foreign Minister of Vietnam from 1954 to 1976, Prime Minister from 1955 to 1987 (post retitled Chairman of the Council of Ministers in 1981). Like many other leaders of the Communist Party of Vietnam, Pham Van Dong was born into privilege, in his case as the son of a civil servant in Hue, the former imperial capital. By 1926, however, he was a convinced supporter of national independence and social revolution. In that year, he joined Ho Chi Minh (see above) at his base in southern China. From 1929 onward, he alternated periods of underground agitation with periods in jail under the French colonial authorities, and in 1945 he was present, as a leading member of the Viet Minh, at Ho's proclamation of the Democratic Republic of Vietnam in Hanoi. He then took a leading role in formulating economic policy until 1954, when he attended the Geneva Conference as Foreign Minister, reluctantly accepting what he was told would be a purely temporary division of his country. During the years of war that intervened between the conference and reunification, he played a crucial role in directing the civilian economy in support of the war effort, and, it appears, in distancing Vietnam from China. After reunification, however, the aging Vietnamese leadership found it increasingly difficult to maintain the centralized, state-directed economy that they had grown accustomed to. As Pham Van Dong himself told Stanley Karnow, "waging a war is simple, but running a country is difficult" (quoted in Karnow's book, cited in Chapter 11, p. 36). He retired one year after the launch of the *doi moi* program of reforms. In 1999, he published an article denouncing the increasing corruption, the persistent poverty, and the abandonment of socialist ideals that seemed to him to be disfiguring the country he had helped to create.

Phibun Songkhram (1897–1964): Prime Minister of Thailand from 1938 to 1944 and from 1948 to 1957. Phibun was originally named Plaek Khittasangkha; the name by which he became famous derives from the noble title he was granted in 1928. As a career soldier, he had spent the years 1924–27 receiving training in France. After his return to the country then known as Siam, he supported Pridi Phanomyong (see below) in launching the revolution of 1932 that brought the absolute monarchy to its end. In 1934, he became Minister of Defense. During his first term as Prime Minister, he imposed the name "Thailand" (in 1939), encouraged a mixture of westernization and nationalism, and forged an alliance with Japan. Once it had become clear that Japan would be defeated, he was removed from office, but he returned to power, at the invitation of the army, four years later. During his second term, he was actively anti-Communist, sending troops to fight in Korea, supporting the British in Malaya, and welcoming the creation of the Southeast Asia Treaty Organization, which had its headquarters in Bangkok. After nine more years of his erratic and corrupt government, the army moved against him once more, and he went into exile in Japan, where he died. Admirers and critics alike have compared Phibun to Kemal Atatürk of Turkey, another soldier turned politician who sought to balance tradition with modernity, and nationalism with westernization; others, however, have dismissed Phibun as an opportunist with little claim on the attention of contemporary Thais.

Pol Pot (1925–98): leader of the Khmer Rouge; Prime Minister of Democratic Kampuchea (Cambodia) from 1975 to 1979. Pol Pot's real name was Saloth Sar; it appears that he assumed the name Pol Pot in 1975. He was also known as "Brother Number One." Born in the province of Kompong Thom, a member of a relatively prosperous peasant family with connections at the royal court, he was educated in the capital, Phnom Penh, and was then sent to Paris in 1948 on a government scholarship. In 1953, he failed examinations in radio engineering and returned home to join what was then the Indochinese Communist Party, linked to the Viet Minh but also active in Cambodia and Laos. In 1960, following the violent death of Tou Samouth, the leader of the Cambodian Communists (probably on Saloth Sar's orders), Saloth Sar became head of a separate Cambodian Workers or Communist Party. In subsequent years, he ensured that it was aligned with China, which gave his movement political support, financial aid, and arms until 1993, and possibly later too. In 1963, Saloth Sar left Phnom Penh to take direct control of the Communist guerillas in northeastern Cambodia, who were soon to be nicknamed the "Khmers Rouges" (Red Khmers) by Sihanouk (see below). Between 1970 and 1975, nominally in coalition with the deposed Sihanouk and with sporadic aid from North Vietnam, which apparently hoped to use the Khmer Rouge against South Vietnam and the United States, Saloth Sar directed the gradual takeover of Cambodia from the military regime of Lon Nol. Victory came in April 1975, when the Khmer Rouge entered the capital. The new regime emptied the cities, imposed forced labor, eliminated "western" education and medicine, persecuted the Cham and Vietnamese minorities, and sent many of its opponents, including increasing numbers from within the Khmer Rouge itself, to be tortured and executed at Tuol Sleng, a former high school in Phnom Penh, and at other centers. Pol Pot then antagonized the government of newly unified Vietnam by launching a campaign to seize "Kampuchea Krom," the region of southern Vietnam that had been part of the medieval Cambodian empire. The Vietnamese invasion of 1978–79 forced Pol Pot and his colleagues to flee back to the jungles of the Northeast, from where they continued to direct campaigns against the new Cambodian government, and against the UN administration installed in 1993. In 1997, rivals within the Khmer Rouge staged a show trial of Pol Pot, not on any charges relating to the events of 1975–79, but in revenge for his ordering the killing of one of his colleagues, Son Sen. Pol Pot was then kept under house arrest until his death.

Pridi Phanomyong (1900–83): Regent of Thailand from 1941 to 1946, Prime Minister in 1946. Pridi, also known by his noble title, Luang Pradist Manudharm, completed a doctorate in law in Paris in 1927 and came under the influence of French socialist thought. Back in Siam, as Thailand was then known, he formed the People's Party, which sought to modernize the country's economic and political systems. In June 1932, Pridi joined with others, including Phibun Songkhram (see above), to carry out the first of Thailand's several military coups. He then went on to take the lead in writing the country's first Constitution. Pridi struggled to maintain his position after the revolution of 1932, in the face of accusations of Communism, linked to his national economic plan, and the rise of the military faction in the People's Party. After a brief period of exile, he rejoined the government, becoming successively Minister of the Interior, of Foreign Affairs, and of Finance. He also founded what is now Thammasat University. He left office in 1941, being opposed to Phibun's plans for an alliance with Japan, and became Regent while the young King Ananda Mahidol studied in Switzerland. In 1942, he accepted US support for the creation of the clandestine Free Thai Movement, which was opposed to the Japanese military presence in Thailand, and in 1944 he took part in the overthrow of Phibun. Two years later, Pridi became the first Prime Minister to enter office as the result of democratic elections, but he was ousted a few months later, following the mysterious death of the King. He left Thailand in 1947, settling first in Singapore, then in China, and finally, from 1970, in Paris. He continued to voice his criticisms of successive Thai military regimes, while his writings in exile displayed an eclectic mixture of elements from Buddhism, Marxism, and liberal democratic thought. Pridi was less flamboyant and less successful than his rival and colleague Phibun, but he too has become a towering figure in the history of modern Thailand.

Rahman, Tengku Abdul (1903–90): Leader of the United Malays National Organization (UMNO) from 1952 to 1970, Prime Minister of Malaya from 1957 to 1963, Prime Minister of Malaysia from 1963 to 1970. As his title indicates, the Tengku was one of the sons of the Sultan of Kedah and his sixth wife, a Thai princess. After studying law in Britain, he joined the civil service in Kedah in 1931. He was among the founders of UMNO in 1946, and became its leader six years later. As an early advocate of cooperation among the parties representing the main ethnic groups in what was then Malaya, he formed the Alliance, which went on to win elections to the Federal Legislative Council in 1955. Rahman then led his country to independence, became the main architect of the wider Malaysian federation in 1963, and successfully pursued policies of peaceful relations with neighboring countries, including an initially hostile Indonesia, and internal harmony, based on maintaining the delicate balance between the majority Malays and the Chinese and Indian minorities. When this balance broke down in riots in Kuala Lumpur in May 1969, Rahman took responsibility, and retired from office the following year. He retained the respect of many Malaysians, however, and his occasional criticisms of the present government, as well as his repeated pleas for social tolerance, accordingly carried some weight.

Sihanouk (1922–): King of Cambodia from 1941 to 1955, and since 1993; Prime Minister from 1955 to 1960; Head of State from 1960 to 1970, from 1975 to 1976, and from 1991 to 1993. Sihanouk, the great survivor of Cambodian politics, was educated in Saigon and Paris. He was selected to become King by the French colonial authorities, who believed that he could be manipulated, thus arousing the lasting resentment of relatives who had expected to succeed the previous King, Sisowath. Sihanouk succeeded in securing Cambodia's independence and, at least for a while, its neutrality. In 1955, he abdicated in favor of his father, Suramarit (who died five years later), founded a party called the

Popular Socialist Community, and proceeded to win several elections, partly by exploiting his charisma as a member of the royal family, partly by manipulating the voting system. Throughout the 1960s, he oscillated between friendliness toward the United States and conciliation of North Vietnam, with the result that both sides in the Vietnam War came to distrust him. After he was overthrown by the military in 1970, Sihanouk made the fateful decision to align himself with the Chinese government and its clients, the Khmer Rouge. He initially served as a powerless figurehead for the regime they established in 1975, only to be deposed in 1976. He eventually made peace with the Khmer Rouge, however, and in 1982 became the figurehead President for their coalition with FUNCINPEC, a party founded in 1981 and led by Sihanouk's son Ranariddh, and other groups opposed to the government installed by the Vietnamese in 1979. The recognition of this coalition by the major powers helped to restore Sihanouk's reputation: in 1991, he was reinstated as Head of State, and in 1993, after serving as President of the Supreme National Council, he became King once again, albeit with very limited powers. He has spent much of his time since then receiving medical treatment in China.

Souphanouvong, Prince (1902–95): President of the Lao Democratic People's Republic from 1975 to 1986. Souphanouvong was born the youngest of the 22 sons of Prince Boun Khong, the most powerful figure at the princely court of Luang Prabang. Since none of the three Lao principalities, then under French rule, had any institutions of postelementary education, Souphanouvong studied in Hanoi, where he met and married a Vietnamese woman, and in Paris. He returned home in 1938 to take up a career as a civil engineer. However, he had already begun to favor independence for a reunited Laos, and in 1945 he joined the Lao Issara (Free Laos) movement, led by one of his brothers, Prince Phetsarath Ratanavongsa, who was briefly Prime Minister in the aftermath of the upheavals of World War II. When the French regained control of the country, Souphanouvong accompanied his brother and other Lao Issara leaders into exile. In 1949–50, the Lao Issara movement split: a more moderate wing, led by yet another son of Boun Khong, Prince Souvanna Phouma, made peace with the French and entered government under French auspices, while Souphanouvong's more radical wing formed the Pathet Lao (Land of the Lao) movement, in collaboration with the Viet Minh. There followed 25 years of civil war, punctuated by ceasefire agreements and shortlived "governments of national unity," and complicated by the political and military incursions of North Vietnam, South Vietnam, and the United States. Throughout these years, Souphanouvong cultivated his image as the "Red Prince," the apparent focus of decision-making within the Pathet Lao, and the commander in chief of its Lao Army of Liberation and Defense. However, by the time the movement seized total power, in 1975, he had conceded most of his authority to Kaysone Phomvihane, who became Prime Minister when Souphanouvong became President of the new republic. He resigned from the presidency after suffering a stroke, and left the Politburo of the Lao People's Revolutionary Party, the dominant force within the Pathet Lao movement, in 1991. As is the case with other Asian Communist leaders, debate continues as to Souphanouvong's character and his contribution to his country's recent history. Was he simply an opportunist politician, deploying radical slogans to satisfy urban intellectuals while relying on his membership of the traditional elite to retain authority among the peasantry? Or did he have, at least initially, a real commitment to Marxist ideals, however attenuated they may have been by later struggles for power within his own movement and beyond it?

Souvanna Phouma, Prince (1901–84): Prime Minister of Laos from 1951 to 1954, from 1956 to 1958, and from 1960 to 1975. Like his half-brother Souphanouvong (see above), Souvanna Phouma studied in Vietnam and France, and was trained as an engineer. He too joined the Lao Issara independence movement, but opted for the faction that sought compromise and reconciliation, rather than the revolutionary wing. During his first period in office, he negotiated independence from France. From this point onward, he was consistent in seeking to foster compromise and in taking a leading role in the repeated negotiations to form coalition governments. In 1973, he overcame the opposition of rightwing politicians and signed a ceasefire with the Pathet Lao. His last government, which included Souphanouvong, left office shortly before the revolution, and he then accepted an appointment as senior adviser to the new government.

Suharto (1921–): President of Indonesia from 1968 to 1998. Suharto was the son of a low-level Javanese official serving in the Dutch colonial administration. Following a brief experience of formal education, he enrolled successively in the Dutch army, one of the police forces created by the Japanese, a wartime paramilitary unit, and the Indonesian national army established in 1945. During the 1950s, he commanded various campaigns in the outer islands, but he was removed from his Central Java command in 1959 because of his business dealings. By 1963, however, he had once again become a major figure in the military, and was entrusted with the command of the Indonesian takeover of West Papua. As a Major General, he was well-placed to lead the military campaign two years later to suppress the Communist Party and curb the powers of President Sukarno. Having replaced Sukarno, first as Acting President from 1967, then as President, Suharto oversaw the stabilization of the economy and the consolidation of an authoritarian political order. By the mid-1990s, however, he had become more isolated, and was being criticized more and more openly for his favors to his children and cronies, the distortion of the economy, and the harsh suppression of dissidents. When the Asian crisis struck the country in 1997–98, an aged Suharto at first responded ineffectively, then transferred power to his Vice President, B.J. Habibie. In 2000, the government of President Abdurrahman Wahid launched a prosecution of Suharto, in connection with his corrupt practices while in office. However, his lawyers, arguing that he was too ill to stand trial, were succeeding in keeping him out of court at the time this book went to press.

Sukarno (1901–70): President of Indonesia from 1945 to 1968. Sukarno was a son of a Javanese schoolteacher of aristocratic descent, and his wife, who was probably Balinese. After high school studies in Surabaya, where he also followed his father in joining the Theosophical Society, and imbibing its eclectic mixture of eastern and western philosophies, Sukarno studied engineering at a technical college in Bandung, graduating in 1926 with a dissertation on harbor design. By then, he had married and divorced his first wife, and married his second: there were five more to come over the years. Sukarno then plunged into the unstable and fragmented world of the independence movement, advocating unity among nationalists, Islamists, and Marxists (a notion that, however much it would have startled both Mohammed and Karl Marx, won widespread support in Indonesia and has analogies throughout Asia). In 1927, Sukarno became chairman of the largest pro-independence group, the Perserikatan Nasional Indonesia (Indonesian National Association, or PNI). As early as 1928, he was predicting that Japan would attempt to control the whole of East Asia, thereby giving Indonesia its opportunity to expel the Dutch. In the event, he was jailed from 1929 to 1931, and exiled to the outlying island of Flores from 1933 until 1942, when his prediction began to come true. Sukarno and his colleagues in the PNI collaborated with the Japanese military administration throughout World War II, to the extent of participating in decisions about the dispatch of forced labor units to other parts of the short-lived "Greater East Asia Co-Prosperity Sphere." In return, he was allowed to retain his control of the nationalist movement. It was Sukarno who proclaimed Indonesia's independence after Japan's defeat in 1945. For the next 22 years, he continued to be the most powerful individual in the new country. However, his freedom of action was limited, first, by the necessities of war against the Dutch, up to 1949; then by the demands of party politicians, who, unlike Sukarno himself, had to submit to elections (though only once, in 1955); and then, during the era of "Guided Democracy," from 1957, by an unstable and eventually disastrous equilibrium between the armed forces and the Communist Party. The military had acquired enormous power and wealth, while the Communists seemed to Sukarno to be useful in his struggles against regional movements for autonomy, as well as against the Islamist parties that had dominated electoral politics. In 1965, when some Communist elements joined with sections of the military to start a coup, the President was forced to favor the military leadership. The ensuing upheavals, which led to the deaths of hundreds of thousands, resulted in the seizure of power by Suharto. In 1967, Sukarno surrendered all his powers; a year later, he resigned his office. By the time of his death, he was effectively under house arrest.

Vo Nguyen Giap (1912–   ): Vietnamese Communist politician and soldier, founder of the People's Army of Vietnam, Deputy Prime Minister of Vietnam from 1976 to 1980. Giap was born into a relatively prosperous and literate family that was able to support his law studies at the University of Hanoi, where his fellow students included Ngo Dinh Diem, the future President of South Vietnam. In 1938–39, Giap, already a member of the Communist Party, wrote, with Truong Chinh, an influential text on the role of the peasantry in revolution, modifying Marxist orthodoxy by allotting them an equal role in partnership with the proletariat, while resisting the tendency, later dominant in most forms of "Asian Marxism," to elevate the peasantry to a leading role. Theory gave way to practice in 1940, when Giap left Hanoi – and his first wife, a fellow Communist who was to die in prison – to join Ho Chi Minh (see above) in China. Giap returned in 1941 to begin organizing armed resistance to the Japanese inside Vietnam. Giap was also largely responsible for the purges of Trotskyists and other dissidents within the Communist Party at this time. As founder and commander of what became the People's Army of Vietnam, Giap secured victory for the Communist-led national liberation movement against the French at Dien Bien Phu in 1954, directed the military activities of the South Vietnam National Liberation Front (Viet Cong) from 1960, organized the Tet offensive in 1968, and laid down the strategy for the overrunning of South Vietnam and national unification, achieved in April 1975. Giap retired from public life in 1982, apparently after opposing the invasion of Cambodia that began in 1978.

Wahid, Abdurrahman (1940–): President of Indonesia since 1999. Wahid comes from a family of prominent Javanese Moslem intellectuals. The Islamic social organization Nahdlatul Ulama (NU, "Revival of Religious Scholars"), often said to be the largest of its kind in the world, was founded by his grandfather and led later by his father, Wahid Hasyim, who was also Minister of Religious Affairs. During the 1960s, Wahid studied at Al-Azhar University in Cairo. During the 1970s, after returning to Jakarta, Wahid became involved with Islamic study groups, marking his emergence as a Moslem intellectual. In 1984, he rose finally to the leadership of the NU, although he continued to write regularly for the middle-class secular newsweekly, *Tempo*. Under Suharto's rule, he bravely demanded political and social change, and in 1991 he became chairman of a prominent discussion group, the Forum for Democracy. The National Awakening Party, which he went on to form in 1998, has an inclusive platform, despite its Islamic credentials, and Wahid has consistently opposed the idea of making Indonesia an Islamic state. In 1999, amid dissatisfaction with the presidential candidacies of Megawati Sukarnoputri and B.J. Habibie, Wahid caused widespread surprise by emerging as the winner. His presidency initially fostered high expectations, but it has been characterized largely by political rivalries and policy failings. In August 2000, Wahid transferred much of his power to his Vice President, Megawati Sukarnoputri.

APPENDICES

Appendix 4
Political and Economic Institutions

Political Systems

The countries of Southeast Asia have had experience of a wide range of political systems, reflecting the variations in their precolonial and colonial legacies, as well as in their development since independence. Four basic types may be distinguished.

Parliamentary democracies: In Cambodia, Malaysia, Singapore, and Thailand, a head of state with limited and largely ceremonial powers presides over a political system dominated by a government formed by the largest party or parties in the lower house of a bicameral legislature or, in Singapore only, the unicameral Parliament. However, there are significant differences of detail among them.

The State of Cambodia (Preah Reach Ana Pak Kampuchea) has experienced a unique series of regime changes and social upheavals in the postwar period. The constitutional monarchy established under French colonial administration, and given a new basis in the Constitution of 1953, served as a vehicle for personal rule by Norodom Sihanouk, first as King, then as Prime Minister and finally as Head of State, until 1970, when a military regime was installed under Lon Nol. His "Khmer Republic" was overthrown in 1975 by the Khmer Rouge, which imposed an extreme Maoist dictatorship under the name of "Democratic Kampuchea." In 1979, a provisional government of the "People's Republic of Kampuchea," financed and defended by the Vietnamese, replaced the Khmer Rouge in most parts of the country. Meanwhile, opponents of this regime, including the Khmer Rouge, formed a coalition government in exile, which occupied Cambodia's seat at the UN and was recognized by most other countries. Cambodia was administered by the UN for a transitional period (1992–93). Under the Constitution of 1993, Sihanouk is King once again – a post he first occupied in 1941 – while the Prime Minister and Council of Ministers are responsible, not to him, but to the National Assembly, the lower house of the legislature, which is elected for five-year terms. Since 1998, there has also been a Senate. The traditional division of the country into 20 largely rural provinces and three municipalities – Phnom Penh, Keb, and Preah Seihanu – has also been restored.

Malaysia has a system of government largely based on the British model, with a Prime Minister and Cabinet chosen from the ranks of the largest party in the House of Representatives (Dewan Rakyat), the lower house of Parliament (Parlimen), which is elected for five-year terms. However, other features of the political system are more distinctive. Instead of a president or a single hereditary monarch, Malaysia has a "paramount ruler" (Yang di-Pertuan Agong), who is elected from among their own number by the hereditary rulers of nine of the 13 states, and serves for five years. The Yang di-Pertuan Agong chooses 43 of the 69 members of the Senate (Dewan Negara), the upper house of Parliament. The other 26 senators are chosen by the legislatures of the 13 states: Johor, Kedah, Kelantan, Negeri Sembilan, Pahang, Perak, Perlis, Selangor, and Terengganu (which all have hereditary rulers); and Melaka, Pulau Pinang, Sabah, and Sarawak (which have governors appointed by the federal government). Under the terms of the agreements that created Malaysia in 1963, Sabah and Sarawak have greater autonomy than the 11 peninsular states, which previously formed the Federation of Malaya. In addition, there are two federal territories directly ruled by the federal government: Wilayah Persekutuan (which contains the capital, Kuala Lumpur) and Labuan.

The Republic of Singapore (Hsin-chia-p'o Kung-ho-kuo in Mandarin, Republik Singapura in Malay, Singapore Kudiyarasu in Tamil) has a President, elected by the unicameral Parliament up to 1993 but directly elected since then. Executive power is in the hands of the Prime Minister and the Cabinet. They are responsible to Parliament, which is elected for five-year terms through a voting system that ensures overwhelming majorities for the ruling People's Action Party. For example, in the most recent elections, held in 1997, it won 65% of the votes cast, yet it took all but two of the elective seats. Parliament also has up to three seats allocated to representatives of opposition parties and up to nine occupied by nominated members; there are limitations on the voting rights of all these special members. Voting is compulsory in Singapore.

Thailand (Muang Thai, "Land of the Thais," or Prathet Thai, "Kingdom of the Thais") has undergone long periods of military or military-dominated rule, but is now governed under the terms of the 1997 Constitution, the 16th to be enacted since 1932. The National Assembly (Rathasapha) comprises a Senate (Wuthisapha), with appointed members who are to be gradually replaced by senators directly elected for six-year terms from 2000 onward; and a House of Representatives (Sapha Phuthaen Ratsadon), with members elected for four-year terms.

Presidential democracies: Both Indonesia and the Philippines have undergone periods of dictatorship in which formal political institutions have either ceased to function or been used merely to endorse decisions taken elsewhere. Both have now returned to the presidential systems with which they began their existence as independent states.

The Republic of Indonesia (Republik Indonesia) is governed under the Constitution promulgated at independence in 1945, abandoned in 1949, restored in 1959, and frequently amended since 1966, most recently in August 2000 (see Appendix 1). Power is shared between the elected House of Representatives (Dewan Perwakilan Rakyat), on the one hand, and a President and Vice President, on the other. Legislators and executive officers alike serve five-year terms. Two special features make Indonesia's legislature unique. First, suffrage extends to all citizens aged 17 or above, as well as to all married citizens whatever their age, but excludes members of the armed forces. Second, in addition to the 462 elected members (425 until 1999), there are 38 appointed representatives of the military (75 until 1999). The two executive officers are formally selected by the People's Consultative Assembly (Majelis Permusyawaratan Rakyat), which comprises the 500 members of the House of Representatives and 500 delegates from the 27 provinces and three special districts (Jakarta, Aceh, and Yogyakarta), the armed forces, and other organizations. The Assembly, which meets once a year, also has the power to amend the Constitution.

The Republic of the Philippines (Republika ng Pilipinas) derives much of its political culture from the United States, with which it had a quasicolonial relationship from 1898 to 1946. A new Constitution, which came into force in 1987, relaunched the system of government that had been corrupted under the regime of Ferdinand Marcos. The President and Vice President are directly elected for six-year terms, and cannot be re-elected. The President, who is both head of state and head of government, appoints his own cabinet without reference to the Congress (Kongreso). The latter has two houses: the House of Representatives (Kapulungan Ng Mga Kinatawan), with 221 members elected for three-year terms; and the Senate (Senado), with 24 members, half of whom are elected in turn for six-year terms. The Philippines departs from the US model in being a relatively centralized state, in which local government is conducted through 75 provinces and 61 cities. Since 1996, however, there have been special provisions for greater autonomy in the mainly Moslem provinces of Mindanao.

One-party states: This remains the standard term for describing the polities of Laos and Vietnam (among other countries), although there are other terms, such as "market Stalinist regimes," that arguably help to distinguish such polities from non-Communist one-party states.

In the Lao People's Democratic Republic (Sathalanalat Paxathipatai Paxaxon Lao), the Lao Front for National Construction, led by the People's Revolutionary Party, remains the only legal political organization. The Front dominates representation in the National Assembly, which is elected for five-year terms, appoints the President (also for a five-year term), and formally approves the President's appointments to the government, the Council of Ministers. The Front also dominates the institutions that control the country's 16 provinces, its capital city, Vientiane, and the "special zone" of Xaisomboun.

Similarly, in the Socialist Republic of Vietnam (Cong Hoa Xa Hoi Chu Nghia Viet Nam), the Fatherland Front, led by the Communist Party, is the only legal political organization. The Party dominates the National Assembly (Quoc Hoi), which is elected for five-year terms, and which chooses the President, the Prime Minister, and the Cabinet from among its own members, also for five-year terms.

Dictatorships: Brunei and Burma are now the only countries in Southeast Asia that lack even the formalities of political competition within representative institutions. In almost every other respect, however, their histories and societies are very different.

Brunei (in full, Negara Brunei Darussalam, "The State of Brunei, Abode of Peace") is formally a constitutional monarchy. Its first and only Constitution, adopted in 1959, established an elected Legislative Council (Majlis Masyuarat Megeri), but its powers were limited after a state of emergency was declared in 1962, and, following the attainment of full independence from the United Kingdom in 1984, it was abolished. There was a brief revival of constitutional politics, with two political parties being officially registered in 1985, but since 1988 only one of these has had legal status, and it barely functions. The hereditary Sultan serves as his own Prime Minister and Defense Minister, and is advised by a variety of bodies, all of which he appoints. These include, most notably, the Council of Ministers and the Religious Council, which is chiefly concerned with administering Sharia law for the Sunni Moslem majority (around two thirds of the population).

The Union of Burma (Pyidaungzu Myanma Naingngandaw) was declared a republic upon independence in 1948. Initially, it retained a parliamentary system of government on the British model, but with a greater degree of centralization than in colonial times. In particular, the Shan States, controlled by their traditional leaders in colonial times, lost their autonomy. The position of the Shan and other minorities has been a major issue in Burmese politics ever since. Seven districts for the Burman majority and seven states for the minorities were delineated in the first years of independence, but they were never fully functional. After the military coup in 1962, they were swept away, along with the established institutions of national government, which were replaced by a People's Assembly (Pyithu Hluttaw) and a Council of Ministers. A new Constitution introduced in 1974 was suspended in 1988 when the military once again took power and established the State Law and Order Restoration Council (SLORC), renamed the State Peace and Development Council (SPDC) in 1997. A new People's Assembly was elected in 1990 and a National Coalition Government was formed by its leading members, but the Assembly has never met and the government was forced to flee to a rebel-held border area. A "constitutional convention" created and supervised by the military regime first met in 1993. The military regime is opposed by the National League for Democracy, the All-Burma Student Democratic Front, and several groups representing ethnic minorities.

East Timor (Timor Lorosae): Since its separation from Indonesia in 1999, East Timor has been under temporary UN administration. Sergio Vieira de Mello, Special Representative of the UN Secretary General and Transitional Administrator, collaborates with the National Consultative Council, which was set up in November 1999 as a joint East Timorese-UNTAET body with 15 members, but restructured in June 2000 as an exclusively East Timorese body with 33 members; with the Transitional Cabinet of four East Timorese and four UNTAET officials, established in July 2000; and with 13 District Advisory Councils, set up in April 2000.

Economic Institutions

Stock exchanges: Five exchanges in the region are full members of the International Federation of Stock Exchanges (based in Paris and known as the FIBV): the Jakarta Stock Exchange in Indonesia, the Kuala Lumpur Stock Exchange in Malaysia, the Philippine Stock Exchange, Inc., in Pasig City, the Stock Exchange of Singapore Ltd., and the Stock Exchange of Thailand, in Bangkok. Another Indonesian exchange, the Surabaya Stock Exchange, is a corresponding member of the FIBV. Two other exchanges have become increasingly important in recent years: the Indonesia NET Exchange on a national scale, and the Singapore International Monetary Exchange Ltd. (Simex) on a global scale. A stock exchange started trading in Ho Chi Minh City in southern Vietnam in October 2000 – the first to be permitted in the country since unification in 1975 – but, like the stock exchanges in Cambodia and Brunei, it does very little business. Laos and Burma do not have officially recognized or regulated stock exchanges.

Labor unions: According to data from the International Labor Organization (ILO), as of 1995 membership of labor unions stood at 22.8% of the nonagricultural labor force in the Philippines, 13.5% in Singapore, 11.7% in Malaysia, 3.1% in Thailand, and 2.6% in Indonesia (then taken to include East Timor). All five of these countries have "tripartite" institutions that bring together representatives of government, employers, and unions, and wield some influence over the setting of wages (including minimum wages), the promotion of greater productivity, and other labor issues. Union activity is extremely limited in Brunei and Cambodia; labor organizations are still largely controlled directly by the state in Laos and Vietnam, although Vietnam enacted a more liberal labor code in 1994; and in Burma, forced labor is commonplace and unions have suffered intensive repression along with other groups opposed to the military regime. Orderly and legal processes of collective bargaining remain relatively undeveloped, except in some sectors in Malaysia, the Philippines, and Singapore, and throughout the region there is widespread disregard even for the minimal "core labor standards" devised and monitored by the ILO.

SEATO and ASEAN

Despite its name, the Southeast Asia Treaty Organization (SEATO), first suggested by the British Prime Minister Winston Churchill and established under the Treaty of Manila in September 1954, had only two member states in the region itself – the Philippines and Thailand – the other members being Australia, France, New Zealand, Pakistan, the United Kingdom, and the United States. A protocol to the founding treaty placed Cambodia, Laos, and the Saigon regime in Vietnam under SEATO's "protection," but the members of the alliance never agreed a joint policy on Indochina in particular or the region as a whole. Pakistan departed in 1972, Thailand and the Philippines suggested in 1975 that the organization was no longer needed, and SEATO was dissolved in June 1977.

By contrast, ever since its foundation under the Bangkok Declaration of August 1967, ASEAN has focused on political and economic goals rather than military cooperation, and its membership has been limited to countries in the region. Its five founding members – Indonesia, Malaysia, Singapore, the Philippines, and Thailand – shared a hostility to Communism in general, and the regimes in Indochina in particular. Nevertheless, from 1979 onward they and the sixth member, Brunei (from 1984), supported the efforts of the Khmer Rouge to regain power in Cambodia, in coalition with other parties, on the grounds that these particular Communists were preferable to the Cambodian government installed by the Vietnamese. In July 1995, following the completion of the UN-brokered peace process in Cambodia, and the introduction of economic reforms in all three states in Indochina, Vietnam joined ASEAN, and its seven members then agreed with Burma, Cambodia, and Laos on a declaration making Southeast Asia a "nuclear-free zone." By 1999, these three countries had also joined ASEAN (see Appendix 1). In 2000, ASEAN adopted the EU concept of the "troika," a group of three ministers, its membership rotating among member states, in an attempt to formalize closer cooperation on common interests.

ASEAN's mechanisms for cooperation and consultation have been extended to cover relations with certain countries outside the region. Papua New Guinea has observer status; China and Russia have been designated "consultative partners," and hold regular summit meetings with ASEAN; and the ASEAN Regional Forum (ARF), founded in June 1994, brings the member states of ASEAN together with Papua New Guinea, Russia, China, and eight "dialogue partners": Australia, Canada, India, Japan, South Korea, New Zealand, the United States, and the European Union. North Korea also sent a delegation to the ARF meeting in July 2000. Since July 1991, there have also been regular "postministerial" conferences between the ASEAN states and, initially, China and Japan, joined later by South Korea. These three countries also took part in the first annual East Asia Cooperation or "ASEAN+3" summit in November 1999.

ASEAN has probably made its greatest impact in the field of economic cooperation. Along with Brunei, which joined in January 1984, the founding members agreed in 1992 to begin progress toward an ASEAN Free Trade Area (AFTA), starting with a common preferential tariff from 1993. Agreements have also been made on common goals to be pursued in relation to environmental protection, and on relations with Northeast Asia (Japan, China, South Korea, North Korea, and Taiwan).

Other International Organizations

Worldwide: The Philippines was one of the founder members of the UN, in 1945; Thailand joined in 1946, Burma in 1948, Indonesia in 1950, Cambodia and Laos in 1955, Malaya in 1957 (and Malaysia by succession in 1963), Singapore in 1965, Vietnam in 1977, and Brunei in 1984. East Timor, illegally occupied by Indonesia between 1975 and 1999, is now under special administration by the UN pending full independence, and does not yet participate in any of the organizations mentioned from this point onward. All the member states of ASEAN now take part in the activities of the UN Economic and Social Commission for Asia and the Pacific (ESCAP), established in 1947 as the UN Economic Commission for Asia and the Far East (ECAFE), as well as those of the IMF and the International Bank for Reconstruction and Development (the World Bank). Seven ASEAN states are full members of the World Trade Organization, while Vietnam is an applicant, and Cambodia and Laos have observer status. Only five of the 10 ASEAN states – Brunei, Cambodia, the Philippines, Thailand, and Vietnam – have ratified the International Covenant on Civil and Political Rights, enacted by the UN in 1966.

All 10 ASEAN countries belong to the Nonaligned Movement, which was founded in 1961 and now has 113 members, and to the Group of 77, which was established in 1967 and retains its original name even though it now has 130 members. Indonesia and Malaysia also belong to another organization aimed at nonaligned development, the Group of 15, founded in 1989; Indonesia belongs to still another such organization, the Group of 19; and the Philippines belongs to the Group of 24, set up in 1972 to promote the interests of developing countries within the IMF. These organizations were created to promote cooperation among developing countries without regard for the Cold War division between the West and the Soviet bloc. However, that division has now been ended, regional organizations have become more important, and many countries have ceased to be "developing" or "less developed" on most definitions of these terms. Accordingly, the effectiveness of these five international forums has decreased.

Malaysia, Singapore, and Brunei, which were all colonized or "protected" by the British, became members of the Commonwealth when they achieved independence. By contrast, in 1948, when Burma achieved independence from the United Kingdom, it left the British Empire and Commonwealth, as the organization was then known, because at that date it was not possible for republics to remain within it. The three states with Moslem majorities, Brunei, Indonesia, and Malaysia, belong to the Organization of the Islamic Conference (OIC), established in 1969, and to the Islamic Development Bank (IDB), established in 1973; the OIC recently granted observer status to the Moro National Liberation Front of the Philippines. Indonesia is a member of the Organization of Petroleum Exporting Countries (OPEC).

Regional and Interregional: All the ASEAN countries except Brunei are member states of the Asian Development Bank (usually ADB, but sometimes AsDB to distinguish it from the African Development Bank or AfDB). This was established by a group of western and Asian nations in 1966 to promote economic cooperation in the world's largest continent.

All except Brunei and Vietnam also take part in the Colombo Plan for Cooperative Economic and Social Development in Asia and the Pacific. This body was established in 1951 to promote development in certain member countries of the Commonwealth, but it is now involved in collaborative projects throughout the Asia-Pacific region.

Seven countries in Southeast Asia – other than Burma, Cambodia, and Laos – also belong to the Asia-Pacific Economic Cooperation (APEC) group, which was founded in 1989 to promote trade and investment in the Asia-Pacific. ASEAN itself has observer status at the annual APEC summit meetings, which started in 1993 and are now attended by the leaders of 21 countries.

Since March 1996, the seven countries that were then members of ASEAN (that is, Southeast Asia excluding Burma, Cambodia, Laos, and East Timor) have taken part, alongside Japan, China, and South Korea, in summit and ministerial conferences with the EU and its 15 member states. These are known as the "Asia-Europe" Meetings (ASEM) – misleadingly, since not even the whole of East Asia or the whole of western Europe is represented. In 2000, Thailand was invited to become a "partner in cooperation" with the Organization for Security and Cooperation in Europe (OSCE), with observer status at its meetings.

Finally, Cambodia, Laos, Thailand, and Vietnam signed an agreement in April 1995 on establishing a Mekong River Commission to coordinate development in the Lower Mekong basin.

APPENDICES

Appendix 5
Ethnic Groups

The Ethnic Composition of National Populations

The total populations and percentages set out below are all estimates as of mid-2000. All the figures should be treated with caution, perhaps especially those for Burma, where no full census has been conducted since 1931, and those for East Timor, where decades of occupation and conflict have hampered the collection of statistics.

Brunei: 336,000; Malays 64%, Chinese 20%, Dayaks 6%, others 10%

Burma: 41.7 million; Burmans 68%, Shan 8.5%, Karen 6.2%, Rakhine 4.5%, Chinese 3%, Mon 2.4%, Chin 2.2%, Indians 1.7%, Kachin 1.4%, Wa 1.4%, others 0.7%

Cambodia: 12.2 million; Khmers 90%, Vietnamese 5%, Chams and others 3%, Chinese 2%

East Timor: around 800,000; Timorese 78%, Indonesians 20%, Chinese 2%

Indonesia: 224 million; Javanese 45%, Sundanese 14%, Madurese 7.5%, other Malays 7.5%, around 350 other groups 26%

Laos: 5.5 million; Lao Lum and Lao Tai (speakers of Tai languages) 65%, Lao Theung (speakers of Mon-Khmer languages) 24%, Lao Sung (speakers of Tibeto-Burman languages) 10%, others 1%

Malaysia: 21.8 million; Malays and other Bumiputras 60%, Chinese 30%, Indian 8%, others 2%

Philippines: 81.2 million; Tagalogs 28%, Cebuanos 24%, Ilocanos 10%, Ilongos 9%, other Malays 24%, Chinese 2%, others 3%

Singapore: 4.2 million; Chinese 76.4%, Malays 14.9%, Tamils 6.4%, others 2.3%

Thailand: 61.2 million; Thais 54%, Laos 28%, Chinese 11%, Malays 4%, Khmers 3%, others 10%

Vietnam: 78.8 million; Vietnamese 88%, Chinese 2%, Tai 2%, Khmers 1%, others 7%

Some Ethnic Majorities and Minorities

What follows is no more than a selection from among the hundreds of ethnic groups in the region. As with the percentages above, great care should be taken over all the figures mentioned below. They are all estimates, based on official sources and standard reference texts. They all fail to take account of intermarriage, religious conversion, and other forms of intermingling between groups and cultures. In Southeast Asia, as elsewhere, claims to ethnic "purity" hardly ever have any basis in fact.

In what follows, the terms "Mon-Khmer," "Tai," "Tibeto-Burman," and "Western Austronesian" are used solely to indicate relationships among languages that are generally recognized as belonging to one or other of these language families. We intend no judgment on the possible relationships among these language families themselves, which are a subject of continuing debate among linguists. It should also be borne in mind that linguistic relationships do not necessarily indicate either ethnic or surviving cultural relationships.

Acehnese: Aceh, in northern Sumatra, Indonesia, was the first territory in Southeast Asia to be conquered by Moslem invaders, in the late 13th century. The Acehnese use a Western Austronesian language that is only distantly related to the languages of neighboring peoples. Their traditional culture was based on a division of roles between men, who traveled, often over very large distances, to take part in trade, and women, who worked the land. Their sultans ruled much of northern Sumatra and parts of Peninsular Malaya until the defeat of the Acehnese by the Dutch in the Aceh Wars (1873–1903); Aceh was then formally incorporated into the Dutch East Indies in 1908. The Acehnese rose against the Dutch during 1942, only to be crushed by the incoming Japanese, but they then resisted the reimposition of Dutch rule after World War II longer than any other part of Indonesia. Between 1953 and 1957 another nationalist rebellion probably won the support of most Acehnese, only to be crushed by the Indonesian armed forces. Since then, Aceh has been governed as a "special autonomous area," but sporadic outbreaks of nationalist protest have led to repeated declarations of martial law, notably since the Free Aceh Movement began its activities in the mid-1970s. (A ceasefire was signed by this movement and the Indonesian government in May 2000, but it has not been entirely effective.) Since the 1980s, the expansion of the oil, natural gas and timber industries, and the tendency for increasing numbers of Acehnese men to remain in other provinces of Indonesia, have had a major impact on Acehnese culture, but most of the 4.3 million or more Acehnese retain a marked devotion to Islam and to their homeland itself.

Arakanese: see Rakhine

Austronesians: This term is used by some scholars to refer to a broad range of peoples in the Asia-Pacific region, including the Malays and the Pacific islanders, although there is disagreement as to whether it is a strictly linguistic term or has ethnic and cultural applications as well. In a narrower sense, and in the context of Southeast Asia, the term is used with reference to those ethnic groups that have traditionally fished, farmed, and traded along the coasts of West Papua, and that speak languages related to the Malay languages spoken in other parts of Indonesia. These characteristics set them apart from the Papuans (see below), the other major ethnic group in West Papua, but are shared with some of the other ethnic groups settled in the province under Indonesia's policy of *transmigrasi* ("transmigration").

Balinese: The Balinese people, who number more than 2.5 million, are the dominant group in Bali itself, but also form minorities on the neighboring islands of Sumbawa and Lombok (see Sasaks), and, since population transfers started in the 1950s, on many other Indonesian islands as well. They share with the Javanese a complex history of conquest by successive Hindu, Buddhist and Moslem rulers, but their culture was also shaped by the island's isolation from the outside world, between around 1600 and the completion of the Dutch conquest in 1908. Except among the minority who are Moslems, this culture is characterized by a caste system adapted from Hinduism, and by three distinctive forms of organization: the *dadia*, a descent group responsible for maintaining the local temple; the *banjar*, which protects and represents each village; and the *subak*, which is responsible for the maintenance and irrigation of the wet-rice paddies. The fact that this culture has largely survived Dutch and Javanese incursion is at least partly due to the transformation of Bali into a major tourist destination, although this has arguably also led to some distortion of Balinese traditions.

Batak: This term refers to a number of ethnic groups originating in the interior of northern Sumatra, including the Karo, the Angkola, the Pakpak, and the Mandailing. They share a form of social organization through the *marga*, traditionally a landowning patrilineal descent group based on exogamy. The *marga* survives even among Batak who have left the land and among those who have moved to other provinces of Indonesia.

Bugis or Buginese: This Malay people, originally concentrated on the coasts and in other lowlying regions of southern Celebes, has a long history of seafaring and trading throughout the archipelago. As a result, Buginese communities can be found almost everywhere in Indonesia and on the coasts of the Kra Peninsula (Malaysia and Thailand).

Bumiputras/Bumiputeras: These are, respectively, the Malaysian and Indonesian forms of a widely used term that means "princes of the land" or "sons of the soil." The alternative term *pribumi* is also used in Indonesia. All three words generally refer to those ethnic groups, mainly of Malay or related stock, considered to be "indigenous" to each of these countries. Hence, the term *pribumi* includes, for example, Buginese and Dayaks (see separate entries), while the term "Bumiputra" includes Malays, the Orang Asli, as the other indigenous peoples of peninsular (West) Malaysia are known, and also the indigenous minority peoples of Sabah and Sarawak. Just as important to those who use them, these terms also imply that Chinese, Indians, and others with more recent antecedents outside these countries, however many generations back, are inherently and inevitably "alien." This in turn has important implications for interethnic relations, and it is not surprising that all three terms have frequently been denounced as racist. In Malaysia, as detailed in Chapter 5 of this book, Bumiputras have special rights guaranteed by the Constitution and by law, and have benefited from affirmative action by governments. In Indonesia, which has a far larger and more varied population, spread over an immense area, regional and religious divisions have tended to have more salience than the nonjuridical but nonetheless pervasive division between Bumiputeras and non-Bumiputeras.

Burmans: As distinct from the "Burmese" – that is, citizens of Burma of varying ethnic backgrounds – the Burmans (or Myanma, or Bama) are the majority group in the country, which they have inhabited for around 2,100 years, mainly in the northern plains. It is generally agreed that they probably originated in what is now Tibet, while their language, which belongs to the Tibeto-Burman language family, and many other aspects of their traditional culture indicates an ancient affinity with the Chinese. In addition, however, there has been strong influence from Indian cultures, ranging from religion – most Burmans are Buddhists – to architecture. Having superseded the Mon (see below) as the dominant group in what is now Burma, in the ninth century CE the Burmans created a kingdom, Pagan, that lasted up to a Chinese invasion in the 13th century. A later Burman kingdom was established in the 15th century and became the basis for the extension of Burman hegemony over neighboring groups; it was this realm that the British took over in three stages (Arakan in 1826, southern or Lower Burma in 1852, northern or Upper Burma in 1885). Although this kingdom paid tribute to the emperors of China until 1874, and the two cultures still have much in common, the Burmans differ most notably from the Chinese in their adherence to the Theravada (Hinayana) tradition of Buddhism.

Cebuanos: The Cebuanos, who number around 19 million, are the largest ethnic group in the Visaya islands, which lie between Luzon and Mindanao in the Philippines. Their language, also called Cebuano, has become the lingua franca of the Visayas.

Cham: This is the name given to a Malay people, now resident in Cambodia and Vietnam, who are descendants of an ethnic group that dominated the Hindu kingdom of Champa. Their realm, founded in around 192 CE, covered much of the Mekong delta and neighboring regions until it was overthrown by Vietnamese invaders in 1471. In Cambodia, the Chams, most of whom were Moslems, were singled out for especially brutal treatment by the Khmer Rouge regime between 1975 and 1978: their distinctive language and customs were prohibited, and around half their number were executed, worked to death, or left to starve. In Vietnam, most Chams remain Hindus, with around one third being Moslems. In both countries, however, the Chams have generally also retained their older traditions of matriarchy and matrilineality.

Chin: This term is used in Burma to refer to members of more than 40 ethnic groups that traditionally have engaged in hunting and fishing in the northwestern highlands and the valley of the Chindwin River. An alternative name for the Chin is Zomi. The Chin, who number around 1.5 million, have close linguistic and cultural affinities with the majority Burmans, as well as with groups in the Indian state of Assam. Most live in the region designated as the Chin State of Burma in 1948. Chin parties and military formations were prominent both in collaborating with the British against the Japanese during World War II, and later in resisting successive Burmese regimes.

Chinese: There are people of Chinese descent in every country in Southeast Asia. The Chinese living in Brunei, who are mostly Hokkien-speakers, constitute around one fifth of the population, but most have not been granted citizenship.

Many of the Chinese traders and workers resident in Burma departed in 1939–41, as the region was being drawn into World War II. Many more departed around 1963, when General Ne Win's regime nationalized most business enterprises. The Chinese remaining in Burma have tended to assimilate with the Burman majority in recent decades, willingly or not, while improved relations with China itself have probably helped to reduce official discrimination against them.

The Chinese in Cambodia probably numbered more than 400,000 by 1970, but there are now only around 60,000, following decades of war and emigration, notably in response to discrimination and nationalization of much of their property under the Khmer Rouge and, on a smaller scale, under the Vietnamese-backed regime that replaced it. Most Chinese in Cambodia have been Teochiu-speakers or Cantonese-speakers, and have been city-dwellers.

In Indonesia and Malaysia, where the Chinese are mostly of Hokkien, Hakka, and Cantonese descent, both governments have long pursued a policy of encouraging those who had Chinese citizenship to renounce it. Most Indonesian and Malaysian Chinese have been city-dwellers: in Indonesia, they were officially forbidden to settle in rural districts up to 1919 and again from the 1960s.

Laos has a relatively small Chinese minority: most emigration from China into Laos has been by non-Chinese minorities (see Lao Sung).

Chinese immigration into the Philippines, mainly but not exclusively from Fujian province, continued up to the early 1950s. There is now a large group of Chinese-Filipino *mestizos*, including, for example, former President Corazon Aquino and the head of the Catholic Church in the Philippines, Jaime, Cardinal Sin.

The Chinese in Singapore, where, uniquely in the region, they form a majority of the population, include Hokkiens, Teochius, Cantonese, Hakkas, Hainanese, Henghuas, and Hokchias. Since 1979, the government has campaigned to promote the use of Mandarin Chinese, which is an official language, in all educational establishments and public services. All other forms of Chinese are officially treated (with blithe disregard for linguistic facts) as "dialects."

The Chinese constitute the largest single ethnic minority in Thailand, and the number of Chinese in the country is probably greater than in any other country in the region, although there are disagreements on this question, notably over the increasing number of people of mixed Chinese-Thai descent. As elsewhere, people of Chinese descent have traditionally been involved in commercial activities in the cities rather than in agriculture. Most still speak one or other form of Chinese. During the 1940s and 1950s, their freedom was restricted and their property put at risk by the military regime's policies of nationalism and anti-Communism. Almost all people of Chinese descent born in Thailand are now Thai citizens, and use standard Thai for at least some purposes. In recent decades, they have generally escaped the hostility that the Chinese minorities in Indonesia and Malaysia have sporadically suffered from.

Finally, in Vietnam, people of Chinese descent, known as the Hoa, still form the largest single ethnic minority, living mainly in the cities of the South. From the 1950s onward, successive regimes based in Saigon imposed Vietnamese names and nationality on most of the Hoa, and made sporadic attempts to reduce and/or co-opt their economic influence. Following unification in 1975–76, and the nationalization of most commercial enterprises in the South in 1978, around 250,000 Hoa left Vietnam, either overland into China or as "boat people." Those ethnic Chinese who remain have continued to suffer discrimination, which tends to vary in intensity and degree of government approval in line with changes in relations with China.

Dayaks: This term (or a variant, "Dyaks") is still commonly used, although it has been contested as inaccurate, to designate more than 200 different ethnic groups, totaling around 1 million people, who inhabit the island of Borneo (or Kalimantan). The island, and the Dayaks, are divided among Brunei, the Indonesian provinces of Kalimantan Barat, Kalimantan Timur, and Kalimantan Selatan, and the Malaysian states of Sabah and Sarawak. Dayak groups are generally regarded as falling into one of six larger groups. Three of these, the Penans, the Klemantans, and the Kenyahs, may be descended from the earliest inhabitants of the island; the other three, the Kayans, the Muruts, and the Ibans, are probably descended from groups that reached it at later stages. Dayak cultures are generally characterized by reliance on swidden agriculture and the use of longhouses shared by several families, apart from the Ibans, also known as the "Sea Dayaks," who generally live on the coasts and are Moslems. Most Dayaks in the Indonesian portions of the island have been either Protestant Christians or adherents of Kaharingan, a syncretistic religion combining elements of shamanism and Hinduism that was not granted official recognition in Indonesia until the 1980s.

Filipino Malays: The Malay peoples of the Philippines, who jointly comprise around 95% of the total population, are conventionally divided into around 100 ethnic groups. Most are mainly Christian – including the four largest groups, the Tagalogs, Cebuanos, Ilocanos, and Ilongos – and inhabit all three parts of the country (Luzon, the Visayas, and Mindanao). Tagalog, the language of the largest single Malay group in the Philippines, is the basis of the official language, Pilipino. The other Malay groups in the country, such as the Maranaos, the Maguindanos or the Sanggils, are mainly Moslem (see also "Moro" below) and are concentrated in Mindanao.

Hmong: The Hmong, a highland people who use a language related to Chinese, live in the Chinese province of Yunnan and in Laos, Burma, Thailand, and Vietnam. They may number as many as 5 million in total. Originating in southern China, groups of Hmong have migrated into Southeast Asia since the late 19th century. In Thailand, the Hmong have attracted the attention of the security forces because their shifting cultivation is said to have contributed to deforestation and because some have cultivated the opium poppy. After the establishment of the Lao People's Democratic Republic in 1975, the Hmong in that country, who had once had their own king and military organization, were held collectively responsible for the collaboration that had developed between some of their leaders and the United States during the Vietnam War. The new regime has been accused by its critics of bringing about the deaths of around 100,000 Hmong, during a campaign of repression that has also seen thousands of Hmong flee into Thailand and other neighboring countries, while others have migrated to California, Minnesota, and other parts of North America.

Indians: There have been Indian traders in Singapore, Thailand, Vietnam, and elsewhere in the region since at least the ninth century, while Hinduism, Buddhism, and a variety of cultural forms first constructed in the subcontinent have had a significant impact on much of the region too. However, most of the Indian minorities in contemporary Southeast Asia were formed during the centuries of British imperialism, when numerous subjects of the Indian Empire, the "Raj," found employment on plantations and in market towns. Hence, while there are people of Indian descent in every country in Southeast Asia, they form significant and distinct minorities in three former British colonial possessions, Burma, Malaysia, and Singapore. (Hence, too, the fact that people from any part of the subcontinent still tend to be called "Indians" in Southeast Asia, despite the attainment of independence by Pakistan, Bangladesh, and Sri Lanka.) Most Singaporean and Malaysian Indians are Tamils, originating in southern India, and it is Tamil that is recognized as one of Singapore's four official languages (alongside English, Mandarin Chinese, and Malay).

Javanese: The 96 million Javanese constitute the largest of the 330 or more ethnic groups in Indonesia. They share the island of Java with the culturally similar but linguistically distinct Sundanese, Madurese, and other groups, but are also to be found throughout Indonesia. Their traditional culture has been grounded in wet-rice cultivation, shaped by successive waves of conquest by Hindu, Buddhist and Moslem rulers, and expressed through kingdoms that fought and succeeded one another from the sixth century to the Java War (1825–30), when the Dutch consolidated their control of the island. However, the traditional aristocracy retains much of its prestige among contemporary Javanese (it produced both Sukarno and Suharto), and Javanese traditions more generally continue to exert considerable influence through the *rukun kampung* (village mutual aid association) or *rukun tetangga* (neighborhood association).

Javanese communities have existed outside Java for centuries, notably in Bali and Borneo (Kalimantan), but increasing numbers of Javanese have also been relocated to other, less overpopulated islands, first by the Dutch colonial authorities in the 1930s, then by the Indonesian government. Those among them who have left Java against their will, as well as many of the poor farmers and traders they have left behind them, may find it hard to recognize descriptions of Indonesia as "the Javanese empire." Nevertheless, this claim has been supported by reference, among other factors, to the sheer weight of numbers of the Javanese within the total population; the derivation of the official language, Bahasa Indonesia, from Javanese (though it lacks that language's complex system of levels of formality); and the continuing predominance of Javanese individuals, mostly from aristocratic backgrounds, in leading positions in society.

Jinghpaw: see Kachin

Kachin: This is the name generally used for a number of ethnic groups that use Tibeto-Burman languages and have traditionally resided in the northern highlands of both Thailand and Burma, parts of southern China, and the valley of the Upper Irawaddy River in Burma. There are many differences among these groups, ranging from traditional clothing to economic activities, but their languages are closely related and they have all traditionally claimed descent from five Kachin clans. The name "Jinghpaw" (or "Theinbaw"), which in Burma is used for the largest of the Kachin groups, is often applied to all Kachin in Thailand. The Kachin have cultural affinities both with the Burmans and with other ethnic groups in southern China. They form the majority in Kachin State, established in 1948, and there was sporadic armed conflict between Burmese government forces and groups seeking greater autonomy for Kachin State from 1961 to 1993, when the largest group, the Kachin Independence Organization, agreed to a ceasefire.

Karen: Members of this ethnic group, using a Tibeto-Burman language, predominantly Christian, and long associated with the eastern highlands of Burma, also inhabit parts of the southern lowlands. There are probably around 4 million Karen in Burma. There are much smaller numbers of Karen in Thailand, where they are also known as Kariang and are probably the most numerous of Thailand's many highland peoples. Opinions differ as to whether the Karen people of Burma includes the group known as the Kayah or Karenni (Red Karen), whose members, mainly resident in the southern part of the Shan Plateau, share the Karen language and much of their culture. During World War II, Karen military units collaborated with the British against the Japanese, arousing the hostility of Burman nationalists. In 1947, Karen political parties boycotted elections to the Constituent Assembly. In 1948, the Assembly decided, without consulting any Karen representatives, to make the areas of Burma in which the Karens are concentrated into two autonomous states, Karen and Karenni. The latter was renamed Kayah in 1951; in the same year, the Karenni leader U Bee Htu Re was murdered by Burmese government agents. Constitutional provisions allowing the Karens to vote on whether to secede from Burma have never been implemented. There have been sporadic armed conflicts between Karen (including Karenni) resistance groups and the central government ever since 1948, notably involving the Karen National Union and its armed wing, the Karen National Liberation Army.

Khmer: The Khmer people, now the majority ethnic group in Cambodia, formed a series of kingdoms starting in the sixth century, embodying their commitment to Theravada (Hinayana) Buddhism and their tradition of wet-rice cultivation. Khmers also form minorities in southern Vietnam, where they are known as the Khmer Krom or "downriver Khmer," and in northeastern Thailand. The Khmer presence in these countries results from two separate upheavals: the defeats of the Khmer empire by the Thai kingdom of Ayutthaya and the growing Vietnamese empire in the 15th century, when the victors took control of large areas mainly inhabited by Khmers; and the civil wars in Cambodia in the second half of the 20th century, when hundreds of thousands of refugees fled from the Khmer Rouge regime in particular.

Khmer Loeu: This term, meaning "highland Khmer," was adopted by the Cambodian government in the 1960s to encompass such groups as the Kuy, the Mnong, and the Stieng, who use Mon-Khmer languages, and the Jarai and the Rade, who use Western Austronesian languages. Most Khmer Leou are concentrated in the northeastern provinces of Cambodia. All were persecuted by the Khmer Rouge regime, which sought to enforce their assimilation with the Khmer majority. Since the 1980s, however, successive governments have been committed, at least in principle, to protecting minority languages and cultures.

Lao Lum: The Lao Lum ("valley Lao") people share a common ethnic and cultural heritage with the Shan of Burma, the Thais, and the Zhuang of southern China; they speak a language closely related to Thai. They constitute the largest ethnic group in Laos, at around 52% of the total population, and are concentrated along the Mekong River and in the cities of that country. However, there are also Lao Lum inhabiting much of the Khorat Plateau and other areas in northeastern Thailand. Indeed, there are probably considerably fewer Lao Lum in Laos than there are in Thailand, where they are known as Thai-Lao or Isan and display varying degrees of assimilation into Thai culture. The first Lao Lum kingdom, Lan Xang, established in 1353, comprised modern Laos and parts of modern Thailand, Cambodia, Burma, and Vietnam, but by the early 18th century this had been reduced through repeated wars, with Burma and Vietnam in particular, and three separate principalities had been established: Luang Prabang, Vientiane, and Champasak. All three then passed under the control of Siam (Thailand), which in turn was forced to cede them to France in 1893. The Lao Lum were traditionally the dominant group in Laos, and the aristocracies of the three principalities were generally Lao Lum, as also are many leading officials in the present government.

Lao Sung: This a term used by the Lao government to refer to the Hmong, the Yao (see separate entries), and other highland groups that are each relatively small in numbers and that all probably originated in southern China. The languages used by these groups belong to the Tibeto-Burman family, rather than the Tai family to which the languages of the Lao Lum and Lao Tai belong (see separate entries). There are also significant differences in culture and historical experience.

Lao Tai: This is a term used by the Lao government, and sometimes represented in English as "tribal Tai"; it refers to a number of ethnic groups – the Tai Dam, the Tai Deng, the Tai Neua, the Tai Phuan, the Phutai, and others – that live mainly in highland districts of Laos, are distinguished by dress, dialect, and other cultural elements, and were traditionally ruled by hereditary nobles. Today, they constitute around 13% of the total population. Like the Lao Lum, with whom other classifications sometimes assimilate them, they speak languages belonging to the Tai family.

Lao Theung: This, yet another official designation used by the Lao government, refers to around 25 separate groups that may (or may not) be largely Malay descendants of the aboriginal people of the region, and that use a variety of Mon-Khmer languages. "Lao Theung" (like "Lao Sung," above) may be misleading, in that it may suggest greater affinities with the Lao Lum majority than in fact exist. However, it is preferable to the older term that it replaced, "Kha," a derogatory word meaning "slaves."

Malays: The Malays, who include the majority populations of Brunei, Indonesia, Malaysia, and the Philippines, as well as minorities in the other countries of Southeast Asia, may be regarded as a single people sharing (at least) a common linguistic heritage, in that all the languages they use belong to the Western Austronesian family. However, there are many significant cultural, economic, and political differences among and within the various Malay peoples, perhaps most importantly in respect of religion: most but not all Indonesian and Malaysian Malays are Moslems, most but not all Filipino Malays are Christians. It is therefore probably more useful to treat the Malay peoples as separate ethnic groups, while noting that the term "Malay" continues to be applied to members of all or any of these groups, notably in the four countries already mentioned as well as in Singapore. See also Filipino Malays.

Maubere: see Timorese

Mien: see Yao

Minangkabau: This highland people, also known as the Orang Padang, have long been concentrated in West Sumatra, Indonesia, forming the largest of the groups of subsistence cultivators in that province. Since the 19th century, however, many Minangkabau have migrated to peninsular Malaya (West Malaysia). Among most Minangkabau even today, Moslem faith is combined with at least some vestiges of their traditionally matrilineal culture, centered on the *suku*, the lineage organization that has traditionally controlled land use.

Moluccans: This is a term of convenience for the numerous peoples spread throughout the 999 islands of the Moluccas (the Indonesian provinces of Maluku and North Maluku). Known to the West for centuries as the Spice Islands, the Moluccas have historically been among the most productive lands in the region, attracting both Moslem and Christian settlers to live alongside Orang Laut and other indigenous groups. Tensions between the two dominant religious groups have erupted in riots and killings on several occasions during the 20th century, notably in the early 1950s, when Indonesia crushed an attempt to create a Republic of South Maluku, and in the late 1990s, in the wake of the financial crisis that swept through Southeast Asia.

Mon: The Mon people of Burma and Thailand, numbering at least 1 million in all, and also known as the Ramang or Tailaing, have linguistic and cultural affinities with the Khmers and the Vietnamese. It is believed that the Mon arrived in what is now southern (or Lower) Burma around 4,000 years ago, probably from what is now Thailand. The first people in Indochina to embrace Buddhism, they established a series of kingdoms, now known as the Dvaravati, across large parts of Burma and Thailand; it was the Mon who founded Rangoon, now the capital of Burma. From the 11th century onward, they were gradually subjugated by incoming Burmans and Thais, although the Mon kingdom of Pegu lasted until 1757. There has since been extensive intermarriage, and a degree of cultural merging, between the Mon and these other two groups. In addition, many Mon have moved from Burma to Thailand as a result of the political upheavals in the former country. As a result of all these developments, while some Mon continue to use the Mon language, others, mainly in Tenasserim and the Sittang valley, use Burmese, and most Mon in Thailand use standard Thai. The New Mon State Party and other resistance groups fought against the Burmese government from 1948, when Mon State was formally recognized, to 1995.

Montagnards: With reference to Vietnam and Cambodia, this French term is used to refer to members of ethnic minorities concentrated in mountain regions; with reference to Laos, it has largely been superseded by "Lao Sung" and "Lao Theung" (see separate entries). In Vietnam, around two thirds of the members of such groups, including the Hmong (see separate entry), live in the North; most of the groups that were traditionally nomadic have been compelled to settle in permanent villages, and the government has settled increasing numbers of ethnic Vietnamese in the highlands. See also Lao Sung and Lao Theung.

Moro: This Filipino word (derived from the Spanish term generally translated as "Moor") refers to the Moslems of the Philippines, around 5% of the total population. They come from diverse ethnic backgrounds, mostly in the South, but have increasingly developed a self-identification as a distinct minority despite their linguistic and cultural differences. This common identity has drawn both on the traditional Filipino institution of the *datu*, the communal leader who settles disputes, and on Moslem reaction to the influx of Christians into the southern provinces, which has increased since the 1950s and has frequently led to disputes over land ownership. The creation of a special autonomous region in Mindanao, in 1990, has satisfied some of the demands of the Moro minority.

Myanma: see Burmans

Naga: There are perhaps 100,000 members of this set of ethnic groups in Burma, mainly living in the highlands close to the border with the Indian state of Nagaland, where most Nagas live. In fact, there are several different Naga groups, each using a different Tibeto-Burman language. Successive Burmese governments have attempted to assimilate them with the Chin peoples who live alongside them (see above).

Negritos: In the Philippines, but also sometimes in relation to Borneo (divided among Brunei, Indonesia, and Malaysia), this Spanish term is used loosely to refer to members of a large number of nomadic groups that inhabit highland areas, and are linguistically and culturally distinct from the majority Malay populations. The Negritos of the Philippines are probably that country's aboriginal inhabitants: their ancestors may have arrived (perhaps across a land bridge) around 25,000 years ago. However, the term can be a misleading one, in that it subsumes peoples who have a wide range of different cultures and, in many cases, have had little or no contact with each other.

Orang Laut: This Indonesian term, frequently translated, with more romanticism than reliability, as "sea gypsies," refers to a variety of coastal peoples, living throughout the archipelago, who are widely believed to be descendants of the aboriginal peoples of the region.

Palaung: The people generally known as Palaung – though they call themselves Ta-ang – live in Laos and Burma, and use a Mon-Khmer language. In Burma, they have traditionally inhabited the Shan Plateau alongside the Shan and the Wa; many Palaung communities traditionally had Shan rulers (*sawbwas*). The two divisions of the Palaung often referred to in Burmese and other sources – the "Shwe" (Gold) Palaung and the "Pale" (Silver) Palaung – derive from differences in traditional clothing rather than any major difference in language or customs. Many Palaung were caught up in the armed conflicts that ravaged Shan State between 1961 and 1991.

Panthay: This dispersed and ethnically varied group, resident mainly in Burma, is also known as the Hui-Hui. It is generally thought that they are descended from Moslem traders who entered China from Central Asia, and perhaps further West, and then migrated into Burma as traders. Most Panthay remain Moslems today.

Papuans: This term refers to a large number and variety of ethnic groups in the highlands of the Indonesian province of West Papua (Irian Jaya), as well as in East Timor and New Guinea. Most of the 1.3 million or more Papuans in West Papua still live in relatively isolated valleys and river basins. The

Dutch did not complete their occupation of the province until as recently as 1927. Nevertheless, and with increasing intensity ever since then, the traditional cultures of some Papuan groups have been damaged or destroyed by contact with Christian missionaries; with settlers brought from other parts of Indonesia under the *transmigrasi* ("transmigration") policy, numbering around 770,000 in a total provincial population of around 2 million; and with multinational enterprises seeking to exploit the province's mineral resources. Accordingly, there has been considerable but sporadic resistance to outside influences, notably by the armed and militant Organisasi Papua Merdeka (Free Papua Organization, or OPM), and calls have been made for secession from Indonesia and, in some cases, union with Papua New Guinea, the independent state that occupies the eastern portion of the same island. There is considerable controversy over the extent of this resistance and the degree to which it represents majority views among either the Papuans or the Austronesian peoples who also live in West Papua.

Pribumi: see under Bumiputra/Bumiputera

Rakhine (or Arakanese): The Rakhine form the majority in the state of Arakan, on the west coast of Burma (there have also been Rakhine in the Chittagong Hills of India since the 19th century). Arakan, which is said to have embraced Buddhism in 146 CE, was not incorporated into the kingdom of Burma until 1784, and was the first part of Burma to be taken over by the British, in 1826 (hence the use in English of certain Rakhine forms, such as "Rangoon," rather than Burman forms such as "Yangon"). Most Rakhine are Buddhists today, while around 25% are Moslems, who are generally known as the Rohingya. Both the majority Rakhine and the Rohingya minority use what is believed to be a dialect of Burmese. The Rohingya have been particularly badly affected by the Burmese government's policies of repressing minority cultures and forcing resettlement of Burmans in minority areas. Accordingly, many have migrated to neighboring Bangladesh in recent decades, although some of these migrants have been repatriated to Burma under an agreement between the two countries signed in 1992.

Rohingya: see Rakhine

Sasaks: The mainly Moslem Sasaks make up around 90% of the population on the Indonesian island of Lombok. (Most of the remaining 10% are Balinese, descendants of former overlords or recent migrants.) Significant differences remain between the "coastal Sasaks" and the "mountain Sasaks," who are frequently described as being less completely converted to Islam and more loyal to pre-Islamic animistic traditions.

Semang: The Semang are an indigenous people who have long inhabited part of the southern mountains of Thailand. Their Western Austronesian language is related to Malay, rather than to the majority Tai languages of the region.

Shan: It is thought that the Shan people arrived in the Shan Plateau, in what is now northern Burma, in the late 13th century; there are also Shan in northern Thailand (where they are known as Ngiaw) and in southwestern China. They use a language belonging to the Tai family, and share a common ethnic and cultural heritage with the Lao, the Thais, and the Zhuang of southern China. (Indeed, it has been argued that the Thais originated as a branch of the Shan.) Over the centuries, the Shan generally abandoned Buddhism for Christianity, and their traditional rulers (the *sawbwas*) formed a total of 34 small states. These were annexed by the British in 1885 and brought into association with the larger British possessions in Burma, but they retained a large measure of local self-rule. The Shan formed the largest single ethnic minority within newly independent Burma, and a single Shan State, in which the Shan were the majority, was formed in 1948, with a promise (never fulfilled) that there would be an opportunity for the Shan to decide whether to secede from Burma or remain within it. In 1959, the *sawbwas* formally submitted to the Burmese government and renounced their traditional powers. The attempts of successive Burmese governments, notably after 1962, to centralize power structures have met with sustained resistance from the Shan, in which the Communist Party of Burma played a major role until its disintegration in 1989. The largest remaining resistance group, the Shan State Nationalities Liberation Organization, signed a ceasefire in 1994. (See also Palaung and Wa.)

Sundanese: The Sundanese, who number at least 21 million, are the second largest ethnic group in Indonesia, where they have traditionally shared the island of Java with the largest group, the Javanese.

Their language, like Javanese, belongs to the Austronesian family, and there are many cultural similarities between the two peoples, but the Sundanese are generally regarded as being more devoted to Islam, and less influenced by older religious beliefs and practices.

Tamils: see under Indians

Tetun: The largest of the 12 ethnic groups that make up the population of East Timor, the Tetun are the source of the language that has served as a lingua franca for most of that population since colonial times. Like eight of the other ethnic groups in the country, the Tetun are users of a Western Austronesian language, and have close ethnic and cultural affinities to other Malay peoples.

Thai: The Thais appear to have arrived in what is now Thailand in the 11th century, having originated in southern China, where many groups with related languages and cultures remain today. They may even once have formed a branch of one such group, the Shan (see above). However, the Thais soon established their own states, in opposition to the Buddhist Mon princes and Hindu Khmer kings previously dominant in the area, and by the late 18th century the Thais were the dominant force in much of Indochina. Pressure from the western powers, which established colonies and protectorates surrounding Thailand, led to a series of drastic reductions in Thai power and territory, but most Thais retain an understandable pride in the fact that their country is the only one in Southeast Asia never to have been colonized. Like many other countries, Thailand is characterized by a linguistic heritage within which one dialect has become the official standard language, without entirely crowding out other dialects. In Thailand, this standard form, Ayutthaya Thai, is the one originating in the delta of the Chao Praya River. Other dialects originating in central Thailand share with it a degree of prestige not generally accorded to other dialects, whether those of the North (including Kham Mu'ang, or Yuan, sometimes regarded as a separate language) or those of the South (which contain elements of Malay vocabulary). In addition, at least one third of the total population of Thailand belong to non-Thai ethnic groups, some of which use languages belonging to the Tai family (Lao, Phutai, Phuan, and Saek, all concentrated in the Northeast), while others speak unrelated languages (notably Chinese, Khmer, and Vietnamese).

Timorese: The people of East Timor, also known as the Maubere, are probably of mixed Malay, Polynesian, and Papuan descent, and comprise 12 ethnic groups. Eight of these (Tetun, Mambai, Tokodede, Kemak, Galoli, Idate, Waima'a, Naueti) use Western Austronesian languages; three (Bunak, Makasae, Fatuluku) are mainly Papuan; and there is a significant group of *mestisu*, persons of mixed Portuguese and Timorese ancestry. Most Timorese are members of the Catholic Church, at least formally, although animism and other precolonial religious traditions also survive. The Timorese have been united by historical experience of Portuguese colonization (from 1520 to 1975) and Indonesian occupation (from 1975 to 1999). See also Tetum.

Toraya: This term is applied to a number of ethnic groups, long resident in central and southern Celebes, that successfully resisted Dutch conquest until 1905, but were then compelled to move from their mountain villages to new settlements in the valleys below. Since the 1950s, when Indonesia began its program of large-scale population transfers to relieve pressure on overpopulated regions, these valleys have also become home to numerous incomers, and there have been occasional outbreaks of hostility between them and Toraya groups.

Vietnamese: The Vietnamese first formed an independent state after rebelling against Chinese rule in around 968 CE, went on to conquer extensive territories inhabited by Cambodians in the 15th century, and established a shortlived "protectorate" over Cambodia in the early 19th century. Yet their emperors were still paying tribute to their Chinese counterparts as late as 1880. Predictably, different versions of these distant historical events, and their implications, continue to be deployed in the rhetoric of nationalists in Vietnam, Cambodia, and China alike. Relations between Vietnamese and Cambodians are also occasionally exacerbated by the fact that most Cambodians adhere to their traditional culture, which was deeply influenced by Indian models, while most Vietnamese are still influenced by Chinese social models. The fact that the Vietnamese are still occasionally referred to as "Annamese" in western accounts reflects the French colonial perception (1862–1945) of the importance of Annam, the central region of the country, in relation to the other two provinces the French created, Cochin China in the South and Tonkin in the North. Nevertheless, Vietnam's society and culture have retained a recognizable unity – notably through the national language, which probably belongs to the Mon-Khmer family but has been deeply influenced by classical Chinese – despite divisions under the French and again during the two Indochina or Vietnam Wars (1946–54 and 1955–75).

However, while Vietnam is less ethnically heterogeneous than any other state in Southeast Asia except Cambodia, it is certainly not homogeneous. By contrast to the other four majority-Buddhist countries in the region, Vietnamese Buddhism belongs to the Mahayana tradition rather than the Theravada (Hinayana) tradition; there has been a significant Catholic minority among the Vietnamese since the 18th century; and Vietnam's population includes more than 50 ethnic minorities (see Chams, Chinese, Hmong, Khmers, Montagnards, Tamils for examples). In turn, the status of the Vietnamese minorities in Cambodia and Laos has improved and declined in line with changes in relations with Vietnam itself, reaching a nadir in Cambodia under the ultranationalist Khmer Rouge regime (1975–79). Finally, there have been Vietnamese in Thailand since the late 18th century, if not earlier: most arrived as, or are descended from, people taking refuge from upheavals and conflicts in Vietnam.

Wa: The Wa people of Burma, who have traditionally inhabited the Shan Plateau alongside the Shan and the Palaung, speak a Mon-Khmer language. It is generally accepted that the Wa are the same people as the Lawa minority that lives in the northwestern highlands of Thailand. However, controversy continues over whether the Wa/Lawa constitute, as many scholars have concluded, a remnant of the aboriginal people of the region, who inhabited these and other highlands long before the Mons, Khmers, and Thais arrived. Between 1961 and the early 1990s, large numbers of Wa took part in the conflict between the Burmese government and groups seeking autonomy for Shan State, some under the leadership of the now defunct Communist Party of Burma.

Yao: The Yao, who call themselves the Mien, but are also known as the Man, are a highland people, originally resident in the Chinese province of Yunnan, who have spread from there into Vietnam, Laos, Burma, and Thailand. Many Yao who once lived in Laos departed for neighboring countries during the decades-long civil war in Laos. Their traditional religion incorporates elements of animism and ancestor worship. There is disagreement as to whether their language is related to Chinese or to Thai. In any case, many Yao speak Chinese.

APPENDICES

Appendix 6
Bibliography

This bibliography is intended to draw the attention of readers to some of the most useful and stimulating English-language books on the countries of Southeast Asia that are currently available. It therefore excludes periodicals and websites, as well as publications in other languages. It should be seen as a supplement to the suggestions for further reading at the end of each chapter.

Reference

Bowman, John. S. (editor), *Columbia Chronologies of Asian History and Culture*, New York: Columbia University Press, 2000

A useful collection of materials on the historical development of the various countries of the continent

Far Eastern Economic Review (editors), *Asia 2001 Yearbook*, Hong Kong: Far Eastern Economic Review, 2000

This is the 42nd edition of an authoritative annual publication that reviews recent events in each Asian state, with additional chapters on international relations and a large amount of statistical data. The weekly magazine that publishes it is also a valuable source of information and debate about developments in the region.

Leifer, Michael, *Dictionary of the Modern Politics of Southeast Asia*, London and New York: Routledge, 1995

Some of the entries in this impressively comprehensive book are already out of date, reflecting the rapidity of change in the region, but this remains a very useful source on the nuts and bolts of political activity in the region, and the social forces and institutions that shape it.

The Region as a Whole

Adas, Michael, *Machines as the Measure of Men: Science, Technology, and Ideologies of Western Dominance*, Ithaca, NY: Cornell University Press, 1989

An important assessment of the role played by forms of knowledge and technology in the culture of imperialism, including in Southeast Asia

Breman, Jan, *Taming the Coolie Beast: Plantation Society and the Colonial Order in Southeast Asia*, New Delhi, Oxford, and New York: Oxford University Press, 1989

This is the definitive study of plantation labor and its relationship to colonial systems of power.

Brown, David, *The State and Ethnic Politics in Southeast Asia*, London and New York: Routledge, 1994

Brown provides the most comprehensive guide now available to the evolution of state strategies in the region and their implications for interethnic relations.

Dixon, Chris, *Southeast Asia in the World Economy*, Cambridge, New York, and Melbourne: Cambridge University Press, 1991

Dixon offers several ways in which the region might be seen: as part of the periphery of the world economy, as part of an Asia-Pacific economy dominated by Japan, as part of an economy centered on the four Asian "tigers," and as a regional economy centered on Singapore.

Field, Graham, *Economic Growth and Political Change in Asia*, London: Macmillan, and New York: St Martin's Press, 1995

This comprehensive survey of the economies and societies of South, Northeast and Southeast Asia contains a great deal of useful information and provocative argument. The numerous comparisons and contrasts drawn among the various political and economic systems of these regions are particularly valuable.

Higgott, Richard, and Richard Robison (editors), *Southeast Asia: Essays in the Political Economy of Structural Change*, London, New York, and Melbourne: Routledge, 1985

Higgott and Robison offer a broad overview of the transformation of Southeast Asia's economies, surveying major changes including industrialization and increased integration with world capital and commodity circuits.

Hirsch, Philip, and Carol Warren (editors), *The Politics of Environment in Southeast Asia: Resources and Resistance*, London and New York: Routledge, 1998

An informative series of recent case studies that chart the emergence of the environment as an issue for public debate in Southeast Asia, and the roles that various political actors play in conflicts over resource use and management

Mackerras, Colin (editor), *Eastern Asia: An Introductory History*, third edition, Melbourne: Longman, 2000

An excellent, comprehensive collection that analyzes the development and contemporary status of East Asia generally, with many sections on Southeast Asia in particular

Mackerras, Colin, and Nick Knight (editors), *Marxism in Asia*, London and Sydney: Croom Helm, 1985

This book has been overtaken by events and, in any case, it contains only three chapters specifically on countries in Southeast Asia (Vietnam, Cambodia, and Indonesia). Nevertheless, it remains a fascinating and thoughtprovoking set of essays on the ways in which a 19th-century European ideology became entangled with 20th-century Asian nationalisms, and was transformed beyond recognition in the process.

Neher, Clark D., *Southeast Asia in the New International Era*, third edition, Boulder, CO: Westview Press, 1998

In this book, each of the nations in the region is discussed with emphasis on political and economic problems and prospects.

Rigg, Jonathan, *Southeast Asia: The Human Landscape of Modernization and Development*, London and New York: Routledge, 1997

A probing and thoughtprovoking assessment of economic progress in the region that gives prominence to indigenous notions of development, and the manner in which different people have responded to the challenges and opportunities of change

Robison, Richard, et al. (editors), *Politics and Markets in the Wake of the Asian Crisis*, London and New York: Routledge, 2000

This collection provides analyses of the impact of the recent regional economic crisis, with case studies of countries in Southeast and Northeast Asia as well as broader thematic treatments.

Rodan, Garry, Kevin Hewison, and Richard Robison (editors), *The Political Economy of Southeast Asia: An Introduction*, second edition, Melbourne, Oxford, and New York: Oxford University Press, 2000

This is an expanded and updated version of an outstanding popular introduction to the political economy of the region, addressing its internal dynamics and its integration with the world system.

Tarling, Nicholas (editor), *The Cambridge History of Southeast Asia*, 4 Volumes, Cambridge, New York, and Melbourne: Cambridge University Press, 1999

A collection of authoritative academic studies on the region

Thomas, Nicholas, *Colonialism's Culture: Anthropology, Travel and Government*, Melbourne: Melbourne University Press, Cambridge: Polity Press, and Princeton, NJ: Princeton University Press, 1994

Using an approach derived from the work of Michel Foucault, Thomas examines how forms of governmentality, in the sense of systems of surveillance and control, were integral parts of the cultures of colonialism.

Specific Countries

Arcilla, Jose S., *An Introduction to Philippine History*, fourth edition, Quezon City: Ateneo de Manila University Press, 1999

This is the latest edition of a useful and wideranging basic text on the complex history of the archipelago.

Booth, Anne (editor), *The Oil Boom and After: Indonesian Economic Policy and Performance in the Soeharto Era*, Singapore, Oxford, and New York: Oxford University Press, 1992

The contributors offer comprehensive discussions of many facets of economic policy-making, and its consequences, under Suharto.

Butler-Diaz, Jacqueline (editor), *New Laos, New Challenges*, Tempe, AZ: Arizona State University Program for Southeast Asian Studies, 1998

This anthology of contributions from diverse disciplines covers many aspects of contemporary Laos, focusing on the many changes that have occurred in both the economy and society. It includes an assessment of the future of Laos.

Carey, Peter (editor), *Burma: The Challenge of Change in a Divided Society*, London: Macmillan, and New York: St Martin's Press, 1997

This is a varied collection of essays addressing the main issues facing the Burmese regime and people at the close of the 20th century.

Chandler, David, *A History of Cambodia*, third edition, Boulder, CO: Westview Press, 2000

This is a highly readable account of Cambodia's history, from the era of "Indianization" through the period of the Angkor empire, French colonialism, and the Sihanouk era, through the brutal years of the Khmer Rouge regime.

Cribb, Robert, and Colin Brown, *Modern Indonesia: A History Since 1945*, Harlow and New York: Longman, 1995

This is a compact historical overview of Indonesia from the closing years of its colonial period up to the middle decade of Suharto's presidency.

Diran, Richard K., et al., *The Vanishing Tribes of Burma*, London: Weidenfeld and Nicolson, and New York: Amphoto Art, 1997

The sensationalist title should not deter readers from a book in which outstanding photographs of people from Burma's numerous minorities are accompanied by comprehensive and reliable texts on their histories, cultures, and current circumstances.

Dunn, James, *Timor: A People Betrayed*, Sydney: ABC Books, 1996

This thorough account of East Timor up to the 1980s is probably the best source on the period surrounding Indonesia's invasion. It includes detailed coverage of relations between the two main independence movements, Fretilin and the UDT, as well as of Australia's role in encouraging the invasion.

Emmerson, Donald K. (editor), *Indonesia Beyond Suharto: Polity, Economy, Society, Transition*, Armonk, NY: M.E. Sharpe, 1999

This collection of essays by leading Indonesianists addresses Suharto's rule and the transition that followed his resignation.

Evans, Grant, *Lao Peasants under Socialism*, New Haven, CT, and London: Yale University Press, 1990

Evans analyzes the early stages of the Lao revolution, and its impact on rural society and development, in a study that remains informative and acute even though it is now somewhat outdated.

Evans, Grant (editor), *Laos: Culture and Society*, Chiang Mai: Silkworm Books, 1999; Singapore: Institute of Southeast Asian Studies, 2000

This collection of 12 research articles addresses such diverse domains as geography, history, the Lao diaspora, ethnicity, religion, ritual, gender, and language and literature. Together, they provide the cultural and social context for understanding the Lao economy and its challenges.

Evans, Grant, and Kelvin Rowley, *Red Brotherhood at War: Vietnam, Cambodia, and Laos since 1975*, revised edition, London and New York: Verso, 1990

This detailed account of the complex relations among the three countries of Indochina, set against the background of US, Soviet and Chinese intervention in their affairs, provides a valuable corrective to conventional accounts based on stereotypes about "national character," Vietnamese "imperialism," or the supposedly uniform nature of "Communism."

Fenton, James, *All the Wrong Places: Adrift in the Politics of the Pacific Rim*, New York: Atlantic Monthly Press, and London: Viking, 1988

This collection of vivid journalistic reports contains short sections on Cambodia, Laos, and South Korea, but the heart of the book comprises eyewitness accounts of the fall of Saigon in 1975 and the EDSA revolution in Manila 11 years later.

Gomez, Edmund, and Kwanme Jomo, *Malaysia's Political Economy: Politics, Patronage, and Profits*, Cambridge, New York, and Melbourne: Cambridge University Press, 1999

The authors examine Malaysia from an economic perspective, paying particular attention to the objectives and effects of major economic policies since the New Economic Policy was introduced in 1971.

Gunn, Geoffrey C., *Language, Power and Ideology in Brunei Darussalam*, Athens, OH: Ohio University Center for International Studies, 1997

This contribution to modernization theory analyzes the links between language and power in Brunei, including the oral tradition, the rise of Islamic literacy, print culture, and mass literacy.

Hewison, Kevin (editor), *Political Change in Thailand: Democracy and Participation*, London and New York: Routledge 1997

A very comprehensive collection of papers covering most aspects of contemporary power relations in Thailand

Hill, Hal, *The Indonesian Economy*, second edition, Cambridge, New York, and Melbourne: Cambridge University Press, 2000

This major and comprehensive work on the Indonesian economy begins with the turn away from import substitution in the 1960s, and includes a section on the causes and effects of the crisis of 1997.

Hill, Michael, and Lian Kwen Fee, *The Politics of Nation Building and Citizenship in Singapore*, London and New York: Routledge, 1995

The authors focus on the impact of Singapore's economic policies and political structures on the social system.

Karnow, Stanley, *Vietnam: A History*, revised and updated edition, New York and London: Viking Penguin, 1991

This book sets the Vietnam War in a broader historical and geographic context, and provides a wealth of detail on many aspects of Vietnam's modern history. It includes numerous photographs, reflecting its origins as an accompaniment to a television series, as well as a chronology of events and biographies of the leading decision-makers.

Kiernan, Ben, *The Pol Pot Regime: Race, Power and Genocide in Cambodia under the Khmer Rouge, 1975-79*, New Haven, CT, and London: Yale University Press, 1996

In this magisterial work on the era of the "killing fields," Kiernan, who has devoted his academic career to research on, and analysis of, the Khmer Rouge, combines interviews with survivors and detailed background information to present a convincing portrait of a genocidal regime.

Kolko, Gabriel, *Anatomy of a War: Vietnam, the United States, and the Modern Historical Experience*, New York: New Press, 1985, reissued 1994 as *Vietnam: Anatomy of a War*, London: Allen and Unwin, 1985; and *Vietnam: Anatomy of a Peace*, London and New York: Routledge, 1997

Professor Kolko's analysis of the Vietnam War, which arguably rivals Stanley Karnow's somewhat more accessible work (see above) as the standard text on the topic, is complemented by his much shorter but equally absorbing discussion of developments since that conflict ended.

Manning, Chris, and Peter van Diermen (editors), *Indonesia in Transition: Social Aspects of* Reformasi *and Crisis*, London: Zed Books, and Singapore: Institute of Southeast Asian Studies, 2000

This is probably the most comprehensive of the several volumes that have already appeared focusing on the period after Suharto, assessing both the brief presidency of B.J. Habibie and the election of Abdurrahman Wahid.

Milne, Robert S., and Diane Mauzy, *Malaysian Politics under Mahathir*, London and New York: Routledge, 1999

Milne and Mauzy examine the effects of Mahathir's tenure on political and economic development in Malaysia, with special focus on ethnic relations, foreign relations, human rights, and the succession.

Robison, Richard, *Indonesia: The Rise of Capital*, Sydney, London, and New York: Allen and Unwin, 1986

Robison presents an excellent discussion of some complex issues of political economy, and in particular the emergence of business networks and activities within Suharto's state apparatus. Much that he describes remains highly relevant 14 years on.

Rotberg, Robert I. (editor), *Burma: Prospects for a Democratic Future*, Washington, DC: Brookings Institution Press, 1998

This collection of essays written for the nonspecialist covers politics, values, foreign policy, the role of the military, the economy, health, and education.

Saunders, Graham, *A History of Brunei*, Kuala Lumpur, Oxford, and New York: Oxford University Press, 1984

This was the first full-length scholarly history of Brunei by a former long-term resident and it is still a valuable source of information.

Smith, Martin, *Burma: Insurgency and the Politics of Ethnicity*, revised edition, London: Zed Books, 1999

> Provides useful materials for understanding the development of the complex and politically crucial relations between the Burman majority and the various ethnic minorities in the country.

Tremewan, Chris, *The Political Economy of Social Control in Singapore*, London: Macmillan, and New York: St Martin's Press, 1994

> Tremewan offers a detailed and absorbing explanation of how and why Singapore's government tries to control the private behavior and thoughts of its citizens.

Thongchai Winichakul, *Siam Mapped*, Honolulu: University of Hawaii Press, 1994

> This award-winning book is a provocative and innovative history of the symbolic emergence of Siam (Thailand) as a nation state.

Turnbull, C.M., *A History of Singapore, 1819-1988*, second edition, Singapore, Oxford, and New York: Oxford University Press, 1989

> This remains probably the best historical account of the years before independence was achieved, and it also offers a relatively unbiased treatment of the first two decades of nation-building.

Vatikiotis, Michael R.J., *Indonesian Politics under Suharto: Order, Development and Pressure for Change*, third edition, London and New York: Routledge, 1998

> This investigation of various aspects of Suharto's rule has been overtaken by events in some respects, but it remains a highly readable account of the main institutions that still figure largely in Indonesia's political landscape.

Warr, Peter G. (editor), *The Thai Economy in Transition*, Cambridge, New York, and Melbourne: Cambridge University Press, 1993

> A comprehensive overview of economic growth and political developments in Thailand since the end of World War II

Wyatt, David K., *Thailand: A Short History*, New Haven, CT, and London: Yale University Press, 1984

> This detailed and readable history of Thailand from ancient times to the 1980s is a fine display of Wyatt's craft.

International Relations

Berger, Mark T., and Douglas A. Borer (editors), *The Rise of East Asia: Critical Visions of the Pacific Century*, London and New York: Routledge, 1997

> A critical look at the rise of East Asia, with important essays on international relations within and between Northeast Asia and Southeast Asia

Chia Siow Yue and Marcello Pacini (editors), *ASEAN in the New Asia*, Singapore: Institute of Southeast Asian Studies, 1997

> A concise survey by scholars from various countries in Southeast Asia who share a concern to raise the profile of their region

Funabashi Yōichi, *Asia Pacific Fusion: Japan's Role in APEC*, Washington, DC: Institute for International Economics, 1995

> This is a useful introduction to the politics behind the formation of the APEC grouping. Although written from a predominantly Japanese perspective, it provides a good analysis of many of the key issues and personalities.

Gill, Ranjit, *ASEAN Towards the 21st Century*, London: ASEAN Academic Press, 1997

> An optimistic account of ASEAN's potential based, perhaps somewhat tenuously, on a review of its modest historical achievements

Henderson, Jeannie, *Reassessing ASEAN*, Oxford and New York: Oxford University Press for the International Institute of Strategic Studies, 1999

> This thoughtful, concise, and refreshingly up-to-date evaluation, reaching qualified conclusions, is aimed at the specialist reader.

Katzenstein, Peter J., and Takashi Shiraishi (editors), *Network Power: Japan and Asia*, Ithaca, NY: Cornell University Press, 1997

> An important collection of essays by leading scholars, including considerations of issues relating to Southeast Asia in particular

Wurfel, David, and Bruce Barton (editors), *Southeast Asia in the New World Order: The Political Economy of a Dynamic Region*, London: Macmillan, and New York: St Martin's Press, 1996

> This collection of papers is mainly concerned with regional economic relations, but it also offers chapters on the region's links with China and Japan.

Index

abangan, 97, 103, 168
ABIM, see Malaysian Islamic Youth Movement
Abu Sayyaf, 176
Aceh, 9, 13, 106, 107, 167, 175, 176–77
Acehnese, 168
acid rain, 183
adat, 168
ADB, see Asian Development Bank
Aditla, 111
Africa, 4, 5
AFTA, see ASEAN Free Trade Agreement/Area
agriculture, 8, 10, 17, 24, 26, 193–203
    and climate changes, 186–87
    and environment, 179, 180
    colonial, 193
    in Brunei, 78, 79, 193
    in Burma, 162, 193, 200–01
    in Cambodia, 140, 193, 196, 202
    in Dutch East Indies, 198
    in East Timor, 115, 193
    in Indonesia, 193, 195, 196, 197, 198–99
    in Laos, 145, 147, 148, 153, 193, 196, 201–02
    in Malaysia, 193, 195, 196, 197, 198, 199
    in Philippines, 193, 196, 197, 200
    in Singapore, 67, 193
    in Thailand, 32, 42, 48, 49, 50, 193, 196, 199–200, 201
    in Vietnam, 122, 127, 128, 129, 130, 193, 196, 201, 203
Aguinaldo, Emilio, 89, 90
aid, 224, 225, 228
    to agriculture, 196
    to Burma, 162
    to Cambodia, 132, 135, 137, 140, 141, 142
    to East Timor, 112, 115–16
    to Laos, 146, 147, 149, 150, 153
AIDS, 49
Air America, 146
air quality, 179, 183
Akashi Yasushi, 136
"Ali Baba" partnerships, 63
aliran, 97, 98, 100, 102, 103, 168, 175
All Indonesia Workers Union, 100
Alliance (Malaya/Malaysia), 59, 60, 62
Al-Ma'unah, 176
Al-Muktahee Billah, Crown Prince, 82
Alorese, 13
Alternative Front (Malaysia), 60
Ambonese, 18, 97

Amedeo, 82–83
Amnesty International, 75
Ampa-Fairly Field, 80
Anand Panyarachun, 46, 49
Ananda Mahidol, King, 43
Anderson Consulting, 83
Angkor, 5, 133
Angkor Wat, 133
animals, 182
ANKI, see Armée Nationalé Khmer Indépendante
Annam, 10, 121, 123, 170
Annamite Mountains, 121
Anti-Imperialist League, 89
Anwar Ibrahim, 57, 59, 60, 61
APEC, see Asia-Pacific Economic Cooperation
Apkindo group, 175
APMT, see Popular Monarchist Association of Timor
Apodeti, 111
Aquino, Benigno ("Ninoy"), 91, 92, 93
Aquino, Corazon ("Cory"), 92, 93, 94, 173
Arabia, 5
Arakhanese, see Rakhine
ARF, see ASEAN Regional Forum
Aris, Michael, 160
Armée Nationalé Khmer Indépendante (Cambodia), 135, 136, 137
armed forces
    of Burma, 158, 159, 160, 161, 162, 163, 201
    of Cambodia, 138, 141
    of China, 220
    of colonial powers, 18
    of Indonesia, 97, 99, 100, 101, 102, 103, 104, 106, 107, 110, 111–12, 113, 114, 117, 118–19, 168, 175, 176, 210
    of Khmer Rouge, 138
    of Laos, 146
    of Portugal, 111
    of Thailand, 42, 43, 44, 45, 46, 47, 161
    of United States, 228
    of Vietnam, 126, 127
ASA, see Association of Southeast Asia
ASEAN, 101, 175, 176, 177, 215, 233, 234, 235, 239
    and Brunei, 80
    and Burma, 162, 163
    and Cambodia, 135, 136, 138, 141, 227
    and China, 221, 227, 233, 234, 235
    and cultural exchanges, 226
    and environment, 189
    and Japan, 227, 235

and Laos, 149, 150
and South Korea, 227, 235
and United States, 233
and Vietnam, 126, 227, 233
establishment of, 219, 227
Free Trade Agreement/Area, 189, 238
Regional Forum, 219, 221, 227, 228, 234, 238
Senior Officials on the Environment, 189
ASEAN+3, see East Asia Cooperation
ASEM, see Asia–Europe Meetings
Asia Group (UN), 227
Asia Pacific Council, 227
Asia, 185 (see also Central Asia, Northeast Asia, South Asia)
Asia–Africa Conference (1955), 12
Asia–Europe Meetings, 219, 227, 228, 235, 236
Asian crisis, 162, 167, 175, 176, 179, 207–15, 227, 228, 232, 234, 236, 237
 and Brunei, 79, 81–82, 85
 and Cambodia, 140
 and Europe, 211, 212
 and IMF, 209, 210, 211, 212–14
 and Indonesia, 97, 101, 104, 107, 207, 208, 209, 210, 211, 213, 214, 215
 and Japan, 207, 211, 212
 and Laos, 151, 153
 and Malaysia, 56, 60, 65, 207, 208, 209, 210, 211, 212, 213, 214, 215
 and Philippines, 93, 208, 209, 215
 and Singapore, 72, 208, 209, 215
 and South Korea, 212
 and Thailand, 47, 207, 208–09, 210, 211, 212, 213, 214, 215
 and United States, 207, 211, 212
 and Vietnam, 129, 130
Asian Development Bank 149, 150, 227
Asian Monetary Fund (proposed), 212, 227
"Asian values," 12, 71, 167, 172, 177, 235
Asian Wall Street Journal, 75
Asia Pacific Economic Cooperation, 101, 237–38
Asia Pacific region, 235
ASOEN, see ASEAN Senior Officials on the Enviornment
ASPAC, see Asia Pacific Council
Asprey Group, 83
Associacão Democratica Intergracão Timor-Leste Australia, see Aditla
Associacão Popular Democratica de Timor, see Apodeti
Associacão Popular Monarquia de Timor, see Popular Monarchist Association of Timor
Association of Southeast Asia (proposed), 227
Association of Southeast Asian Nations, see ASEAN
Astra International, 106
Atauro, 111
Attorney General of Malaysia, 58
Aung San, 159, 160
Aung San Suu Kyi, 33, 160, 161, 162, 163
Australia, 94, 136, 148, 227, 234, 236, 237–38, 239
 and East Timor, 110, 111, 114, 116, 238
 and Five-Power Defense Arrangements, 220
 and Indonesia, 116, 238
 and Laos, 149
 and West Timor, 110

authoritarianism, 58, 171, 175, 233
 in Brunei, 79–80
 in Indonesia, 102, 210
 in Singapore, 74–76
Ayutthaya, 5, 9
Azahari, Sheikh Ahmad, 79

Bahasa Indonesia, see Indonesian language, Malay language
Bahasa Malay, see Malay language
baht, 32, 41, 49, 151, 209
Bali, 13, 117, 168, 198
Balibo, 111
"bamboo network," 171
Banda Sea, 8
Bandung, 97
Bandung Conference (1955), 12
Bangkok, 45, 49, 170, 172, 182, 183
Bangkok Airport, 73
Bangladesh, 201
bangsa Filipino, 174
bangsa Malaysia, 64
bangsa Melayu, 171
bangsa Moro, 174
Bank Bumiputra, 64
Bank for International Settlements, 211
Bank of Thailand, 49
banks and banking, 207, 208, 211, 212, 214
 in Indonesia, 101, 106
 in Indonesia, 209
 in Philippines, 209
 in Singapore, 67
Bao Dai, 124, 125
Bapak Pembangunan, 104
Barbosa, Duarte, 19
Bardez, Félix-Louis, 19
Bargain of 1957 (Malaysia), 59, 61–62
Barisan Alternatif, see Alternative Front
Barisan Nasional, see National Front
Barisan Socialist Party of Singapore, 74
Barus, 4
Bataks, 168
Batam Island, 73
Batavia, 9, 15, 98 (see also Jakarta)
Batjan, 8
Batu Gade, 111
Baucau, 115, 117
Bawazier, Fuad, 106
Belo, Bishop Carlos, 113
Bengal, Bay of, 5
Bengkulu, 186
Beraja, 79
"Berkeley Mafia," 99
Bhineka tunggal ika, 167
Bhumibon Adulyadej, King, 42, 44, 45, 51, 52
BIA, see Brunei Investment Agency
biodiversity, 179, 182
BLDP, see Buddhist Liberal Democratic Party
"boat people," see migration from Vietnam
Bonifacio, Andres, 88
Bontec, 169
Borneo Bulletin, 81, 83, 84

Borneo, 15, 64, 87, 168, 174, 179, 196, 198
    colonial, 13, 14, 16
    premodern, 6
Bowring Treaty (1855), 10, 42
Brando, Marlon, 41
Brantas River, 6
Brazil, 212, 214
Britain, see United Kingdom
Brunei, 24, 27, 31, 78–85, 180, 193
    and China, 221
    and Indonesia, 79
    and Japan, 79, 80, 224, 225
    and Malaysia, 169
    and South Korea, 80, 224
    and Spratly Islands, 121
    and United Kingdom, 78, 79
    and Vietnam, 201
    premodern, 12
Brunei Coldgas, 80
Brunei Economic Council, 80, 82
Brunei Investment Agency, 82–83, 84
Brunei LPG, 80
Brunei People's Party, 79
Brunei Shell Petroleum, 80
Brunei Shell Tanker, 80
BSPP, see Burma Socialist Program Party
Buddhists and Buddhism, 8, 172
    in Burma, 157–58, 169
    in Cambodia, 133
    in Laos, 146, 148, 153, 154
    in Philippines, 87
    in premodern period, 7
    in Thailand, 44, 170
    in Vietnam, 8, 123
Buddhist Liberal Democratic Party (Cambodia), 137, 138
Buencamino, Felipe, 89
BULOG, 106
Bumiputras, 59, 60, 61, 62, 63, 64, 171, 199, 213
Bureau of Non-Christian Tribes (colonial Philippines), 89
bureaucracy
    in Brunei, 79
    in Burma, 158, 160
    in Indonesia, 101, 107
    in Philippines, 89
    in Thailand, 42, 43, 44, 45, 46, 47, 50
Burgos, Father, 88
Burma, 14, 27, 29, 33, 151, 157–63, 167, 168, 169–70, 171, 181, 188, 189, 193, 200–01, 222, 224, 226, 235
    and ASEAN, 162, 163
    and Canada, 162
    and China, 162, 163, 201, 220, 222, 225
    and Commonwealth, 25
    and Europe, 162
    and India, 201, 228
    and Japan, 158–59, 162, 225
    and Laos, 145
    and Singapore, 72
    and South Korea, 162
    and Thailand, 161
    and United Kingdom, 158, 159
    and United States, 162
    colonial, 10, 16, 17, 19
    premodern, 5, 9, 13

Burma Socialist Program Party, 159
Burmans, 158, 169, 170

Caday, Leonides, 177
Caetano, Marcello, 110
California, 94
Cambodia, 27, 33, 132–42, 148, 167, 170, 140, 183, 189, 193, 196, 202, 222, 224, 226, 235
    and ASEAN, 135, 136, 138, 141, 227
    and China, 132, 135, 136, 141
    and East Timor, 114
    and France, 123, 124, 133, 141
    and IMF, 139
    and India, 135
    and Japan, 133, 220, 225
    and Laos, 132, 133, 145
    and Soviet Union, 132, 136, 141
    and Thailand, 132, 133, 134, 135, 136, 138
    and UN, 132, 135, 136, 137, 141
    and United States, 132, 133–34, 135, 136, 141, 202
    and Vietnam, 121, 125, 128, 132, 133, 134, 135, 136, 141, 142, 202
    and World Bank, 139
    colonial, 10, 14, 16, 19
    premodern, 5, 231
Cambodian People's Party, 136–37, 138, 139, 140–41, 142
Canada, 94, 148, 162, 187, 227
Cao Dai, 123
capital controls (Malaysia), 57, 60, 61, 210, 212, 214
Capital Issues Committee (Malaysia), 64
capitalism, 180, 231, 232, 234, 236, 238
    and agriculture, 193, 202, 203
    and imperialism, 13
    in Indonesia, 98, 99, 100, 101
    in Laos, 153, 154
    in postcolonial period, 26
carbon dioxide, 183, 186
carbon monoxide, 183
Carrascalao, Mario, 115
cars, 31, 182, 183
cash crops, 193, 194–95, 198–99, 203
cassava, 199, 200
Catholic Church
    in East Timor, 113, 117
    in Indonesia, 168
    in Philippines, 15, 87, 88, 91, 92, 93, 169
    in Vietnam, 123
Celebes, 13, 168, 198
censorship
    in Brunei, 84
    in Burma, 159
    in Singapore, 76
    in Thailand, 51
Center for International and Development Studies, in Indonesia, 105
Central Asia, 3
Central Provident Fund (Singapore), 70
chaebols, 223
Chakri dynasty, 170
Champasak, 146
Chams, 133
Changi Airport, 69, 70, 73
Chao Praya delta, 168, 199

chat, 170
Chatichai Choonhaven, 46
Chea Sim, 135
chemicals, 224
Cheuang Somboukham, 152
Chin, 158, 169
China, 33, 34, 70, 74, 126, 174, 219, 220–21, 223, 225, 226, 227, 228
   and ASEAN, 221, 227, 233, 234, 235
   and Asian crisis, 214
   and Australia, 237
   and Brunei, 221
   and Burma, 162, 163, 201, 220, 222, 225
   and Cambodia, 132, 135, 136, 141
   and East Asia Cooperation, 239
   and Indonesia, 221, 224
   and Laos, 145, 146, 147, 148, 152, 153, 220, 222
   and Malaysia, 221, 225, 226
   and Paracel Islands, 121, 221
   and Philippines, 87, 221
   and Singapore, 225, 226
   and Spratly Islands, 121
   and Thailand, 136, 225
   and Vietnam, 8, 121, 122, 123, 124, 125, 220, 221, 222
   cultural exchanges with, 226
   premodern, 3, 4, 5, 6, 7, 9
   trade with, 224–25
"Chinatowns," 98
Chinese characters, 175
Chinese language
   in Cambodia, 133
   in Indonesia, 175
   in Singapore, 68, 73
Chinese people, 10, 26, 28, 72, 171, 176, 226
   and China, 220, 223
   and Javanese, 18
   in Brunei, 79
   in Burma, 29, 158
   in Cambodia, 133, 226
   in colonial period, 15, 16, 24
   in Dutch East Indies, 18
   in Indonesia, 15, 29, 98, 100, 101, 104–05, 168, 172, 174, 226
   in Malaya, 18
   in Malaysia, 59, 61, 62, 63, 64, 168, 169, 171, 172, 174, 226, 232
   in Philippines, 29, 172, 173
   in Singapore, 15, 29, 68, 69, 70, 71, 169, 171, 172
   in Thailand, 29, 42, 43, 170, 172–73, 223
   in Vietnam, 122, 226
chlorofluorocarbons, 183
Chola, 6
Cholon, 122
Christian Conference of Asia, 75
Christians and Christianity
   in Burma, 157
   in Cambodia, 133
   in Indonesia, 97, 168, 174, 175, 177
   in Philippines, 8, 173, 174
   in Vietnam, 123
   (see also Catholic Church, Protestants)
Chuan Likphai, 47
Chulalongkorn, King, 42

chum, 185
CIA, see United States Central Intelligence Agency
CIC, see Capital Issues Committee
cities, 15, 182
   in Cambodia, 134
   in colonial period, 19
   in Philippines, 87
   in Vietnam, 122, 123, 130
citizenship
   of Brunei, 79
   of China, 174–75, 226
   of Philippines, 173
   of Thailand, 173
city states, 71
civil service, see bureaucracy
Clark Air Base, 30
Clark, Helen, 238
classes, social,
   and hazards, 187
   in Indonesia, 98
   in Philippines, 94, 200
   in Thailand, 50, 52
climate, 186–87
Clinton, Bill, 149, 212
cloves, 3, 8
CNRM, see National Council of Maubere Resistance
CNRT, see National Council of Timorese Resistance
coal, 16
Cochin China, 10, 121, 123, 170
cocoa, 196, 199
coconuts, 194, 200
coffee, 112, 116, 152, 194, 201, 202
Cold War, 12, 30, 124, 136, 146, 167, 176, 219, 227, 233, 234
colonialism, 12, 13, 29, 168, 180, 238
   agriculture under, 193
   and development, 14
   and nationalism, 95
   and postcolonial economies, 24–25
colonization, 8–10, 12, 231–32, 235
Commission of National Integration (Philippines), 173
Commonwealth, 25
communalism, 172
Communism, 26, 27, 170, 225, 226, 232, 233
   and China, 220
   in Dutch East Indies, 19
   in French Indochina, 19
   in Indonesia, 30
   in Laos, 146, 154
   in Malaya, 30, 56, 67, 169, 171, 196
   in Philippines, 30, 93, 94, 200
   in Singapore, 74, 76
   in Thailand, 43, 45, 136
Communist Party
   of China, 126
   of Indonesia, 98, 99, 103, 168, 175
   of the Philippines, 90, 91, 200
   of Singapore, 67
   of Thailand, 46, 173
   of the Soviet Union, 126
   French, 134
   in Vietnam (Workers Party), 125, 126–27, 128, 201
Community Development Department (Thailand), 44
computer assembly, 74

Index

concubines, 18
Conference of Rulers (Malaysia), 57
Conference on the Human Environment (1972), 188
Confrontation, see *Konfrontasi*
Confucianism, 123, 172
Congress (Philippines), 91
Conselho Nacional de Resistencia Maubere, see National Council of Maubere Resistance
Conselho Nacional de Resistencia Timorense, see National Council of Timorese Resistance
Consolidated Fund (Brunei), 81
Constitution(s)
    of Brunei, 78, 81
    of Burma, 159, 160, 162, 163
    of Cambodia, 133, 137, 141
    of Indonesia, 98, 103
    of Laos, 150
    of Malaysia, 58, 61–62, 169, 171
    of Philippines, 91, 93
    of Singapore, 171
    of Thailand, 42, 44, 46, 47, 188
    of Vietnam, 126, 127
Constitutional Council (Cambodia), 141
construction, 82, 83, 140
coolies, 15, 16
cooperatives (Laos), 147, 148, 201, 202
copper, 31
copra, 194
copyright infringement, 74
coral reefs, 182, 185, 187
Cordillera (Philippines), 90
corruption, 19, 213
    in Brunei, 84
    in Burma, 160
    in Cambodia, 133, 137, 139, 140, 142
    in Indonesia, 101–02, 105, 106, 113, 209, 211
    in Laos, 153
    in Malaysia, 57, 64
    in Thailand, 41–42, 43, 46, 47
    in Vietnam, 130
Council for Security and Cooperation in the Asia Pacific, 221
Council for the Popular Defense of the Proclamation of the Democratic Republic of East Timor, 115
Council of Trust for the Indigenous People (Malaysia), 64
CPP, see Cambodian People's Party
crime, 226
    in East Timor, 118
    in Indonesia, 97
    in Laos, 154
    in Singapore, 73–74
crisis, see Asian crisis
"crony capitalism," 46, 184, 213, 234
    in Cambodia, 137
    in Indonesia, 105, 107
    in Malaysia, 57, 61
Crown Agents, 82
CSCAP, see Council for Security and Cooperation in the Asia Pacific
Cuba, 126, 147
cukong, 175
"cultivation system" or "culture system" (Dutch East Indies), see *Cultuurstelsel*

cultural exchanges, 225–26
Cultuurstelsel, 14, 195
currency traders, 209–10

Danang, 123
DAP, see Democratic Action Party
Dayaks, 13, 168, 174
debt, 212, 213
    and Asian crisis, 211
    and development, 189–90
    in Thailand, 209
    of Burma, 162
    of Cambodia, 141
    of Indonesia, 215
    of Malaysia, 56, 215
    of Philippines, 215
    of Thailand, 215
    of Vietnam, 130
deforestation, see forests and forestry
Democrat Party (Thailand), 43, 47
Democratic Action Party (Malaysia), 60, 61, 65
Democratic Association for the Integration of East Timor into Australia, see Aditla
dengue fever, 117
dependistas, 25
devaluation
    and trade, 211
    in Brunei, 81–82
    in Cambodia, 137
    in Indonesia, 210, 215
    in Laos, 151
    in Malaysia, 209, 210, 215
    in Philippines, 215
    in Thailand, 49, 209, 210, 215
Development Fund (Brunei), 81
development, economic, 207
    and colonialism, 14
    and hazards, 187
    and investment, 214
    in East Timor, 115–16, 118
    in Laos, 146, 147, 151, 152, 153
    in Malaysia, 56
    in Thailand, 49–50
    (see also maldevelopment, underdevelopment)
Dewan Negara, see Senate (Malaysia)
Dewan Rakyat, see House of Representatives (Malaysia)
Dienbienphu, 124
Díli, 111, 112, 113, 114, 115, 118
doi moi, 33, 127, 128, 148, 201
dollar,
    Brunei, 81–82
    Singapore, 81
    US, 127, 129, 208
Dominican order, 110
Dong Cong San Vietnam, see Communist Party in Vietnam
dong, 128, 129
Downer, Alexander, 116, 238
"dragon economies," 56, 129
droughts, 186, 201
drugs,
    abuse of, 74
    trade in, 72, 160, 161
Duch, see Kang Kek Ieu

Dutch companies, see Netherlands, companies from
Dutch East Indies, 10, 12–13, 14, 16, 19, 20, 170, 195, 198 (see also Indonesia, Netherlands)
Dutch language, 17
dwifungsi, 102, 168, 175

EAEC, see East Asian Economic Caucus
East Asia Cooperation, 219, 227, 235, 236, 238, 239
East Asian Economic Caucus (proposed), 227, 236
East India Company, British, 67
East Timor, 107, 110–19, 175, 176, 177, 193, 196
    and Australia, 110, 111, 114, 116, 238
    and Cambodia, 114
    and Indonesia, 105, 110, 111–12, 113, 114, 115, 116, 117, 118–19, 211
    and Japan, 110
    and Malaysia, 114
    and New Zealand, 114
    and Philippines, 114
    and Portugal, 113, 114, 117
    and Singapore, 112, 114
    and South Korea, 114
    and Thailand, 114
    and UN, 113, 114, 238
    and United Kingdom, 114
    and United States, 114
    and World Bank, 117
    colonial, 18, 115, 116
    premodern, 12
Economic Development Board (Singapore), 69
economic growth, see growth, economic
Economic Planning Unit (Brunei), 81
Economic Planning Unit (Malaysia), 58, 65
Economist, 75
EDSA revolution, 91, 92–93, 95
education, 225
    in Brunei, 79
    in Burma, 157, 158, 159, 160
    in Cambodia, 134, 139, 140
    in East Timor, 112, 117–18
    in Indonesia, 175
    in Laos, 145, 147, 148, 150, 154
    in Philippines, 90
    in Singapore, 68–69, 73, 172
    in Thailand, 173
    in Vietnam, 122, 123, 201
El Niño southern oscillation, 179, 182, 186
elections
    in Brunei, 79
    in Burma, 159–60
    in Cambodia, 132, 136–37, 141
    in Indonesia, 103, 105
    in Laos, 150
    in Malaysia, 62, 63, 210
    in Philippines, 91, 92
    in Singapore, 75–76
    in Thailand, 47
    in Vietnam, 127
Electoral Commission (Thailand), 47
electrical goods, 32
electricity, 140, 151, 152, 153, 181
electronics, 67, 223
Elf Aquitaine, 80
"Emergency" (Malaya), 30, 196

emigration, see under migration
emissions of pollutants, 183
employment, see labor force
energy consumption, 182–83
English language, 17
    in Cambodia, 133
    in East Timor, 115
    in Malaysia, 62
    in Philippines, 87, 95
    in Singapore, 68
Englund, George H., 41
Enrile, Juan Ponce, 92
ENSO, see El Niño southern oscillation
environment, 49, 179–90
Epifanio de los Santos Avenue, see EDSA revolution
EPU, see Economic Planning Unit
equities, see stock markets
erosion, 180
Estrada, Joseph, 93–94
"Ethical Policy," 18, 102
ethnic groups and ethnicity, 167, 171, 177
    and agriculture, 196
    in Burma, 16, 158, 168, 169
    in Cambodia, 133
    in colonial period, 18–19
    in East Timor, 110
    in Indonesia, 97, 98, 100, 167, 168, 174, 175, 176, 210, 213
    in Laos, 145, 153, 154
    in Malaysia, 62, 64, 169, 174, 175, 232
    in Philippines, 89, 169, 173–74
    in Singapore, 70–71, 171–72
    in Thailand, 167–68, 172–73
    in Vietnam, 122, 167
EU, 227, 235–37, 238
Eurasians, 172
Europe, 222, 231, 232, 235, 236, 238, 239
    and Asian crisis, 211, 212
    and Burma, 162
    and Japan, 32
    and Singapore, 68
    colonization by, 8–9, 12
    in colonial period, 194
    premodern, 3, 4, 6
    trade with, 30
European Economic Community, 30
Europeans in colonial period, 16, 18, 24
exchange rates, 207–08, 209, 210, 211, 212, 214
export-oriented industrialization, see industrialization, export-oriented
export-processing zones, 32
export-promoting industrialization, see industrialization, export-oriented
exports, 208
    and Asian crisis, 211
    and manufacturing, 207
    from Brunei, 82
    from Burma, 201
    from Cambodia, 140
    from East Timor, 115, 116
    from Indonesia, 210, 224
    from Laos, 146, 149, 150, 151
    from Malaysia, 56
    from Singapore, 68

from Thailand, 32, 48, 210, 224
from Vietnam, 201
of commodities, 31
to China, 224
to Northeast Asia, 224
to South Korea, 224
extraterritoriality, 10

Fa Ngum, King, 145
Faifo, 123
Falintil, 111–12, 113, 115
"familism," 104
famine, 193
   in Cambodia, 135
   in Vietnam, 128, 129
Far Eastern Economic Review, 75
farmers
   and Communism, 196
   in Cambodia, 202
   in colonial period, 15, 24
   in Laos, 147–48, 201, 202
   in Malaysia, 64
   in Philippines, 90
   in Thailand, 44, 45, 47, 51
   in Vietnam, 127, 201
Federal Land Development Authority (Malaysia), 196, 199
federalism, 57, 102
Federation of Filipino-Chinese Chambers of Commerce (Shang Zong), 173
FELDA, see Federal Land Development Authority
fertility rate (Laos), 153
fertilizers, 180, 185, 194, 196, 197, 199, 201
FIC, see Foreign Investment Committee
Fiery Cross Reef, 221
"Filipino," 89, 95
Filipino language, 95
financial services, 207–15
   in Brunei, 80
   in Singapore, 222
   in Thailand, 48, 49
fisheries, 181–82, 184–85, 193, 194
   in Brunei, 78, 80, 83
   in Burma, 181
   in East Timor, 116
   in Indonesia, 181, 185
   in Philippines, 181, 185
   in Vietnam, 181
Five-Power Defense Arrangements, 220
floods, 186, 201
Florinese, 168
"flying geese theory," 71, 74, 214
forced labor, 17
   in Burma, 33
   in Cambodia, 134
   in Thailand, 42
Foreign Investment Committee (Malaysia), 64
forest fires, 179, 182, 186, 189
Forest People's Conference, 188
forests and forestry, 3, 4, 5, 6, 179, 180, 182, 183, 184, 193, 194, 203, 223
   and indigenous peoples, 188
   in Brunei, 79, 180
   in Burma, 160
   in Cambodia, 140
   in colonial period, 17
   in Indonesia, 31, 184, 222
   in Laos, 152, 202
   in Malaysia, 184
   in Philippines, 31, 180, 184
   in Singapore, 67
   in Thailand, 48, 184, 199
Formosa, Strait of, 221
France, 136, 148, 152, 231
   and Cambodia, 14, 133, 141
   and Indochina, 10, 13, 14, 25, 123, 124, 133, 170
   and Laos, 14, 16, 146, 147, 149
   and Thailand, 18, 170
   and traditional monarchies, 16
   and Vietnam, 14, 123, 124, 125, 130
   attitudes to imperialism in, 18
   in postwar world, 25
Free Aceh Movement, 176–77
Free Papua Organization, 174, 177
Free Thai movement, 43
free trade zones, 32
French language, 17, 133
Fretilin (Frente Revolucionária de Timor L'Este Independente), 111, 112, 113, 114–15
Friendship Bridge, 154
Front Unie Nationale pour Camboge Indépendante, Neutrale, Pacifique et Coopérative, 137, 138, 139, 141, 142
Fujianese, 172
Funan, 5
FUNCINPEC, see Front Unie Nationale pour Camboge Indépendante, Neutrale, Pacifique et Coopérative
Furnivall, John, 18

Galman, Rolando, 92
GAM, see Free Aceh Movement
gambling, 19
Gandhi, Indira, 58
Garuda Indonesia, 179
gas, natural,
   exported to Japan, 224
   in Brunei, 78, 79, 80, 81, 85
   in East Timor, 116
   in Indonesia, 209, 222
   in Vietnam, 130
GATT, see General Agreement on Tariffs and Trade
gazetting, 75
gems, 160
General Agreement on Tariffs and Trade, 227
Geneva Conference and Accords (1954), 123, 124, 125, 133, 146
Gerakan Rakyat Malaysia, see Malaysian People's Movement
Gerakanan Aceh Merdeka, see Free Aceh Movement
Germany, 18, 149, 150
Gilbert Global Equity, 106
global warming, 186, 189
globalization, 167, 180, 189, 214, 224, 228, 239
   and Laos, 154
   and Thailand, 51
Glodok, 175
God's Army, 161
Goh Chok Tong, 75–76

Goh Keng Swee, 69
Golkar, 103, 105
Gomez, Father, 88
Government of Singapore Investment Corporation, 70, 72
Greater East Asia Co-Prosperity Sphere, 12, 223
Greece, 5
Green Forum (Philippines), 188
green revolution, 193, 196–98, 199, 200, 201
greenhouse gases, 186
Group of Seven, 211, 212
growth, economic, 24, 29, 167
   and capital cities, 182
   and environment, 187, 190
   and exports, 207, 208
   and import-substituting industrialization, 30
   and investment, 214
   in Brunei, 82
   in Indonesia, 99, 101, 104
   in Laos, 150, 153
   in Malaysia, 62, 63, 65
   in Philippines, 31
   in Singapore, 67, 71, 72
   in Thailand, 32, 48, 49, 173
   in Vietnam, 129, 130
"Guided Democracy," 27, 103, 168
"Guided Economy," 27, 99, 100, 103
Gusmão, José Alexandre, 112, 113, 115

Habibie, B.J., 101, 105, 113, 114, 175
Hakim, Prince, 84
Hakka, 172
Halmahera, 8
Hamid, Prince, 83
Hamilton and Inches, 83
Hanoi, 121, 124, 130
Hassan, Mohammed "Bob," 175
Hassanal Bolkiah, Sultan, 78, 79, 80, 81, 83, 84, 106
Hawaii, 94
Hawke, Bob, 237
hazards, 179, 185–87, 189, 190
headhunting, 89
health and environment, 179, 183
health care, 225
   in Brunei, 79
   in Burma, 159
   in Cambodia, 139, 140
   in East Timor, 112
   in Laos, 150
   in Philippines, 90
   in Singapore, 69
   in Vietnam, 201
Heng Samrin, 135
heroin, 161
Herzog, Chaim, 176
High Court (Brunei), 83, 84
highlanders (Philippines), 89, 90
Hindus and Hinduism, 172, 219
   in Cambodia, 133
   in Indonesia, 168
   in Philippines, 87
   in premodern period, 7, 8
Hitam, Musa, 59
Hmong, 173

Ho Chi Minh, 124
Ho Chi Minh City, 121, 122, 130
Ho Chi Minh Trail, 124
Hoang Sa, see Paracel Islands
Hobson, J.A., 13
Hoi An, 123
Honasan, Greg, 93
Hong Kong, 32, 70, 71, 94, 150, 185, 219, 222, 223, 226, 227
   and Asian crisis, 214
   and Vietnam, 223
   companies from, 32
   trade with, 224
Hong Leong, 174
Horta, José Ramos, 113, 115, 116
House of Representatives (Malaysia), 57
House of Representatives (Thailand), 46, 47
housing, 207
   in Brunei, 79
   in Singapore, 68, 69, 71, 209
Housing and Development Board (Singapore), 69
Howard, John, 238
Hsaya San uprising, 19
Hsisha, see Paracel Islands
Htoo, Johnny and Luther, 161
Hue, 121
Huk rebellion, 90, 92
human rights, 236
   in Cambodia, 132, 136, 137, 139
   in Indonesia, 211
   in Singapore, 74–76
Hun Sen, 135, 137, 138, 141
hutan, 183
"hyperdependence," 146

IBRA, see Indonesia Bank Restructuring Agency
Ieng Mouly, 138, 140
Ieng Sary, 134, 141
illiteracy, see literacy
ilustrados, 173
IMF, 233, 234, 235, 236
   and Asian crisis, 211, 212–14, 227
   and Cambodia, 139
   and Indonesia, 105, 210, 213
   and Laos, 151
   and Malaysia, 57, 210
   and Philippines, 209
   and Thailand, 48, 49, 51
immigration, see under migration
imperialism, 12–20
import substitution, see industrialization, import-substituting
imports, 208
   from China, 224
   from South Korea, 224
   into Indonesia, 210
   into Laos, 146, 149, 150, 151
   into Philippines, 28
   into Thailand, 210
   into Vietnam, 201
income distribution, 182
   in Malaysia, 62
   in Thailand, 49
   in Vietnam, 122

income per capita
  in Burma, 157
  in Cambodia, 139, 140
  in Laos, 145, 146, 150
  in Singapore, 67
  in Thailand, 157
  in United States, 157
  in Vietnam, 122, 130
indenture systems, 15, 16
Independent National Khmer Army, see Armée Nationale Khmer Indépendante
India, 74, 219, 220, 228
  and Burma, 201
  and Cambodia, 135
  premodern, 3, 4, 5, 6, 9
  religious and cultural influence from, 7, 8
  under British rule, 67, 158
Indian Ocean, 162
Indians, 10
  in Burma, 29, 158
  in colonial period, 24
  in Malaya, 15, 16
  in Malaysia, 59, 63, 232
  in Singapore, 172
indigenism, 28–29
indigenous peoples, 188
indijenas, 115
"Indios Bravos," 95
indios, 95, 173
Indochina, 25, 30, 33, 132, 146, 170, 225
  and France, 123, 124, 133
  and United States, 13
  colonial, 10, 13, 14, 16
  (see also Cambodia, Laos, Vietnam)
"Indologists," 17
Indo-Malayan region, 182
Indo-Malesian region, 182
Indonesia, 24, 25, 27, 28, 29, 30, 31, 32, 60, 74, 80, 97–108, 136, 151, 162, 167, 168, 171, 172, 174–75, 176–77, 179, 181, 182, 184, 185, 186, 187, 188, 189, 193, 195, 196, 197, 207, 208, 209, 210, 211, 213, 214, 215, 222, 223, 225, 226
  and ASEAN, 177
  and Australia, 116, 238
  and Brunei, 79
  and China, 221, 224
  and East Timor, 105, 110, 111–12, 113, 114, 115, 116, 117, 118–19, 211
  and IMF, 105, 210, 213
  and Japan, 98, 224
  and Malaysia, 56, 103, 171, 176
  and "Maphilindo," 227
  and Netherlands, 168
  and Philippines, 87
  and Singapore, 73, 226
  and South Korea, 224
  and Spratly Islands, 121
  and Taiwan, 223
  and United States, 30, 99, 103, 220
  and Vietnam, 201
  colonial, 10, 12–13, 15, 97, 98, 102, 232
  predicted future of, 24
  (see also Dutch East Indies)
Indonesia Bank Restructuring Agency (IBRA), 106

Indonesian Democracy Party of Struggle, 105
Indonesian Environmental Forum, see WALHI
Indonesian language, 115, 117
Indonesian National Party, 103, 168
"Indonesians" (Philippines), 89
industrialization, 25–28, 33
  and agriculture, 195
  and East Timor, 118
  and investment, 33, 34
  and pollution, 182–83
  export-oriented, 31, 32, 48, 56, 69, 207, 222, 234
  import-substituting, 27–28, 30, 31, 44, 48, 56, 69, 99, 208, 222
  in Indonesia, 99, 100, 104
  in Malaysia, 56
  in Singapore, 69
  in Thailand, 44, 48
infant mortality
  in Cambodia, 139
  in Singapore, 68
inflation, 212
  in Burma, 162
  in Cambodia, 141
  in Indonesia, 99
  in Laos, 148, 150, 152
  in Singapore, 68
  in Vietnam, 129
insurance, 83
"integralism," 98, 100
Integration Struggle Soldiers, 114
intellectual property, 74
Intengan, Father Romeo, 91
Interfet, see International Force for East Timor
Internal Security Council (Singapore), 74
International Force for East Timor, 110, 114
International Labor Organization, 162
Intramuros, 88
investment, 226
  and Asian crisis, 213
  and economic development, 214, 215
  and exchange rates, 208
  and export-oriented industrialization, 31, 32
  and hazards, 187
  and import-substituting industrialization, 28
  and manufacturing, 207, 223
  by Japanese companies, 32, 212
  from Northeast Asia, 222–23, 224, 228
  from United States, 234
  in Brunei, 80, 81, 82
  in Burma, 33, 160, 161, 162, 163, 222
  in Cambodia, 137, 140, 141, 222
  in colonial period, 24, 25
  in East Timor, 118
  in Indonesia, 32, 99, 100, 101, 106, 107, 108, 209, 210, 222
  in Laos, 148, 149, 150, 151, 152, 153, 222
  in Malaysia, 32, 222
  in Philippines, 32, 222
  in Singapore, 27, 68, 69, 70, 72, 74, 76, 222
  in Thailand, 48, 199, 210, 222
  in Vietnam, 33, 127, 128, 129, 130, 222, 223
  industrialization and, 33, 34
Irian Jaya, see West Papua
Irish Republic, 114

irrigation, 180, 181, 186, 199, 201, 202
Isa, Pehin, 83
Isan region, 170, 173
Islam, see Moslems and Islam
Israel, 176
Italy, 94

Jakarta, 9, 98, 104, 175, 177, 182, 183, 209 (see also Batavia)
Jakarta Stock Exchange, 101, 174
Japan, 48, 73, 94, 207, 214, 219, 220, 221, 222, 224, 226, 227, 228, 232, 233
    aid from, 225
    and ASEAN, 227, 234, 235
    and Asian crisis, 207, 211, 212
    and Asia-Pacific Economic Cooperation, 237
    and Brunei, 79, 80, 224, 225
    and Burma, 158–59, 162, 225
    and Cambodia, 133, 220, 225
    and East and West Timor, 110
    and "flying geese" theory, 71
    and Indonesia, 98, 224, 225
    and Laos, 146, 149, 150, 152, 225
    and Malaya, 56
    and Malaysia, 223, 225
    and Philippines, 87, 90, 225
    and Singapore, 68, 225
    and Thailand, 212, 224, 225
    and Vietnam, 124, 224, 225
    companies from, 31, 32, 100, 184, 208, 212, 223, 224
    cultural exchanges with, 225–26
    economic growth in, 32
    in postwar world, 25
    in World War II, 12
    premodern, 9
    reconstruction of, 26
    trade with, 30
Jasra Elf, 80
Jasra International Petroleum, 80
Java, 98, 168, 174, 175, 193, 198
    colonial, 12, 14, 16, 19, 195
    premodern, 6–7, 9, 231
Java Sea, 6, 9
Javanese language, 17
Javanese people, 16, 97, 102, 103, 167, 168
    and Chinese, 18
    in outer islands of Indonesia, 196
    in Sumatra, 15–16
Jayavarman II, King, 133
Jefri, Prince, 81, 82, 83, 84
Jehovah's Witnesses, 75
Jerudong Hotel, 83
Jesuits, 91
Jeyaretnam, J.B., 70, 74–75
Jintanagan Mai, see New Economic Mechanism
Johor, 9, 17, 72, 181
Johor Bahru, 73
Joint Secretariat of Functional Groups, see Golkar
Judicial Committee of the Privy Council, 75, 84
judiciary
    in Cambodia, 142
    in East Timor, 118
    in Malaysia, 58
    in Singapore, 75

juries (Singapore), 75
Jurong Industrial Estate, 69

Kachin, 158, 169
Kalimantan, see Borneo
Kamphoui Keoboualapha, 152
kampones, 115
Kang Kek Ieu, 141
karaoke, 154
Karen, 18, 158, 161, 168, 169, 170
Karim, Dato Abdul Rahman, 82
Karunga, 169
Kaysone Phomvihane, 147, 150, 153
Keating, Paul, 237, 238
kekeluaragaan, 104
Kelantan, 176
kenaf, 199
Keppel Corporation, 70
kerajaan, 171
Khieu Samphan, 134, 141
Khin Nyunt, 160
Khmer language, 133
Khmer People's National Liberation Front (Cambodia), 135, 136
Khmer Rouge, 132, 134–35, 136, 137, 138, 141–42, 148, 196, 220
Khmers, 122, 133
Khon Muang, 173
Khun Sa, 161
kinship politics, 94
kip, 149, 151, 152
Klibur Oan Timur Aswain, see Kota
Konfrontasi, 56, 103, 171
Kong Le, 146
kongsi, 15
Konsortium Perkapalan Berhad, 61
Korea, North, 126, 157, 219, 226, 227, 233
Korea, South, 32, 34, 207, 210, 214, 219, 222, 223, 225, 226, 228
    and ASEAN, 227, 235
    and Asian crisis, 212
    and Brunei, 80, 224
    and Burma, 162
    and East Timor, 114
    and "flying geese" theory, 71
    and Indonesia, 224
    and Laos, 151, 152
    and Singapore, 68
    and Vietnam, 220
    cultural exchanges with, 226
    trade with, 224
Korean Peninsula, 221
Korean War, 26, 98
Kota, 111
KPB, see Konsortium Perkapalan Berhad
KPNLF, see Khmer People's National Liberation Front
Kra Isthmus, 4
Kra Isthmus Canal project, 73, 76
Kriangsak Chamanand, 46
Kuala Lumpur, 29, 62, 69, 169, 171, 177, 183
Kuching, 179
Kukrit Pramoj, 41, 45
Kunming, 150
Kupang, 115

Kwik Kian Gie, 105, 106
kyat, 159, 162

labor, forced, see forced labor
labor force, 26, 31, 223
   in agriculture, 193, 196, 197, 198
   in Cambodia, 140
   in colonial period, 15–16
   in East Timor, 117, 118
   in Indonesia, 100, 104
   in Singapore, 67, 68, 69, 73
   in Thailand, 48, 50
   in Vietnam, 123, 201
   industrialization and, 33, 34
labor standards, 236
labor unions
   in Cambodia, 139
   in Philippines, 91
   in Singapore, 71
   in Thailand, 42, 43, 44, 45, 46, 49, 51
   in Vietnam, 126
Laguna, 198
Laguna de Bay, 181
lakes, 181
Lan Xang, 145–46
land, 180, 183
   and green revolution, 197
   in East Timor, 118
   in Java, 198
   in Malaya/Malaysia, 180
   in Philippines, 90
   in Thailand, 47, 48, 180, 199, 200
   in Vietnam, 201
landmines (Cambodia), 139, 202
landowners, 15, 197, 200
languages, 17, 167, 183
   in Burma, 158
   in Cambodia, 133
   in East Timor, 110, 115
   in Indonesia, 97
Lao Issara, 146
Lao language, 147, 168
Lao Lum, 145
Lao Patriotic Front, 146
Lao people, 133, 145
Lao People's Revolutionary Party, 146, 154
Lao Sung, 145
Lao Theung, 145
"Laoism," 145, 154
Laos, 27, 33, 126, 145–54, 167, 170, 188, 189, 193, 196, 201–02, 222, 224, 226, 235
   and ASEAN, 149, 150
   and Asian Development Bank, 149, 150
   and Australia, 149
   and Burma, 145
   and Cambodia, 132, 133, 145
   and China, 145, 146, 147, 148, 152, 153, 220, 222
   and Cuba, 147
   and France, 123, 124, 146, 147, 149
   and Germany, 149, 150
   and IMF, 151
   and Japan, 146, 149, 150, 152, 225
   and Malaysia, 151, 152
   and Nordic countries, 149
   and Russia, 147
   and South Korea, 151, 152
   and Soviet bloc, 148
   and Soviet Union, 146, 147, 149, 150
   and Switzerland, 149
   and Thailand, 145, 146, 148, 149, 150, 151, 152, 154
   and UN Development Program, 150
   and United States, 146, 148, 149
   and Vietnam, 121, 145, 146, 147, 148, 152
   and World Bank, 149, 150, 151
   and World Trade Organization, 150
   colonial, 10, 14, 16
   premodern, 9
Lascar Jihad, 177
Latin America, 212
lead, 183
Lee and Lee, 72
Lee Hsien Loong, 71, 72, 176
Lee Kuan Yew, 27, 67, 72, 73, 74, 171
Leiden, University of, 17
Lenin, V.I., 13, 20
liberalization, economic, 208, 212, 214, 215, 228
   and environment, 189
   in Burma, 160, 161
   in Cambodia, 33
   in Indonesia, 101, 209, 210
   in Laos, 33, 148, 149, 202
   in Singapore, 70
   in Thailand, 44, 48, 49
   in Vietnam, 33
liberalization, political, 150
Libya, 174
life expectancy
   in Burma, 157
   in Cambodia, 139
   in Laos, 145, 146
   in Singapore, 68
   in Thailand, 157
   in United States, 157
Lim Sioe Liong, 175
literacy
   in Cambodia, 139
   in Laos, 145, 146, 147
   in Singapore, 68, 69, 73
   in Vietnam, 122–23
Litigan, 176
livestock, 203
loans, 207, 208, 211, 213
   from Japan, 224
   from Northeast Asia, 224
   in Indonesia, 209
   to Thailand, 48
logging, see forests and forestry
Lon Nol, 134
longyi, 163
"Look East" policy, 57
lowlanders (Philippines), 89, 90
LPRP, see Lao People's Revolutionary Party
Luang Prabang (city), 150
Luang Prabang (principality), 146
lumber, see forests and forestry
Luzon, 197, 198

Mabini, Apolinario, 89
Madura, 174
Madurese, 168, 174
Magellan, Ferdinand, 8
Magsaysay, Ramon, 90
Mahathir Mohamad, 56, 57, 59, 61, 63, 64, 65, 176, 210, 224, 227, 236
maize, 199, 200
Majapahit, 6, 7
Majlis Amanah Rakyat, see Council of Trust for the Indigenous People
Makarim, Zacky Anwar, 114
Makian, 8
Malacca, 7, 9, 67
Malacca, Straits of, 5, 6, 10, 73, 179, 220
Malari riots, 100
malaria, 17, 117, 186
Malay Chinese Association, see Malaysian Chinese Association
Malay language
  in colonial period, 17
  in Indonesia, 168
  in Malaysia, 62
  in Singapore, 68
Malay Peninsula, 4, 5, 6, 10
Malaya, 12, 13, 14, 15, 16, 17, 18, 24, 25, 30, 171, 196
  and Japan, 56
  and Singapore, 68, 69
  and United Kingdom, 56, 169, 194, 232
Malayan Indian Congress, see Malaysian Indian Congress
"Malayans" (Philippines), 89
Malays
  in Malaya, 18
  in Malaysia, 29, 59, 60, 61, 62, 63, 64, 168, 169, 171, 199, 232
  in Singapore, 70–71, 172, 176
  in Thailand, 173
Malaysia, 24, 25, 28, 29, 31, 32, 56–65, 74, 80, 162, 167, 168, 169, 171, 172, 174–75, 176, 179, 180, 181, 182, 184, 186, 187, 188, 193, 195, 196, 197, 198, 199, 207, 208, 209, 210, 212, 213, 214, 215, 222, 223, 226, 232
  and ASEAN, 177
  and Asia-Pacific Economic Cooperation, 238
  and Brunei, 169
  and China, 221, 225, 226
  and East Asia Cooperation, 239
  and East Timor, 114
  and Five-Power Defense Arrangements, 220
  and IMF, 57, 210
  and Indonesia, 56, 103, 171, 176
  and Japan, 223, 225
  and Laos, 151, 152
  and "Maphilindo," 227
  and Philippines, 174, 176
  and Singapore, 67, 69, 72–73, 169, 176
  and Spratly Islands, 121
  and Taiwan, 223
  and Thailand, 176
  and United States, 30, 220
  and Vietnam, 201
"Malaysia Incorporated," 57

Malaysian (formerly Malay) Chinese Association (MCA), 59, 60, 62, 169, 171
Malaysian (formerly Malayan) Indian Congress (MIC), 59, 60, 169, 171
Malaysian Chinese Organizations Election Appeals Committee, 61
Malaysian Islam Party, 59, 60, 61, 65, 171
Malaysian Islamic Youth Movement, 59
Malaysian People's Movement, 60
maldevelopment, 49–50
Maliana, 111
mangroves, 181
Manila, 9, 95, 182, 183, 227
Manila Declaration, 221
manufacturing, 26, 29–30, 179, 224
  and agriculture, 195
  and Asian crisis, 211
  and "flying geese" theory, 71, 214
  and import-substituting industrialization, 30
  and investment, 207, 208, 223
  Chinese people and, 28
  in Burma, 33, 201
  in Cambodia, 140
  in Indonesia, 32, 104, 213
  in Laos, 147
  in Malaysia, 28, 32
  in Philippines, 31
  in Singapore, 27, 28, 67, 70, 74, 222
  in Thailand, 28, 32, 48, 49, 50
  in Vietnam, 129
  Japanese companies and, 31
"Maphilindo," 227
MARA, see Council of Trust for the Indigenous People
marble, 116
Marcos, Ferdinand, 91, 92, 93, 95, 171, 173, 174, 224
Marcos, Imelda, 91, 93, 94
marine environment, see sea
Martaban, 13
martial law
  in Indonesia, 103
  in Philippines, 91–92, 93, 95, 173
Marx, Karl, 20
Marxism, 153, 172
Masyumi, 103, 168, 175
maternal mortality (Cambodia), 139
Maugham, W. Somerset, 16
Maung Aye, 160
MCA, see Malaysian Chinese Association
Medan, 179
media
  in Brunei, 84
  in Burma, 159, 161
  in Cambodia, 138
  in Indonesia, 113
  in Singapore, 75
Mekong Delta, 5, 10, 15, 121, 124, 133
Mekong River, 121, 132–33, 146, 149, 154
Mekong River Commission, 137, 138
Melaka, see Malacca
Melanesians (East Timor), 110
Mello, Sergio Vieira de, 118
Merauke, 186
"meritocracy," 69, 73

Index

mestizos
  in East Timor (*mestisu*), 18, 115
  in Philippines, 17, 89, 90, 173
methane, 183, 186
Metro Manila, see Manila
Mexico, 88
MIC, see Malaysian Indian Congress
Michelin, 16
Middle East, 3, 4, 5, 7, 68, 94, 176
migration,
  and agriculture, 196
  between Northeast and Southeast Asia, 226, 228
  from Burma, 226
  from Laos, 148, 226
  from Philippines, 33, 73, 94, 226
  from Singapore, 73
  from Thailand, 48
  from Vietnam, 122, 128, 226
  in colonial period, 15, 226
  to Australia and New Zealand, 237
  to China, 226
  to East Timor, 112
  to Hong Kong, 226
  to Japan, 226
  to Singapore, 69, 73
  to Taiwan, 226
  to Thailand, 43
  within Cambodia, 196
  within Indonesia, 174, 196
  within Malaysia, 196, 199
  within Vietnam, 15, 196
MILF, see Moro Islamic Liberation Front
military, see armed forces
militias (East Timor), 114, 118–19
Minahassans, 97
Minangkabau, 168
Mindanao, 91, 169, 173, 176, 200
mining, 24, 28, 98, 106, 213
Mischief Reef, 221
Misuari, Nur, 91
Mitsubishi Corporation, 80
Miyazawa bonds, 212
Miyazawa Ki'ichi, 212
MNLF, see Moro National Liberation Front
modernity, 13, 17, 20, 90, 154, 231
Mohammed, Prince, 81
Molinaka Party, 137
Moluccas, 4, 9, 107, 168, 177
Mon, 158, 169
monarchy
  in Brunei, 79, 80
  in Burma, 17
  in Cambodia, 133, 137, 170
  in colonial period, 16
  in Laos, 146, 170
  in Malaysia, 57, 60, 62
  in Thailand, 41, 42, 46, 51, 170
Mongkut, King, 10
Mongolia, 219, 227
Mongols, 6
Monivong, King, 133
monks, 44, 45, 134, 163
monsoon, 3, 186
"moral economy," 202

Moro Islamic Liberation Front, 176
Moro National Liberation Front, 91, 174
Moros, see Moslems and Islam in Philippines
mortality, see infant mortality, maternal mortality
Moslems and Islam, 172, 176, 219
  in Brunei, 79
  in Burma, 157
  in Cambodia, 133
  in Indonesia, 98, 105, 168, 174, 175, 176, 177
  in Malaysia, 62, 171, 175, 176
  in Philippines, 87, 89, 90, 91, 93, 94, 169, 173–74, 176, 200
  in premodern period, 7, 8
  in Thailand, 43, 170, 173, 176
Motir, 8
motor vehicles, 223
multinational corporations, 69, 223
Mummadiyah, 175, 177
murder (Singapore), 74
Musi River, 5
"Myanmar," 158
Myanmar Alin, 160

Nadhlatul Ulama, 175
NAFTA, see North American Free Trade Agreement
Nansha, see Spratly Islands
Nasser, Gamal Abdel, 133
"Nation, Religion, and Monarchy," 42, 45, 170
National Assembly (Cambodia), 136, 137, 138, 139, 140, 141, 142
National Assembly (Vietnam), 126, 127
National Awakening Party (Indonesia), 175
National Consultative Council (East Timor), 114
National Corporation, see Pernas
National Council of Maubere Resistance (East Timor), 112
National Council of Timorese Resistance, 113, 114, 115, 116, 117, 118
National Development Plans (Brunei), 79, 81
National Development Policy (Malaysia), 63, 64
National Economic and Social Development Board (Thailand), 50
National Economic Development Authority (Philippines), 27
National Economic Development Board (Thailand), 27
National Election Committee (Cambodia), 141
National Equity Corporation, see Permodalan Nasional Berhad
National Front (Malaysia), 59, 60, 61, 169
National League for Democracy (Burma), 33, 159, 160, 162, 163
National Peacekeeping Council (Thailand), 46, 47
National Trades Union Congress (Singapore), 74
National Union Front for Political Affairs (East Timor), 115
National Unity Party (Burma), 159, 160
nationalism, 12, 25, 26, 172
  and economic activity, 29
  and industrialization, 27
  Chinese people and, 28
  in colonial period, 20
  in Europe, 14
  in Indonesia, 107, 168, 175, 222
  in Laos, 146

in Malaysia, 199
in Philippines, 88, 94–95, 169
in Thailand, 42, 43
nationalization
   in Burma, 159
   in Indonesia, 25, 27, 99, 198
nation-building, 167, 170–75, 177
nature, attitudes to, 183
NDP, see National Development Policy
Ne Win, 27, 159, 160
Negritos, 89
Negros, 200
NEM, see New Economic Mechanism
Neo Lao Hak Sat, 146
neocolonialism, 12
neoliberalism, 50, 52
NEP, see New Economic Policy
Netherlands, 231
   and Indonesia, 9, 10, 12–13, 14, 97, 98, 102, 168, 232
   and Timor, 110
   and trade, 13
   attitudes to imperialism in, 18
   companies from, 25, 27, 99, 198
   in postwar world, 25
New Economic Mechanism (Laos), 148–50, 152
New Economic Policy (Laos), 147–48
New Economic Policy (Malaysia), 27, 29, 58, 60, 61, 62–63, 63–65, 105, 171, 199
New Economic Zones (Vietnam), 196
New Light of Myanmar, 161
"New Order," 27, 30, 168, 175, 198
New People's Army (Philippines), 90–91, 200
New Zealand, 227, 236, 237–38, 239
   and East Timor, 114
   and Five-Power Defense Arrangements, 220
Newbridge Capital, 106
newly industrialized countries, 24, 48, 222, 223, 224
Ngo Dinh Diem, 125
Nguyen Ai Quoc, see Ho Chi Minh
NICs, see newly industrialized countries
Nike, 152
"Nine Dragons," 133
nitrogen oxide, 183
NLD, see National League for Democracy
Nonaligned Movement, 12, 97
Nordic countries, 149
Norodom Ranariddh, see Ranariddh, Prince
Norodom Sihanouk, see Sihanouk
North America, 68, 194, 222
North American Free Trade Agreement, 237
North Borneo Company, 14
Northeast Asia, 185, 219–28, 233, 234, 237
NPA, see New People's Army
NU, see Nadhlatul Ulama
Nu, U, 159
Nuon Chea, 134, 141

Oc-eo, 5
oil, 31, 220, 221
   exported to Japan, 224
   in Brunei, 27, 78, 79, 80, 81, 82, 84, 85
   in Burma, 14
   in East Timor, 116
   in Indonesia, 31, 32, 98, 100, 101, 209, 222
   in Singapore, 67, 73
   in Vietnam, 130
Omar Ali Saifuddin III, Sultan, 78, 79
Onn, Tun Hussein, 57, 59
OPEC, see Organization of Petroleum-Exporting Countries
Operation Cold Store (Singapore), 74
opium, in Dutch East Indies, 18, 44, 161
OPM, see Free Papua Organization
Orang Asli, 196
"organicism," 98, 100
Organisasi Papua Merdeka, see Free Papua Organization
Organization of Petroleum-Exporting Countries, 80, 100
Organization of the Islamic Conference, 80
Orwell, George, 19
Osaka Gas Company, 80
outer islands of Indonesia, 99, 103, 167, 174

pa thuan, 183
Pacific Dunlop, 16
Pagan (city), 163
Pagan (polity), 5
Pakpahan, Mochtar, 175
Palembang, 6
Palestin, Sulaiman, 59
palm oil, 179, 180, 194, 196, 198, 199, 203
Pancasila, 100, 101, 103, 168, 175
PAP, see People's Action Party
Papua New Guinea, 182, 184
Papuans, 13
Paracel Islands, 121, 221
Paris Conferences and Agreement (1989 and 1991), 132, 136
Parliament (Malaysia), 57–58
Parliament (Philippines), 93
Partai Komunis Indonesia, see Communist Party of Indonesia
Partai Nasional Indonesia, see Indonesian National Party
Parti Islam SeMalaysia, see Malaysian Islam Party
Parti Rakyat Brunei, see Brunei People's Party
Partido Trabalhista, see Workers Party of East Timor
parties, political,
   in Brunei, 79
   in Burma, 160
   in Cambodia, 137, 139, 142
   in East Timor, 111, 115
   in Indonesia, 97, 103, 106, 171, 175
   in Malaysia, 59, 60, 169, 171
   in Singapore, 74, 76, 171
   in Thailand, 43, 47
PAS, see Malaysian Islam Party
Pasukan Perjuang Integrasi, see Integration Struggle Soldiers
Pathet Lao, 146
patronage, 184
   in Indonesia, 101, 105, 106
   in Malaysia, 57, 59, 61
   in Philippines, 92, 94
   in Thailand, 43–44, 46
Pattani, 173
Pattani United Liberation Organization, 176
PDI-P, see Indonesian Democracy Party of Struggle

Peasant Federation of Thailand, 45
peasants, see farmers
peceklik, 193
Pedra Branca, 176
Pegu, 9
pela gandong, 177
Penang, 9, 67, 74
"People Power" movement, 173 (see also EDSA revolution)
People's Action Party (Singapore), 69, 70, 71, 74, 75, 76, 171
People's Army of Vietnam, see armed forces of Vietnam
People's Assembly (Burma), 159
People's Consultative Assembly (Indonesia), 114
People's Council, see Volksraad
People's Party (Thailand), 42, 43
People's Sovereignty Party (Indonesia), 105
pepper, 194
Perak, 176
peranakan, 172
perestroika, 148
Permodalan Nasional Berhad, 64
Pernas, 64
Persian Gulf, 220
Pertamina, 80, 100
Peru, 88
pesticides, 180, 194, 197
petrochemicals, 31, 80
petroleum, see oil
Petronas, 80
Pham Van Dong, 124
Phibun Songkhram, 27, 42–43, 44, 172
Philippines, 24, 25, 27, 28, 29, 30, 31, 32, 87–95, 162, 167, 169, 170–71, 172, 173–74, 177, 179, 180, 181, 182, 183, 184, 185, 188, 189, 193, 196, 197, 200, 208, 209, 215, 222, 225, 226
  and China, 87, 221
  and East Asia Cooperation, 239
  and East Timor, 114
  and IMF, 209
  and Indonesia, 87
  and Japan, 87, 90
  and Malaysia, 174, 176
  and "Maphilindo," 227
  and Southeast Asia Treaty Organization, 220
  and Spratly Islands, 121
  and Taiwan, 87
  and United States, 30, 94, 220, 222
  and Vietnam, 201
  colonial, 10, 12, 15, 17, 19, 24–25, 87–90, 169, 200, 231
  premodern, 7–8, 87
Phillips Petroleum, 116
Phnom Penh, 132, 133, 134, 138, 139–40
Phra Prayom Kulyano, 50
pirating of intellectual property, 74
pisantren, 175
PKI, see Communist Party of Indonesia
PKP, see Communist Parties of the Philippines
planning, economic, 26, 27, 148, 149, 195
plantations, 10, 194
  and Communism, 19
  and forest fires, 179
  in Brunei, 78
  in Cambodia, 202
  in colonial period, 16, 17, 24
  in Indonesia, 97, 98
  in Malaysia, 25, 199
  in Sumatra, 16
  in Vietnam, 15
plants, 182
Plaza Accords (1985), 32, 48
pluralism, 18, 102
PNB, see Permodalan Nasional Berhad
PNI, see Indonesian National Party
poisons, 185
Pol Pot, 125, 134, 138, 141, 196
political parties, see parties
pollution, 33, 179, 181–82, 183, 189
Polynesians (East Timor), 110
"popular consultation," see referendum
Popular Democratic Association of Timor, see Apodeti
Popular Monarchist Association of Timor, 111
Popular Socialist Community of Cambodia, 133
Popular Timorese Party, 115
population
  of Burma, 158
  of Laos, 145
  of Singapore, 67
  of Vietnam, 121
population density, 180
  and agriculture, 194
  of Laos, 145, 153
  of Thailand, 145
  of Vietnam, 122, 145
population growth
  and environment, 189
  in Laos, 153
  in Vietnam, 123
Portugal, 231
  and East Timor, 12, 110–11, 113, 114, 115, 116, 117
  and Malacca, 9
  and trade, 13
  and Vietnam, 123
  attitudes to imperialism in, 18
Portuguese language, 17, 115, 117
Portuguese Telecommunications Company, 117
Potsdam Conference (1945), 124
poverty
  and agriculture, 193, 195, 196, 197
  and environment, 189
  and hazards, 187
  in Cambodia, 135
  in East Timor, 115
  in Java, 195, 198
  in Laos, 146
  in Malaysia, 64
  in Singapore, 67, 73
  in Vietnam, 122, 123, 127, 128, 130
PPI, see Integration Struggle Soldiers
PPT, see Popular Timorese Party
Prachathipok, King, 42
Prem Tinsulanonda, 46, 172
pribumi, 29
Pridi Phanomyong, 43
primam, 175
Prime Minister's Department (Malaysia), 58, 65

privatization
    in Indonesia, 101
    in Laos, 149
    in Malaysia, 63, 65
    in Singapore, 70
Privy Council, see Judicial Committee of the Privy Council
priyayi, 16, 20, 97, 103
Prosperous Labor Union, 175
prostitution, 15, 17, 19
Protestants, 123, 168
PST, see Timorese Socialist Party
P.T. Denok, 116
PT Telkom, 117
Ptolemy, 3
PULO, see Pattani United Liberation Organization

QAF Holdings, 81
Quan Doi Nhan Dan Viet Nam, see armed forces of Vietnam
Quezon, Aurora, 91–92
Quezon, Manuel, 92

racism, 29
Raffles, Sir Stamford, 67
Rahman, Tengku Abdul, 57, 59, 60, 171
rai-na'in, 115
rainfall, 186, 199 (see also acid rain)
Rais, Amien, 175, 177
Raj, see India under British rule
Rakhine, 158
rakyat, 84
Ramos, Fidel, 93, 209
Ramos, Narciso, 92
Ranariddh, Prince, 137, 138, 139, 141
Rangoon, 158, 160, 163
Ratchaburi, 161
Razak, Tun Abdul, 57, 59, 60, 171
real estate, 207, 208, 209
Red River, 121, 124
referendum (East Timor, 1999), 113, 114, 176
reforestation, see forests and forestry
refugees
    from Burma, 161
    from Cambodia, 134, 135, 136, 138
    from East Timor, 114
    from Laos, 148
    from Vietnam, 226
religion, 167, 171
    in Burma, 157
    in Cambodia, 133, 134
    in Indonesia, 97, 174, 176–77
    in Malaysia, 174
    in Philippines, 88, 89
    in Singapore, 71, 75, 172
    in Vietnam, 123
Renetil, 112
rentier economy (Brunei), 78, 79, 82, 85
Resistencia Nacional dos Estudantes de Timor L'Este, see Renetil
resources, 179, 180, 183–85, 187, 188, 189
Revolutionary Front for an Independent East Timor, see Fretilin
Rhodes, Father Alexandre de, 123

Riau, 17
Riau and Johor, Sultan of, 67
Riau archipelago, 72
rice, 193, 194, 202–03
    and climate changes, 186
    and green revolution, 197
    from United States, 26
    in Brunei, 78
    in Burma, 14, 162, 200–01
    in Cambodia, 202
    in Java, 6, 195, 198
    in Laos, 145, 151, 152, 153, 202
    in Malaysia, 199
    in Philippines, 200
    in premodern period, 6, 8
    in Thailand, 31, 42, 48, 200
    in Vietnam, 128, 129, 201
    trade in, 10
riel, 137, 140
ringgit, 209, 210
Rizal, Jose, 88, 92, 95
roads, 153, 199
Robert Kuok group, 174
Roman empire, 3, 4, 5
Romasz Starweski, 83
Royal Dutch Shell, 78, 80, 116
royal family (Brunei), 80–81, 84, 85
Royal Navy (United Kingdom), 68
royalists (Thailand), 43
rubber, 194, 203, 223
    and forest fires, 179
    in Brunei, 78
    in Burma, 200
    in Cambodia, 202
    in colonial period, 15, 24
    in Dutch East Indies, 198
    in Indonesia, 31
    in Malaya, 14, 180
    in Malaysia, 31, 180, 196, 199
    in Singapore, 67
    in Sumatra, 16
    in Thailand, 31
    in Vietnam, 201
    trade in, 10
Rukunegara, 58
rupiah, 98, 105, 209, 210, 213
Russia, 212, 214, 228
    and Laos, 147
    and Vietnam, 130

Sabah, 14, 64, 67, 169, 174, 176, 182, 184
Sabat Alam, 188
Saigon, 121, 125
Salim group, 175
Salleh Abbas, Tun, 58
Saloth Sar, see Pol Pot
Sam Rainsy, 139, 141
Sam Rainsy Party, 141, 142
Samane Vignaket, 154
Sammkakhi Tham, 47
Samudra-Pasai, 7
sang chat, 172
Sanskrit, 8, 17
Santa Cruz Cemetery, 112–13

santri, 98, 103, 168
Sarawak, 64, 67, 78, 169, 182, 184
Sarit Thanarat, 44, 172
Sasono, Adi, 105
Saudi Arabia, 94
saving, 207, 208, 209
Saw Maung, 159
sawah agriculture, 194, 195
schools, see education
Schurz, Carl, 89
sea, 180, 181–82, 187
  law of the, 221
  levels, 186
SEATO, see Southeast Asia Treaty Organization
SEDCs, see State Development Corporations
Sein Lwin, 159
Sekretariat Bersama Golongan Karya, see Golkar
semiconductors, 32
Senate (Malaysia), 57
Senate (Thailand), 43, 46, 47
service industries, 207
  in Cambodia, 140
  in Indonesia, 104
  in Singapore, 67, 222
  in Thailand, 49
  in Vietnam, 129–30
17th parallel, 124
sex industry, 49, 152, 154
sexuality, 19
Shan, 158, 169, 170
Shan states, 17
Shang Zong, see Federation of Filipino-Chinese Chambers of Commerce
Shanghai, 223
Shell, see Royal Dutch Shell
"Shellfare," 79
shipbuilding, 31, 67, 68
Shipping Consortium Limited, 61
Shwedagon Pagoda, 163
Siam, see Thailand
Siem Reap, 133
Sihanouk, 27, 132, 133, 134, 135, 136, 137
Sijori, 72–73
Sikhs, 172
silk, 4, 5
Silk Road, 4
Sin, Jaime, Cardinal, 92, 173
Singapore, 24, 27, 28, 29, 31, 32, 67–76, 94, 167, 171–72, 176, 179, 181, 193, 208, 209, 215, 222, 226, 227
  and ASEAN, 177
  and Burma, 72
  and China, 225, 226
  and East Timor, 112, 114
  and Europe, 68
  and Five-Power Defense Arrangements, 220
  and Indonesia, 73, 226
  and Japan, 68, 225
  and Malaya, 68, 69
  and Malaysia, 32, 67, 69, 72–73, 169, 176
  and Middle East, 68
  and North America, 68
  and South Korea, 68
  and United Kingdom, 67, 68
  and United States, 30, 74, 220, 222
  and Vietnam, 201
  and World Bank, 68
  colonial, 10, 15
  companies from, 32
Singapore Airlines, 70
Singapore Straits, 68
Singapore Technologies Group, 72
Singapore Telecom, 72
Singapore Trades Union Congress, 74
Sinhalese, 172
Sipidan, 176
Sirik Matak, 134
Sirindhorn, Princess, 51
Sisavang Vong, King, 146
Sison, Jose Maria, 90, 91
SLORC, see State Law and Order Restoration Council
small and medium-sized enterprises, 82, 101
smallpox, 17
smuggling, 72, 127, 201
Social Democratic Party (East Timor), 115
Social Democrats (Philippines), 91
Soeryadjaya, Edwin, 106
Soewandi, Rini, 106
sōgō shōsha, 184
Solo, 175
Solomon Islands, 182, 184
Son Sann, 135, 137
Son Sen, 134, 138, 141
Sons of the Mountain Warriors, see Kota
Sorong, 186
Souligna Vongsa, King, 145, 146
Souphanouvong, Prince, 146
South Asia, 185
South China Sea, 121, 133, 220, 221
Southeast Asia Treaty Organization, 220, 227
Souvanna Phouma, Prince, 146
Soviet bloc
  and Laos, 148
  and Vietnam, 124, 126, 127
Soviet Union, 26, 33, 148, 162, 220, 227, 232, 233
  and Cambodia, 132, 136, 141
  and Laos, 146, 147, 149, 150
  and Vietnam, 124, 128, 129, 130, 135
Spain, 94, 231
  and Philippines, 10, 12, 15, 17, 87–89, 90, 95, 169, 200
Spanish language, 88, 95
Spanish–American War, 88
SPDC, see State Peace and Development Council
spices, 3, 4, 5, 6–7, 8, 9, 87, 194
Spratly Islands, 121
SPSI, see All Indonesia Workers Union
Srivijaya, 5–6
stamps, 146, 152
State Development Corporations (Malaysia), 64
state enterprises
  in China, 222, 223
  in Indonesia, 100, 101
  in Laos, 149, 152
  in Malaysia, 62, 63, 64
  in Singapore, 70
  in Thailand, 43, 46
  in Vietnam, 128, 129

state intervention, 27, 31
State Law and Order Restoration Council (Burma), 33, 159, 160, 161, 201
State Peace and Development Council (Burma), 33, 160, 161, 162, 163
state-building, 167, 194
steel, 31
Stock Exchange of Singapore, 72
stock markets, 72, 101, 174, 207, 208, 209, 213
Stockholm Conference (1972), 188
Straits Settlements, 67
students, 46, 99–100, 104, 112
Subic Bay Naval Base, 30
Suchinda Krapyrayoon, 47
Sudarsono, Juwono, 107
Sufis, 7
Sufri, Prince, 81
sugar, 15, 24, 31, 194, 199, 200
Suharto, 27, 99, 100–01, 102, 103–04, 105, 106, 107, 108, 113, 168, 171, 175, 179, 189, 213
Suharto family, 101, 175
suicide (Singapore), 74
Sukarno, 12, 27, 98, 99, 100, 102, 103, 104, 105, 133, 168, 171
Sukarnoputri, Megawati, 105
Sulawesi, see Celebes
sulfur dioxide, 183
Sulu, 87, 174, 200
Sumatra, 4, 5, 6, 7, 9, 13, 16, 179, 198
Sumatrans, 167
Sunda Shelf, 182
Sunda Straits, 5
Sundanese, 168
Supreme Court (Malaysia), 58
Suqiu, see Malaysian Chinese Organizations Election Appeals Committee
Surakarta, 17
Surin Pitsuwan, 176
Suu Kyi, see Aung San Suu Kyi
swidden agriculture, 194, 202
Switzerland, 149

Ta Mok, 141
Tagalog language, 95
"Taglish," 95
Taiwan, 32, 94, 174, 185, 219, 220, 222, 225, 226, 227, 228
  and "flying geese" theory, 71
  and Indonesia, 223
  and Malaysia, 32, 223
  and Philippines, 87
  and Spratly Islands, 121
  and Vietnam, 223
  companies from, 32
  cultural exchanges with, 226
  trade with, 224
Takeshita Noboru, 226
Tam Giao, 123
taman, 183
Tamil language, 68
Tamils, 172
Tan, Tony, 224
Tanaka Kakuei, 100
Taoism, 123

taw, 183
taxation
  and agriculture, 194
  and forestry, 184
  in Brunei, 79
  in Cambodia, 141
  in colonial period, 14, 15, 19
  in Dutch East Indies, 18, 195
  in East Timor, 112
  in Singapore, 70
  in Vietnam, 127
Tay Ninh, 123
tea, 152, 194, 201
telecommunications, 83, 117, 140
television, 154
Telstra, 117
temperature changes, 186
Teochew, 172
Ternate, 8
Tetun Dili language, 115
Tetun language, 115, 117
textiles, 32
  in Brunei, 82
  in Indonesia, 32, 100
  in Laos, 152
  in Singapore, 67
  in Thailand, 48
TFET, see Trust Fund for East Timor
thaba-wa, 183
Thai language, 173
Thailand, 17, 18, 24, 27, 28, 29, 31, 32, 41–52, 73, 74, 157, 162, 167–68, 170, 176, 179, 180, 181, 182, 183, 184, 186, 187, 188, 193, 196, 199–200, 201, 207, 208–09, 210, 211, 212, 213, 214, 215, 222, 223, 226
  and Burma, 161
  and Cambodia, 132, 133, 134, 135, 136, 138
  and China, 136, 225
  and East Timor, 114
  and France, 170
  and IMF, 48, 49, 51
  and Japan, 212, 224, 225
  and Laos, 145, 146, 148, 149, 150, 151, 152, 154
  and Malaysia, 176
  and Northeast Asia, 224
  and Southeast Asia Treaty Organization, 220
  and United Kingdom, 170
  and United States, 30, 43, 44, 45, 220
  and Vietnam, 30, 45
  and World Bank, 44, 48, 49, 51
  avoidance of colonization by, 10, 12, 13
  predicted future of, 24
  premodern, 5, 42, 231
Thailand, Gulf of, 5, 132, 181
Thais Help Thais Fund, 50
Thais in Singapore, 73
thammachaat, 183
Thammasat University, 45
Than Shwe, 160
Thanin Kraivixien, 45, 46
Thanom Kittikachorn, 44–45
That Luang, 150
Thatcher, Margaret, 58
Thebaw, King, 17

Index

Tidore, 8
"tiger economies," 32, 33, 56
timber, see forests and forestry
Timor, 5, 110 (see also East Timor, West Timor)
Timor Gap, 116
Timorese Democratic Union, see União Democratica de Timor
Timorese Social Democratic Association, 111
Timorese Socialist Party, 115
tin, 14, 15, 16, 24, 25, 31
Tito, Josip Broz, 133
tobacco, 15, 16, 194, 199
Tokyo Electric Power Company, 80
Tokyo Gas Company, 80
Tonkin, 10, 121, 123, 170
Tonle Sap, 132–33
tourism, 179
   between Northeast and Southeast Asia, 226, 228
   in Brunei, 80
   in Burma, 33, 162
   in Cambodia, 140
   in East Timor, 116–17
   in Laos, 149, 150–51, 152
   in Singapore, 67
   in Thailand, 48
trade unions, see labor unions
trade, international, 13, 223, 226
   and colonization, 12
   and devaluation, 211
   and environment, 180, 189
   and exchange rates, 208
   Chinese people and, 28
   in colonial period, 24–25
   in premodern period, 3–10
   of Burma, 162
   of Indonesia, 211
   of Laos, 148, 150, 152, 202
   of Malaysia, 31
   of Philippines, 88
   of Singapore, 68, 70, 71, 72, 74
   of Thailand, 211
   of Vietnam, 127, 128, 129–30
   precolonial, 219
   with Australia and New Zealand, 237
   with China, 224
   with Europe, 235
   with Northeast Asia, 223–25, 228
   with other regions in 1965, 30
   with South Korea, 224
   with United States, 235
transmigrasi (transmigration), see migration within Indonesia
transportation, 67, 70
trials (Singapore), 75
Trisakti University, 104
Truong Sa, see Spratly Islands
Trust Fund for East Timor, 116
typhoons, 186
typhus, 17

UDT, see União Democratica de Timor
Umar, Pehin Abdul Aziz Awang, 84
UMNO, see United Malays National Organization

UN, 227
   Advance Mission in Cambodia, 136
   and Cambodia, 132, 135, 136, 137, 141
   and East Timor, 113, 238
   and Vietnam, 126
   Assistance Mission to East Timor, 110, 114
   Convention on the Law of the Sea, 221
   Development Program, 150
   General Assembly, 135
   Security Council, 111, 114
   Transitional Administration in East Timor, 114, 116, 117, 118
   Transitional Authority in Cambodia, 136, 137, 138, 141, 142
UNAMET, see UN Assistance Mission to East Timor
UNAMIC, see UN Advance Mission in Cambodia
underdevelopment, 25, 190
unemployment
   and agriculture, 196
   in Brunei, 81
   in Burma, 162
   in Cambodia, 140
   in East Timor, 117, 118
   in Malaysia, 62
   in Singapore, 67
   in Vietnam, 130
Unesco, 150
União Democratica de Timor, 111, 112, 113, 114
Unicef, 117
unions, see labor unions
United Kingdom, 222, 231
   aid from, 225
   and Brunei, 12, 78, 79
   and Burma, 158, 159
   and East Timor, 114
   and Five-Power Defense Arrangements, 220
   and Malaya, 10, 12, 13, 14, 17, 24, 56, 169, 194, 232
   and migration from India, 15
   and Singapore, 10, 67, 68
   and Thailand, 18, 42, 170
   and traditional monarchies, 16, 17
   and Vietnam, 124
   attitudes to imperialism in, 18
   in postwar world, 25
United Malays National Organization (Malaya/Malaysia), 56, 58–61, 62, 63, 65, 169, 171, 174, 210
United States, 157, 221, 227, 228, 231, 232–35, 236, 238, 239
   Agency for International Development, 44, 117
   aid from, 225
   and ASEAN, 233
   and Asian crisis, 207, 211, 212, 213, 214
   and Burma, 162
   and Cambodia, 132, 133–34, 135, 136, 141, 202
   and East Timor, 114
   and Indochina, 13
   and Indonesia, 30, 99, 103, 220
   and Japan, 32
   and Laos, 146, 148, 149
   and Malaysia, 30, 220
   and Philippines, 30, 12, 17, 19, 88–90, 94, 95, 169, 200, 220, 222

and Singapore, 30, 74, 220, 222
and Thailand, 30, 43, 44, 45, 220
and Vietnam, 30, 33, 123, 124, 125, 220
attitudes to imperialism in, 18
Central Intelligence Agency, 133, 146
companies from, 25, 32
exports from, 26
in postwar world, 25
Operations Mission in Thailand, 44
trade with, 30, 224
Treasury, 213
"unity In diversity," 167, 168
UNTAC, see UN Transitional Authority in Cambodia
UNTAET, see UN Transitional Administration in East Timor
US News and World Report, 75
USAID, see United States Agency for International Development

Vajiralongkorn, Crown Prince, 51
Vajiravudh, King, 42, 170
Vientiane (city), 146, 151
Vientiane (principality), 146
Viet Cong, 125
Viet Minh, 124
Vietnam, 27, 33, 121–30, 133, 149, 167, 170, 181, 182, 183, 193, 196, 201, 203, 222, 223, 225, 226
  and ASEAN, 126, 227, 233
  and Brunei, 201
  and Cambodia, 121, 125, 128, 132, 133, 134, 135, 136, 141, 142, 202
  and China, 8, 121, 122, 123, 124, 125, 220, 221, 222
  and France, 123, 124, 125, 130
  and Hong Kong, 223
  and India, 228
  and Indonesia, 201
  and Japan, 124, 224, 225
  and Laos, 121, 145, 146, 147, 148, 152
  and Malaysia, 201
  and Paracel Islands, 121, 221
  and Philippines, 201
  and Portugal, 123
  and Russia, 130
  and Singapore, 201
  and South Korea, 220
  and Soviet bloc, 124, 126, 127
  and Soviet Union, 124, 128, 129, 130, 135
  and Spratly Islands, 121
  and Taiwan, 223
  and Thailand, 30, 45
  and UN, 126
  and United Kingdom, 124
  and United States, 30, 33, 123, 124, 125, 220
  colonial, 10, 14, 15, 16
  premodern, 5, 7, 8, 9, 231
Vietnamese alphabet, 123
Vietnamese language, 133
Vietnamese people, 16, 133, 137
Visayas, 200
"Vision 2020," 57
Vo Nguyen Giap, 124
Volksraad, 102, 170

wages, 224
  in Burma, 200
  in Indonesia, 74
  in Malaysia, 74
  in Singapore, 67, 71, 72, 74
  in Vietnam, 130
Wahid, Abdurrahman, 105–06, 107, 175, 177
WALHI, 188
"Washington consensus," 213
water, 180–81
  and agriculture, 194, 196, 197
  and climate changes, 186
  in Cambodia, 139, 140
  in Malaysia, 181
  in Philippines, 181
  in Singapore, 73, 181
  in Thailand, 200
waterpumps in Laos, 151
wealth distribution in Malaysia, 64
West Papua, 13, 103, 106, 107, 167, 168, 174, 175, 177, 188, 196
West Timor, 110, 111, 114, 115, 118
Whitlam, Gough, 237
Wilhelmina, Queen, 18
Wiranto, General, 114
women
  in Burma, 160
  in colonial period, 18
  in East Timor, 112
  in Laos, 153
  in Philippines, 87, 91
  in Thailand, 49
Women's Association (Vietnam), 126
wood, see forests and forestry
Woodside Petroleum, 116
Worcester, Dean, 89
Workers Party (East Timor), 111
Workers Party (Singapore), 75
Workers Party (Vietnam), see Communist Party in Vietnam
workers, see under labor
workforce, see labor force
World Bank, 26, 227, 233
  and agriculture, 196
  and Asian crisis, 213
  and Cambodia, 139
  and East Timor, 117
  and Laos, 149, 150, 151
  and Singapore, 68
  and Thailand, 44, 48, 49, 51
World Trade Organization, 150, 189, 227

Xanana, Kay Rala, see Gusmão, José Alexandre

Yang Di-Pertuan Agong, 57, 62
"Yangon," 158
yen, 32, 48, 208, 223, 224, 225
Yugoslavia, 162

Zambales, 181
Zamboanga del Sur, 181
Zamora, Father, 88
Zhou Enlai, 133